THE LABYRINTH IN NUCLEAR STRUCTURE

Related Titles from AIP Conference Proceedings

704 Tours Symposium on Nuclear Physics V: Tours 2003
Edited by M. Arnould, M. Lewitowicz, G. Münzenberg, H. Akimune, M. Ohta, H. Utsunomiya, T. Wada, and T. Yamagata, March, 2004, 0-7354-0177-2

698 Intersections of Particle and Nuclear Physics, 8th Conference; CIPANP 2003
Edited by Zhoreh Parsa, February 2004, CD-ROM included, 0-7354-0169-1

681 Proton Emitting Nuclei: Second International Symposium; PROCON 2003
Edited by Enrico Maglione and Francesca Soramel, September 2003, 0-7354-0150-0

656 Frontiers of Nuclear Structure
Edited by Paul Fallon and Rod Clark, March 2003, 0-7354-0116-0

644 Exotic Clustering: 4th Catania Relativistic Ion Studies; CRIS 2002
Edited by Salvatore Costa, Antonio Insolia, and Cristina Tuvè, November 2002, 0-7354-0099-7

638 Mapping the Triangle: International Conference on Nuclear Structure
Edited by Ani Aprahamian, Jolie A. Cizewski, Stuart Pittel, and N. Victor Zamfir, November 2002, 0-7354-0093-8

610 Nuclear Physics in the 21st Century: International Nuclear Physics Conference, INPC 2001
Edited by Eric Norman, Lee Schroeder, and Gordon Wozninak, April 2002, 0-7354-0056-3

594 Hadrons and Nuclei: First International Symposium
Edited by Il-Tong Cheon, Taekeun Choi, Seung-Woo Hong, and Su Houng Lee, November 2001, 0-7354-0037-7

561 Tours Symposium on Nuclear Physics IV: Tours 2000
Edited by M. Arnould, M. Lewitowicz, Yu. Ts. Oganessian, H. Akimune, M. Ohta, H. Utsunomiya, T. Wada, and T. Yamagata, April 2001, 1-56396-996-3

549 Intersections of Particle and Nuclear Physics: 7th Conference, CIPANP2000
Edited by Zohreh Parsa and William J. Marciano, December 2000, 1-56396-978-5

518 Proton Emitting Nuclei: PROCON'99—First International Symposium
Edited by Jon C. Batchelder, May 2000, 1-56396-937-8

512 Nuclear Physics at Storage Rings: Fourth International Conference: STORI99
Edited by Hans-Otto Meyer and Peter Schwandt, June 2000, 1-56396-928-9

495 Experimental Nuclear Physics in Europe: ENPE 99, Facing the Next Millennium
Edited by Berta Rubio, Manuel Lozano, and William Gelletly, November 1999, 1-56396-907-6

481 Nuclear Structure 98
Edited by C. Baktash, September 1999, 1-56396-858-4

To learn more about these titles, or the AIP Conference Proceedings Series, please visit the webpage http://proceedings.aip.org

THE LABYRINTH IN NUCLEAR STRUCTURE

International Conference on
The Labyrinth in Nuclear Structure,
an EPS Nuclear Physics Divisional Conference

Crete, Greece   13-19 July 2003

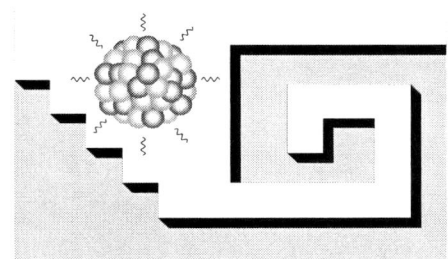

EDITORS

Angela Bracco
Università di Milano and INFN
Milan, Italy

Constantine A. Kalfas
NCSR Demokritos, Institute of Nuclear Physics
Athens, Greece

SPONSORING ORGANIZATIONS
EPS Nuclear Physics Division
NCSR Demokritos
INFN and University of Milan
NTU Athens
Greek Ministry of Culture

Melville, New York, 2004
AIP CONFERENCE PROCEEDINGS ■ VOLUME 701

Editors:

Angela Bracco
Department of Physics
University of Milan
and INFN sez. Milano
via Celoria, 16
20133 Milan
ITALY

E-mail: Bracco@mi.infn.it

Constantine A. Kalfas
Institute of Nuclear Physics
National Center for Scientific Research "DEMOKRITOS"
Agia Paraskevi
153 10 Athens
GREECE

E-mail: kalfas@inp.demokritos.gr

Authorization to photocopy items for internal or personal use, beyond the free copying permitted under the 1978 U.S. Copyright Law (see statement below), is granted by the American Institute of Physics for users registered with the Copyright Clearance Center (CCC) Transactional Reporting Service, provided that the base fee of $22.00 per copy is paid directly to CCC, 222 Rosewood Drive, Danvers, MA 01923. For those organizations that have been granted a photocopy license by CCC, a separate system of payment has been arranged. The fee code for users of the Transactional Reporting Service is: 0-7354-0174-8/04/$22.00.

© 2004 American Institute of Physics

Individual readers of this volume and nonprofit libraries, acting for them, are permitted to make fair use of the material in it, such as copying an article for use in teaching or research. Permission is granted to quote from this volume in scientific work with the customary acknowledgment of the source. To reprint a figure, table, or other excerpt requires the consent of one of the original authors and notification to AIP. Republication or systematic or multiple reproduction of any material in this volume is permitted only under license from AIP. Address inquiries to Office of Rights and Permissions, Suite 1NO1, 2 Huntington Quadrangle, Melville, N.Y. 11747-4502; phone: 516-576-2268; fax: 516-576-2450; e-mail: rights@aip.org.

L.C. Catalog Card No. 2004101189
ISBN 0-7354-0174-8
ISSN 0094-243X
Printed in the United States of America

Contents

Preface xi

Nuclear Physics and Astrophysics at the ISAC Radioactive Beams Facility 1
    J. M. D'Auria (for the DRAGON Collaboration)

Nuclear Spectroscopy Using Radioactive Ion Beams from the HRIBF 10
    A. Galindo-Uribarri

Role of Surface Vibrations in the Pairing Gap of Atomic Nuclei 19
    G. Gori, F. Barranco, P. F. Bortignon, R. A. Broglia, G. L. Colò, and E. Vigezzi

Recent Experimental Results from the NSCL on the Structure of Exotic Nuclei 26
    T. Glasmacher, D. Bazin, C. M. Campbell, J. A. Church, D. C. Dinca, A. Gade, W. F. Mueller, H. Olliver, B. M. Sherrill, and K. L. Yurkewicz

Nuclear Structure around the N=16 Subshell Closure 31
    O. Sorlin, M. Stanoiu, Z. Dombrádi, F. Azaiez, B. A. Brown, M. Belleguic, D. Sohler, M. G. Saint Laurent, M. J. Lopez-Jimenez, Y. E. Penionzhkevich, G. Sletten, N. L. Achouri, F. Becker, J. C. Angélique, C. Borcea, C. Bourgeois, J. M. Daugas, Z. Dlouhý, C. Donzaud, J. Duprat, Z. Fülöp, D. Guillemaud-Mueller, S. Grévy, F. Ibrahim, A. Kerek, A. Krasznahorkay, M. Lewitowicz, S. Leenhardt, S. M. Lukyanov, S. Mandal, P. Mayet, H. Van der Marel, W. Mittig, J. Mrázek, F. Negoita, F. de Oliveira-Santos, Z. Podolyák, F. Pougheon, M. G. Porquet, H. Savajols, Y. Sobolev, C. Stodel, J. Timár, and A. Yamamoto

Rotational Bands in Neutron-Rich $^{160-162}$Ho 39
    D. Escrig, A. Jungclaus, B. Binder, A. Dietrich, T. Härtlein, H. Bauer, C. Gund, D. Pansegrau, D. Schwalm, D. Bazzacco, G. de Angelis, E. Farnea, A. Gadea, S. Lunardi, D. R. Napoli, C. Rossi-Alvarez, and C. Ur

Systematic AMD+GCM Study of Structure of Carbon Isotopes 44
    G. Thiamova, N. Itagaki, T. Otsuka, and K. Ikeda

Exotic Cluster Structure in Light Neutron-Rich Nuclei 49
    N. Itagaki and S. Aoyama

New Lines of Research with the MAGNEX Large-Acceptance Spectrometer 54
    F. Cappuzzello, A. Cunsolo, A. Foti, A. Lazzaro, S. E. A. Orrigo, J. S. Winfield, M. C. Allia, and C. Nociforo

Nuclear Structure Studies with GEANIE at the LANSCE/WNR Facility 60
    N. Fotiades, P. E. Garrett, E. Tavukcu, M. Devlin, R. O. Nelson, J. A. Becker, W. Younes, and L. A. Bernstein

^{8}Li(α,n)^{11}B at Big Bang Temperatures: Neutron Counting with a Low Intensity ^{8}Li Radioactive Beam 68
    S. Cherubini, P. Figuera, A. Musumarra, C. Agodi, R. Alba, L. Calabretta, L. Cosentino, A. Del Zoppo, A. Di Pietro, L. Lamia, L. Pappalardo, M. G. Pellegriti, R. G. Pizzone, A. Rinollo, C. Rolfs, S. Romano,

C. Spitaleri, F. Strieder, S. Tudisco, and A. Tumino

Single-Particle Structure of Neutron-Rich Nuclei 73
J. A. Cizewski, K. L. Jones, J. S. Thomas, D. W. Bardayan,
J. C. Blackmon, C. J. Gross, J. F. Liang, D. Shapira, M. S. Smith,
D. W. Stracener, R. L. Kozub, C. D. Nesaraja, U. Greife, R. J. Livesay,
and Z. Ma

A Modular and Compact Multidetector System Based on Monolithic Telescopes 77
P. Figuera, F. Amorini, G. Cardella, A. Di Pietro, G. Fallica,
A. Musumarra, M. Papa, G. Pappalardo, F. Rizzo, W. Tian,
S. Tudisco, and G. Valvo

Octupole Correlations in High-K States in ^{240}Pu 82
L. Amon, S. M. Mullins, B. Akkus, and M. N. Erduran

Collective and Single-Particle Properties of Neutron-Rich Nuclei Investigated via Reactions of Relativistic Radioactive Beams 87
T. Aumann

The Giant Dipole Resonance at Very High and Very Low Excitation Energies 95
F. Camera

A Study of the Jacobi Shape Transition in Light, Fast Rotating Nuclei with the EUROBALL IV, HECTOR, and EUCLIDES Arrays 104
A. Maj, M. Kmiecik, M. Brekiesz, J. Grębosz, W. Męczyński, J. Styczeń,
M. Ziębliński, K. Zuber, A. Bracco, F. Camera, G. Benzoni, B. Million,
N. Blasi, S. Brambilla, S. Leoni, M. Pignanelli, O. Wieland, A. Airoldi,
B. Herskind, P. Bednarczyk, D. Curien, E. Farnea, G. de Angelis,
D. R. Napoli, J. Nyberg, M. Kicińska-Habior, C. M. Petrache, D. Petrache,
N. Dubray, J. Dudek, and K. Pomorski

Spectroscopy of Light Exotic Nuclei Using Nuclear Break-Up 112
D. Cortina-Gil, J. Fernandez-Vazquez, T. Aumann, T. Baumann,
J. Benlliure, M. J. G. Borge, L.V. Chulkov, U. Datta Pramanik, C. Forssén,
L. M. Fraile, H. Geissel, J. Gerl, F. Hammache, K. Itahashi, R. Janik,
B. Jonson, S. Mandal, K. Markenroth, M. Meister, M. Mocko,
G. Münzenberg, T. Ohtsubo, A. Ozawa, Y. Prezado, V. Pribora, K. Riisager,
H. Scheit, R. Schneider, G. Schrieder, H. Simon, B. Sitar, A. Stolz,
P. Strmen, K. Sümmerer, I. Szarka, and H. Weick

Protein Folding and Non-conventional Drug Design: A Primer for Nuclear Structure Physicists 117
R. A. Broglia, G. Tiana, and D. Provasi

Gamma-Ray Spectroscopy Studies at GANIL: Status and Perspectives 127
G. de France

Nuclear Moments of Nuclei in the Neighborhood of the Neutron Drip Line 135
H. Ueno, K. Asahi, H. Ogawa, D. Kameda, H. Miyoshi, A. Yoshimi,
H. Watanabe, K. Shimada, W. Sato, K. Yoneda, N. Imai, Y. Kobayashi,
M. Ishihara, and W.-D. Schmidt-Ott

RISING Status Report 143
M. Górska (for the RISING Collaboration)

Structure of Proton-Neutron Multiplets Around ^{132}Sn 149
    A. Gargano, L. Coraggio, A. Covello, and N. Itaco

Chaotic Dynamics in Warm Rotating Nuclei 157
    M. Matsuo, E. Vigezzi, S. Leoni, A. Bracco, G. Benzoni, T. Døssing,
    B. Herskind, G. B. Hagemann, A. Lopez-Martens, and T. L. Khoo

Order to Chaos Properties of the Decay-out Gamma Rays from Superdeformed Bands 164
    T. Døssing, A. P. Lopez-Martens, T. L. Khoo, T. Lauritsen, and S. Åberg

Nuclear Alignment in Projectile Fragmentation as a Tool for Moment Measurements 169
    G. Georgiev, I. Matea, J. M. Daugas, M. Hass, R. Astabatyan, L. T. Baby,
    D. L. Balabanski, G. Bélier, D. Borremans, F. de Oliveira Santos,
    G. Goldring, H. Goutte, P. Himpe, M. Lewitowicz, S. Lukyanov, V. Méot,
    G. Neyens, Y. E. Penionzhkevich, O. Roig, and M. Sawicka

Coulomb Breakup of Neutron-Rich Oxygen Isotopes 174
    C. Nociforo, R. Palit, P. Adrich, T. Aumann, K. Boretzky, D. Cortina-Gil,
    U. Datta Pramanik, T. W. Elze, H. Emling, H. Geissel, M. Hellström,
    N. Iwasa, K. L. Jones, J. V. Kratz, R. Kulessa, Le Hong Khiem,
    A. Leistenschneider, G. Münzenberg, P. Reiter, C. Scheidenberger,
    H. Scheit, H. Simon, K. Sümmerer, S. Typel, E. Wajda, W. Walus, and
    H. Weick

Continuum Response and Reaction in Neutron-Rich Be Nuclei 179
    T. Nakatsukasa, M. Ueda, and K. Yabana

Spectroscopy of Very Heavy Elements 184
    R. Julin

Proton Emitter Studies Using the Argonne Fragment Mass Analyzer 192
    P. J. Woods

Tracking Dissipation in Capture Reactions 200
    T. Materna, V. Bouchat, V. Kinnard, F. Hanappe, O. Dorvaux, C. Schmitt,
    L. Stuttgé, K. Siwek-Wilczynska, Y. Aritomo, A. Bogatchev,
    E. Prokhorova, and M. Ohta

Deformation of the Very Neutron-Deficient Rare-Earth Nuclei Produced with the SPIRAL ^{76}Kr Radioactive Beam and Studied with EXOGAM + DIAMANT 208
    N. Redon, A. Prévost, D. Guinet, P. Lautesse, M. Meyer, B. Rossé,
    O. Stézowski, P.J. Nolan, C. Andreoiu, A. J. Boston, M. Descovich,
    A. O. Evans, S. Gros, J. Norman, R. D. Page, E. S. Paul, G. Rainovski,
    J. Sampson, G. de France, J. M. Casandjian, C. Theisen, J. N. Scheurer,
    B. M. Nyakó, J. Gál, G. Kalinka, J. Molnár, Z. Dombrádi, J. Timár,
    L. Zolnai, K. Juhász, A. Astier, I. Deloncle, M. G. Porquet, R. Wadsworth,
    P. Raddon, Y. Lee, A. Wilkinson, P. Joshi, J. Simpson, D. Appelbe, D. Joss,
    R. Lemmon, J. Smith, D. Cullen, A. Brondi, G. La Rana, R. Moro,
    E. Vardacci, and M. Girod

Chaos in the Nucleus: SD Decay-out and Masses 213
    S. Åberg, T. Døssing, H. Olofsson, I. Ragnarsson, C. Andreoiu,
    C. Fahlander, and D. Rudolph

Decay-out of ^{151}Tb Yrast Superdeformed Band and Shape Coexistence 222

    G. Duchêne, J. Robin, A. Odahara, T. Byrski, F. A. Beck, P. J. Twin, P. Bednarczyk, D. Curien, S. Courtin, O. Dorvaux, B. Gall, P. Joshi, A. Nourreddine, E. Pachoud, I. Piqueras, J. P. Vivien, K. Zuber, N. Adimi, D. E. Appelbe, A. Bracco, B. Cederwall, D. M. Cullen, S. Ertück, G. de France, S. L. King, A. Korichi, K. Lagergren, S. Leoni, G. Lo Bianco, A. Lopez-Martens, S. Lunardi, B. Million, E. S. Paul, C. Petrache, N. Redon, A. Saltarelli, J. Simpson, O. Stézowski, and R. Venturelli

Determining the Excitation Energy, Spin, and Parity of Levels in the Superdeformed Bands of ^{152}Dy 230

    T. Lauritsen, M. P. Carpenter, R. V. F. Janssens, T. L. Khoo, P. Fallon, B. Herskind, D. G. Jenkins, F. G. Kondev, A. Lopez-Martens, A. O. Macchiavelli, D. Ward, K. S. Abu Saleem, I. Ahmad, R. Clark, M. Cromaz, T. Døssing, J. P. Greene, F. Hannachi, A. M. Heinz, A. Korichi, G. Lane, C. J. Lister, P. Reiter, D. Seweryniak, S. Siem, R. C. Vondrasek, and I. Wiedenhöver

Investigation of Dipole Bands in the ^{142}Gd Region with EUROBALL 238

    R. M. Lieder, A. A. Pasternak, and E. O. Podsvirova

Precise ft-Value Measurement for the Superallowed $0^+ \rightarrow 0^+$ β Decay of ^{22}Mg 244

    V. E. Iacob, J. C. Hardy, M. Sanchez-Vega, R. G. Neilson, A. Azhari, C. A. Gagliardi, V. E. Mayes, L. Trache, and R. E. Tribble

Beta Decay of ^{76}Sr Using the Total Absorption Spectrometer "Lucrecia" at ISOLDE-CERN 252

    E. Nácher, A. Algora, B. Rubio, J. L. Taín, M. J. G. Borge, D. Cano-Ott, S. Courtin, P. Dessagne, D. Escrig, L. M. Fraile, W. Gelletly, A. Jungclaus, G. Le Scornet, F. Maréchal, C. Miehé, E. Poirier, and O. Tengblad

Recent Results at the $N=Z$ Line with GASP and EUROBALL 257

    E. Farnea

AGATA 265

    D. Bazzacco

Study of Superdeformation in the $A \approx 60$ Mass Region: High Resolution γ-Ray Spectroscopy at EUROBALL IV with the Recoil Filter Detector and the EUCLIDES Charged Particle Detector 273

    J. Dobaczewski, J.P. Vivien, K. Zuber, P. Bednarczyk, T. Byrski, D. Curien, G. de Angelis, O. Dorvaux, G. Duchêne, E. Farnea, A. Gadea, B. Gall, J. Grębosz, R. Isocrate, A. Maj, W. Męczyński, J. C. Merdinger, A. Prévost, N. Redon, J. Robin, O. Stézowski, J. Styczeń, and M. Ziębliński

Angular Momentum Population in Projectile Fragmentation 280

    Z. Podolyák, K. A. Gladnishki, J. Gerl, M. Hellström, Y. Kopatch, S. Mandal, M. Górska, P. H. Regan, H. J. Wollersheim, K.-H. Schmidt, and O. Yordanov (for the GSI-ISOMER Collaboration)

What Is the Signature of $T=0$ np Pairing in Rotating Nuclei? 285

    A. L. Goodman

Competing Decay-out Mechanisms of the Yrast Superdeformed Band in ^{59}Cu 289

    C. Andreoiu, C. Fahlander, D. Rudolph, T. Døssing, I. Ragnarsson, and S. Åberg

Nuclei at Extreme Deformations..295
    P. Fallon

Hyperdeformed Shapes and Jacobi Transitions in ^{126}Ba..................303
    B. Herskind, G. B. Hagemann, G. Sletten, T. Døssing, C. Rønn Hansen,
    S. Ødegård, H. Hübel, P. Bringel, A. Bürger, A. Neusser, A. K. Singh,
    A. Al-Khatib, S. B. Patel, A. Bracco, S. Leoni, F. Camera, G. Benzoni,
    P. Mason, A. Paleni, B. Million, O. Wieland, P. Bednarczyk, F. Azaiez,
    T. Byrski, D. Curien, O. Dakov, G. Duchêne, F. Khalfallah, B. Gall,
    I. Piqueras, J. Robin, J. Dudek, N. Rowley, N. Redon, F. Hannachi,
    J. N. Scheurer, J. N. Wilson, A. Lopez-Martens, A. Korichi, K. Hauschild,
    J. Roccaz, S. Siem, P. Fallon, I.-Y. Lee, A. Goergen, B. M. Nyakó,
    A. Algora, Z. Dombrádi, J. Gál, G. Kalinka, D. Sohler, J. Molnár, J. Timár,
    L. Zolnai, K. Juhász, A. Maj, M. Kmiecik, M. Brekiesz, J. Styczeń,
    K. Zuber, J. C. Lisle, B. Cederwall, K. Lagergren, A. O. Evans,
    G. Rainovski, G. de Angelis, G. La Rana, R. Moro, W. Gast, R. M. Lieder,
    E. Podsvirova, H. Jäger, C. M. Petrache, and D. Petrache

Nuclear Clusters and Structure in Light Nuclei..........................316
    T. Kokalova, W. von Oertzen, S. Thummerer, H.G. Bohlen, M. Milin,
    A. Tumino, G. de Angelis, E. Farnea, M. Axiotis, N. Marginean,
    D.R. Napoli, S.M. Lenzi, C. Ur, M. Rousseau, and P. Papka

The Neutron Facility at NCSR "Demokritos"—Implementation in the Case of the ^{232}Th(n,2n) Reaction...............................324
    R. Vlastou, C. T. Papadopoulos, G. Perdikakis, M. Kokkoris, C. A. Kalfas,
    S. Kossionides, D. Karamanis, and P. A. Assimakopoulos

Studies around A ~ 100 Using Binary Reactions..........................329
    P. H. Regan, A. D. Yamamoto, C. Y. Wu, A. O. Macchiavelli, D. Cline,
    J. F. Smith, S. J. Freeman, J. J. Valiente-Dobón,
    R. S. Chakrawarthy, M. Cromaz, P. Fallon, A. Hayes, H. Hua,
    S. D. Langdown, I-Y. Lee, C. J. Pearson, Z. Podolyák, R. Teng, and
    C. Wheldon

The Radioactive Ion Beam Project SPES at LNL..........................334
    A. Pisent and A. Bracco

The SPIRAL2 Project at GANIL...343
    H. Savajols, A.C.C. Villari, and W. Mittig (for the SPIRAL2 Group)

Status of the EXCYT Facility at INFN-LNS..............................354
    G. Cuttone, R. Alba, L. Calabretta, L. Celona, F. Chines, L. Cosentino,
    P. Finocchiaro, A. Grmek, S. Gammino, M. Menna, G. E. Messina,
    G. Raia, S. Passarello, M. Re, D. Rifuggiato, A. Rovelli, S. Russo,
    G. Schillaci, V. Scuderi, and E. Zappalà

Heavy Ion Radiative Capture: ^{12}C(^{12}C, γ)..............................361
    D. G. Jenkins, B. R. Fulton, J. Pearson, C. J. Lister, M. P. Carpenter,
    S. Freeman, N. Hammond, R. V. F. Janssens, T. L. Khoo, T. Lauritsen,
    E. F. Moore, A. H. Wuosmaa, P. Fallon, A. Görgen, A. O. Macchiavelli,
    M. McMahan, M. Freer, and F. Haas

Chiral Bands and Triaxiality..366
    C. M. Petrache

Supersymmetry in Nuclei..375
    A. Algora

Chirality in the A ∼ 100 Region..383
    P. Joshi, D. G. Jenkins, P. M. Raddon, A. J. Simons, R. Wadsworth,
    T. Wilkinson, D. B. Fossan, K. Starosta, C. Vaman, J. Timár, Z. Dombrádi,
    A. Krasznahorkay, J. Molnár, D. Sohler, L. Zolnai, A. Algora, E. S. Paul,
    G. Rainovski, J. Gizon, A. Gizon, P. Bednarczyk, D. Curien, G. Duchêne,
    N. Fotiades, and J. N. Scheurer

Author Index...389

Preface

This volume contains the invited and contributed papers presented at the International Conference on "The Labyrinth in Nuclear Structure," held at Aghia Pelaghia in Crete, Greece, 13–19 July 2003 (EPS Nuclear Physics Divisional Conference). It was jointly organized by the I N P of the National Center for Scientific Research DEMOKRITOS, INFN, and the University of Milano.

The traditional goal of a conference in this series is to offer the opportunity to discuss the progress and future perspectives in the field of nuclear structure. In particular, the presentations of the 30 invited talks and the 30 oral contributions focused on recent achievements in experimental and theoretical nuclear physics, which are related to the activities with large detector arrays, installed at facilities offering beams of stable and unstable ions. The topics that were covered during the week represent the main lines of investigations of the nuclear structure laboratories around the world. Results concerning the physics of nuclei far from stability in the direction of both the proton and neutron drip lines, together with a number of interesting astrophysical applications, were presented. New insights into nuclear structure given by recent results on exotic and superdeformed nuclei populated at the very high spins were discussed, as was the problem of the search for hyperdeformed nuclear shapes. Special attention was also given to the problem of order to chaos transition in nuclei, to the understanding of the origin of the pairing force, and to the investigation of giant resonance states. It is important to stress that time has been devoted to the presentation of new facilities for radioactive beams, presently under discussion or under construction that will provide intense beams of unstable nuclei in different mass regions. In addition, developments in the instrumentation for gamma-ray spectroscopy were presented. The concluding roundtable discussion centered on the perspectives opened by such new facilities as well as by the new instrumentation.

A major success of the meeting was the lively discussion based on the new ideas and on the exciting experimental and theoretical results presented at the different sessions. We warmly thank the speakers and the participants for the high quality of the presented talks and for the very active participation.

We would like to thank the international advisory committee for their valuable suggestions: Sven Aberg, Faisal Azaziez, Jolie Cizewski, Giacomo De Angelis, Antonio Del Zoppo, George Dracoulis, Sidney Gales, Bill Gelletly, Hubel Grawe, Herbert Hubel, Jan Jolie, Rauno Julin, Teng Lek Khoo, Marek Lewitowich, Marcello Pignanelli, Hiroyuki Sagawa, John Sharpey-Shafer, Alan Shotter, John Simpson, Geirr Sletten, Michael Thoennessen, Wolfram von Oertzen, David Ward, and Phil Woods.

We are also deeply grateful to the organizing committee, Leo Skouras, Rosa Vlastou, Costas Papadopoulos, and Michael Kokkoris. We thank them especially for their work and help prior to and during the conference.

Angela Bracco
Costas Kalfas

Nuclear Physics and Astrophysics at the ISAC Radioactive Beams Facility

John M. D'Auria[*] for the DRAGON Collaboration

Department of Chemistry, Simon Fraser University, Burnaby, B.C. V5A 1S6, CANADA

Abstract. The ISAC (Isotope Separator and Accelerator) Radioactive Beams laboratory has been in operation for about 4 years. An upgrade in the RB upper energy and mass range will be operational in 2 years and funding for a further upgrade in available number of RB will be submitted this year. There are a number of major experimental facilities now available with physics results produced. A key new facility in the area of nuclear astrophysics, the DRAGON (Detector of Recoils And Gammas Of Nuclear reactions) facility, has finished its first major study, namely a direct measurement for the first time of the ^{21}Na(p,γ)^{22}Mg reaction, a reaction believed to play a key role in the production of ^{22}Na during novae explosions. This report will provide a review of the experimental facilities now available at ISAC, along with details of this first published result from the DRAGON facility.

INTRODUCTION

ISAC

ISAC is a radioactive beams facility of the ISOL (Isotope Separation On-Line) approach, using the intense (<100 µA), 500 MeV p$^+$ beam from the TRIUMF (Vancouver, Canada) cyclotron to bombard a thick (~19 cm) cylindrically shaped, target vessel, containing a target material of choice, inducing spallation, fragmentation and fission reactions. This material is chosen to optimize the production and rapid release of a high quantity of the desired element/isotope, usually operated at high temperatures (<1800 C). At present only a surface ionization source has been used leading to primarily alkali elemental beams, however an ECR ion source is in the testing phase, and a laser ion source is in the design phase. A table (Table 1) of some of the beam intensities achieved with a surface ion source is given below; p$^+$ intensities up to 45 µA have been used.

Table 1. Some ISAC RB Intensities

Isotope	Target	Yield
^{8}Li	Ta	8 x 10^8 pps
^{11}Li	Ta	2 x 10^4 pps
^{21}Na	SiC	9.5 x 10^9 pps
^{74}Rb	Nb	1.3 x 10^4 pps
^{79}Rb	Nb	4.6 x 10^9 pps
^{160}Yb	Ta	8.4 x 10^9 pps

An artist's conception of the ISAC facility is given in Fig. 1. The crucial TIS (target and ion source) are mounted on the bottom of an 8 m plug, which is removable using specially designed, precision crane system. Thick, concrete walls to contain the high levels of radiation, during intense beam irradiations, surround this. The crane is used to remove the plug system to a hot cell to remove and replace target and ion source systems. There are two (east and west stations), essentially identical target irradiation areas to allow for fast TIS changes. All radiation damageable material is located near to the top of the plug.

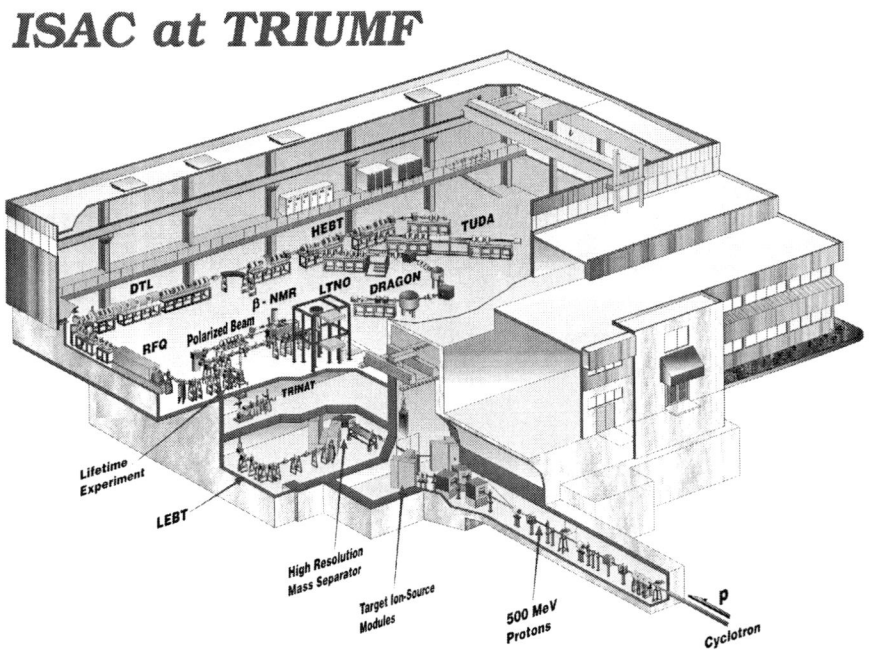

FIGURE 1. Artist's representation of the ISAC I Radioactive Beams Facility

The extracted ion beam (< 60 keV) is initially mass analyzed in a pre-separator magnet for initial clean-up and then, mass analyzed again with a high resolution magnet ($\Delta M/M \sim 10{,}000$), providing a final high purity, 1^+ beam of the isotope of interest. This beam is then directed using primarily electrostatic elements to experimental facilities located in a stopped beam area (LEBT) or into the HEBT area using a specially designed LINAC accelerator for experiments in nuclear astrophysics. This latter beam line will also service the new ISAC II area, presently being installed.

The LINAC consists of a room temperature RFQ LINAC, accelerating the 2 keV/u, 1^+ radioisotopic beam (with $q/m < 1/30$) to 150 keV/u, a foil stripper to increase q, and a room temperature drift tube LINAC (DTL) with $q/m < 1/6$, resulting in an accelerated beam ranging in energy from 0.15 to 1.5 MeV/u. This is then delivered to DRAGON and TUDA (see below).

ISAC II

In 2000 funding was approved to increase the mass and energy range of the accelerated ISOL beam. The projected goal will result in beams with energies above the coulomb barrier (and up to 15 MeV/u depending on the mass) and for energies with A < 150. It is expected that beams with A<60 and up to about 4.3 MeV/u will be available by 2005. These will be achieved by using a charge state booster (CSB), following mass separation, to increase the q of the beam injected into the RFQ, and then to use superconducting (SC) LINAC sections following the DTL to achieve this initial beam energy. With additional funding, further SC sections will be used in a different configuration to achieve the desired higher energy (see Fig. 2). The ISAC II building is now available for occupancy. Additional details on ISAC (I and II) can be found elsewhere (www.triumf.ca/isac/isac.html).

EXPERIMENTAL FACILITIES: PRESENT AND PLANNED AT ISAC (I AND II)

Overview

Initially ISAC was built for scientific programs, namely nuclear astrophysics, fundamental symmetries and condensed matter physics. However a fourth area of nuclear structure was added given the potential scientific productivity of ISAC. Table 2 gives a simplified view of the scientific area of interest of the present facilities present or planned for ISAC I and II.

The layout of the ISAC I experimental hall is shown in Fig. 2. The TRINAT and the RT (Long Lived Radioisotope Production) facilities are located in the mezzanine and beam floor areas, respectively and are not shown in Fig. 2.

Table 2. Experimental Facilities Present or Planned for ISAC (I and II)

	Area	Facility	Status
ISAC 1			
	Nuclear Astrophysics	DRAGON	Operational
		TUDA	Operational
	Nuclear Structure	8 Pi	Operational
		Sceptar	Commissioned
		GPS	Operational
	Fundamental Symmetries	TRINAT	Operational
		TITAN	Planned
	Condensed Matter	Beta NMR	Operational
		LTNO	Commissioning
ISAC II			
	Nuclear Structure	TIGRESS	Design
	and Astrophysics	EMMA	Planning
	Nuclear Reactions	HERACLES	Planned

LEBT facilities (present and planned): A summary

TRINAT

TRINAT (TRIUMF Neutral Atom Trap) was assembled some years ago to perform studies of fundamental symmetries using the novel concept of trapping neutral radioactive atoms using laser beams. A collection of up to 10^{10} atoms can be suspended freely without support material in a volume of mm^3 and their decay emissions measured using appropriate detection systems. At present the program has concentrated on studies of beta-neutrino asymmetries searching for the presence of scalar particles. A study setting a limit on the mass of the heavy neutrino has been completed and further details of the system can be found elsewhere (http://www.triumf.ca/welcome/trinat.html).

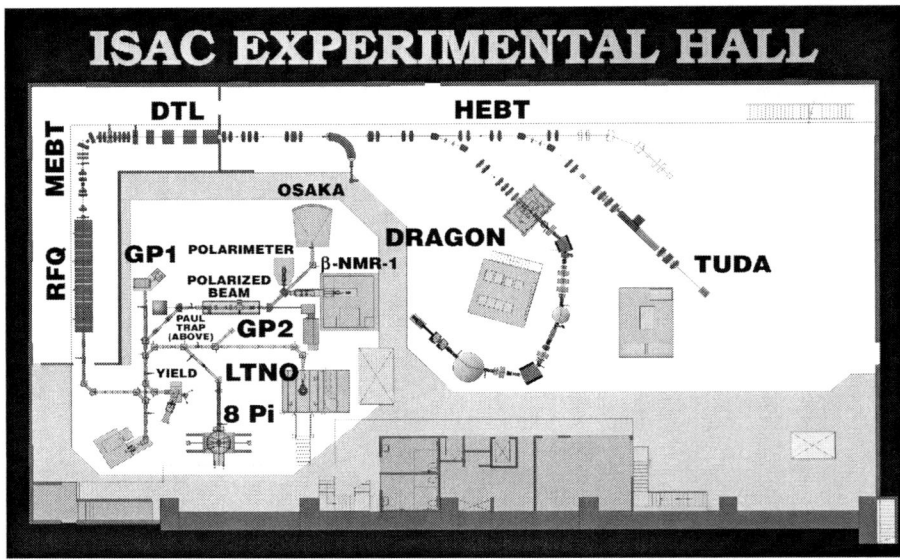

FIGURE 2. Layout of ISAC I Experimental Hall

LTNO

A low temperature refrigerator system was obtained from the Oak Ridge National laboratory to perform studies of interest both for nuclear physics and condensed matter physics. Operating at a temperature of microkelvins, nuclei can be oriented in space and properties of either the stopping material or the nuclei themselves can be measured depending upon the goal of the study. This facility is still being commissioned.

GPS

GPS or General Purpose Spectroscopy facilities are available at several locations in the LEBT area. Armed with appropriate gamma, beta, conversion electron, and particle detectors along with fast moving radioactivity tape transport systems (either in vacuum or in air systems), nuclear spectroscopy studies and studies focused on fundamental symmetries have been pursued. Of particular interest were studies of the Z=N nuclide, ^{74}Rb. Both the lifetime and the branching ratio of superallowed, Fermi beta decaying nuclide have been measured with very high precision (~0.01%) and will play a role in a test of the Standard Model of ElectroWeak Interactions. Other isotopes such as ^{62}Ga will also be pursued.

TITAN

A newly proposed and now funded facility, TITAN (TRIUMF Ion Trap for Atomic and Nuclear science), will be used to measure with high precision, masses of exotic nuclei. The basic concept of TITAN is that of the Penning Ion Trap but TITAN will utilize multicharged ions, which in principle should allow it to perform such studies with a precision significantly higher than other facilities. It consists of four main components in series: a linear radio frequency (RFQ) trap, for cooling and bunching of the beam, an electron beam ion trap (EBIT), for rapid charge breeding, a Wien velocity filter, for selecting the isotope in the charge state of choice, and a Penning trap, for high precision mass measurements. Positioned on a mezzanine in the LEBT region of the ISAC Hall, it is expected to be operational by the end of 2004.

Beta NMR

This consists of a β-detected nuclear magnetic resonance (β-NMR) facility including a radioactive beam polarizer using an in-flight optical pumping technique for studying materials and phenomena of interest in condensed matter science.

OSAKA Polarization Spectroscopy

An array of plastic scintillator detectors (provided by Osaka University) for neutron TOF measurements in β-n-γ studies of polarized ^{11}Li; one study has been completed and submitted for publication on the levels of ^{11}Be.

HEBT facility descriptions for nuclear astrophysics

Two facilities are operational for studies of problems in nuclear astrophysics using the accelerated radioactive beam. TUDA (TRIUMF University of Edinburgh Detector Array) is used to study fusion reactions of low Z RB leading to particle emission while DRAGON (Detector of Recoils And Gammas Of Nuclear reactions) is designed to measure radiative proton and alpha capture reactions. DRAGON is described in more detail below.

Facilities Planned for ISAC-II

TIGRESS

Funding was approved in 2003 for a new state-of-the-art gamma-ray detector array with high efficiency and high energy resolution. Named TIGRESS (TRIUMF-ISAC Gamma- Ray Escape Suppressed Spectrometer), it will initially have 12 active detector elements, with each being a high-purity germanium (HPGe) multi-crystal high-resolution system of the clover variety. Each clover will have eight-fold segmented contracts, surrounded by reconfigurable bismuth germanante (BGO) and CsI(Tl) movable suppression shields; the latter to achieve either high geometric efficiency or high suppression. The granularity of the array will be dramatically increased through 32-fold segmentation of the outer electrical contacts of the clover detectors. Each detector module will be surrounded by a bismuth-germanate (BGO) Compton suppression shield to provide a high efficiency veto of events in which γ rays scatter out of the HPGe detector. Signals from both the HPGe detectors and the BGO suppressors will undergo fast waveform digitization and digital signal processing. A key aspect of this processing is to provide ~2 mm position resolution for the first γ-ray interaction location within the HPGe crystals. The position sensitivity will effectively eliminate the degradation of the γ-ray energy resolution due to Doppler-broadening effects that occurs when detectors subtending large solid angles are used in fast beam experiments. In the close packed configuration TIGRESS will provide a factor of 400 increase in γ-γ coincidence efficiency for 1 MeV γ rays compared to the current detector array, the 8π spectrometer. The improvement will be even greater for the higher-energy γ rays typical of astrophysically important radiative capture reactions. a sub-unit is expected to be ready for studies on ISAC-II by 2006.

EMMA

A versatile variable mode recoil spectrometer named EMMA (Electro Magnetic Mass Analyzer) is required at ISAC-II to detect the exotic heavy products of fusion-evaporation reactions in coincidence with prompt gamma radiation detected in TIGRESS modules surrounding the target and/or to be used in conjunction with the light particle multi-detector array for elastic and inelastic scattering and transfer reaction in inverse kinematics (*e.g.* (d,p) on neutron-rich nuclei near the r-process path). A workshop was held at TRIUMF in the summer of 2002 to identify the physics goals and a second workshop is planned for Dec. 2003 to begin to define the detailed requirements for the spectrometer. It is anticipated that a proposal for funding the spectrometer will be sought in 2004

HERACLES 4π Multi-detector Array

The HERACLES 4π multi-detector array has been used to study a variety of heavy-ion reaction phenomena with stable beams up to A~60 and energies $E \leq 50$ MeV/u. The array consists of 144 charged particle detectors covering the angular range from 3.5 to

140 deg. Four rings of 16 plastic phoswich detectors subtend the region from 3.3 to 24 deg followed by two rings of 16 CsI detectors coupled to phototubes and a 48 element CsI ball coupled to photodiodes. Modifications to the array are required to optimize the performance for the lower energy beams from ISAC-II.

Nuclear Astrophysics at ISAC

The DRAGON Facility: An Overview

As indicated above, the DRAGON facility was constructed to measure radiative proton and alpha capture reactions using accelerated RB with A < 30 and with energies from 0.15 to 1.5 MeV/u. Following commissioning studies aimed at determining the operations parameters of the facility, a major study of the rate of the ^{21}Na(p,γ)^{22}Mg reaction at energies of importance to the production of the isotope, ^{22}Na, in novae explosions was completed.

The Facility

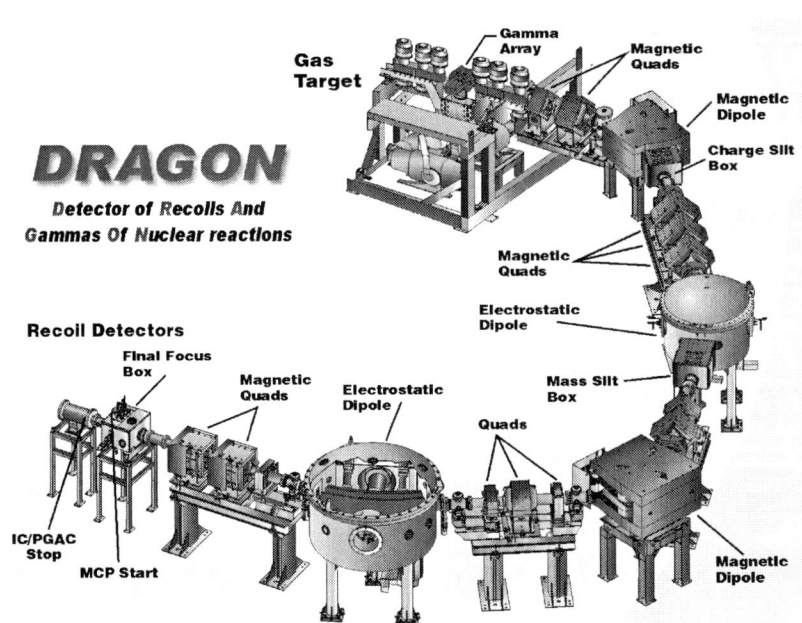

FIGURE 3. Schematic 3-d representation of the DRAGON facility at ISAC.

Figure 3 displays a schematic representation of the DRAGON facility. The facility is composed of windowless gas target, surrounded by a 30 unit, BGO gamma array to detect prompt gammas from the radiative capture reactions. Recoiling reaction products entered the 20 m long, electromagnetic separator (along with beam particles of similar momentum). Following selection of one charge state in the first magnetic dipole, separation of recoils from beam particles is achieved in the first electric dipole (~2% energy difference). At present a silicon strip detector is used at the focal plane for separated reaction products either singly or coincident with reaction gammas. Suppression factors as high as 10^{13} have been measured. The facility was commissioned by measuring resonance strengths of known resonances. The energy of the beam is measured and NRM probe with the first magnetic dipole, following calibration using resonances of known energies. Further information can be found at http://www.triumf.ca/dragon.

The $^{21}Na(p,\gamma)^{22}Mg$ Reaction

This reaction is believed to be important in the production and destruction of ^{22}Na as part of the hot Ne-Na cycle. The production of this isotope is a benchmark of our understanding of a nova explosion but as of yet, it has not been observed using gamma-ray observatories. Various studies have explored the rationale for this lack of any observation but one problem is that the ^{21}Na(p,γ)^{22}Mg reaction had never been measured directly. The DRAGON facility has now measured the resonance strengths of all known resonances up to E_{cm}= 0.821 MeV in ^{22}Mg including those important for novae using an intense beam of ^{21}Na delivered by ISAC. With these data a revised reaction rate for the production of ^{22}Na was calculated.

$E_x = 5.714$ MeV State in ^{22}Mg

Calculations of the Gamow window indicate that the state at E_x= 5.714 MeV will be the dominant contributor (as compared to higher resonances and DC) to the ^{21}Na(p,γ)^{22}Mg reaction at all nova temperatures from 0.2 to 0.35 GK, while the states at 5.962 and 6.046 may contribute to a limited extent at higher temperatures. A beam of ^{21}Na with intensities up to 5 x 10^8 s^{-1} was delivered to DRAGON hydrogen gas target (4.6 Torr); a total of ~10^{13} Na atoms. Data taking was done both in singles and coincidence (prompt reaction gammas and recoiling reaction products). The beam energies were measured by adjusting the field of MD1 so as to position the beam on the ion-optical axis at an energy dispersed focus. From parameters of MD1, it was possible to calculate the beam energy and these were confirmed by measuring a number of known resonances with stable beams (discussed earlier). The lower panel of fig. 4 shows the yield curve for one of these studies, the ^{24}Mg(p,γ)^{25}Al reaction, demonstrating our agreement (inflection point of 214.4±0.5 keV) with the literature resonance energy of 214±0.1 keV. As shown in the upper panel, we find the energy (inflection point) for the ^{21}Na(p,γ)^{22}Mg resonance to be 205.7±0.5, and not 212 keV the difference between the Q value and the level excitation energy, 5713.9 ± 1.2 keV. Given that the latter value is based upon a direct gamma de-excitation measurement of the 5713.9 keV level, this disagreement is most likely explained by an incorrect mass excess for ^{22}Mg; our data imply a value of -403.2 ± 1.3 keV rather than -396.8 keV.

The yield (5.76 ± 0.88 x 10^{-12} per incident ion) for the mid-target data point (Fig. 4) was used to calculate a (ωγ) of 1.03 ± 0.16_{stat}±0.14_{sys} meV.

The effect of these results was to enhance the calculated stellar reaction rate at a particular temperature as compared to some previous estimates. Using a new model of a nova outburst involving an O-Ne white dwarf of 1.25 solar mass, computing for the onset of accretion up to the explosion and ejection stages by means of a spherical hydrodynamic code, results in a lower amount (20%) of the key isotope, ^{22}Na. This arises because the enhanced reaction rate leads to more ^{22}Na (from ^{22}Mg decay) which is consumed at peak temperatures by proton capture. With these results, there is now a firmer basis for the predictions of observing the expected gamma-ray signature (1.275 MeV) γ-ray from a nova outburst using the new INTEGRAL spectrometer.

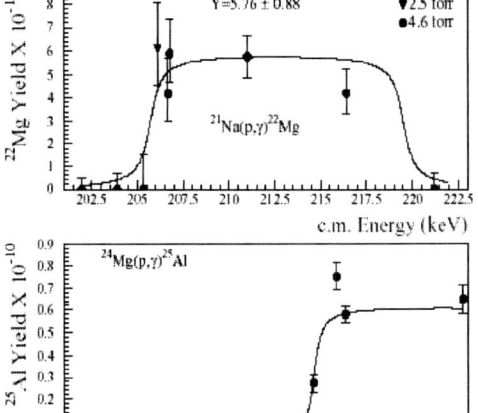

FIGURE 4.
The upper panel displays the thick target yield data for the ^{21}Na(p,γ)^{22}Mg reaction, with the solid line showing the nominal target thickness for 4.6 Torr. Only the point at 212 keV/u was used to calculate the resonance strength.

Yield of the ^{24}Mg(p,γ)^{25}Al reaction for the resonance at E_{cm} = 214 keV, used for beam energy calibration, is displayed in the lower panel. Statistical errors only are displayed in both.

Nuclear Spectroscopy Using Radioactive Ion Beams from the HRIBF

A. Galindo-Uribarri

Physics Division, Oak Ridge National Laboratory, Oak Ridge, TN 37831 USA

Abstract. Exciting opportunities in the study of nuclei far from stability in both the neutron and proton rich side are opening with the recent availability of radioactive ion beams (RIBs) at energies above the Coulomb barrier at the Holifield Radioactive Ion Beam Facility (HRIBF). These RIBs provide a unique opportunity for a whole class of measurements that could never before be realized. A recent highlight has been the acceleration of "pure" beams of fission fragments such as ^{82}Ge ($T_{1/2}$=4.6s) and ^{132}Sn ($T_{1/2}$=40s). These semi-magic and doubly-magic nuclei are important benchmarks within the chart of nuclides, because they are constraints for the shell-model parameter sets. We are currently developing the required experimental tools and specialized techniques for studies in nuclear astrophysics, reaction spectroscopy, and nuclear structure research with RIBs. I will discuss some of the challenges encountered with examples from selected topical areas with which I have been involved.

INTRODUCTION

Radioactive ion beams (RIBs) both in the proton and the neutron rich side are now available for nuclear structure, reactions and astrophysics research from HRIBF at Oak Ridge National Laboratory (ORNL). HRIBF is the first facility to produce post-accelerated beams of heavy neutron-rich nuclei opening important physics opportunities. RIB generation at the HRIBF is based on the well-established Isotope Separation On-Line (ISOL) technique where nuclear reaction products produced in thick target materials are released and rapidly ionized. An ISOL facility requires the use of two accelerators. The first accelerator, the driver, is the Oak Ridge Isochronous Cyclotron (ORIC), a K=100 cyclotron used to produce intense beams of light ions to bombard the production target. Perhaps the most important element is the RIB injector, which in turn consists of various components including the production target, the RIB ion source and two mass separation stages. The post-accelerator consists of the folded-geometry 25-MV Tandem accelerator, which is the highest operating voltage electrostatic accelerator in the world. The HRIBF 25 MV Tandem Accelerator is capable of producing beams of 0.1 up to 14 MeV per nucleon for light nuclei, and up to 5 MeV per nucleon for heavier masses.

The production of a RIB species contrasts sharply with the production of the ions for other accelerators used in today's main areas of research in nuclear physics such as an electron machine (one beam) or a relativistic heavy ion accelerator (a few species).

In general, the development of each RIB beam is a research project in itself and constitutes one of the main challenges of RIB physics. There are now more than 120 RIBs available at HRIBF of at least 10^3 ions per second on target and energies up to 3 MeV/nucleon. Perhaps the list of RIB beams should be expanded, as we have been able to do experiments such as mass measurements with as few as 20 particles per hour. The RIBs complement the approximately 80 stable ion beams (SIBs) that are also available at HRIBF.

Production of p-rich nuclei is done using n, 2n, pn, and alpha-n reaction channels. For example ^{17}F and ^{18}F were produced in a HfO_2 fiber target using $^{16}O(d,n)^{17}F$ and $^{16}O(\alpha, pn)^{18}F$ reactions [1]. However, the majority of the RIBs available at HRIBF are on the n-rich side. They originate from the fission spectra of ^{nat}U bombarded with protons of energy of about 42 MeV. Very recently the ORIC proton beam energy has been increased from 42 MeV to greater than 52 MeV [2]. The expected result is that the fission-fragment yield for neutron-rich RIB production will nearly double due to the increase in proton energy.

The production target for n-rich nuclei consists of UC deposited on a reticulated-vitreous-carbon-fiber (RVCF). In order to avoid excessive losses through radioactive decay of short-lived nuclei, prompt diffusion release and transport of the release product to the ion source in a time period commensurate with the life-time of the species is required. A vigorous program of research is being conducted at HRIBF to study various properties that affect the diffusion release from ISOL production targets and transport properties. Various target/ion sources are being designed to operate with the highest possible efficiency [3-5]. Compared to conventional stable heavy-ion beams (SIBs), RIB intensities are several orders of magnitude lower, particularly for those further from stability. Therefore it is most important also to optimize the beam transport in every stage to provide enough RIB intensity for physics. Generally when using SIBs, losses on transmission efficiency are easily compensated with, for example, more intensity coming out of the ion source. In the case of RIBs, the operation of every single component of the transport system must be optimized as the cumulative losses can make the difference of whether an experiment is successful or not. The field of RIB Physics demands a closer interaction between experimenters and accelerator operations staff to fully exploit these exotic beams.

EXPERIMENTAL CHALLENGES OF RADIOACTIVE ION BEAMS

More than 25 RIB experiments have been done at HRIBF, using beams of ^{7}Be, ^{17}F, ^{18}F, ^{75}Cu, ^{77}Cu, ^{78}Cu, ^{79}Cu, ^{78}Ge, ^{80}Ge, ^{82}Ge, ^{83}Ge, ^{84}Ge, ^{85}Ge, ^{86}Ge, ^{83}Se, ^{84}Se, ^{118}Ag, ^{126}Sn, ^{128}Sn, ^{130}Sn, ^{132}Sn, ^{133}Sn, ^{134}Sn, ^{136}Sn, ^{132}Te, ^{134}Te and ^{136}Te with intensities from 10^8 to 0.005 ions/sec. This wide range of intensities, purity and energies require a variety of detection schemes and sometimes the "non-conventional" use of detectors. Current low energy RIB projects at HRIBF include:

- Elastic and quasielastic reactions
- Resonant reactions
- 2p emission
- Transfer reactions
- COULEX using neutron-rich RIBs
- g-factor measurements
- Near Barrier Fusion
- Breakup
- Mass measurements
- Decay studies
- Instrumentation

RIBs also bring with them their own problems such as low intensities, and potentially high backgrounds in the gamma-ray and charged particle detectors caused by the radioactivity of the beam. The latter effects can be mitigated with good quality beams (Figure 1) and with specially designed detection systems. The 25MV-Tandem is an excellent tool for delivering high quality beams in a wide range of energies.

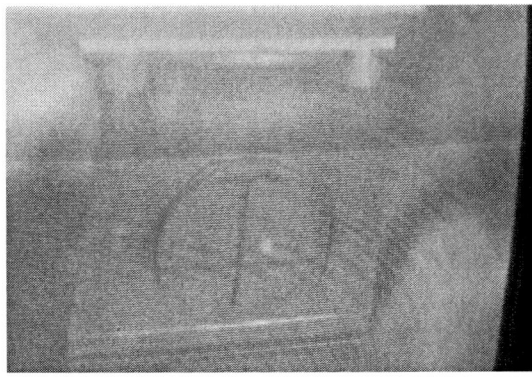

FIGURE 1. Picture taken directly from a TV monitor of the beam spot for a RIB A=82 on an imaging diagnostics screen made of Al_2O_3:Cr. The beam intensity was of 10^5 pps at 220 MeV and the beam spot diameter less than 1.5 mm. With careful tuning, RIBs of excellent quality can be delivered which is of particular importance to avoid large background from stopped and scattered beam in transport components, slits, and chamber.

Radioactive Ion Beams are often isobar cocktails. Removal of all electrons to separate the interfering isobaric ions by the beam transport system has been used for light beams such as ^{17}F [6]. Isobar suppression factors larger than 10^7 have been achieved with this technique commonly used in Accelerator Mass Spectrometry (AMS) [7]. The isobar separator in the injection side of the Tandem has a resolving power of $M/\Delta M \sim 20000$; however, the isobaric separation actually achieved depends not only on isobaric mass differences but also on the beam emittance, and energy spread. It was recognized at ORNL [8] that the sulfide-formation provided an efficient and selective method for isobar enhancement of Sn and Ge isotopes. This chemical method has allowed the production of purified beams of ^{82}Ge ($T_{1/2}$=4.6s) and ^{132}Sn ($T_{1/2}$=40s). These semi-magic and doubly-magic nuclei are important benchmarks within the chart of nuclides. Figure 2 shows the beam composition measured with a

Bragg detector for a RIB of mass A=82 used in a recent experiment to measure the B(E2) of the first 2^+ state of ^{82}Ge. Axial ion chambers or Bragg curve detectors developed for a research program of AMS have become important beam diagnostic tools at HRIBF to determine and monitor the isobaric content of RIBs. Projectile Z identification using X-rays is another method that has been used at HRIBF to monitor the beam composition. However in order to be able to quantify with confidence the beam composition we are currently studying experimentally the X-ray yield dependence on various parameters including projectile and target Z, projectile energy, target thickness, and projectile charge state, for energies above 3 MeV per nucleon relevant to COULEX, fusion-evaporation, and AMS studies at HRIBF.

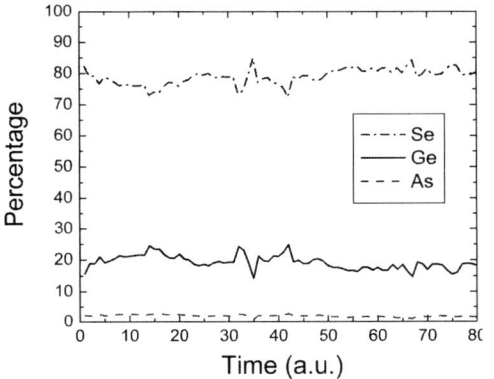 

FIGURE 2. Beam composition for a mass A=82, 220 MeV RIB monitored as a function of time (left) using a Bragg curve detector (right) located downstream from the target at zero degrees. Isobaric resolution was determined to be roughly one FWHM per Z at Z=50.

EXPERIMENTS WITH LIGHT PROTON-RICH RADIOACTIVE ION BEAMS

Study of Resonant States and Observation of Simultaneous Two-Proton Emission

We have been studying resonant states in light nuclei with radioactive beams using a fast and efficient method described in [6]. It involves the use of inverse kinematics, thick targets, large acceptance detectors with high-granularity and high-energy resolution double-sided silicon strip detectors (DSSDs), and a carbon-foil microchannel plate (MCP) timing/position detector to suppress the background associated with RIBs and to measure the beam profile. The advantage of this technique is that an excitation function can be obtained at a single bombarding energy with high efficiency. The excitation functions of elastic scattering are extracted from the measured energy spectra and scattering angles of the recoiling light-charged particles. The target is thick enough to stop the beam, but thin enough to let the recoils escape. Fully stripping the ions and separating the interfering isobars ions by the beam transport system produce pure RIBs from HRIBF. Radioactive ion beams of $^{17}F^{+9}$ were used to study several resonant states in ^{18}Ne [9] including the astrophysically important 4.52 MeV 3^+ state [10] (E_{cm} = 0.6 ± .05 MeV, Γ = 18 ± 2 keV), and the 6.15 MeV 1^- state (E_{cm} = 2.22 ±.01 MeV, Γ = 50 ± 5 keV) of relevance to the 2p emission process.

Radioactive ion beams of 55 MeV ^{11}C from the BEARS project at LBNL [11], a "portable" experimental setup and the thick-target technique were used to study low lying resonances in ^{12}N via elastic scattering with a ^{11}C radioactive beam [12]. If resonances dominate at low energies the reaction $^{11}C(p,\gamma)^{12}N$ then it is important to study properties of the resonant states of ^{12}N such as spin, parity and total width. In addition to its importance in astrophysics, the study of ^{12}N is also of considerable interest for nuclear structure. Very light nuclei have also received much recent attention in nuclear structure in connection with mixing of valence states between the p and the sd shell and the melting of the N=8 neutron shell closure.

The same efficient techniques led to the discovery at ORNL of the simultaneous two-proton emission from resonant states in ^{18}Ne populated using the exotic radioactive beam ^{17}F [9, 13] at a bombarding energy of 44 MeV. This elusive mode of decay might provide a new powerful tool to study how nucleons interact inside the nucleus. The nucleus ^{18}Ne makes a good place to look for this phenomenon as sequential single proton decay, $^{18}Ne \rightarrow ^{17}F+p \rightarrow ^{16}O+p+p$, is not possible for ^{18}Ne states up to about 6.5 MeV. Furthermore, the state involved (E* = 6.15 MeV) is unbound with respect to two-proton emission by 1.6 MeV. The p-p angular correlation data along with p-p relative energy distributions gave evidence for the simultaneous 2 proton emission. These results have received considerable interest [14, 15]. New results using targets of CH_2 and ZrH and a radioactive beam of ^{17}F at 60 MeV have confirmed our initial findings. A future experiment with larger solid angle coverage detection will provide better data for p-p angular correlation and relative energy distributions.

Polarization

Elastic scattering of protons (polarized and unpolarized) from light nuclei has provided important information on the spin, parity and isospin assignments of the resonant states involved. These measurements have consisted mainly of excitation functions at a few angles. Even for these conventional techniques the measurement of excitation functions of cross sections and analyzing powers with small energy steps are very time consuming. More efficient methods are desirable for the study of resonances using the limited RIB intensities available. One natural extension to our measurements of excitation functions of cross sections with the thick-target is to explore the determination of spin observables such as analyzing powers. Together with a low temperature group from Paul Scherer Institute we are exploring [16] the development of a polarized thick target of hydrogen or deuterium, which together with large arrays of DSSD's that are being developed at HRIBF will permit efficient measurements of analyzing powers. A possible additional advantage is to make the target a scintillator to serve as an "active" target [17,18] helping to eliminate background events.

EXPERIMENTS WITH NEUTRON-RICH RADIOACTIVE IONS BEAMS

COULEX of Te, Sn and Ge Isotopes

Approximately 275 stable nuclei are known and close to 3000 of about 7000 different nuclei thought to exist that can decay radioactively have been created and studied in the laboratory. Only about ten nuclei are doubly magic. Above $A \geq 56$ only ^{208}Pb is on the stability line and thus accessible to be studied by different probes. The existence of doubly closed shell nuclei is characterized by increased stability that reflects on number of isotopes, mass, binding energy, one-neutron or two-neutron separation energy. They also have relatively higher excitation energy of its first excited state. Of considerable interest is the development of RIBs in the region centered on the doubly magic nucleus ^{132}Sn. Low energy Coulomb excitation of RIBs in inverse kinematics will allow precise mapping of the evolution of collectivity, deformation and phase transitions along isotopic chains. The successful development of neutron-rich RIBs at HRIBF has provided the opportunity to study reduced electric quadrupole transition probabilities [B(E2)] in nuclei far from stability and therefore

FIGURE 3. View of HyBall with 95 CsI(Tl) photodiode detectors, mounted inside the CLARION γ-ray spectrometer. Each one of the 11 Clover HPGe detectors has BGO Compton shields.

stringently tests current nuclear structure theories. A series of Coulomb excitation experiments in inverse kinematics have resulted in the measurement of the B(E2) value of the lowest 2^+ excited states in a number of even-even Sn and Te isotopes [19, 20]. The experiments required the detection of γ-rays using the CLARION Ge detector array in coincidence with target recoils detected in the HyBall array [21] of 95 CsI(Tl) crystals (See Figure 3). RIBs of ~ 3 MeV/nucleon Te and Sn isotopes were used to bombard a natural 1 mg/cm^2 carbon foil. Ratios of Coulomb excitation to Rutherford scattering cross sections were determined as described in [19]. The measured B(E2) allowed a systematic analysis of quadrupole strength across the N=82 shell closure for the Te isotopes. An interesting feature that emerged from this analysis was the abnormally low B(E2) value (0.10 e^{2}b^2) of the first 2^+ state in ^{136}Te. A comparison of these results with Quasiparticle Random Phase Approximation calculations [22] suggests that weak neutron pairing in ^{136}Te can explain this anomaly.

We have measured the B(E2) for the first excited 2^+ states in the doubly closed-shell (N=82, Z=50) nucleus ^{132}Sn and the two-neutron nucleus ^{134}Sn. These experiments were done with the ORNL-MSU-TAMU BaF$_2$ Array. The large efficiency of this detector is useful to detect high energy gammas such as the 4.04 MeV transition from the first 2^+ state of ^{132}Sn and to deal with beams of very weak intensities such as the ~3000 ions/s available for ^{134}Sn. Beam and target-recoils were detected in an S2 model "CD"-type DSSD [23] with 48 radial strips and 16 azimuthal sectors. The ^{132}Sn experiment was performed using 470 MeV and 495 MeV ions incident on a 1.3 mg/cm^2 ^{48}Ti target. These energies however are above the "safe-energy" limit the detection on the CD permits, using the scattering angle to select only events with a "safe" distance of closest approach. For the ^{134}Sn experiment a 1.0 mg/cm^2 ^{90}Zr target was used and a lower "safe" energy of 400 MeV. Preliminary results give a B(E2; $0^+ \rightarrow 2^+$) value for the two-neutron nucleus ^{134}Sn similar to the value for the two-hole nucleus ^{130}Sn while the doubly magic ^{132}Sn exhibits a much larger B(E2); behavior similar to the one shown by ^{208}Pb.

Measurements of the B(E2) values for the first excited 2^+ states in ^{78}Ge, ^{80}Ge and ^{82}Ge [24] presented a major experimental challenge and were made possible by the experimental techniques and equipment available at HRIBF. The $2^+ \rightarrow 0^+$ γ-ray transitions of ^{78}Ge and ^{78}Se (contaminant) differ only by 5 keV, and the corresponding transition for ^{80}Ge and ^{80}Se (contaminant) differ by 7 keV. Correction for the Doppler broadening of the γ-rays was possible to about 3.9 keV FWHM with position information of the electrically segmented Clovers and using the detected target recoils in the HyBall. Even with this correction capability the first attempt to measure the B(E2) for ^{80}Ge was not possible since the beam was strongly contaminated by ^{80}Se and the ^{80}Se B(E2) value is larger than the predicted value for ^{80}Ge. The definite measurement had to make use of the chemical method to purify the beam using the GeS molecule [8]. For the measurement of the B(E2) value for the semi-magic ^{82}Ge nucleus we used the BaF$_2$-CD setup and the sulfide purification. The A=82 RIB beam of 5 x 10^4 ions/s consisted of 78.9 % ^{82}Se, 19.2 % ^{82}Ge and 1.9 % ^{82}As.

Mass Measurements of Unstable Nuclei near ^{78}Ni

Masses of neutron-rich nuclei, produced by fission, have been investigated [25]. The mass of some nuclides, which is so far only known from theoretical predictions, has been determined for the first time. The measurements were performed using a simple setup consisting of a position-sensitive channel plate detector [26] and a Bragg detector mounted near the focus of the energy-analyzing magnet of the 25-MV tandem accelerator. Measurements of mass differences were performed by measuring the position and Z of each ion in a mixed beam composed of unknown- and known-mass isobars. Beams of exotic nuclei such as ^{77,78,79}Cu at 3 MeV/nucleon were used at rates of 1.5 ions/s, 0.15 ions/s and 20 ions/h, respectively although they were limited by the overall rate of 5000 ions/s in the gas detector.

INSTRUMENTATION

The HyBall CsI(Tl) detector array in its standard form has thick absorbers in front of the detectors. We are in the process of building a new partial HyBall dedicated for RIB experiments that will cover up to 90° with minimum absorbers. The advantage of using this "bare" HyBall is a lower detection threshold as a consequence of lower energy losses of the target-like recoils. This will allow the use of heavier targets.

CONCLUSIONS

We can look forward to an invigorating time in the next few years when new techniques will be developed to deal with the wide field of physics opened by the availability of RIBs. This paper is dedicated to Jerry Garrett who in his last years devoted most of his time to planning and promoting the new RIB facility at Oak Ridge.

ACKNOWLEDGMENTS

I wish to thank the staff of HRIBF for their dedication in the production of RIBs. This talk is based on work done in collaboration with various individuals: C. Baktash, J. Batchelder, J.R. Beene, B. Fuentes, J. Gomez del Campo, C.J. Gross, M.L. Halbert, P.A. Hausladen, Y. Larochelle, J. F.Liang, P.E. Mueller, E. Padilla-Rodal, D.C. Radford, D. Shapira, D.W. Stracener, J. P. Urrego, R.L. Varner and C.-H. Yu.

Oak Ridge National Laboratory is managed by UT-Battelle, LLC for the U.S. Department of Energy under Contract No. DE-AC05-00OR22725.

REFERENCES

1. Welton, R.W., Nucl. Phys. **A701**, 452c (2002).
2. Mallory M., in HRIBF Newsletter, http://www.phy.ornl.gov/hribf/usersgroup/news/jul-03/jul-03.html#A.
3. Alton, G.D., et al., Nucl. Instrum. Methods **B 170**, 515 (2000).
4. Carter, H.K., et al., Nucl. Instrum. Methods **B 126**, 166 (1997).
5. http://www.phy.ornl.gov/hribf/usersgroup/news/jan-03/jan_03.html#RA1.
6. Galindo-Uribarri, A., et al., Nucl. Instrum. Methods **B172**, 647 (2000).
7. Kubik P.W., G. Korschinek, E. Nolte, Nucl. Instrum. Methods **B1**, 51 (1984).
8. Stracener, D.W., Nucl. Instrum. Methods **B204**, 42 (2003).
9. Gomez del Campo, J., et al., Phys. Rev. Lett. **86**, 43 (2001).
10. Bardayan, D.W., et al., Phys. Rev. Lett. **83**, 45 (1999).
11. Powell, J., et al., Nucl. Instrum. Methods **A455**, 452 (2000).
12. Galindo-Uribarri, et al., in Proc.Frontiers of Nuclear Structure, Berkeley, California, P.Fallon and R.Clark, Eds., p.323 (2003); AIP Conf.Proc. 656 (2003).
13. Galindo-Uribarri, A., et al., Nucl. Phys. **A682**, 363c (2001).
14. Woods, P., Science 2001 February 9; 291: 995-997 (in Perspectives).
15. Brown, B.A., F.C. Barker, D.J. Millener, Phys. Rev. **C65**, 051309 (2002).
16. Galindo-Uribarri, A., in Exploratory Workshop on Polarized Radioactive Beams and Polarized Targets, Strasbourg, 6-7 March 2003, http://sbgat252.in2p3.fr/polar03/.
17. van den Brandt, B., et al., Nucl. Instr. Methods **A446**, 592 (2000) and in Exploratory Workshop on Polarized Radioactive Beams and Polarized Targets, Strasbourg, 6-7 March 2003, http://sbgat252.in2p3.fr/polar03/.
18. Galindo-Uribarri, A., et al., Nucl. Instrum. Methods **A301**, 457 (1991).
19. Radford, D.C., et al., Phys. Rev. Lett. **88**, 222501 (2002).
20. Radford, D.C., et al., in proceedings of RNB6 to be published in Nucl. Phys. **A**.
21. Galindo-Uribarri, A., et al., to be published in Nucl. Instrum. Methods.
22. Terasaki, J., et al., Phys. Rev. **C 66**, 054313 (2002).
23. Micron Semiconductor Ltd. 1 Royal Buildings, Marlborough Road, Lancing, Sussex, BN15 8UN, England..
24. Padilla E, Ph.D. Thesis UNAM (2004) and Padilla, E., et al. Rev. Mex. Fis., **48 S2**, 93 (2002).
25. Hausladen P.A., et al., to be published in Nucl. Instrum. Methods.
26. Shapira, D., et al., Nucl. Instr. Methods **A454**, 409 (2000).

Role of surface vibrations in the pairing gap of atomic nuclei.

G. Gori*, F. Barranco†, P. F. Bortignon*, R. A. Broglia***, G. L. Colò* and E. Vigezzi*

*Dipartimento di Fisica, Universitá degli Studi di Milano, via Celoria 16, 20133, Milano, Italy and INFN, Sezione di Milano, via Celoria 16, 20133, Milano, Italy
†Departamento de Fisica Aplicada III, Escuela Superior de Ingenieros, Camino de los Descubrimientos s/n, 41092 Seville, Spain.
**The Niels Bohr Institute, University of Copenhagen, Blegdamsvej 17, 2100 Copenhagen Ø, Denmark

Abstract. Medium polarization effects induce an effective pairing interaction between nucleons which renormalize the nucleon-nucleon bare potential. We assess the importance of these renormalization processes concerning pairing correlations in both exotic nuclei as well as nuclei lying along the stability valley.

INTRODUCTION

Nuclear Field Theory (NFT) is a systematic approach to the description of single-particle and collective degrees of freedom of atomic nuclei and of their interweaving. The basic elements entering the theory are the bare masses of nucleons and the bare nucleon-nucleon interaction [1][2][3]. These elements define the mean-field properties of the system (k-mass and mean field potential), as well as the collective vibrations of this potential, particle-vibration coupling vertices, dressed particles (ω-mass, cf. Fig. (1a)), induced interaction (medium polarization, cf. Fig. (1b)), vertex corrections (cf. e.g. [4][5][6] and references therein), etc.. In the present contribution we shall concentrate our attention on the consequence polarization effects have on the pairing channel.

POLARIZATION EFFECTS AND PAIRING IN STABLE NUCLEI.

A first assessment of the importance polarization effects have concerning pairing correlation in nuclei is obtained by solving the BCS equation making use of the pairing matrix elements associated with the process depicted in Fig. 1(b).

In Fig. 2 we show the results [1] associated with Ca-isotopes in comparison with the

---

[1] To be noted that the induced interaction accounts in this case not only for the isotopic effect but also for almost all of the absolute value of the pairing gap. This is related to the fact that in the calculations reported in Fig. 2 one is using the bare nucleon mass instead of the k-mass (cf. next section).

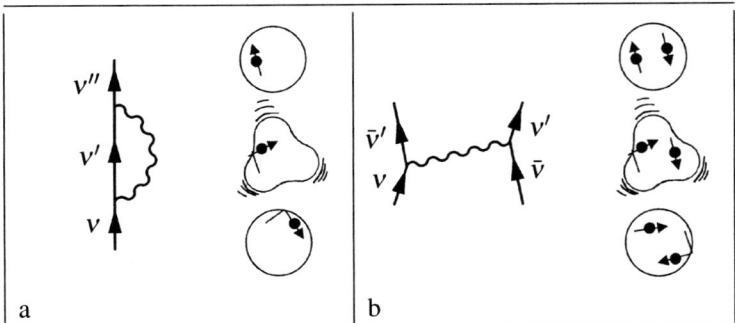

FIGURE 1. (a): on the left, we show the Feynman diagram associated to the self-energy process while on the right the schematic representation of a nucleon emitting and then reabsorbing a phonon through (virtual) anelastic processes (self-energy). (b): on the left, the Feynman diagram associated to the interaction mediated by the medium polarization while on the right the schematic representation of a nucleon emitting a phonon which is then reabsorbed by a second nucleon.

experimental odd-even mass difference (cf. ref. [6]). To be noted that the isotopic dependence of the pairing gap mirrors the trend of the sequence of the β_2 and β_3 deformation parameters controlling the particle-vibration coupling strength, a dependence which was also found important in the discussion of the mean-square radius of Ca-isotopes [7]. The importance of this result is underscored by the fact that while an empirical parametrization of the mass formula exists which can account for the masses of 2000 nuclear species with a root mean square deviation smaller than 1 MeV, subtle isotopic effects seem to be the major stumbling block for further improvements [8].

A STEP FORWARD: THE SOLUTION OF THE DYSON EQUATION.

In a recent work different observables of ^{120}Sn have been calculated making use of NFT, within the framework of Dyson equation, treating on equal footing self-energy, vertex corrections and induced interaction effects. In this work, nucleons are allowed to move in a self-consistent field calculated within Hartree-Fock theory [9]. The nucleons were allowed to interact through the Argonne potential and through the exchange of phonons. Figure 3 shows the state-dependent pairing gap for levels close to the Fermi energy. It can be seen that the bare Argonne nucleon-nucleon potential accounts only for the 50% of the experimental findings, while the inclusion of the self-energy, induced interaction and vertex corrections processes account for the rest. The lowest quasi-particle spectrum of ^{120}Sn predicted by theory is shown in Fig. 4. The experimental energies obtained from the analysis of the low energy spectrum of ^{119}Sn and ^{121}Sn, are compared with the results of a simple Hartree-Fock (HF) calculation, with the results of a HF+Bogoliubov calculation with Argonne potential, and with the results obtained by properly taking into account the NFT renormalization and dressing effects. The inclusion of these processes improves the agreement between theory and experimental data. The

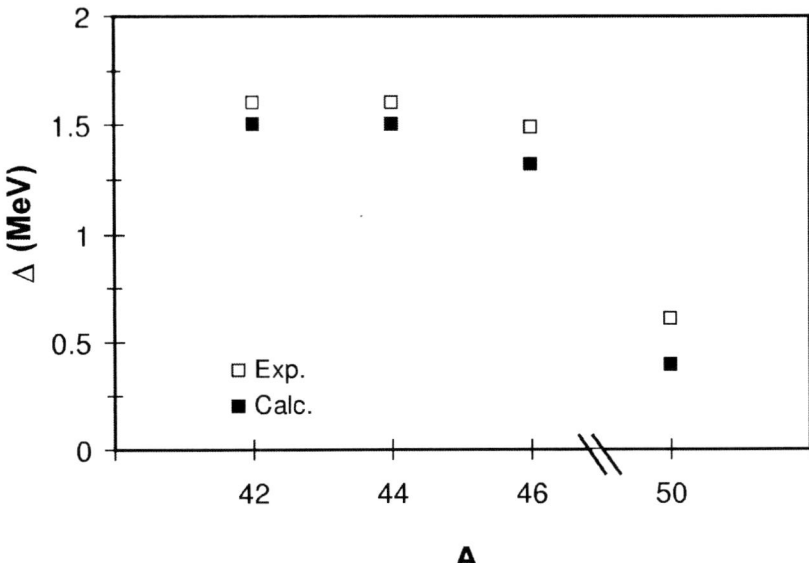

FIGURE 2. For four members of the calcium isotopic chain, the figure compares the average of Δ_k around the Fermi energy with the odd-even-mass-difference [10] obtained from the experimental binding energies [11].

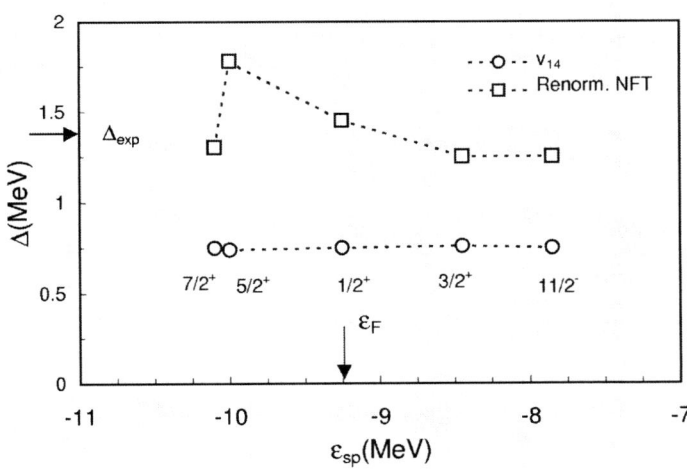

FIGURE 3. State-dependent pairing gap plotted as a function of single-particle energies for five levels around the Fermi energy ε_F. The circles are associated to a calculation which uses only the Argonne potential while the squares are results obtained by using Argonne plus the coupling of nucleons to collective vibrations. Also shown is the value of the experimental pairing gap.

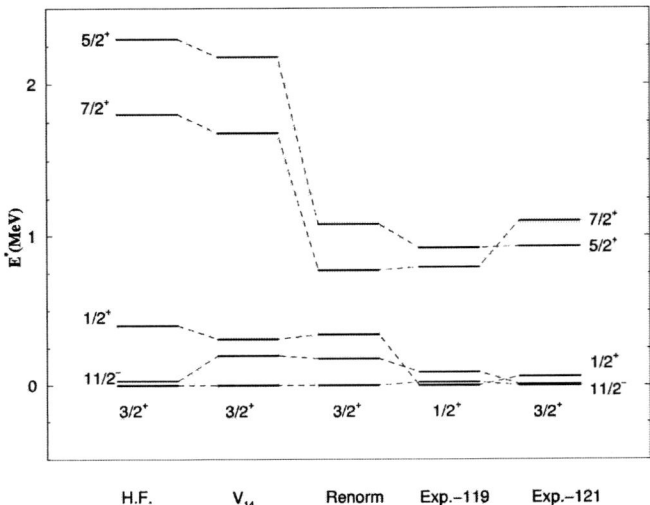

FIGURE 4. The experimental low-lying quasi-particle spectrum of ^{119}Sn and ^{121}Sn compared with different theoretical predictions obtained from a mean-field calculation (HF), a HFB calculation using Argonne potential (V_{14}), and taking into account coupling to collective vibrations (Renorm.)

same improvement is obtained in the case of the properties of the first 2^+ excited state: by including renormalization processes associated to the exchange of phonons between particle-hole excitations, good agreement between theory and experiment is obtained both for the energy and for transition probability of the collective states (cf. Table 1).

TABLE 1. Energies and transition probabilities of the lowest 2^+ excited state of ^{120}Sn calculated through RPA calculations using a Gogny (RPA(Gogny)) or Skyrme force (RPA(Sly4)) compared with the calculation which includes the renormalization associated to exchange of phonons (RPA+renorm.) and experimental data.

	$\hbar\omega_{2_+}$(MeV)	B(E2↑) (e^2fm^4)
RPA(Gogny)	2.9	660
RPA(Sly4)	1.5	890
RPA+renorm.	0.9	2150
Exp.	1.2	2030

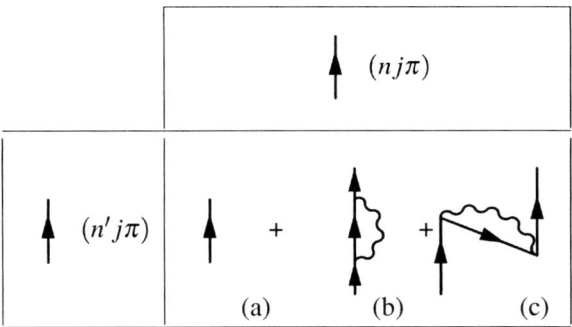

FIGURE 5. Schematic representation of the effective matrix used in the Bloch-Horowitz perturbation theory to calculate the eigenvalues of ^{11}Be. An arrowed line pointing upwards (downwards) indicates a particle (hole), while a wavy line indicate a collective vibrational state

THE CASE OF NEUTRON-RICH LIGHT NUCLEI.

In what follows we apply NFT, to the description of the spectrum of exotic nuclei. In particular we analyze the case of Beryllium and Lithium neutron-rich isotopes [12][13]. As an example we discuss in some detail the case of $^{11}_{4}$Be and $^{12}_{4}$Be. An analogous discussion can be carried out for $^{10}_{3}$Li and $^{11}_{3}$Li.

One of the features which characterizes the unusual properties displayed by ^{11}Be is that this nucleus shows a sequence of valence levels which is inverted with respect to the sequence associated to a standard mean-field potentials. In the case of ^{11}Be the level $1/2^+$ is more bound than the $1/2^-$ level. To be noted is the extremely small one neutron separation energy displayed by this nucleus (0.5 MeV). Furthermore, when one neutron is added to ^{11}Be, the system gains 2 MeV of binding energy, testifying to the importance pairing correlations have in correlating the two neutrons moving around ^{10}Be core.

In the calculation carried out, ^{11}Be and ^{12}Be are considered as systems obtained by adding one and two neutrons outside ^{10}Be respectively. These valence nucleons may interact with the core through the particle-vibration coupling mechanism, namely through the exchange of phonons. The effects of the particle-vibration coupling are taken into account within the Bloch-Horowitz theory (cf. [5]). That is the eigenvalues of the correlated system are obtained by diagonalizing the effective energy-dependent matrix depicted in Fig. 5 built on a basis in which only single-particle states appear explicitly.

To describe the ground state of ^{11}Be, the effective matrix was calculated considering configurations with $J^\pi=1/2^+$. In a mean-field representation, we started from a $2s_{1/2}$ state with energy in the continuum. Due to the coupling with states given by the product of particles in $d_{5/2}$ levels and a 2^+ phonon, the $1/2^+$ state gains energy and the lowest eigenvalue of the matrix turns out to have an energy of -0.48 MeV. Associated to this eigenvalue we obtain a ground state in which the valence nucleon spends 87% of the time in a pure $s_{1/2}$ single-particle state and 13% of the time in more complicated configurations (given by the product of a $d_{5/2}$ single-particle state and a 2^+ phonon). As far as the first excited state of ^{11}Be is concerned, it is known to be of $1/2^-$ character.

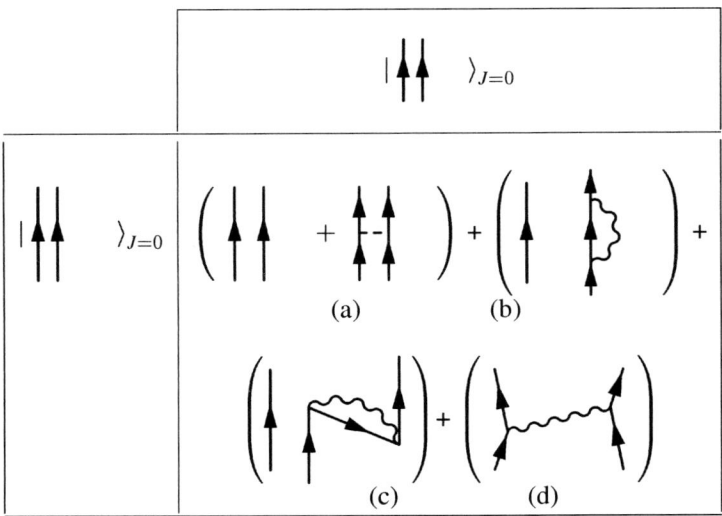

FIGURE 6. Schematic representation of the effective matrix used in the Bloch-Horowitz perturbation theory to calculate the eigenvalues of ^{12}Be. The dashed horizontal line represents the bare (Argonne v_{14}) nucleon-nucleon potential. Pairs of nucleons are coupled to angular momentum $J = 0$.

Thus, the energy matrix was worked out using an explicit basis of $p_{1/2}$ states. The lowest single-particle level calculated using a Woods-Saxon potential with standard parametrization (cf. [10]) is bound by 3.12 MeV, while the renormalized $1/2^-$ state is shifted upwards to -0.27 MeV due to Pauli principle corrections arising from the coupling of the $p_{1/2}$ with the 2^+ low-lying vibration. Summing up, the combined effects of self-energy and Pauli-principle blocking associated to the particle-vibration coupling explain the parity inversion observed in ^{11}Be.

To describe ^{12}Be we have used a basis made out of two-particle states, two particles and one phonon, three particles-one hole and one phonon. All these three types of configurations are coupled to zero angular momentum. The effects of the particle-vibration coupling in ^{12}Be are calculated by diagonalizing an effective matrix built on a basis of two particles coupled to zero angular momentum (cf. Fig. 6). These matrix elements include self-energy effects as in the case of ^{11}Be as well as induced interaction processes. To be noted that also the contribution arising from the action of the bare Argonne nucleon-nucleon potential has been taken into account. The corresponding contribution is, in this case, small, a fact associated with the low angular momentum content of the single-particle space available to the two neutrons to correlate.[2]

Diagonalizing the resulting matrix, one obtains a lowest eigenvalue equal to -3.58 MeV, to be compared with the experimental two-particle separation energy of -3.67

---

[2] The situation is quite different from that found in nuclei lying along the stability valley, where the pairing energy was a result of a combined action of the induced interaction ($\approx 50\%$) and of the bare nucleon-nucleon interaction ($\approx 50\%$).

MeV. The induced interaction between the two neutrons stabilizes the system leading to 2 MeV of extra binding energy.

CONCLUSIONS

The interweaving of nucleons and collective vibrations lead to (quantitatively) important renormalizations of the pairing interaction in nuclei. This interaction has to be treated explicitly to achieve a quantitative description of the nuclear system both in nuclei lying along the stability valley as well as in exotic nuclei.

REFERENCES

1. D.R. Bes et al., Nucl.Phys. A260(1976)1;
2. D.R. Bes et al., Nucl.Phys. A260(1976)27;
3. D.R. Bes et al., Nucl.Phys. A260(1976)77
4. C. Mahaux et al., Phys.Rep. 120 (1985)1;
5. P.F. Bortignon et al., Phys.Rep. 30C (1977)305;
6. F. Barranco et al., Phys. Rev. Lett. 83(1999)2147;
7. F. Barranco and R.A. Broglia, Phys.Lett.B 151B(1985)90;
8. F. Tondeur et al., Phys. Rev. C62(2000)24308
9. F. Barranco et al., nucl-th/0304049
10. A. Bohr and B.R. Mottelson, Nuclear Structure Vol. I, Benjamin, New York (1969);
11. G. Audi and A.H. Wapstra, Nucl. Phys. A595(1995)409;
12. G. Gori et al., nucl-th/0301097
13. F. Barranco et al., Eur.Phys.J. A11(2001)385;

Recent Experimental Results from the NSCL on the Structure of Exotic Nuclei

T. Glasmacher*, D. Bazin†, C.M. Campbell*, J.A. Church*, D.C. Dinca*, A. Gade*, W.F. Mueller†, H. Olliver*, B.M. Sherrill* and K.L. Yurkewicz*

*National Superconducting Cyclotron Laboratory and Department of Physics and Astronomy, Michigan State University, East Lansing, MI 48824, USA.
†National Superconducting Cyclotron Laboratory, Michigan State University, East Lansing, MI 48824, USA.

Abstract. The Coupled Cyclotron Facility at the National Superconducting Cyclotron Laboratory at Michigan State University provides a large variety of new isotopes previously inaccessible and others at rates sufficienct for in-beam spectroscopy. This talk presents some of our recent results elucidating the structure of exotic nuclei.

After a general overview of scientific highlights from the first two years of operation particular results from several in-beam gamma-ray spectroscopy experiments in the vicinities of neutron numbers $N=16$, $N=20$, $N=28$ in the $\pi(sd)$ shell around ^{56}Ni will be discussed. Inelastic scattering experiments with gamma-ray detection on light and heavy targets have determined specific transition matrix elements and excited state energies. One- and two-particle nucleon knockout reactions were used to investigate the wave functions of specific states and to deduce corresponding spectroscopic factors.

INTRODUCTION

Properties of exotic nuclei are now routinely studied at rare isotope facilities. Facilities based on in-flight fragmentation allow experimenters to offset low beam rates with thick targets to retain standard luminosity in the experiment. The use of γ-rays to indicate inelastic scattering is now a well-established technique [1, 2, 3, 4] and requires the use of position-sensitive photon detectors. These can establish the first interaction point of the emitted γ-ray, and thus, together with some knowledge about the emission point, the emission angle relative to the scattered beam particle. With this information the photon's energy in the moving frame can be established properly. The final energy resolution of the Doppler reconstructed γ ray is primarily dictated by the velocity spread of the beam at the time of γ-ray emission (particularly related to the velocity change in the target), the finite opening angle of the individual detectors, and the intrinsic energy resolution of the detector itself.

At the Coupled Cyclotron Facility at the National Superconducting Cyclotron Laboratory we have completed experiments on intermediate-eneryg Colomb excitation and knock-out reactions with the Segmented Germanium Array (SeGA) as a γ-ray detector.

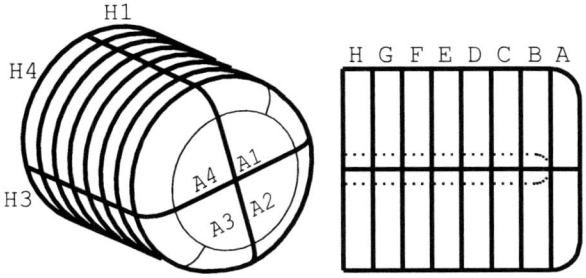

FIGURE 1. Schematic views of the crystal segmentation for a SeGA detector.

THE SEGMENTED GERMANIUM ARRAY (SEGA)

The Segmented Germanium Array (SeGA) consists of eighteen 32-fold segmented germanium detectors [5]. This array is specifically designed for in-beam γ-ray studies of fast exotic beams at the NSCL, where typical secondary beam velocities are $\beta \approx 0.3 - 0.4$. The individual crystals are coaxial 70% relative efficiency n-type closed-end germanium crystals. Each crystal is electronically segmented into eight one-centimeter wide sections along the symmetry axis and each is divided further into four quadrants for a total of 32 segments (see Fig. 1). In this manner one obtains a very high degree of granularity along the axis of symmetry, consequently, when the detector is operated with this axis perpendicular to the ray to the target one obtains a 0.6 cm (FWHM) position resolution for γ-ray interactions. For events with segment multiplicity larger than one, an algorithm is applied to deduce the first interaction point from all interaction points.

The SeGA detectors have a dual readout of the central contact of each crystal and a single readout from each of the external 32 segments. The energy gain matching of the segments is done in an automated way using the algorithm described in Ref. [6]. SeGA has been used in several experiments at the NSCL [7, 8, 9].

EXPERIMENTAL RESULTS

To illustrate the utility of SeGA for in-flight spectroscopy with fast beams, we discuss two recent measurements. The first is a study on the structure of nuclei around the doubly-magic ^{56}Ni [9]. In these experiments the reduced transition matrix elements to the first excited states of several nuclei, ^{52}Fe and ^{54}Ni in particular, were observed from intermediate-energy Coulomb excitation. These results provide information about the structure of Fe and Ni isotopes just below the $N = 28$ shell closure.

The second measurement that will be presented is two-proton knockout from the intermediate-energy exotic beam, ^{28}Mg [7]. The experiment employs two-proton knockout on neutron-rich nuclei as a direct reaction, and thus a tool from which one can obtain structure information about neutron-rich nuclei far from the valley of β stability. In both experiments, SeGA was located at the target position of the S800 high-resolution magnetic spectrograph [10].

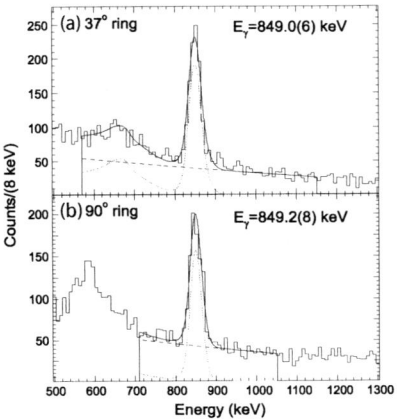

FIGURE 2. Energy spectra observed with SeGA in coincidence with ^{52}Fe recoils detected in the S800 spectrograph. Panel (a) shows the sum of detectors in the 37° ring and (b) the detectors in the 90° detectors. The curves are from GEANT simulations.

The nucleus ^{56}Ni is of interest because it is the heaviest double-magic $N = Z$ nucleus presently available for detailed study. Experiments to measure the reduced transition strength from the ground state to the first excited state in ^{56}Ni have resulted in $B(E2;0^+ \rightarrow 2^+) = 600(120)$ e^2fm^4 from an inelastic proton scattering experiment [11] and 580(70) e^2fm^4 observed from intermediate-energy Coulomb excitation [12]. These results can be compared to the value for ^{58}Ni, 695(20) e^2fm^4 [13]. This rather small drop in the reduced transition rate from ^{58}Ni to ^{56}Ni is inconsistent with what would be expected for a good double-magic nucleus. For example, the $B(E2 \uparrow)$ for ^{40}Ca ($Z = 20, N = 20$) is 100(10) e^2fm^4 [13], which compared to the value in ^{42}Ca (400(10) e^2fm^4 [13]) is significantly smaller.

An experiment was untaken at the Coupled Cycoltron Facility (CCF) and the NSCL to further study the nucleus ^{56}Ni as well as neighboring lighter nuclei ^{54}Ni and ^{52}Fe. To produce the radioactive beams, a 140 MeV/nucleon ^{58}Ni beam produced in the K500 and K1200 cyclotrons was directed on to a thick ^{9}Be ($\approx$ 400 mg/cm^2) fragmentation target located at the entrance to the A1900 fragment separator through which the fragments are selected and identified event-by-event at beam energies of 86 MeV/nucleon for ^{56}Ni and 70 MeV/nucleon for both ^{54}Ni and ^{52}Fe. These nuclei are then transported to the target position of the S800 spectrograph where a secondary target of 184 mg/cm^2 (in the case of ^{56}Ni) or 257 mg/cm^2 (for ^{52}Fe and ^{54}Ni) ^{197}Au is positioned. Surrounding the target position was SeGA with fifteen detectors installed.

The S800 spectrograph was set so that the scattered incident beam was centered at its focal plane, while the γ-ray decay of the inelastically scattered components of the beam were detected in SeGA. An event-by-event correlation was made between the detected γ rays and the coincident scattered nucleus.

Figure 2 shows two Doppler-corrected germanium spectra in coincidence with scattered ^{52}Fe isotopes. The gray lines in both spectra are the results of a GEANT3 [14] simulation of the array response for the 849-keV $2^+ \rightarrow 0^+$ transition in ^{52}Fe. The dot-

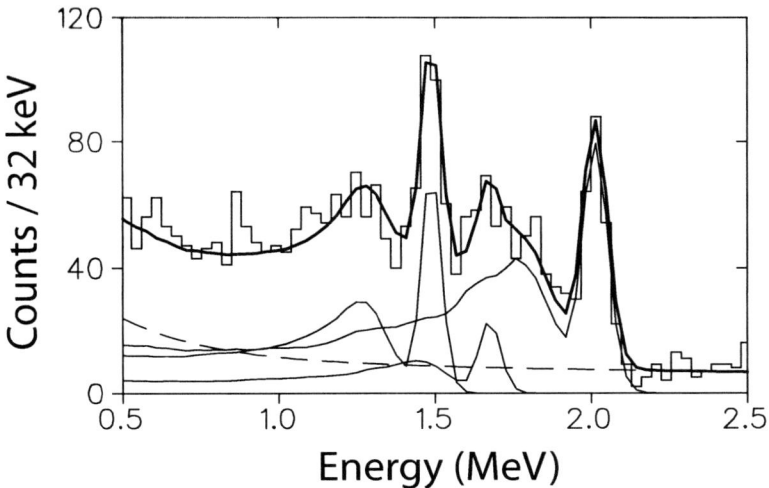

FIGURE 3. Doppler-corrected SeGA energy spectra correlated with ^{26}Ne residues from the two-proton knockout of ^{28}Mg. Definitions of the lines in this spectrum are presented in the text.

ted lines are exponential functions approximating the continuous background, while the black lines correspond to the summed contribution of the two functions.

Using stopped sources for absolute normalization and GEANT3 simulations to model the boost of the in-flight γ rays, one can determine the total γ-ray yield from SeGA, relative to the number of detected ^{52}Fe nuclei to determine the Coulomb excitation cross section. We calculate a B($E2\uparrow$) value from the resulting cross section measurment. The analysis of ^{52}Fe, as well as for ^{54}Ni, and other neighboring nuclei is in progress.

For the two-proton knockout from ^{28}Mg a primary beam of ^{40}Ar was accelerated to 140 MeV/nucleon in the coupled cyclotrons and directed onto a thick ^{9}Be target to create ^{28}Mg at an average energy of 88.3 MeV/nucleon. The ^{28}Mg was then selected in the A1900 fragment separator and identified event-by-event at the A1900 focal plane.

The intermediate-energy beam is then delivered to the target position of the S800 spectrograph where a secondary target of 375 mg/cm^2 ^{9}Be was located. The target was surrounded by SeGA and the S800 spectrograph was set to accept ^{26}Ne residues that result from the knockout of two protons from the ^{28}Mg beam. The ^{26}Ne nuclei are then identified event-by-event in by their energy loss and position in the focal plane of the S800.

The γ rays resulting from this two-proton knockout reaction are detected in SeGA and correlated with the ^{26}Ne detected in the S800 focal plane. The resulting Doppler-corrected γ-ray spectrum is shown in Fig. 3. In this figure, the histogram represents the experimental observation, while the three thin solid lines correspond to GEANT3 [14] simualations of three known γ rays at 1.47, 1.66, and 2.02 MeV [15, 16, 17], the dashed line corresponds to an expontential function that approximates the continuous background, and the thick solid line is the summation the three simulated γ-ray spectra and the background.

From the analysis of the narrow parallel momentum distribution of the ^{26}Ne residues in the S800 focal plane, it can be seen that the knockout of two protons from ^{28}Mg occurs as a direct reaction. Consequently, by comparison to an appropriate reaction theory it is possible to extract spectroscopic information. From the γ-ray yields one can obtain spectroscopic information on individual excited states. Because this reaction takes one farther from the valley of β stability, the two-proton knockout reaction opens the possibility to study the structure of nuclei much closer to the neutron-drip line than previously feasible. A more detailed description of this reaction, as well as a discussion of the theoretical interpretation is provided in Bazin et al. [7].

SUMMARY

With the development of an array of highly-segmented germanium detectors, it has become possible to perform in-flight γ-ray spectroscopy experiments on intermediate energy beams with unprecedented γ-ray energy resolution. Presented here are examples of two techniques in which SeGA was used for the detection of γ rays emitted from fast-moving exotic beams. SeGA used in conjunction with a high-resolution magnetic spectrograph (S800) to detect the reaction residues in coincidence represents a powerful combination for in-beam γ-ray studies.

ACKNOWLEDGMENTS

The work presented here is supported by the U.S. National Science Foundation under Grant numbers PHY-9875122 and PHY-0110253.

REFERENCES

1. T. Glasmacher, Annu. Rev. Part. Sci. 48 (1998) 1.
2. P.G. Hansen and B.M. Sherrill, Nucl. Phys. A693 (2001) 133.
3. H. Iwasaki et al., Phys. Lett. B 522 (2001) 227.
4. M. Belleguic et al., Nucl. Phys. A682 (2001) 136c.
5. W.F. Mueller et al., Nucl. Instr. and Meth. A 466 (2001) 492.
6. Z. Hu et al., Nucl. Instr. and Meth. A 482 (2002) 715.
7. D. Bazin et al., Phys. Rev. Lett. 91 (2003) 012501.
8. A. Gade et al., Phys. Rev. C 68 (2003) 014302.
9. K.L. Miller, PhD. dissertation, Michigan State University (2003).
10. D. Bazin et al., Nucl. Instr. and Meth. B 204 (2003) 629.
11. G. Kraus et al., Phys. Rev. Lett. 73 (1994) 1773.
12. Y. Yanagisawa et al., AIP Conf. Proc. 455 (1998) 610.
13. S. Raman et al., At. Data and Nucl. Data Tables 78 (2001) 1.
14. GEANT v.3.21. CERN library long writeup, Technical report W5013, CERN, 1994.
15. B.V. Pritychenko et al., Phys. Lett. B 461 (1999) 322; 467 (1999) 309(E).
16. A.T. Reed et al., Phys. Rev. C 60 (1999) 024311.
17. O. Sorlin et al., Nucl. Phys. A685 (2001) 186c.

Nuclear Structure around the N=16 subshell closure

O. Sorlin*, M. Stanoiu*, Zs. Dombrádi[†], F. Azaiez*, B.A. Brown**, M. Belleguic*, D. Sohler[†], M.G. Saint Laurent[‡], M.J. Lopez-Jimenez[‡], Y.E. Penionzhkevich[§], G. Sletten[¶], N.L. Achouri[∥], F. Becker[‡], J.C. Angélique[∥], C. Borcea[††], C. Bourgeois*, J.M. Daugas[‡], Z. Dlouhý[‡‡], C. Donzaud*, J. Duprat*, Zs. Fülöp[†], D. Guillemaud-Mueller*, S. Grévy[∥], F. Ibrahim*, A. Kerek[§§], A. Krasznahorkay[†], M. Lewitowicz[‡], S. Leenhardt*, S.M. Lukyanov[§], S. Mandal[¶¶], P. Mayet[¶¶], H. Van der Marel[§§], W. Mittig[‡], J. Mrázek[‡‡], F. Negoita***, F. de Oliveira-Santos[‡], Zs. Podolyák[†††], F. Pougheon*, M.G. Porquet[‡‡‡], H. Savajols[‡], Y. Sobolev[§], C. Stodel[‡], J. Timár[†] and A. Yamamoto[†††]

*IPN, IN2P3-CNRS, Orsay, France
[†]Institute of Nuclear Research, Debrecen, Hungary
**NSCL, East Lansing, USA
[‡]GANIL, Caen, France
[§]FLNR, JINR, Dubna, Russia
[¶]NBI, University of Copenhagen, Denmark
[∥]LPC, Caen, France
[††]IFIN Bucharest Magurele, Romania
[‡‡]Nuclear Physics Institute, Řež, Czech Republic
[§§]Royal institute of Technology, Stockholm, Sweden
[¶¶]GSI Darmstadt, Germany
***IFIN Bucharest Magurele, Romania
[†††]Department of Physics, University of Surrey, Guildford, UK
[‡‡‡]CSNSM, Orsay, France

Abstract.
In-beam gamma-spectroscopy using fragmentation reactions of both stable and radioactive beams have been performed in order to study the structure of excited states in the neutron-rich ^{18,20}C, $^{21-24}$O and ^{23,25}F isotopes. For the produced fragments, γ-ray energies, intensities and $\gamma\gamma$ coincidences have been measured. Based on this information new level schemes are proposed up to the neutron separation energy for some of these isotopes. The systematics of the 2^+ energy has been extended up to ^{20}C in the C isotopic chain. Level schemes are suggested for ^{21}O and ^{22}O. The non observation of any γ-ray from ^{23}O and ^{24}O suggests that their excited states lie above the neutron decay thresholds. The study of F isotopes provides hint of the possible change in nuclear structure between the F and O isotopic chains. These results, together with the size of the N=14 and N=16 shell gaps in C,O,F isotopes are discussed in the framework of shell model calculations.

In this paper we shall report on experiments probing the structure of the N=12 to N=16 C, O and F nuclei by the use of fragmentation reactions and γ-ray spectroscopy. A special emphasis will be made to the existence of the N=14,16 subshell closures and their evolution between C and F nuclei.

EXPERIMENTAL TECHNIQUE

In order to obtain information on the structure of neutron rich nuclei in the proton-sd shell region, in-beam γ-ray spectroscopy using the projectile fragmentation reactions has been used. Two experiments using the fragmentation of ^{36}S stable beam have been performed at GANIL. Both experiments are based on coincidence measurements between the projectile like fragments and their prompt γ-decay. A common γ-detector array has been used in both experiments. It consists of 74 BaF$_2$ detectors placed at a mean distance of 30 cm from the target which cover a major part of the solid angle around it. The relatively high velocity of the produced projectile fragments (v/c$\simeq$0.35) requires to apply proper Doppler shift corrections to the detected gamma rays. The produced fragments were selected and identified through the SPEG spectrometer by the use of time of flight, energy loss, and focal-plane position information.

The first experiment consisted in using the fragmentation of a 77.5 MeV·A ^{36}S beam with an intensity of about 1 pnA on a thin Be target (2.77 mg/cm^2) located at the entrance of SPEG, the energy loss spectrometer at GANIL. The SPEG was tuned to the $A/Z = 2.75$ mass to charge ratio. As a result, the nuclei $^{21-23}$O were transmitted. In addition to the BaF$_2$ array, four high resolution Ge detectors were used at the most backward angles. In the following this experiment will be refereed to as 'single step fragmentation' (SSF) experiment.

For the second experiment, we used a primary beam of ^{36}S delivered by the two GANIL cyclotrons at an energy of 77.5 MeV·A and an intensity of 400 pnA on a carbon target of 348 mg/cm^2 thickness placed in the SiSSi fragment separator. The produced nuclei were selected through the ALPHA spectrometer using a 130 mg/cm^2 Al wedge. The magnetic rigidity of the ALPHA spectrometer and the optics of the beam line were optimized for the transmission of a secondary beam mainly composed of ^{24}F, ^{25,26}Ne, ^{27,28}Na and ^{29,30}Mg fragments with energies varying from 54 MeV·A up to 65 MeV·A. An 'active' target composed of a plastic scintillator (103 mg/cm^2) sandwiched by two carbon foils of 51 mg/cm^2 each was used at the same dispersive focus of SPEG as the Be target in the first experiment. The plastic scintillator part of the 'active' target was used to identify in an event by event basis the incoming nuclei through energy loss and time of flight measurements. The fragments induced by reactions of the secondary beam are collected and identified at the SPEG focal plane which was optimized for the ^{24}O products (A/Z=3). This method which has also been used at RIKEN [8] and GSI [9] is called the 'double step fragmentation' (DSF) method because it uses two consecutive fragmentation reactions in order to populate excited states of the nuclei of interest.

In the case of the SSF method the ^{36}S beam intensity has been limited to $6 \cdot 10^9$ pps in order to reduce the counting rate in the individual γ-ray detectors to a reasonable rate of $2 \cdot 10^4$ s^{-1}. With the DSF technique, a secondary beam made of species close to the nuclei of interest was used with an intensity of $8 \cdot 10^4$ s^{-1}. Consequently, ^{24}O was produced with almost a factor 10 higher intensity than in the SSF, and the counting rates of individual γ-ray detectors was reduced by two orders of magnitude allowing for much better working conditions.

EXPERIMENTAL RESULTS AND DISCUSSIONS

Oxygen isotopes

The ^{21}O nucleus has been studied earlier by transfer reactions [3] where the existence of four excited states has been revealed. The spin and parity of the ground state has been recently suggested to be 5/2$^+$ [10]. The ^{21}O isotope has been produced with relatively high statistics in both the SSF and DSF experiments. Based on γ-ray coincidences from the BaF$_2$ detectors, we are able to build the level scheme of ^{21}O as shown in Fig. 1, which is compared to the previous work of Catford et al. [3]. The USD Shell model calculations of Brown [11, 12] are in excellent agreement with experiment.

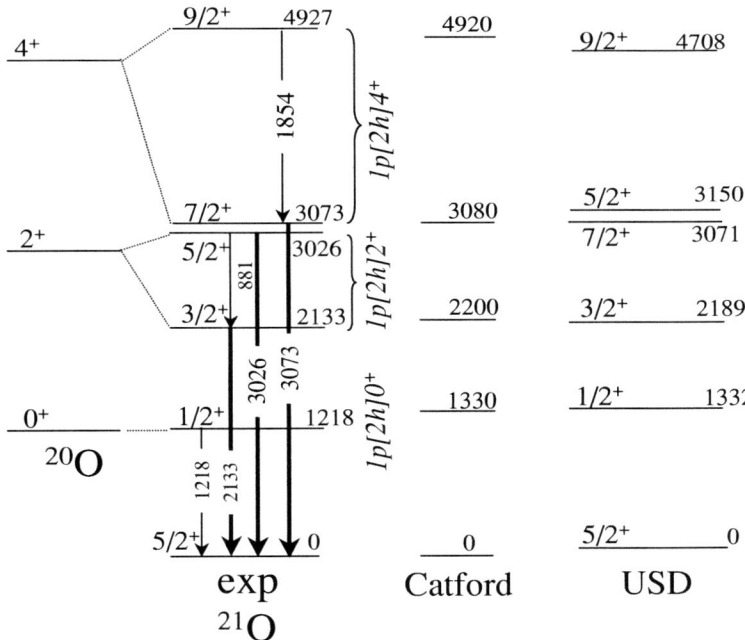

FIGURE 1. Level scheme of ^{21}O from the present experiment. Comparison is made with the work of Catford et al. [3] and with the USD Shell Model predictions [12]. Configurations in terms of particle-hole excitations are indicated.

Relative to the dominant closed d$_{5/2}$ shell for ^{22}O ground state which will be called 0p-0h, the g.s. structure of ^{21}O can be interpreted in term of the $0p-1h$ 5/2$^+$ configuration. The $1p-2h$ excited states are based upon coupling of the s$_{1/2}$ (particle) orbit to the $(d_{5/2})^{-2}$ 2h state 0$^+$, 2$^+$ and 4$^+$ to give the ^{21}O excited states 1/2$^+$, (3/2$^+$,5/2$_2^+$) and

($7/2^+$, $9/2^+$), respectively. The spacing of the $2h$ states is related to the low-lying state of ^{20}O such as 0^+ (gs), 2^+ (1.64 MeV) and 4^+ (3.57 MeV).

We have already reported two gamma rays of 3190(15) and 1380(10) keV [4] corresponding to gamma decay of excited states in ^{22}O. The first one was identified to be the 2^+ to 0^+ transition whereas the second one was found to correspond to the decay from a higher excited state to the 2^+ state. However, no spin or parity assignment was made for this state. Recently Thirolf et al. [5] have confirmed the energy and measured the electromagnetic matrix element B(E2) for the 2^+ state in ^{22}O via inelastic scattering.

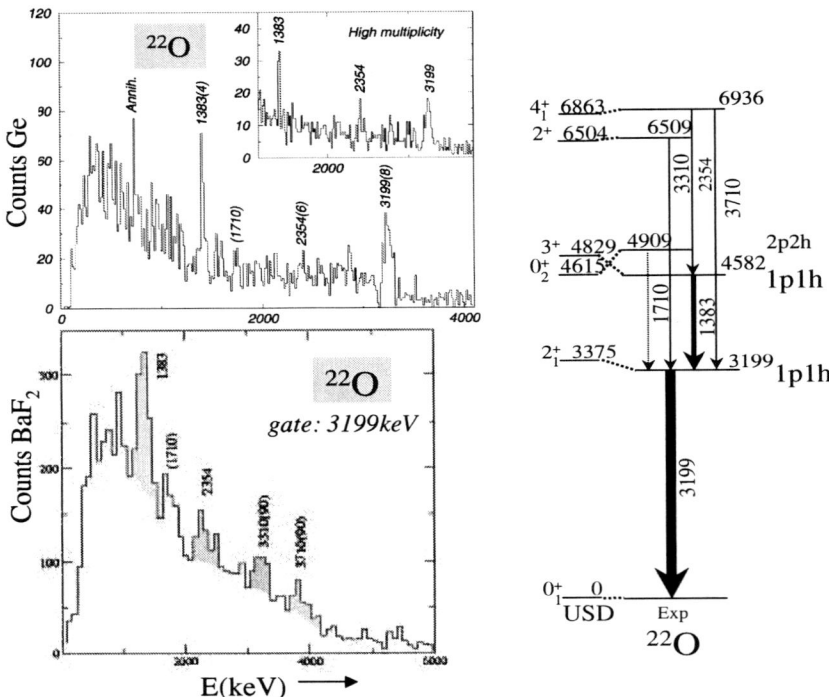

FIGURE 2. Left top: gamma-rays spectrum of ^{22}O from the Ge detectors. The high multiplicity spectrum is shown in the inset. Left bottom: gamma-rays spectrum from the BaF$_2$ detectors with the requirement of a 3.2 MeV γ-ray in coincidence. Right: Deduced level scheme of ^{22}O is shown with the sd shell model calculations.

The γ-ray spectrum of ^{22}O have been obtained with the information from the SSF experiment. The high resolution spectrum cumulated during the SSF experiment from the Ge detectors is shown in Figure 2. The levels above the first 2^+ state have been found from the BaF$_2$ spectrum in which the coincidence relation with the 2^+ to 0^+ 3.2 MeV transition was required.

According to shell model calculations, the full sd-shell wave function for ^{22}O ground state is 77% $0p - 0h$, 22% $2p - 2h$ and 1% $4p - 4h$ and the 2^+ energy is 3.38 MeV.

Relative to the dominant $0p-0h$ ground state the excited states in ^{22}O are $1p-1h$ with one particle in the $s_{1/2}$ or $d_{3/2}$ orbits coupled to one hole in the $d_{5/2}$ orbit. The lowest of these are the 2^+ and 3^+ states which are dominated by the $s_{1/2}$ orbit. The energies of these states are split by the residual $1p-1h$ interaction. Relative to the ESPE (Effective Single particle Energy) gap of 4.3 MeV the 2^+ state is pushed down to 3.4 MeV and the 3^+ state is pushed up to 4.8 MeV. The $(2J+1)$ weighted average of 4.2 MeV is close to the ESPE gap [13].

The information for ^{23,24}O nuclei is very limited due to experimental difficulties to reach nuclei close to the drip lines. Their S_n-values are known to be 2.7(1)MeV and 3.7(4) MeV, respectively [14]. The ground state spin and parity of ^{23}O has been recently identified by E. Sauvan et al. [10] as $1/2^+$, value which is in good agreement with the USD sd-shell model calculations. In the DSF experiment a total number of 19620 and 7000 nuclei of ^{23}O and ^{24}O have been produced, respectively. The obtained γ-ray spectra from the BaF$_2$ detector array are presented in Fig. 3. Both exhibit a relatively low statistics and a structureless component above 600 keV despite a fairly large amount of ^{23}O nuclei produced. The broad energy peak found at around 500 keV is attributed to the contribution from the 511 keV line and the 477 keV line from the ^{7}Li target-like fragment emitted almost at rest. This peak becomes thinner when the Doppler correction is not applied, confirming that it does not belong to the decay of excited state in O isotopes in flight. We, therefore, tentatively conclude that no bound excited state exist in these nuclei.

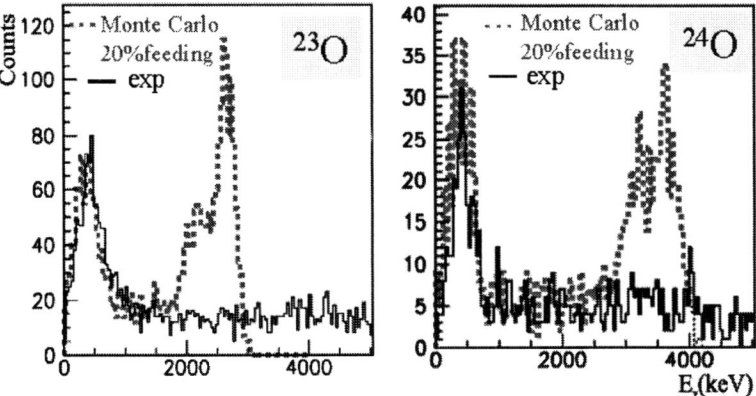

FIGURE 3. Left: gamma-rays spectrum of ^{23}O from the BaF$_2$ detectors array with Doppler correction. The corresponding Monte Carlo simulated spectrum is shown with a dashed line. Right: Same for ^{24}O.

In order to prove that the ^{23,24}O nuclei don't have any bound excited state, we have performed a Monte Carlo simulation of the BaF$_2$ array spectrum obtained from a 2.7 MeV (3.7 MeV) γ-ray emitted by a source moving with the velocity of the ^{23}O (^{24}O) fragments. The S_n values in ^{23}O and ^{24}O are 2.7 MeV and 3.7 MeV, respectively. No γ-ray is expected above this value. The lowest value of the direct feeding intensity of the first excited state being of the order of 30% in all the cases in the double step experiment, a safe value of 20% was taken for the simulation. For comparison, the simulated spectrum with the proper Doppler correction is shown in Fig. 3 together with the one obtained from our experiment. This comparison shows that if a bound excited state existed in ^{23}O, a γ line should be clearly seen in the measured spectrum. Similar conclusion holds true for ^{24}O.

Relative to the $0p-0h$ model for ^{22}O, the ^{23}O levels are $1p-0h$ levels and $2p-1h$ ones. Of these only the lowest $1p-0h$ $s_{1/2}$ (the g.s. $1/2^+$) is predicted to be bound and this agrees with the present experiment. The first excited state $5/2^+$ state at 2.72 MeV is the lowest state which is dominated by $2p-1h$ configuration. This is predicted to be very near the one-neutron decay threshold, and its non-observation in this experiment indicates that it is in fact above the neutron decay threshold. The $1p-0h$ state which is dominated by $d_{3/2}$ is the $3/2^+$ predicted to be at 3.28 MeV and thus also unbound to neutron decay. The $s_{1/2}$ state which is relatively strongly bound in ^{23}O provides another magic nucleus for ^{24}O. Relative to a dominant $0p-0h$ wave function for the ^{24}O ground state, the excited states are $1p-1h$; $d_{3/2}$-$s_{1/2}^{-1}$ with $J = 1^+$ and 2^+ and $d_{3/2}$-$d_{5/2}^{-1}$ with $J = 1^+$-4^+. Of these the lowest is the 2^+ predicted to be at 4.18 MeV and is close to the neutron-decay threshold of 3.7(3) MeV. The non observation of any γ transitions in ^{24}O is consistent with theory.

The importance of the high (unbound) energy for the $d_{3/2}$ orbit is that all of the nuclei beyond ^{24}O where one or more nucleons goes into the $d_{3/2}$ orbit are unbound – ^{24}O is at the drip line. The $d_{3/2}$ must be sufficiently unbound so that the pairing energy is not enough to bind ^{26}O (which has not been observed). With the USD interaction in the full sd-shell ^{25}O is unbound to one-neutron decay by 0.77 MeV and ^{26}O is bound to one-neutron decay by 1.77 MeV and bound to two-neutron decay by 1.00 MeV. Thus to make ^{26}O unbound to two-neutron decay, either the $d_{3/2}$ ESPE for ^{24}O has to be at least 0.5 MeV higher or the pairing energy in the $d_{3/2}$ has to be reduced by at least 1.0 MeV (or a combination of these).

Fluorine isotopes

The ^{22}O and ^{24}O nuclei can be used as cores to model the ^{23}F and ^{25}F nuclei, respectively. We expect in these nuclei $1p-0h$ $d_{5/2}$ and $s_{1/2}$ proton configurations for their g.s. and first excited state. In addition to this a multiplet of levels corresponding to $d_{5/2} \otimes 2^+$ couplings should be observed at the 2^+ energy of the core, i.e. around 3.2 MeV for ^{23}F. The level scheme of ^{23}F extracted from the SSF experiment exhibits three high energy transitions ranging from 2.9 to 3.8 MeV which correspond to the decay of the Yrast members of the multiplet to the g.s. We observe a similar behaviour in the case of ^{25}F, i.e. a group of high energy transitions at a mean value of about 3.5 MeV. The fact that we can model the F isotopes in the weak coupling framework from ^{22}O and ^{24}O cores witnesses that the nuclear structure has not changed dramatically when adding a proton to the O core. We could then question why the F isotopes are bound with up to 20 neutrons and above, while $^{25}O_{17}$ is already unbound by about 0.77MeV. Part of the answer is provided by the $\pi d_{5/2}$-$\nu d_{3/2}$ interaction [15] which binds the $\nu d_{3/2}$ orbital in the Z>8,N>16 nuclei. Gamma-rays of up to 657 keV are observed in the $^{26}_{9}$F energy spectrum [16]. So there is a ~1.4 MeV single particle energy gain when the proton is added to the $d_{5/2}$. This agrees with the USD shell model predictions in which a multiplet of states with J=1,4 is bound at a centroid value of 0.46 MeV in ^{26}F.

Carbon isotopes

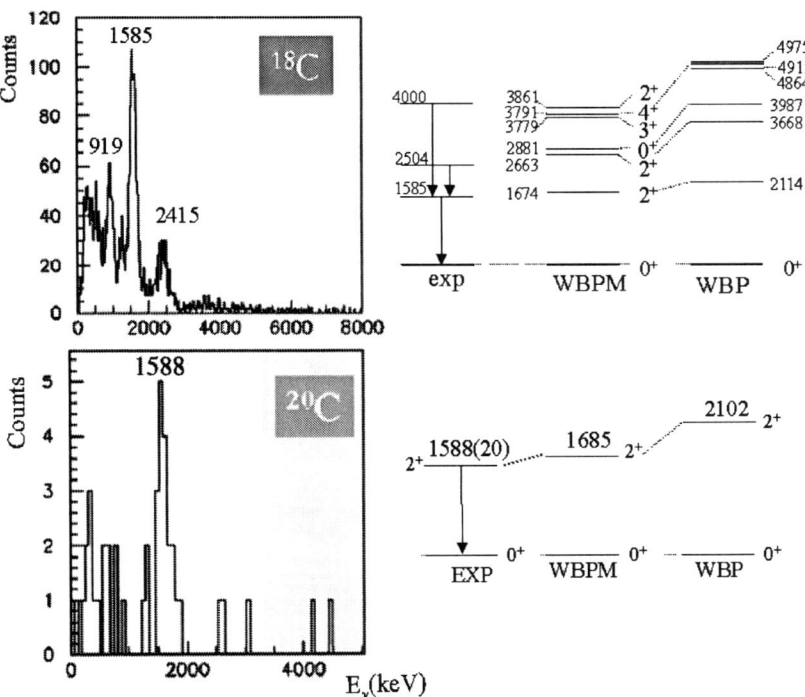

FIGURE 4. Top: gamma-rays spectrum of ^{18}C from the BaF$_2$ detectors. The deduced level scheme is shown adjacent to the spectrum. It is compared with shell model calculation using the WBP and WBPM interactions (see text for details). Bottom: same for ^{20}C.

The spectra observed for ^{18}C and ^{20}C are shown in Fig 4 with the corresponding level schemes. The first excited state of ^{20}C is observed for the first time and new excited states above the first 2^+ are found in ^{18}C. The experimental values are compared to shell model calculations using the WBP interaction [18]. This interaction uses the same Two Body Matrix Elements (TBME) as those which reproduce the O nuclei successfully. We observe that the experimental spectrum is shrunk by about 25% as compared to theory. To obtain a good agreement between theory and experiment (see Figure.4), it is necessary to multiply the TBME in C by a factor 0.75. The corresponding calculation is referenced as WBPM (M meaning modified). Indeed, the S_n values of the C nuclei are about half those of the O isotones. Consequently the radii of the neutron valence orbitals are larger in the C isotones. The matrix elements of the two-body interaction scales approximately as the inverse of the squared radius. The scaling factor applyed in the TBME of C isotopes account for their loosely bound character as compared to O isotopes. This argument is hitherto qualitative and the TBME will be calculated in a forthcoming future in the C isotopic chain.

The energies of the 2^+ states in the C and O isotopes are remarkably similar and constant at a value of about 1.6 MeV from N=10 to N=12. At N=14, they diverge suddenly, the energy of the 2^+ state of ^{20}C remains constant whereas it is two times larger in ^{22}O. From this trend, we deduce that the N=14 gap no longer exists in the C isotopes. Consequently the $d_{5/2}$ (6 neutrons) and $s_{1/2}$ (2 neutrons) orbitals are almost degenerated, and the addition of up to six neutrons from ^{16}C does not affect the energy of the first 2^+ state which is mainly composed of ph neutron excitations in these two orbitals. Since the $d_{5/2}$ and $s_{1/2}$ are closely packed, a large subshell effect is expected at N=16 in the C chain, which would make the first 2^+ of ^{22}C probably unbound with respect to neutron emission.

ACKNOWLEDGMENTS

The experiment using in-beam γ spectroscopy with fragmentation reactions benefit from the smooth running of the accelerator by the GANIL crew. This work has been supported by the European Community contract N° HPRI-CT-1999-00019, from OTKA-D34587,T38404,T42733, PICS(IN2P3) 1171, INTAS 00-0043, RFBR N96-02-17381a and NSF PHY-0244453 grants, as well as from Bolyai Janos Foundation.

REFERENCES

1. O. Tarasov et al., Phys. Lett. **B 409**, (1997) 64.
2. H. Sakurai et al., Phys. Lett. **B 448**, (1999) 180.
3. W.N. Catford et al., Nucl. Phys. **A 503** (1989) 263.
4. M. Belleguic et al., Nucl. Phys. **A 682**, (2001) 136c.
5. P.G. Thirolf et al., Phys. Lett. **B 485**, (2000) 16.
6. A. Ozawa et al., Phys. Rev. Lett. **84**, (2000) 5493.
7. Z. Dlouhy et al., Proc Int. Nucl. Phys. Conf. INPC 2001, ed. by E. Norman et al., AIP Conf. Proc 610 (2002) 736.
8. K. Yoneda et al., Phys. Lett. **B 499** (2001) 233.
9. see the contributions of Th. Aumann, D. Cortina-Gil, and C. Nociforo in the present volume.
10. E. Sauvan et al., Phys. Lett. **B 491**, (2000) 1.
11. B. A. Brown and B. H. Wildenthal, Ann. Rev. of Nucl. Part. Sci. **38**, 29 (1988).
12. www.nscl.msu.edu/~brown/sde.htm
13. B. A. Brown, Prog. in Part. and Nucl. Phys. **47**, 517 (2001).
14. G.Audi and A.H.Wapstra, Nucl. Phys. **A595**, (1995)409.
15. T. Otsuka et al., Phys. Rev. Lett. **87**, 082502 (2001).
16. M. Stanoiu, Thesis work GANIL T03 01, Université de Caen (2003).
17. M. Stanoiu et al. Proc Int. Conf. NS2002, Eur. Phys. J. A, in press.
18. E. K. Warburton and B. A. Brown, Phys. Rev. C **46**, 923 (1992).

Rotational bands in neutron-rich $^{160-162}$Ho

D. Escrig*, A. Jungclaus†*, B. Binder**, A. Dietrich**, T. Härtlein**, H. Bauer**, Ch. Gund**, D. Pansegrau**, D. Schwalm**, D. Bazzacco‡, G. de Angelis‡, E. Farnea‡, A. Gadea‡, S. Lunardi‡, D. R. Napoli‡, C. Rossi-Alvarez‡ and C. Ur‡

Instituto de Estructura de la Materia, C.S.I.C., E-28006 Madrid, Spain
†*Departamento de Física Teórica, Universidad Autónoma de Madrid, E-28049 Madrid, Spain*
**Max-Planck-Institut für physik, D-69029 Heidelberg, Germany*
‡*I.N.F.N, Laboratori Nazionale di Legnaro, I-35020 Legnaro, Italy*
Dipartimento di Fisica dell' Universita di Padova, Italy

Abstract. We have studied the high spin states in $^{160-162}$Ho in order to investigate the properties of the rotational bands and their dependence on the single particle orbits involved. The reaction ^{158,160}Gd(^{7}Li,xn) at 56 MeV were used to produce the Ho isotopes of interest. In all three Ho isotopes the known rotational bands have been significantly extended. New band-crossings have been observed for the first time in this work.

INTRODUCTION

Since the discovery of the backbending effect in the early seventies [1] much effort has been spent to study experimentally and theoretically this rotational behavior. The case of ^{161}Ho was the first proof *'that backbending can occur in rotational bands in odd A-nuclei'* leading to the conclusion that *'the $h_{11/2}$ proton is not strongly involved in the mechanism producing the backbending in this region'* [2].

In this work we studied the high-spin properties of the neutron-rich nuclei ^{160,161,162}Ho (Z=67,N=93-95) in the rare earth region. Whereas a lot of experimental information is available about the high-spin structure of the neutron deficient nuclei in this region, for the more neutron-rich isotopes this information is scarce due to the difficulty to populate these nuclei with heavy-ion induced fusion-evaporation reactions. However incomplete fusion reactions (also called massive-transfer reactions) have been recently shown [3] to be a good spectroscopic tool to gain access to high spin states in relatively neutron-rich nuclei near stability.

We report here on the investigation of the backbending effect in the rotational bands of $^{160-162}$Ho and the role played by the single-particle orbital occupied by the odd proton in this behavior. To do this we have compared the results obtained for the Ho-isotopes with the neighboring Dysprosiums isotones (Z=66, N=93-95) [4, 5].

Information about the experiment and the reaction is given in section 2. The data analysis is explained in section 3 and in section 4 the backbending plots for all three isotopes as well as the discussion of their properties are presented. In this contribution we have omitted the discussion of the level schemes which can be found in [6].

EXPERIMENTS

The experiment was performed in the Laboratori Nazionali di Legnaro. Excited states in $^{160-162}$Ho were populated using the reactions $^{158,160}Gd(^7Li,xn)$, x= 5 or 6, at a beam energy of 56 MeV. The beam was delivered by the XTU tamdem accelerator and directed onto targets of thicknesses of 3.7 mg/cm^2 (^{158}Gd) and 3.7 mg/cm^2 (^{160}Gd).

The γ radiation was detected using the GASP detector array. GASP is composed of 40 Compton-suppressed HpGe detectors with a total photopeak efficiency of 3%. The 80-element BGO inner ball was used as a multiplicity filter and the 40 ΔE-E telescopes Silicon ball ISIS to detect charged particles. More details about the experimental set-up are explained elsewhere [7].

All events with at least three coincident γ-rays in the Ge detectors or two in the Ge plus one particle detected in the Si ball were recorded on tape with the additional condition of multiplicity three or higher in the BGO ball.

DATA ANALYSIS AND RESULTS

As mentioned before the $^{160-162}$Ho isotopes are the pure neutron channels of the ^{7}Li+^{158,160}Gd reactions. In this reactions also some Dy isotopes were strongly produced with the emission of either a proton, a deuteron or a triton. Whereas charged particle gated $\gamma-\gamma$ matrices were used for the analysis of the Dy isotopes [4, 5], the $\gamma-\gamma$ matrices sorted in anti-coincidence with the silicon detector were not clean enough to allow for a detailed analysis of the Ho isotopes. Therefore $\gamma-\gamma-\gamma$ cubes were built to study the pure neutron channels.

The analysis of this cube data lead to a significant extension of the levels schemes in all three isotopes.

In ^{160}Ho [8] the bands with proton-neutron configurations $\pi\, 7/2^-$ [523] $\nu\, 5/2^+$ [642], $\pi\, 7/2^+$ [404] $\nu\, 5/2^-$ [523] and $\pi\, 7/2^-$ [523] $\nu\, 11/2^-$ [505] were extended up to spin 28^-, 17^- and 24^+ respectively. The ground state band was previously known [8] only up to spin 11^+ and has been increased up to spin 24^+ in the present work.

The sequences of the $7/2^+$ [404], $1/2^+$ [411] and $1/2^-$ [541] bands in ^{161}Ho [9] have been extended up to spin $51/2^-, 35/2^+, 51/2^+$ and $41/2^+$ respectively.

Finally, for the ^{162}Ho case, the only known rotational band [10], built on the $\pi\, 7/2^-$ [523] $\nu\, 5/2^+$ [642] proton-neutron orbitals has been extended up to the 28^- state.

Detailed discussion of all results obtained can be found in reference [6].

Backbending in the odd-even ^{161}Ho

In figure 1 the spin alignment plots as a function of rotational frequency are shown for all rotational bands in ^{157,159,161}Ho (Z=67, N=90-94). ^{157,159}Ho are included for completeness. In order to investigate the role of the unpaired proton in the backbending, which in the rare earth region is produced by the antipairing of an $i_{13/2}$ neutron pair [2], the alignment plots of the Yrast band in the neighboring ^{156,158,160}Dy isotopes (Z=66, N= 90-94) are included in figure 1 (small dots in left column).

As a result of our analysis we were able to determine for the first time the total spin gain for the ground state band, which is based on the $7/2^-[523]h_{11/2}$ proton orbital. This gain is measured to be around $8\hbar$ and the critical rotational frequency is $\hbar\omega_c \approx 0.28\ MeV$. From figure 1 (left column) we can observe that this band has a similar behavior for all three Ho isotopes and also very similar to the Yrast band in the Dy isotopes. This fact indicates that the unpaired proton does not play important role in this case.

In the $7/2^+[404]g_{7/2}$ and $1/2^+[411]d_{3/2}$ bands upbends have been observed for the first time at a critical rotational frequency of $\hbar\omega_c \approx 0.28\ MeV$ as in the ground state band.

Finally, for the band built on the proton $1/2^-[541]h_{9/2}$ orbital we observed upbending at a rotational frequency of $\hbar\omega_c \approx 0.35\ MeV$. This different behavior can be explained because in this configuration the orbital occupied by the unpaired proton has a sharp positive slope, minimization of energy leads to a change in nuclear deformation and therefore a change in rotational properties.

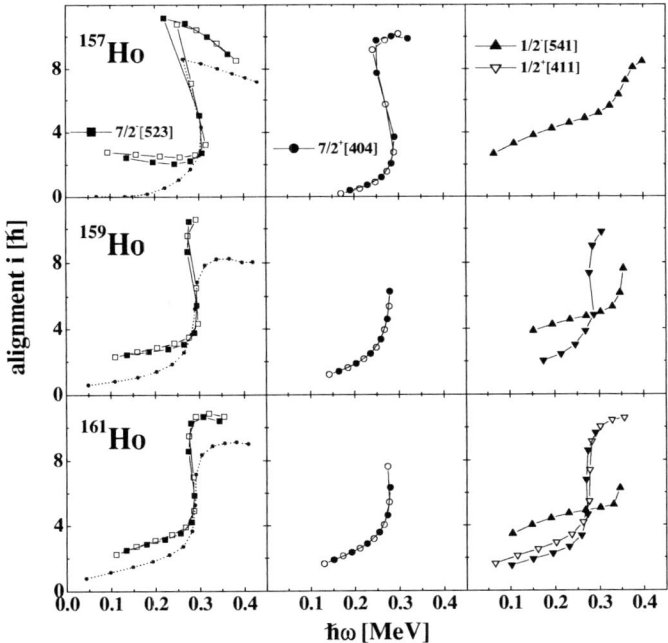

FIGURE 1. Alignment plots for the rotational bands in ^{157}Ho (top row), ^{159}Ho (middle row) and ^{161}Ho (bottom row). States of signature α=+1 and α=-1 are drawn with filled and open symbols respectively. In the left column the alignment of the Yrast bands in the even neighbors ^{156,158,160}Dy is shown for comparison (in small dots), see text for explanation.

Backbending in the odd-odd ^{160,162}Ho

The odd-odd ^{160,162}Ho isotopes have neutron numbers N=93 and N=95 (Z=67) respectively. We will compare them to the isotones ^{159}Dy$_{93}$ and ^{161}Dy$_{95}$ which have the same neutron configuration and with ^{159}Ho isotope which has the same proton configuration, see figure 2.

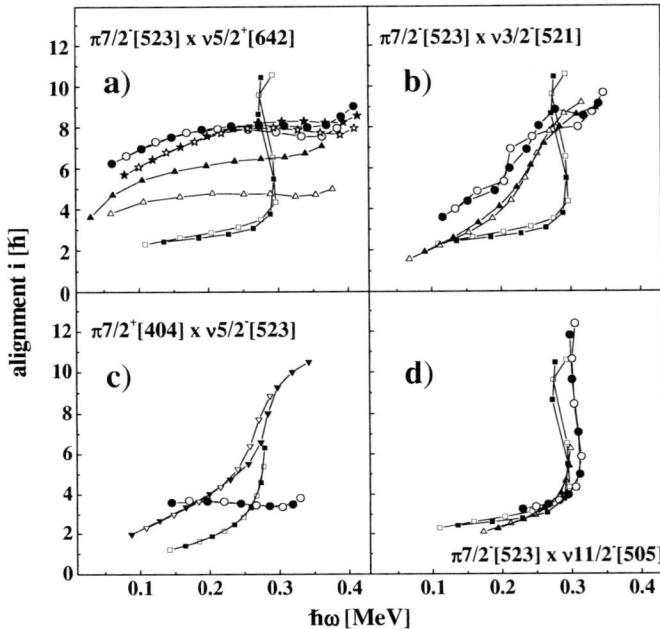

FIGURE 2. Alignment plots for the four rotational bands in ^{160}Ho (circles) and the $\pi\,7/2^-$ [523] $\nu\,5/2^+$ [642] band in ^{162}Ho (stars). The following bands in neighboring nuclei are also shown: a) 7/2$^-$[523] in ^{159}Ho and 5/2$^+$[642] in ^{159}Dy, b) 7/2$^-$[523] in ^{159}Ho and 3/2$^-$[521] in ^{159}Dy, c) 7/2$^+$[404] in ^{159}Ho and 5/2$^-$[523] in ^{161}Dy, d) 7/2$^-$[523] in ^{159}Ho and 11/2$^-$[505] in ^{159}Dy. In all cases squares symbols refer to Dy isotopes and triangle symbols to ^{159}Ho. Filled symbols refer to states of signature $\alpha = +1$ and open symbols to the ones with $\alpha = -1$.

In figure 2.a, the experimental aligned angular momentum is shown for the $\pi\,7/2^-$ [523]$h_{11/2}$ $\nu\,5/2^+$ [642]$i_{13/2}$ bands in ^{160}Ho and ^{162}Ho. The bands in ^{159}Ho and ^{159}Dy based in the single-particle orbit are also plotted. Whereas backbending is observed for the 7/2$^-$[523]$h_{11/2}$ band in ^{159}Ho no backbending appears in the 5/2$^+$[642]$i_{13/2}$ band in ^{159}Dy (taken from [5]). In this latter isotope the antipairing of the $i_{13/2}$ neutron pair is blocked by the unpaired neutron which is located in the same shell. Consequently, also for the 7/2$^-$ [523] ν 5/2$^+$ [642] bands in ^{160}Ho and ^{162}Ho no crossing is found.

The $\pi\,7/2^-\,[523]\,\nu\,3/2^-\,[521]$ and $\pi\,7/2^-\,[523]\,\nu\,11/2^-\,[505]$, figure 2.b and 2.c, bands in ^{160}Ho are found to have similar properties as the corresponding neighbor ^{159}Dy with an upbend at a rotational frequency of $\hbar\omega_c \approx 0.28\,MeV$.

In figure 2c the alignment plot for the $\pi\,7/2^+\,[404]g_{7/2}\,\nu\,5/2^-\,[523]f_{7/2}$ band in ^{160}Ho is shown. Although backbending is observed for the band based on both single particle configurations no backbending is found for the ^{160}Ho case. This behavior is rather surprising because no blocking effect is expected.

SUMMARY AND CONCLUSION

The rotational behavior of the neutron-rich $^{160-162}$Ho nuclei has been investigated at high spin via gamma spectroscopy. The reactions $^{158,160}Gd(^7Li,xn)$ were used to produce the Ho isotopes. We have shown that ^{7}Li induced reactions allow to populate high spin states above the band-crossing region. In each of the three isotopes, the known rotational bands have been extended up to significantly higher spin. Backbending phenomena have been observed for the first time in some of the bands.

REFERENCES

1. Johnson, A., and Sztarkier, J., *Phys. Lett.*, **34**, 34 (1971).
2. Grosse, E., Stephens, F., and Diamond, R., *Phys. Rev. Lett.*, **32**, 74 (1974).
3. Dracoulis, G. e. a., *Appl. Phys. Letters*, **85**, 1192–1202 (1997).
4. Jungclaus, A., and et al, *Phys. Rev. C*, **66**, 014312 (2002).
5. Jungclaus, A. e. a., *Phys. Rev.C*, **67**, 034302 (2003).
6. Escrig, D., Jungclaus, A., and et al, *Rotational bands in neutron-rich* $^{160,161,162}Ho$, Eur. Phys. J., submitted, 2003.
7. Bazzaco, D., *The gamma array spectrometer GASP*, International Conference on Nuclear Structure at High Angular Momentum, Otawa, Canada, 1992.
8. Drissi, S., and et al, *Nucl. Phys.A*, **600**, 63 (1996).
9. Funke, L., and et al, *Nucl. Phys.A*, **170**, 593 (1971).
10. Firestone, R., *Table of Isotopes*, John Wiley & Sons, Inc., 1996.

Systematic AMD+GCM Study of Structure of Carbon Isotopes

G. Thiamova[*,†], N. Itagaki[*], T. Otsuka[***] and K. Ikeda[**]

[*]*Department of Physics, University of Tokyo, Hongo, Tokyo 113-0033, Japan*
[†]*Nuclear Physics Institute, Czech Academy of Sciences, Prague-Rez, Czech Republic*
[**]*The Institute of Physical and Chemical Research (RIKEN), Wako, Saitama, 351-0198, Japan*

Abstract. The structure of low-lying states of the carbon isotopes is investigated using the extended version of the Antisymmetrized Molecular Dynamics (AMD) Multi-Slater Determinant model. We can reproduce reasonably well many experimental data for carbon isotopes ^{12}C-^{22}C. A special approach is adopted for ^{15}C to better describe the tail of the wave function.

INTRODUCTION

In this work we apply the extended version of the Antisymmetrized Molecular Dynamics (AMD) approach [1] to perform a systematic analysis of the carbon isotopes. The main advantage of the AMD method [2] is that it is free from any model assumptions concerning wave functions.

The motivation for introducing the extended version of AMD is as follows; one Slater determinant has been shown to be not enough for the description of a halo system or a system with a neutron skin. An attempt to improve the description by superposing several Slater determinants did not lead to a substantial improvement.

The method adopted here corresponds to the combination of AMD and the Generator Coordinate Method (GCM). The initial GCM basis functions are constructed in such a way that they correspond to a certain value of a properly chosen physical quantity and by changing this quantity a lot of Slater determinants with different intrinsic structure are prepared. In our case r.m.s. radii are constrained.

AMD+GCM FRAMEWORK

The total wave function of the system is described as a superposition of J^π projected AMD wave functions. Their spatial part is expressed by a Gaussian wave packet in coordinate representation,

$$\psi_i(r) = (\frac{2\nu}{\pi})^{3/4} \exp[-\nu(r - \frac{Z_i}{\sqrt{\nu}})^2 + \frac{1}{2}Z_i^2], \quad (1)$$

$$\propto \exp[-\nu(r - R_i)^2 + \frac{i}{\hbar}K_i \cdot r], \quad (2)$$

where complex parameters $Z_i = \sqrt{v}R_i + \frac{i}{2\hbar\sqrt{v}} K_i$ represent centers of the Gaussian wave packets and v is the width parameter, fixed to $v = \frac{1}{2b^2}$, $b = 1.6$ fm.

The diagonal elements of the Hamiltonian-matrix become a function of the parameter Z,

$$E(Z,Z^*) \equiv \frac{<\Phi^{\pm}(Z)|\hat{H}|\Phi^{\pm}(Z)>}{<\Phi^{\pm}(Z)|\Phi^{\pm}(Z)>}. \qquad (3)$$

The parameters Z are optimized by solving a frictional cooling equations

$$\frac{dZ_i}{d\tau} = -\frac{\partial E}{\partial Z_i^*} + \eta \frac{\partial O}{\partial Z_i^*}, \quad \frac{dZ_i^*}{d\tau} = -\frac{\partial E}{\partial Z_i} + \eta \frac{\partial O}{\partial Z_i}, \qquad (4)$$

The r.m.s. constraint is kept constant by introducing a Lagrange multiplier η. The expectation value of the Hamiltonian (E) decreases as development of the imaginary time τ, since the τ derivative of E is always negative. Finally, the angular momentum is projected and the diagonalization of the Hamiltonian matrix is done.

The Hamiltonian and the effective nucleon-nucleon interaction used is the same as in ref. [3].

RESULTS

The comparison of the calculated and experimental binding energies is shown in Fig. 1. The binding energy of ^{12}C is smaller partially due to the stronger LS force which suppresses the 3α like component in the ground state wave function. Also, the Majorana parameter $M=0.6$ has been fitted to the binding energy of ^{16}O and it produces underbinding of ^{12}C. The overbinding observed for ^{20}C can be also related to the used LS force.

In ^{15}C the $2s_{1/2}$ orbital is below the $1d_{5/2}$ orbital. This fact is clearly observed as an abnormal ground state spin parity $J^{\pi} = 1/2^+$ of this nucleus. The lowering of the s orbital is due to the halo formation. The halo is formed since the orbital with lowest angular momentum gains energy by extending its wave function. For ^{15}C we obtained the ground state spin $5/2^+$ (Fig. 1). This may be due to strong LS force which pushes the $d_{5/2}$ orbit down but also the s-orbit is not expressed well by a single Gaussian. Better description of the s-orbit will produce a substantionally lower $1/2^+$ state, as is shown later.

The ^{12}C and ^{14}C nuclei correspond to neutron sub-closed ($N=6$) and closed ($N=8$) shell, respectively, and both the calculated and experimental 2^+ energies are rather high (Fig. 2). However, the calculated 2^+ energy of ^{14}C (8.32 MeV) is higher than the experimental value (7.01 MeV). One of the reasons is most probably larger spin-orbit splitting of the $1p_{1/2}$ and $1p_{3/2}$ spin-orbit partners which brings the dominant proton configuration $(1p_{1/2})^3(1p_{3/2})^1$ higher in energy. Another support for this argument comes also from higher $3/2^-$ state (6.49 MeV) than the experimental value (3.68 MeV) in ^{13}C (Fig. 4a).

On the other hand, the spin-orbit interaction plays an important role in describing the 2_1^+ state in ^{12}C. In the cluster model calculations the level spacing between the 0_1^+

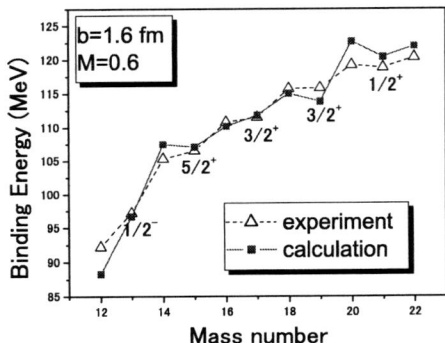

FIGURE 1. Binding energies of the carbon isotopes.

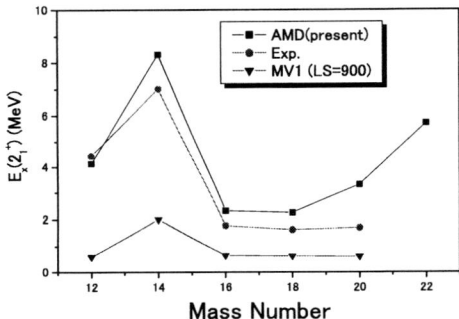

FIGURE 2. The excitation energies of the 2_1^+ states of the even-even carbon isotopes.

and 2_1^+ states was always underestimated. In the present approach this level spacing is reproduced because the LS force describes the dissolution of the α clusters in the ground state [4]. The values obtained with the MV1 force are taken from [5].

From the larger 2_1^+ energy in ^{20}C it seems that a $(1d_{5/2})^6$ sub-shell closure appears. On the contrary, the experimental values of the 2_1^+ energies of ^{16}C, ^{18}C and ^{20}C are almost the same, suggesting the $2s_{1/2}$ and $1d_{5/2}$ orbits are almost degenerate in these isotopes. In our case, the large spin- orbit splitting of the $1d_{5/2}$ and $1d_{3/2}$ orbits brings the $1d_{5/2}$ orbit lower in energy and the sub-shell closure may develop.

The experimental 2_1^+ energy of ^{22}C is not known yet. In our calculation a high value (5.7 MeV) is obtained. If this sharp increase at ^{22}C is measured, this would be a strong evidence for the $N = 16$ new magic number.

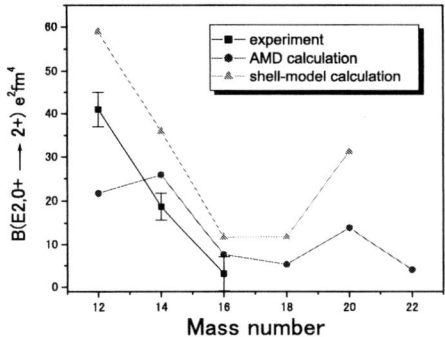

FIGURE 3. The electromagnetic transition strength

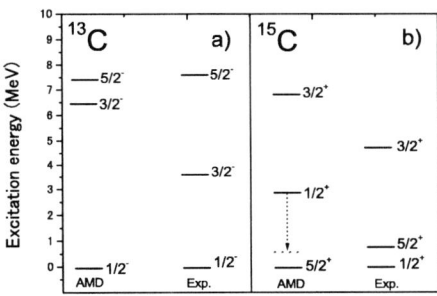

FIGURE 4. The energy levels of ^{13,15}C. The energy of $1/2^+$ obtained by P+MRO - dashed.

The comparison of the experimental $B(E2)$ values with those calculated within the shell model [6] with effective charges and the present approach where bare charges are used is shown in Fig. 3. The $B(E2)$ values for the ^{16}C and ^{18}C isotopes are very small because protons constract almost closed shell configuration. In the shell-model calculations it is because of small neutron effective charges for weakly bound neutrons. The B(E2) value for ^{16}C has been measured recently [7] and it is indeed very small. In ^{20}C the proton contribution to the $B(E2)$ value becomes again larger, because the contribution of protons increases.

We have already mentioned that the ground state spin of ^{15}C was not reproduced. To take into account one of the mechanisms responsible for the development of halo structure, namely the fact that the s-orbit can extend its wave function very far to decrease its kinetic energy, we adopt a projection + multiple relative orientation (between the core and the valence neutron) technique (P+MRO). First, the core wave functions (^{14}C) are generated and afterwards on each of them several wave functions of the last valence neutron are superposed, each of them corresponds to a different relative orientation with respect to the core. The angular momentum projection is performed for the total system. When a large number of basis wave functions is adopted (5 wave functions for 3 different r.m.s. constraints for ^{14}C and for each of them 10 wave functions for the valence neutron) the $1/2^+$ state is 0.6 MeV above $5/2^+$ (Fig. 4b).

CONCLUSION

In this paper we have presented systematic calculations for ^{12}C-^{22}C. Large number of quantities have been calculated for the even-even isotopes. The calculated binding energies are in reasonable agreement with the experimental data. The systematic comparison of the binding and 2_1^+ energies of the even-even isotopes with the experimental data reveals the importance of the spin-orbit term of the effective interaction. Specifically, the calculated energy of the 2_1^+ state in ^{14}C and the observed sub-shell closure in ^{20}C suggest the spin-orbit term should be weaker.

The neutron magic number $N = 8$ is reflected by the large 2_1^+ energy of ^{14}C. Very large 2_1^+ energy of ^{22}C supports the idea of $N = 16$ neutron magic number.

The calculated $B(E2)$ values show good tendency, similar to this obtained by shell model calculations with reduced effective neutron charges. The advantage of the AMD method is that no effective charges have to be used, because the changes of neutron and proton distribution are authomatically described by the model. Recently measured very small $B(E2)$ value for ^{16}C is successfully reproduced by our model.

In case of the ^{15}C isotope good description of the tail of the wave function is important. It has been done by applying a projection + multiple relative orientation technique and a much lower $1/2^+$ state has been obtained.

ACKNOWLEDGMENTS

This work was supported in part by Grant-in-Aid for Scientific Research (13740145) from the Ministry of Education, Science and Culture. The financial support from the Japanese Society for Promotion of Science is acknowledged.

REFERENCES

1. N. Itagaki and S. Aoyama, Phys. Rev. C **61**, 024303 (2000).
2. Y. Kanada-En'yo and H. Horiuchi, Phys. Rev. C **54**, R468 (1996).
3. N. Itagaki and S. Okabe, Phys. Rev. C **61**, 044306 (2000).
4. Y. Kanada-En'yo , Phys. Rev. Lett. **81**, 5291 (1998)
5. Y. Kanada-En'yo and H. Horiuchi, Prog. Theor. Phys. Suppl. **142**, 205 (2001)
6. R. Fujimoto, PhD thesis, University of Tokyo, 2002
7. N. Imai and Z. Elekes, private communication

Exotic cluster structure in light neutron-rich nuclei

N. Itagaki[*] and S. Aoyama[†]

[*]*Department of physics, University of Tokyo, Hongo, Tokyo 113-0033, Japan*
[†]*Information Processing Center, Kitami Institute of Technology, Kitami 090-8507, Japan*

Abstract. A new method for calculating the low-lying states of light nuclei is proposed: Antisymmetrized Molecular Dynamics (AMD) – Superposition of Selected Snapshots (AMD triple-S). In addition to the cluster features of the core nucleus, the properties of the wave of valence nucleons are well expressed in terms of a superposition of selected AMD wave functions. For ^{6}He, the binding energy and the halo nature are reproduced using effective interactions determined from a phase-shift analysis. For the deformed core cases, in ^{12}Be, the newly observed $J = 0$ state at 2.24 MeV is analyzed to be the second 0^+ state.

INTRODUCTION

The cluster feature of light nuclei has been studied for more than four decades [1], and theoretical investigations have further proceeded to systems beyond $4N$ nuclei: a cluster structure with valence neutrons. Recently, the identification of such states become one of the main subjects concerning the structure of unstable nuclei. For example, the appearance of the cluster rotational band structure in the excited states of ^{10}Be and ^{12}Be and the breaking of the magic number ($N = 8$) in ^{12}Be due to clustering have attracted much attention, just as finding the so-called halo structure in light nuclei (^{6}He, ^{11}Li, ^{11}Be, ^{14}Be, $\cdots$).

From the theoretical side, Antisymmetrized Molecular Dynamics (AMD) has been proposed and extensively applied to calculating the properties of light neutron-rich nuclei [2]. In AMD, each single particle is described as one local Gaussian (G_i) characterized by Gaussian-center parameter (z_i). The shape of the nucleus and the configuration of nucleons are determined by solving the cooling equation for the parameters $\{z_i\}$. The appearance and disappearance of the cluster structure can be discussed without any assumptions. The wave function can be easily projected to the eigen state of the parity and angular momentum, and the levels and electromagnetic properties can be easily discussed.

However, the valence neutron(s) in the halo state is spread widely in space with a low density, and the single AMD wave function is not sufficient to correctly reproduce the density distribution; since each nucleon is expressed as one local Gaussian wave packet, whereas the halo orbit has been known to have a Yukawa tail. Therefore, superposition of the AMD wave functions is necessary. Also, if the core nucleus is deformed, the projection of the angular momentum (J) for the core becomes important, and J projection only for the total system is not enough to correctly reproduce the binding energy of the valence nucleon(s) from the threshold energy. In other words, the J projection of the core nucleus and for the relative motion between the core and the valence nucleon(s) are necessary.

Therefore, in weakly bound systems, expressing each valence nucleon not as a localized particle at a fixed position, but as a "wave" is required. To more precisely describe the "wave" nature of the nucleons, we applied the Molecular-orbit approach to Be and C isotopes, and revealed the motion of the valence neutrons around the α-α core [3]. This method has succeeded in classifying the observed spectra in terms of the configurations of the valence neutrons around clusters. However, the number of Slater determinants becomes huge, when the number of valence nucleons increases. When each valence nucleon is expressed as a linear combination of 8 Gaussians, a system with 4 valence nucleons is expressed as a linear-combination of $8^4 = 4096$ Slater determinants, even if the degree of freedom of the core is fixed.

To overcome these problems, we propose a new AMD approach, namely, AMD – Superposition of Selected Snapshots (AMD triple-S) [4]. In this approach, the superposed AMD wave functions are randomly generated, and we select important wave functions from among AMD wave functions. In this case, we use the idea of the Stochastic Variational Method (SVM). To make the basis states more effective, we apply the cooling method and optimize the parameters of the Gaussian centers of the single-particle wave functions. If we optimize both the real and imaginary parts of the Gaussian centers, the obtained wave functions of the basis states (Slater determinants) become essentially identical; if no local minimum exists. Therefore, only the imaginary part of the Gaussian centers are optimized. Optimization of the imaginary part enables us to correctly include the spin-orbit contribution. An AMD wave function whose imaginary part of the Gaussian centers are optimized can be understood as being a snapshot of the nucleus: the position of each nucleon is fixed, but the expectation value of the momentum operator is non-zero, and the spin-orbit contribution is expressed. By taking other snapshots, in other words, by generating another basis state, we can see other aspects of the nucleus at different times, and by superposing selected important snapshots (basis states), we can obtain a real picture of the nucleus.

FRAMEWORK

The total wave function is fully antisymmetrized and is given by a superposition of the basis states (Slater determinants $\{\Psi_k\}$) with coefficients $\{c_k\}$:

$$\Phi = \sum_k c_k P^J_{MK} \Psi_k, \qquad (1)$$

$$\Psi_k = \mathscr{A}[(\psi_1 \chi_1)(\psi_2 \chi_2) \cdots]_k. \qquad (2)$$

Projection onto a good angular momentum is performed by the projection operator P^J_{MK}, and the coefficients $\{c_k\}$ are determined by diagonalizing the Hamiltonian matrix after this projection. Each Slater determinant (Ψ_k) consists of A nucleons and each nucleon

($\psi_i \chi_i$ $i = 1 \sim A$) has a Gaussian form,

$$\psi_i = \left(\frac{2\nu}{\pi}\right)^{\frac{3}{4}} \exp[-\nu(\vec{r} - \vec{z}_i/\sqrt{\nu})^2 + \vec{z}_i^2/2], \tag{3}$$

where $\{\vec{z}_i\}$ are complex parameters and $\{\chi_i\}$ represent the spin-isospin eigen functions. The oscillator parameter ($b = \frac{1}{\sqrt{2\nu}}$) is set equal to 1.46 fm, which is common for all nucleons to exactly remove the center-of-mass kinetic energy. This single-particle wave function satisfies the following relations:

$$<\vec{r}> = \frac{1}{\sqrt{\nu}} Re[\vec{z}_i], \tag{4}$$

$$<\vec{p}> = 2\sqrt{\nu}\hbar\, Im[\vec{z}_i]. \tag{5}$$

The real and imaginary parts of $\vec{z}_i$ represent the expectation values of the position and momentum of the single particle. When we assume the presence of an α-cluster(s), it is expressed by assuming a common $\vec{z}_i$ value for four nucleons (proton spin-up, proton spin-down, neutron spin-up, and neutron spin-down).

For each Slater determinant (Ψ_k), the values of $\{\vec{z}_i\}$ are randomly generated in the framework of SVM, but we optimize imaginary part of these parameters before the angular momentum projection by using the frictional cooling method in AMD:

$$\frac{d\vec{z}_i}{d\tau} = -Im\left[\frac{\partial E}{\partial \vec{z}_i^*}\right]i, \quad \frac{d\vec{z}_i^*}{d\tau} = -Im\left[\frac{\partial E}{\partial \vec{z}_i}\right]i. \tag{6}$$

During this cooling process, the parity of the Slater determinant is projected.

The Hamiltonian operator ($\hat{H}$) has the following form:

$$\hat{H} = \sum_{i=1}^{A} \hat{t}_i - \hat{T}_{c.m.} + \sum_{i>j}^{A} \hat{v}_{ij}, \tag{7}$$

where a two-body interaction ($\hat{v}_{ij}$) includes the central part, the spin-orbit part and the Coulomb part. For the central part, we use the following Volkov No.2 effective $N-N$ potential, where $W = 1 - M$, $M = 0.60$ and $B = H = 0.125$. For the spin-orbit term, we introduce the G3RS potential. The original Volkov No.2 potential gives the bound state for the n-n system, but is eliminated by introducing B and H parameters.

APPLICATION TO LIGHT NUCLEI

We discuss the applicability of the AMD triple-S method in ^{6}He, a famous nucleus with two-neutron halo. In Fig. 1, the 0^+ energy convergence is shown. If the number of trial wave functions exceeds 29, the energy becomes lower than the energy of ^{4}He (-27.57 MeV), and the energy converges at -28.56 MeV. (Exp. B.E. = 0.98 MeV from the ^{4}He+n+n threshold, Cal. B.E. = 0.99 MeV). We check the convergence for other trial

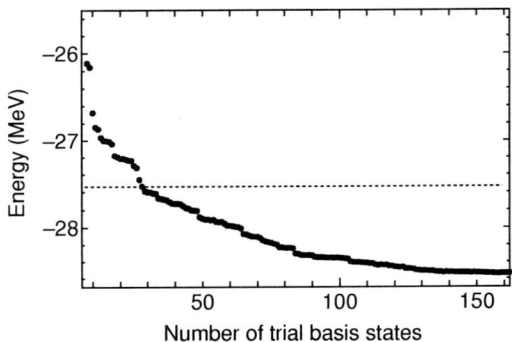

FIGURE 1. Energy convergence of the ground 0^+ state of ^{6}He. The horizontal axis is the number of trial basis states. When the inclusion of one new basis state decreases the sum of the energies of the ground 0^+, the second 0^+, and the third 0^+ states by more than 0.05 MeV, the new basis state is incorporated. The dotted line shows the energy of the ^{4}He+n+n threshold.

basis functions, which is randomly generated. All converged energies are smaller than -28.5 MeV and the most bound one is -28.56 MeV. The r.m.s. radius of the ground state is 2.37 fm, which agrees well with the experimental data. If we adopt the wave function of single Slater determinant AMD, whose real and imaginary parts of $\{\vec{z}_i\}$ are optimized, the energy of the ground 0^+ state is calculated at -21.82 MeV. Therefore, the superposition of many basis states is absolutely necessary.

The 2^+ state, which is an unbound state is calculated at -26.67 MeV ($E_x = 1.89$ MeV) within the bound state approximation, using the same basis functions for the ground state and projecting them to 2^+. Experimentally, it is observed at $E_x = 1.80$ MeV with a decay width of 113 keV. Combining the present AMD triple-S method and the method of Analytic Continuation in the Coupling Constant (ACCC) to correctly calculate the energy and width of the resonance state is promising and powerful. The calculated value by ACCC with the present basis set is $E_x = 1.86$ MeV ($E_r = 0.875$ MeV) with a decay width of 115 keV (the detail will be given elsewhere).

Next, we discuss the structure of the ^{12}Be nucleus using the $\alpha+\alpha+4n$ model. The breaking of the $N = 8$ magic number in ^{12}Be has been widely discussed, and experimental evidence for the breaking exists. We also discussed that the configuration of $(1/2^-)^2$ and $(1/2^+)^2$ for the last two valence neutrons is almost degenerate in energy, and the sd-shell component strongly mixes in the ground state. Recently, the 1^- state has been observed at 2.7 MeV; this presence of low-lying 1^- state is considered to be evidence of the degeneracy of the $1/2^-$ and $1/2^+$ single-particle orbits. Furthermore, a candidate for the second 0^+ state has been observed at 2.24 MeV [5], which is a counterpart of the ground 0^+ state: a state with different amplitudes for the $(1/2^-)^2$ and $(1/2^+)^2$ configurations from those of the ground state. However, experimentally, the parity of the state cannot be fully determined. Therefore, strictly speaking, a possibility that the observed state may be 0^- exists. It is naively considered that 0^- is higher than 1^- in energy; since the spin-spin interaction between the last two valence neutrons contributes attractively to the 1^- state and repulsively to the 0^- state. Here, we examine whether the 0^- state ap-

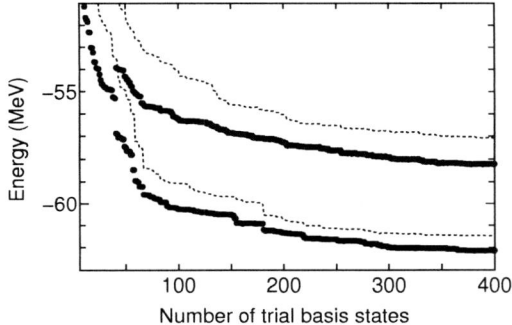

FIGURE 2. Energy convergence of the two lowest 0^+ states of ^{12}Be. The model space is α-α-$4n$ and the distance between two α-clusters is set to 3 fm, the optimal value for the ground state. The horizontal axis is the number of trial basis states. The solid circles are the result of AMD triple-S, and the dotted lines show the result using basis states whose imaginary part of the Gaussian-centers are not cooled. When the trial functions are not incorporated 25 times continuously, the calculation is stopped.

pears much higher than the 1^- state, and whether the observed state must be the second 0^+ state.

We apply AMD triple-S to ^{12}Be. The α-α distance is set to 3 fm, optimal for the ground state. The convergence of the two lowest 0^+ states is shown in Fig. 2. They are compared with results, whose basis states are not optimized by applying the cooling equation to the imaginary part of the Gaussian centers. The difference is very large for both states; therefore, solving the cooling equation is important, especially for the excited state; when the number of the valence neutrons increases. Here, both basis states with total spins of $S_z = 0$ and $S_z = 1$ for the valence neutrons are alternately generated. The ground and second 0^+ states converge to -62.14 MeV and -58.22 MeV, respectively, and the newly observed 1^- state appears at -57.43 MeV, while the 0^- state appears at -54.24 MeV. The 0^- state is calculated to be much higher than the second 0^+ and 1^- states; therefore, it is quite likely that the observed $J = 0$ state at 2.24 MeV is the second 0^+ state. To obtain the correct energy spectra, superposing the wave functions with respect to the α-α distance is necessary; because the deformations of the first and second 0^+ states are slightly different.

[1] D.M. Brink, Proc. Int. School of Phys. ENRICO FERMI, course 36, 247, edited by C. Bloch, (Academic press, New York and London, 1966)
[2] Y. Kanada-En'yo, H. Horiuchi, and A. Ono, Phys. Rev. C 52, 628 (1995).
[3] N. Itagaki and S. Okabe, Phys. Rev. C 61, 044306 (2000).
[4] N. Itagaki and S. Aoyama, Phys. Rev. C in press
[5] S. Shimoura et al., Phys. Lett. B560, 31 (2003).

New lines of research with the MAGNEX large-acceptance spectrometer

F. Cappuzzello[1], A. Cunsolo[1,2], A. Foti[2,3], A. Lazzaro[1,2], S.E.A. Orrigo[1,2], J.S. Winfield[1], M.C. Allia[1,2], C. Nociforo[4]

1. INFN – Laboratori Nazionali del Sud, Catania, Italy
2. Dipartimento di Fisica e Astronomia, Università di Catania, Catania, Italy
3. INFN - Sezione di Catania, Catania, Italy
4. GSI, Darmstadt, Germany

Abstract: A new design of magnetic spectrometer, MAGNEX, is under construction for the INFN-LNS, Catania. It is primarily intended for use with the Tandem-accelerated radioactive beams from the EXCYT facility. Both projects are expected to be completed by early 2004. Considering the broad choice of both stable and radioactive beams that will be available at the LNS, the spectrometer provides new opportunities for, e.g., spectroscopy of weakly bound nuclei by direct reactions, reaction mechanisms with large isospin and nuclear astrophysics.

INTRODUCTION

Historically the technique of magnetic spectrometry for nuclear physics experimentation has enjoyed a considerable advantage with beams from Tandem accelerators. The coupling to the excellent emittance of these beams allows the full exploitation of the high energy and mass resolving power of a magnetic spectrometer. In addition the possibilities to measure at very forward angles (including zero degrees), to compensate for the energy dependence of the reaction products with scattering angle, and to suppress part of the background through a $B\rho$ analysis strongly enhance the attraction of such devices. Benchmark examples like the Enge Split-Pole [1] or the Q3D [2] of the 1960s and 70s, and many others, indicate the durability of magnetic spectrometers for studies of nuclear spectroscopy and reaction mechanisms at low bombarding energy. A limitation of these devices has been the low acceptance (not more than 10 msr solid angle), mainly due to the high order aberrations generated for large solid angles and momentum bites. Recently, with the advent of the ray-reconstruction techniques, the performances of magnetic spectrometers in terms of acceptance have been upgraded (~20 msr for the Osaka LAS [3] and the NSCL S800 [4]) and new scenarios have been consequently opened.

The concept and layout of the MAGNEX spectrometer has been described in refs. [5-7]. In brief, it is a large-acceptance device (50 msr solid angle and ±10% momentum bite) based on a vertically focussing quadrupole and 55° bend-angle dipole. The angles and profiles of the dipole entrance and exit pole faces are used to correct partly the aberrations in the ion-optics. Further corrections are performed in software by a ray-reconstruction technique, resulting in an expected average momentum and mass resolution of about 1/2000 and 1/200 respectively. The ray-reconstruction is based on a solution to high orders of the equation of motion using, in our case, the program COSY INFINITY of Berz [8,9]. Detailed measured field maps in five vertical planes are used. A position-sensitive timing detector (PSD) between the target and quadrupole gives both the angle of the scattered particles and a start signal for the time-of-flight. The focal plane detector (FPD) measures positions and angles as well as providing particle identification information [10], allowing also a remarkably low energy threshold (~ 0.5 MeV/u).

Presently the spectrometer is in an advanced step of construction, the end of which is foreseen for the beginning of 2004. In this paper we focus on some of the experimental opportunities opened by the advent of MAGNEX at the LNS, regarding its use with radioactive ion beams (RIB) from the ISOL-type EXCYT facility [11].

DIRECT REACTIONS WITH EXCYT RIB'S

Direct two-body reactions are essential tools to study the structure of nuclei [12]. Spectroscopic information regarding the ground as well as the excited states of nuclei is accessed in a consistent manner, thus giving stringent conditions to nuclear structure theories. This information is often accomplished if highly monochromatic beams such as Tandem Van de Graaff ones are accelerated and magnetic spectrographs are used for the detection of the ejectiles. The future availability of Tandem post accelerated RIB's from EXCYT coupled to the unique features of MAGNEX will allow the extension of these studies to the challenging field of nuclear structures far from stability. The precise plan for future experiments is obviously dependent on what RIB's are developed for EXCYT.

One interesting nucleus that may be studied with a ^{9}Li beam and the (d,p) reaction in inverse kinematics is the unbound ^{10}Li. The interest is in the location of the $2s_{1/2}$ and $1p_{1/2}$ resonances, which is an important stepping-stone to a better understanding of the classic two-neutron halo nucleus ^{11}Li [13]. As another example, we consider the use of a ^{14}O beam for the two-proton stripping reaction (^{14}O,^{12}C). This reaction has a typically small negative, or even positive, Q-value, and would be a powerful spectroscopic tool to investigate proton-rich nuclei. The idea is that the favourable Q-value matching would enhance the reaction cross section sufficiently to overcome the inherently weak beam intensity and possibly excite previously unobserved states. One disadvantage of the (^{14}O,^{12}C) reaction compared with the lighter ion reactions is the loss of energy-resolution because of straggling and energy-loss differences in the target. However, preliminary simulations for the MAGNEX spectrometer show that a target thickness of 250 µg/cm^2 would lead to a resolution of about 300 keV in the excitation spectrum. Even with such a relatively thin target, one might still obtain an acceptable count rate: a few per hour in a peak for which the cross section were ~ 0.5 mb/sr, with a ^{14}O beam intensity of 10^5 pps.

A powerful tool to obtain information on light nuclei as well as on the isovector nucleon-nucleon interaction is the (^{7}Li,^{7}Be) charge exchange reaction [14-16]. These studies may be extended toward the neutron drip-line by the use of the foreseen ^{8}Li first RIB from the EXCYT facility [17]. We can use this beam to study the neutron-rich nucleus ^{8}He. The proposed experiment is ^{7}Li(^{8}Li,^{7}Be)^{8}He, i.e., the ^{8}Li beam bombards a ^{7}Li target and ^{7}Be reaction products are detected in the focal plane of the spectrometer. One of the interests in ^{8}He is that the ground state has a significantly extended matter distribution compared to ^{4}He, and it is considered to have a α-like core surrounding by 4 neutrons. The current consensus is that ^{8}He is not a halo nucleus, but instead has a *neutron skin* [18-20], even though this conclusion has not yet firmly established. Even more controversial is the discussion about low lying excited states. In the most recent evaluation of mass 8 nuclei, three excited states in ^{8}He (S_{2n} = 2.14 MeV) are listed at ~ 3.1, 4.36 and 7.16 MeV [21] with evidence for another at 6.03 MeV. In fact, it is not clear whether the 2^+ resonance observed at about 3.1 MeV refers to a single state or to a pair of states (see ref. [21] and references therein). The exact location of this 2^+ resonance have major consequences on the role of spin-orbit interaction on the structure of nuclei at large isospin.. Besides the possibility of exciting bound states in the continuum, in a similar manner as for ^{11}Be [14-16], the low-lying level structure of ^{8}He could be more firmly established. In particular with a high resolution spectrum from a generally unselective reaction as (^{7}Li,^{7}Be), one expects to better analyze the shape of the resonance observed around 3 MeV. An advantage of the proposed experiment is that due to the inverse kinematics of the reactions carbon or oxygen impurities or backing material in the target are not important as sources of background in the spectra. In fact it would be unlikely to produce ^{7}Be nuclei from a ^{8}Li beam. Thus the well known problem to have pure (self-supporting) lithium targets is avoided.

For the above ^{8}He experiment a complete simulation accounting for the ray reconstruction technique [9] is presented in the following. In the simulation a beam of 57 MeV incident energy and of 1.6 × 10^6 pps intensity is assumed, according to the preliminary predictions for the EXCYT facility [17]. From the systematics for light nuclei of the (^{7}Li,^{7}Be) reaction at 57 MeV, the cross section for the Gamow-Teller (GT) transition from the ^{8}Li$_{gs}$(2^+) ground to the ^{8}He(2^+) excited state is estimated to be about 100 μb/sr at forward angles. The weak intensity of the beam and the small cross section put severe constraints on the solid angle and target thickness, in order to perform the experiment in a reasonable time. For this purpose the large solid angle (50 msr) of MAGNEX, connected to the precise reconstruction of the scattering angle and the effective compensation of the kinematic effect, are key elements for the feasibility of this experiment. The kinematic broadening of the peaks in the energy spectra due to the large scattering angle interval (from 1° to 15°) is more than 2 MeV when the spectrometer aperture is fully open. Under these conditions the counting rate per each micron of target thickness is about 0.8 counts / 1 μm × hour, while the effect of target on the energy resolution is about 28 keV / 1 μm. To limit the contribution of the target to the energy resolution to within 150 keV a counting rate of about 6 counts / hour is achievable. This means the spectrometer should be kept in the same conditions for at least 100 hours to get enough counts in the GT transition peak. In Fig. 1 the initial conditions of the simulations are shown for a sample of 11000 particles distributed over 12 simulated levels, which corresponds to about 150 hours of measurement. The ground and the known excited states at 2.7, 3.6, 6.03, 7.16 MeV are visible. A broad resonance known at 4.5 MeV is also included. In the simulation states at 2.7 and 3.6 MeV are both considered to exist, since the experimental goal is to resolve them. For each ^{8}He excitation two lines are present in the spectra, arising from either the population of the ground or the 0.429 MeV bound excited state of the ^{7}Be ejectiles. The strong kinematic effect is evident in the plot, appearing as a noticeable curvature of the kinetic energy lines as a function of the scattering angle. In the right panel of Fig. 1 the projected kinetic energy spectrum is shown, emphasizing the need to measure the scattering angle with good precision. It is important to bear in mind that any possible angular segmentation of the data is hindered by the very low counting rate in the actual experiment.

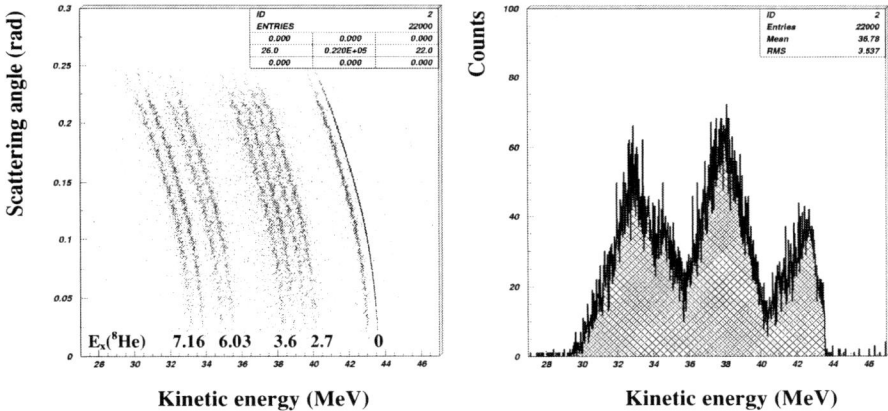

FIGURE 1. Left panel; initial conditions (after the target) for the simulation of the ^{7}Li(^{8}Li,^{7}Be)^{8}He reaction at 57 MeV. Right panel; initial energy spectrum.

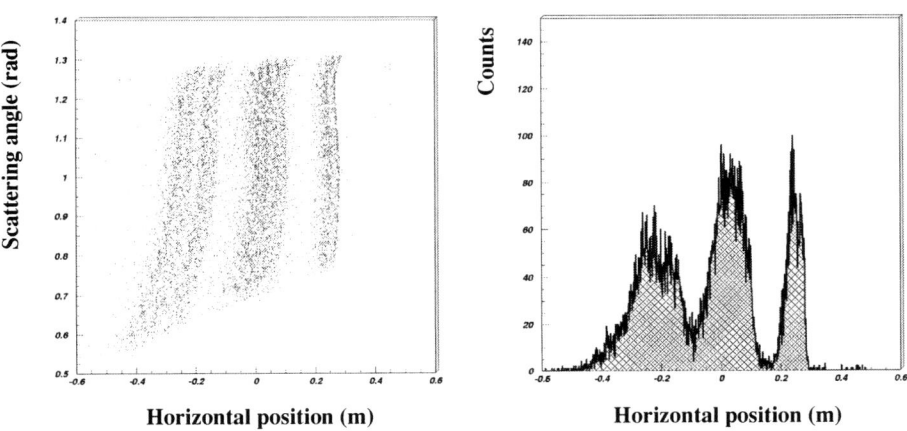

FIGURE 2. Left panel; scatter plot at the focal plane for the ^{7}Li(^{8}Li,^{7}Be)^{8}He reaction at 57 MeV. Right panel; focal plane position spectrum.

The distribution of particles along the focal plane after tracking through the spectrometer is shown in Figure 2. In the simulations all the active and dead layers are included realistically. To reduce the kinematic effect, the detector has been shifted to the predicted location of the focal plane (as allowed by MAGNEX), and the quadrupolar and sextupolar surface correction coils are used. The scatter plot and the one dimensional spectrum give an idea of the difficulties with the large acceptance condition if trajectory reconstruction is not employed, even with an optically-refined spectrometer such as MAGNEX [7]. The position resolution is obviously not enough to distinguish the peaks at 2.7 and 3.6 MeV

In Figure 3 the result of the application of the ray reconstruction method is shown. The order of reconstruction has been set to the 11th. In the figure the reconstructed kinematic plot clearly indicates the power of this technique in compensating both the kinematic effect and the effects of residual aberrations that were observed in Figure 2. The excitation energy spectrum shows the clear separation of the hypothetical peaks in the region of interest around 3 MeV. The resolution of the peaks (FWHM ~ 120 keV) comes almost entirely from the effect of target thickness, which is unavoidable, and is not limited by the instrument itself.

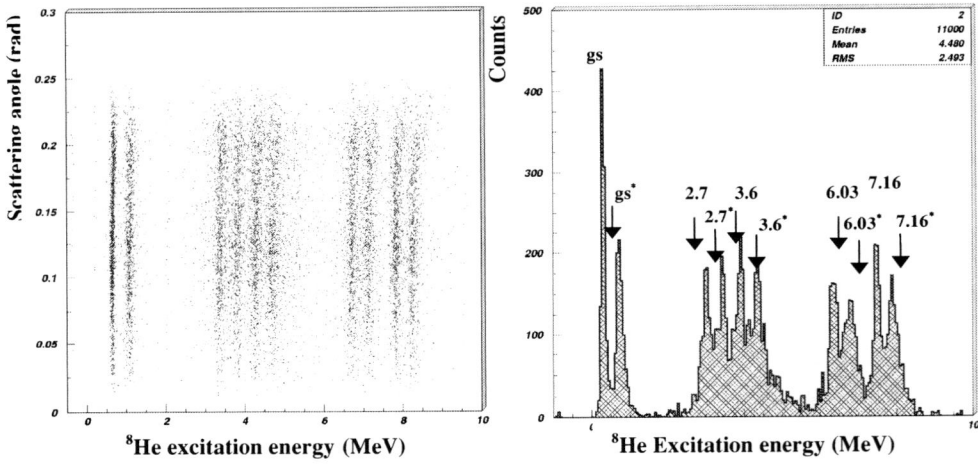

FIGURE 3. Reconstructed scatter plot (left) and energy spectrum (right) at the target for the ^{7}Li(^{8}Li,^{7}Be)^{8}He reaction at 57 MeV. The 11th order algorithm of Cosymag has been used.

REFERENCES

1. Spencer, J.E. and Enge, H.A, *Nucl. Instr. and Meth.* **49**, 181 (1967).
2. Enge, H.A., *Nucl. Instr. and Meth.* **162**, 161 (1979).
3. Okihana, A., et al., *RCNP Osaka Annual Report*, 1994.
4. Bazin, D., et al., *Nucl. Instr. and Meth.* B **204**, 629 (2003).

5. Cunsolo, A., et al., Proceedings Workshop *Giornata EXCYT*, Catania pp. 143-161 (1996), Proceedings Workshop *II Giornata EXCYT*, Catania pp. 71-80 (1997).
6. Cunsolo, A., et al., Proceedings to 9th *International Conference on Nuclear Reaction Mechanisms*, Varenna, p. 661 (2000).
7. Cunsolo, A., et al., *Nucl. Instr. and Meth.* A **481**, 48 (2002); **484**, 56 (2002).
8. Berz, M., *Nucl. Instr. and Meth.* A **298**, 473 (1990).
9. Lazzaro, A., Ph.D. thesis, University of Catania, (2003), Lazzaro, A. et al., Proceedings 7th *International Computational Accelerator Physics Conference*, East Lansing, Michigan, Oct. 2002 (Berz, M., ed., IOP Publishing, in press).
10. Cunsolo, A., et al., *Nucl. Instr. and Meth.* A **495**, 216 (2002).
11. Ciavola, G., et al., *Nucl. Phys. A* **616**, 69c (1997).
12. Satchler, G.R., *Direct Nuclear Reactions*, edited by Elliott, R.J., Krumhansl, J.A., and Wilkinson, D.H., Oxford: Clarendon Press, 1983.
13. Bonaccorso, A., Proceedings to 10th *International Conference on Nuclear Reaction Mechanisms*, Varenna, (2003).
14. Cappuzzello, F., et al., *Phys. Lett.* B **516**, 21 (2001) and *Nucl Phys. A*, in press.
15. Nociforo, C., Ph.D. thesis, Univ. of Catania (2002) and Proceedings XXXVII Zakopane School of Physics, 2002, *Acta Physica Polonica B* **34**, 2387 (2003).
16. Orrigo, S.E.A. et al., Proceedings to 10th *International Conference on Nuclear Reaction Mechanisms*, Varenna, (2003).
17. Cuttone G., Workshop on forthcoming facilities at LNS, Catania, March 2003 and Proocedings to this Conference.
18. G.D. Alkhazov, G.D., et al., *Phys. Rev. Lett.* **78**, 2313 (1997).
19. Karataglides, S., et al., *Phys. Rev. C* **61**, 024319 (2000).
20. Korsheninnikov, A.A., et al., *Nucl. Phys. A* **617**, 45 (1997).
21. Kelley, J.H., et al., *Energy levels of light nuclei A = 8, preliminary version #1*, TUNL Nuclear Data Evaluation Group, February 2002.

Nuclear structure studies with GEANIE at the LANSCE/WNR facility

N. Fotiades*, P. E. Garrett[†], E. Tavukcu**, M. Devlin*, R. O. Nelson*, J. A. Becker[†], W. Younes[†] and L. A. Bernstein[†]

Los Alamos National Laboratory, Los Alamos, New Mexico 87545, USA
[†]*Lawrence Livermore National Laboratory, Livermore, California 94550, USA*
**North Carolina State University, Raleigh, North Carolina 27695, USA*

Abstract. Recent results pertaining to nuclear structure from neutron-induced reactions on ^{90}Zr, ^{193}Ir, ^{196}Pt and ^{238}U are presented. The data were taken using the GEANIE spectrometer comprised of 26 high-purity Ge detectors with 20 BGO escape-suppression shields. The broad-spectrum pulsed neutron source of the Los Alamos Neutron Science Center's WNR facility provided neutrons in the energy range from 0.6 to 200 MeV. The time-of-flight technique was used to determine the incident neutron energies. Results from shell model calculations for ^{90}Zr and from IBM-2 calculations for ^{196}Pt are generally in good agreement with the observed spectrum of excited states.

INTRODUCTION

The $(n,n'\gamma)$ reaction at low neutron energies is a non-selective process. Hence, low-spin off-yrast states can be populated and studied in this type of reaction easier than in heavy-ion induced reactions.

The combination of a high-energy-resolution suppressed-germanium-detector array (GEANIE) with the "white"-neutron source at the LANSCE/WNR facility enables the detection of γ-rays in neutron-induced reactions over a wide range of incident neutron energies. The primary goal of these experiments is to determine the partial cross sections for production of these γ-rays, but these experiments also provide additional results on the nuclear structure of the isotopes studied. Recent results on mainly off-yrast states of ^{90}Zr, ^{193}Ir, ^{196}Pt and ^{238}U as studied in $(n,n'\gamma)$ reactions with GEANIE are presented here.

EXPERIMENTS

Discrete γ-rays have been measured for isotopes populated in $(n,xn\gamma)$ reactions on ^{90}Zr, ^{193}Ir, ^{196}Pt and ^{238}U as a function of incident neutron energy in four separate experiments. The Ge γ-ray spectrometer GEANIE [1] (GErmanium Array for Neutron-Induced Excitations) and neutrons from the "white" source neutron beam at the LANSCE/WNR (Los Alamos Neutron Science Center/Weapons Neutron Research) facility [2] were used.

GEANIE is comprised of 11 Compton-suppressed planar Ge detectors (Low Energy Photon Spectrometers - LEPS), 9 Compton-suppressed coaxial Ge detectors and 6 unsuppressed coaxial Ge detectors. The energy of the neutrons was determined using the time-of-flight technique. The neutron flux on target was measured with a fission chamber [3], located upstream on the GEANIE neutron flight path. The fission chamber consists of two thin ^{235}U and ^{238}U foils. The known ^{235,238}U(n,f) cross sections are used to deduce the neutron flux [3].

Two-parameter data were acquired for each detector consisting of time-of-flight relative to a fast beam-pickoff signal from the LANSCE/WNR proton accelerator and gamma-ray pulse height from the Ge detectors. Escape suppression was implemented by vetoing the associated Ge signals with the BGO signals in the hardware. Absolute detector efficiencies were determined using a variety of calibrated gamma-ray reference sources. Electronic "dead-times" were measured using scalers and corrections were applied to the data. During the experiments the data were stored on magnetic tapes for subsequent off-line analysis. Excitation functions for all the observed γ-rays were determined.

Symmetrized, two-dimensional matrices, constructed from the γ-γ data, were used to investigate the coincidence relationships between the γ rays. The placement of the newly observed γ rays in the level schemes was based on the combination of the γ-γ coincidence results with the deduced γ-ray excitation functions. The level-scheme systematics in neighboring isotopes was also used in some cases to further support the placement of these transitions.

As configured during the experiments described here, GEANIE covered too few unique detector angles (with respect to the neutron beam) to accurately determine angular distributions of the detected transitions. For ^{90}Zr [4] and ^{196}Pt [5] the angular distributions of the new transitions, necessary for spin and parity assignment to the new levels, were determined in separate $(n,n'\gamma)$ experiments using the Ge-detector spectrometer and monoenergetic-accelerator-produced neutrons at the University of Kentucky van de Graaff accelerator facility.

RESULTS AND DISCUSSION

The results from each experiment are discussed briefly in the following four subsections:

$$^{238}\text{U}(n,n'\gamma)^{238}\text{U}$$

As an example of how easily an $(n,n'\gamma)$ reaction can populate off-yrast states, a partial level scheme of ^{238}U is shown in Fig. 1 consisting of previously known transitions and levels [6] observed also in the present work. The absolute partial cross sections for the production of all transitions included in Fig. 1, as well as for several other transitions (a total of 45) from the ^{238}U$(n,xn\gamma)$ ($x \leq 4$) reaction channels, were reported in Ref. [7].

The ground state band 2^+ state together with the first three off-yrast 2^+ states of ^{238}U, forming part of the first and second β-vibrational bands and the γ-vibrational band, are

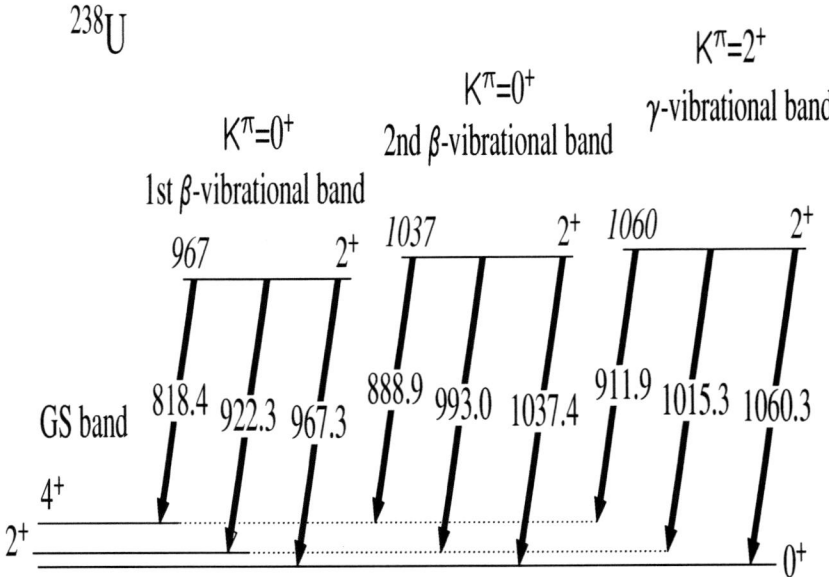

FIGURE 1. Previously known partial level scheme of ^{238}U with the first four 2^+ states of ^{238}U as observed in the GEANIE experiment. Transition and excitation energies are given in keV. The interpretation of the levels is also included.

shown in Fig. 1. Moreover, ten additional previously known off-yrast states with spins from 0 to $6\hbar$ were observed [7].

A similar experiment has been recently performed with GEANIE with a ^{232}Th target [8] to study the vibrational bands in this isotope.

^{196}Pt$(n,n'\gamma)^{196}$Pt

Despite a great deal of previous work including $(n,n'\gamma)$ studies, a total of 13 new levels, below $E_x = 3$ MeV, and 24 new transitions were assigned to ^{196}Pt in this experiment [5]. A partial level scheme of ^{196}Pt including two of the new levels and five of the new transitions is shown in Fig. 2. The absolute partial cross sections for the production of all transitions included in Fig. 2 are reported in Ref. [9].

With six bosons, ^{196}Pt has been shown [10] to be a good example of the SO(6) limit of the interacting boson model (IBM). The results of the IBM-2 (which distinguishes

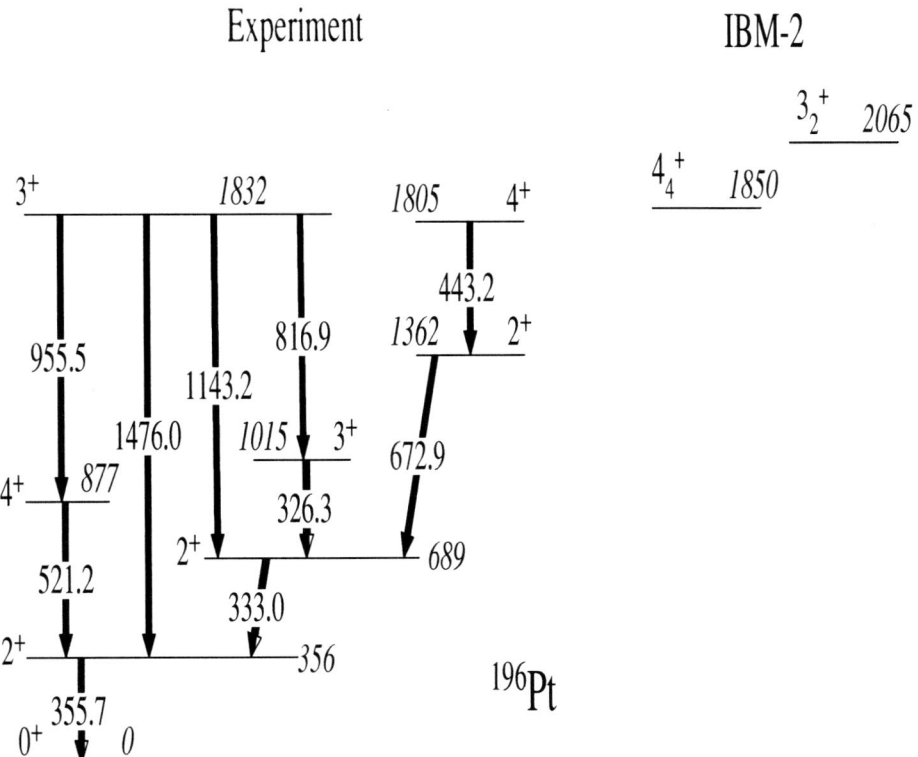

FIGURE 2. Partial level scheme of ^{196}Pt as deduced from the GEANIE experiment. Transition and excitation energies are given in keV. The 1805- and 1832-keV levels, as well as the transitions deexciting these levels, were observed for the first time. The results from IBM calculations for the new levels are also included.

between proton and neutron bosons) calculations [5] for the two new states in Fig. 2 are in good agreement with the experimentally observed ones.

$$^{193}\text{Ir}(n,n'\gamma)^{193}\text{Ir}$$

The data analysis of this experiment is still in progress. So far, a total of 7 new levels, below $E_x = 1.5$ MeV, and 10 new transitions have been assigned to ^{193}Ir. A partial level scheme of ^{193}Ir including two new levels and two new transitions is shown in Fig. 3 together with the previously known levels above the 80 keV, $11/2^-$, $\tau = 10.5$ days isomer [6]. The 19/2⁻ spin and parity assignment of the newly observed 1025-keV level is supported by the γ-γ coincidences, the excitation function of the 546.8 keV transition and the systematics of the corresponding 19/2⁻ states in the lighter odd-mass Ir isotopes [11]. The absolute partial cross sections for the production of all transitions included in Fig. 3 are reported in Ref. [11].

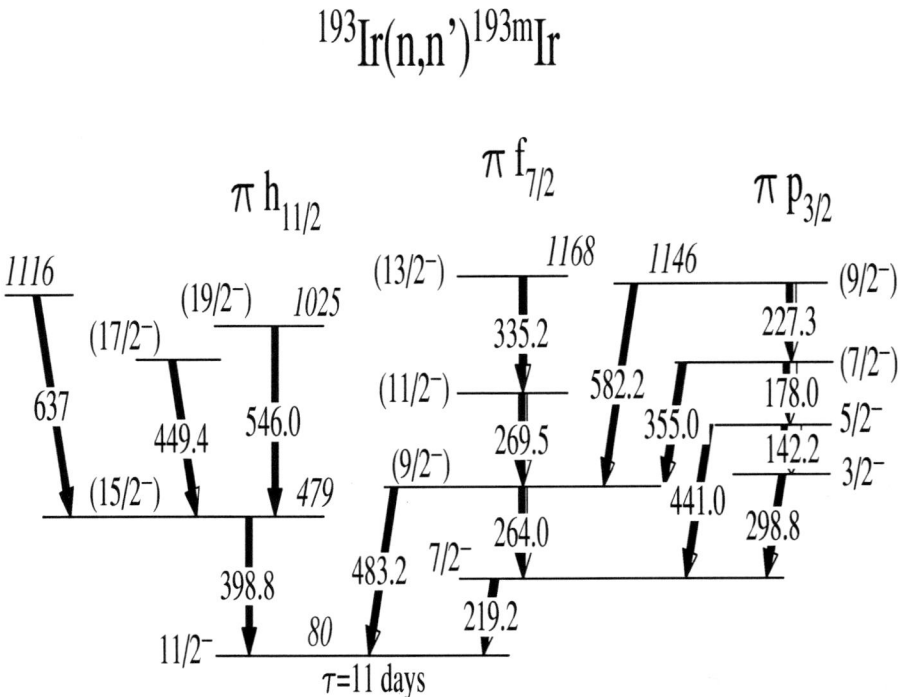

FIGURE 3. Partial level scheme of ^{193}Ir, above the 80 keV, 11/2⁻, $\tau = 10.5$ days isomer, as observed in the GEANIE experiment. Transition and excitation energies are given in keV. The 1025- and 1116-keV levels, as well as the transitions deexciting these levels, are observed for the first time. The configuration of the bands is also given, where possible.

As an example of the quality of the γ-γ data obtained with GEANIE, a spectrum obtained from a gate on the 398.8 keV transition in Fig. 3 is shown in Fig. 4. As can be seen in Fig. 4, while counts are limited, generally, the gated spectra have greatly reduced backgrounds. Moreover, additional neutron energy gates can be introduced to further clean up the resulting spectra in cases where contamination is present.

An experiment will be performed with GEANIE in the near future with a ^{197}Au target [12] to establish the yet unknown structure above the 409 keV, 11/2⁻, $\tau = 8$ sec isomer [6] in this isotope, expected to be similar to the structure observed in Fig. 3.

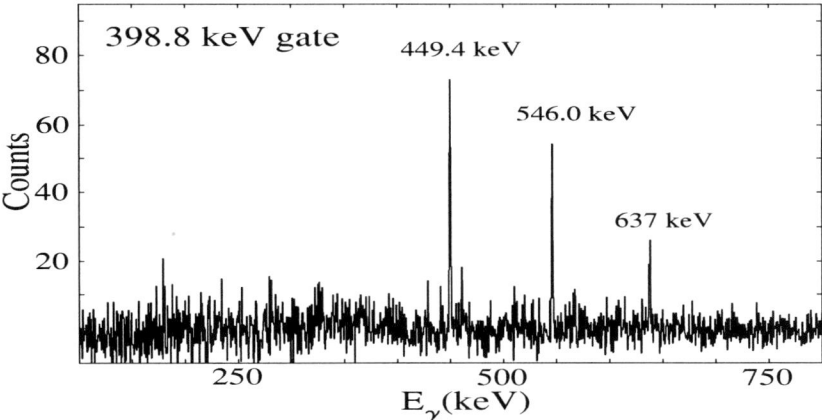

FIGURE 4. Spectrum in coincidence with the 398.8 keV transition in Fig. 3 from the γ-γ data. Since the spectrum is rather clean the use of an additional neutron energy gate has been deemed unnecessary.

$$^{90}\text{Zr}(n,n'\gamma)^{90}\text{Zr}$$

A total of 20 new levels, below $E_x = 6$ MeV, and 122 new transitions were assigned to ^{90}Zr in this experiment [4]. A partial level scheme of ^{90}Zr including two of the new levels and seven of the new transitions is shown in Fig. 5. The absolute partial cross sections for the production of all transitions included in Fig. 5 are reported in Ref. [13].

Shell model calculations, assuming a ^{88}Sr closed core and employing a space spanning twelve protons in the $0f_{5/2}$, $1p_{5/2}$, $1p_{1/2}$, and $0g_{9/2}$ orbits, reproduce the spectrum of excited states very well [4], as can be seen, for instance, for the newly observed 3^+ state in Fig. 5. In the case of the 5^- state in Fig. 5, the calculation predicts three different 5^- states in this excitation energy region, so, it is not clear which one (or, possibly, which combination of these states) can be used for the interpretation of the observed state.

SUMMARY

In conclusion, off-yrast states were studied in ^{90}Zr, ^{193}Ir, ^{196}Pt and ^{238}U using neutron-induced reactions and γ-ray spectroscopy. The pulsed neutron beam from the LANSCE/WNR facility, combined with the GEANIE spectrometer, is a unique and powerful facility for such studies. Several new γ rays were observed and were placed in the level schemes using both the deduced γ-ray excitation functions and the γ-γ coincidence results. Calculations, using a shell model for ^{90}Zr, and IBM-2 for ^{196}Pt, are generally in good agreement with the observed excitation energy spectra.

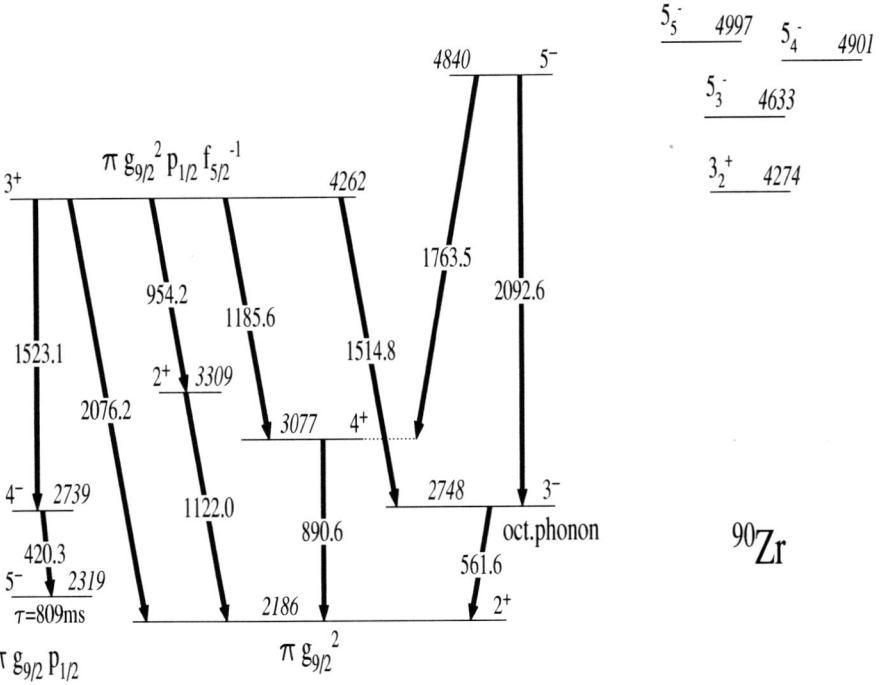

FIGURE 5. Partial level scheme of ^{90}Zr as deduced from the GEANIE experiment. Transition and excitation energies are given in keV. The 4262- and 4840-keV levels, as well as the transitions deexciting these levels, were observed for the first time. The results from shell model calculations for the new levels are also included.

ACKNOWLEDGMENTS

This work was performed under the auspices of the U.S. Department of Energy by the University of California, Los Alamos National Laboratory under contract no. W-7405-ENG-36 and Lawrence Livermore National Laboratory under contract no. W-7405-ENG-48, and benefitted from the use of the Los Alamos Neutron Science Center supported under contract no. W-7405-ENG-36.

REFERENCES

1. J. A. Becker, R. O. Nelson, *Nuclear Physics News International* **7**, 11 (1997).
2. P. W. Lisowski, C. D. Bowman, G. J. Russell, S. A. Wender, *Nucl. Sci. Eng.* **106**, 208 (1990).

3. S. A. Wender, S. Balestrini, A. Brown, R. C. Haight, C. M. Laymon, T. M. Lee, P. W. Lisowski, W. McCorkle, R. O. Nelson, W. Parker, *Nucl. Instr. Meth.* A **336**, 226 (1993).
4. P. E. Garrett, L. A. Bernstein, J. A. Becker, W. Younes, E. M. Baum, D. P. DiPrete, R. A. Gatenby, E. L. Johnson, C. A. McGrath, S. W. Yates, M. Devlin, N. Fotiades, R. O. Nelson, B. A. Brown, Phys. Rev. C **62**, in press (2003).
5. E. Tavukcu, L. A. Bernstein, K. Hauschild, J. A. Becker, P. E. Garrett, C. A. McGrath, D. P. McNabb, W. Younes, P. Navratil, R. O. Nelson, G. D. Johns, G. E. Mitchell, and J. A. Cizewski Phys. Rev. C **65**, 064309 (2002).
6. R. B. Firestone, V. S. Shirley, C. M. Baglin, S. Y. Frank Chu, and J. Zipkin, *Table of Isotopes*, Wiley, New York (1996)
7. N. Fotiades, G. D. Johns, R. O. Nelson, M. B. Chadwick, M. Devlin, W. S. Wilburn, P. G. Young, D. E. Archer, J. A. Becker, L. A. Bernstein, P. E. Garrett, C. A. McGrath, D. P. McNabb, W. Younes, submitted to Phys. Rev. C (2003).
8. P. E. Garrett, W. Younes *et al.*, LLNL report, to be submitted.
9. E. Tavukcu, L. A. Bernstein, K. Hauschild, J. A. Becker, P. E. Garrett, C. A. McGrath, D. P. McNabb, W. Younes, M. B. Chadwick, R. O. Nelson, G. D. Johns and G. E. Mitchell, Phys. Rev. C **64**, 054614 (2001).
10. J. A. Cizewski, R. F. Casten, G. J. Smith, M. L. Stelts, W. R. Kane, H. G. Börner, W. F. Davidson, Phys. Rev. Lett. **40**, 167 (1978).
11. N. Fotiades, R. O. Nelson, M. Devlin, M. B. Chadwick, P. Talou, J. A. Becker, L. A. Bernstein, P. E. Garrett, D. P. McNabb, W. Younes, LANL report, to be submitted (2003).
12. N. Fotiades, R. O. Nelson, M. Devlin, J. A. Becker, L. A. Bernstein, W. Younes, P. E. Garrett, LANL proposal No 2003503, (2003).
13. P E Garrett, J A Becker, L A Bernstein, W E Ormand, W. Younes, R O Nelson, M Devlin, N Fotiades, Bull. Am. Phys. Soc. **47**, No 2, 161 (2002), and P. E. Garrett *et al.*, LLNL report, to be submitted.

^{8}Li(α,n)^{11}B at Big Bang Temperatures: Neutron Counting With a Low Intensity ^{8}Li Radioactive Beam

S. Cherubini*, P. Figuera†, A. Musumarra†**, C. Agodi†, R. Alba†, L. Calabretta†, L. Cosentino†, A. Del Zoppo†, A. Di Pietro†, L. Lamia†, L. Pappalardo†, M.G. Pellegriti†**, R.G. Pizzone†, A. Rinollo†**, C. Rolfs*, S. Romano†**, C. Spitaleri†**, F. Strieder*, S. Tudisco†** and A. Tumino†**

*Institut fær Physik mit Ionenstrahlen, Ruhr-UniversitŻt Bochum, Bochum, Germany
†Istituto Nazionale di Fisica Nucleare - Laboratori Nazionali del Sud, Catania, Italy
**Dipartimento di Metodologie Fisiche e Chimiche per l'Ingegneria, Universit? di Catania, Catania, Italy

Abstract. The cross section of ^{8}Li(α,n)^{11}B is very important for the study of primordial nucleosinthesys models. In this paper we report on the production of a ^{8}Li beam via the ^{7}Li(d,p)^{8}Li reaction at the Laboratori Nazionali del Sud. Also, a novel experimental technique for measuring the reaction ^{8}Li(α,n)^{11}B at energies of astrophysical interest has been implemented and tested.

INTRODUCTION

In the framework of the so called Inhomogeneous Big Bang Model (IBBM) primordial nucleosynthesis proceeds via the reaction chain

$$^1H(n,\gamma)^2H(n,\gamma)^3H(d,n)^4He(t,\gamma)^7Li(n,\gamma)^8Li(\alpha,n)^{11}B(n,\gamma)^{12}B(\beta v)^{12}C.$$

The bottleneck of this reaction chain is constituted by the ^{8}Li(α,n)^{11}B process that involves the radioactive ^{8}Li nucleus (half-life 840 msec). The precise measurement of this cross is hence an important goal in nuclear astrophysics. Various experiments aimed at the measurement of this cross section but there are discrepancies in their results [1, 2, 3, 4, 5].

Different approaches where used in these works. In [1] the inverse reaction ^{11}B(n,α)^{8}Li was measured in the energy range E_n=7.6 - 12.6 MeV. In the experiments described by [2, 3] a ^{8}Li beam was used to bombard a ^{4}He target and the produced neutrons were detected inclusively. The first of these measurements could measure the ^{8}Li(α,n)^{11}B only down to 1.5 MeV in the center of mass, while the interesting range for primordial nucleosynthesis (roughly 0.3-0.8 MeV) was reached by the second experiment. The results of these two measurements agree in the common energy region but they show an enhancement of a factor of 5 or more with respect to those of [1]. This discrepancy has been partially explained by observing that in the inverse experiment only the ground state of ^{11}B contributed to the observed cross section.

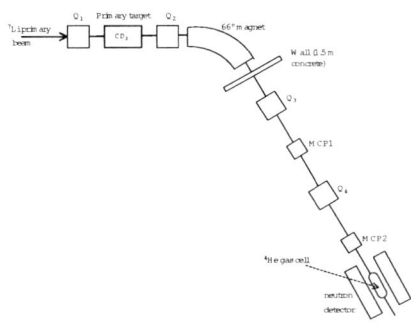

FIGURE 1. Experimental setup schematic diagram

Attempting to clarify the situation an exclusive measurement was performed [5], but the results of this work suffer from the poor statistics that was possible to accumulate in the experiment.

In order to add information on this important process, we planned an experiment at LNS that aims at the direct measurement of the $^{11}B(n,\alpha)^8Li$ reaction at astrophysical energies using a ^8Li beam.

EXPERIMENTAL SETUP

The setup used in the experiment is sketched in Figure 1. The 15 MV tandem of INFN-LNS in Catania provided a ^7Li beam of E_{lab} = 24.6 MeV with a current of about 100 pnA. The beam was focused by a quadrupole doublet (Q1) onto a deuterated solid target, where ^8Li nuclides were produced via the reaction $d(^7Li,p)^8Li$ (Q = -0.19 MeV). The ^8Li nuclides and the ^7Li projectiles left the solid target predominantly in the 3^+ charge state. The ^8Li nuclides were then momentum-filtered from the intense ^7Li primary beam by a 66° magnet in combination with a quadrupole doublet (Q2) and focused onto a ^4He gas cell by 2 quadrupole doublets (Q3, Q4). In the gas cell, the $^4He(^8Li,n)^{11}B$ reaction was initiated and identified by the produced neutrons.

Beam production and transport.

The primary target was a 1 mg/cm2 thick CD_2 foil sandwiched between carbon layers of 30 $\mu g/cm2$ thickness. After this target, the ^7Li ion beam had an energy of 23.7 MeV. Since the forward kinematic solution in the ^8Li production process was in momentum p too close to that of the ^7Li beam ($p(^8Li)/p(^7Li)$ = 0.98), we used the backward kinematic solution ($p(^8Li)/p(^7Li)$ = 0.66) for the momentum-filtering in the magnet. After the magnet, the ^{8}Li nuclides (backward kinematic solution) had an energy of $E_{lab}(^8Li)$ = 10.3±0.4 MeV, where the quoted uncertainty reflects the momentum acceptance of the magnet. The focusing of the beam was made in two steps. In a first step, a ^{7}Li beam, with an energy such that it emerged from the primary target with the same rigidity as the ^{8}Li nuclides, was produced and used as a "pilot beam" through the beam line up to the gas cell, using quartzes and Faraday cups as monitors. The last quartz and the last Faraday cup were installed at a respective distance of 20 and 50 cm from the entrance foil of the gas cell. In the second step, a 24.6 MeV ^{7}Li primary beam was used to induce the $d(^7Li,p)^8Li$. The associated ^{8}Li nuclides produced in the primary target where then tagged in the beam line using 2 micro-channel-plate detectors (MCP1 and MCP2). The upstream MCP had a 40 mm diameter active area (with a 0.8 micron thick mylar foil) in order to catch a large part of the ^{8}Li nuclides, while the downstream MCP had a 25 mm diameter active area (with a 30 $\mu g/cm^2$ thick C foil), which was smaller than the 40 mm diameter of the downstream gas cell; the MCP2 was at a distance of 20 cm from the gas cell. With a 5 m distance between the MCP's, the ^{8}Li nuclides could be clearly identified by their time-of-flight (TOF) between the two MCP's and resolved well from the leaky ^{7}Li projectiles. The ^{8}Li rate was used in a final tuning of the elements of the beam line; it turned out that the optimum parameters of the beam optical elements were nearly identical to the settings found with the ^{7}Li pilot beam. The measured ^{8}Li rate was on average about 150 nuclides/s. The TOF spectrum allowed to measure absolutely the ^{8}Li energy which was consistent with the expected value of 10.3 MeV.

^{4}He gas cell and neutron detector

The gas cell is 4 cm in diameter, 20 cm in length, and it has an entrance window consisting of a 5 μm thick Ni foil, which could sustain a 4He gas pressure of 150 mbar (cleanliness of the 4He gas = 99.9999%). A Si detector (100 mm2 active area) was placed at the end of the gas cell to measure the energy loss of the ^{8}Li nuclides in the Ni foil and in the ^{4}He gas. Taking into account the energy losses in the foil and in the gas itself, the mean ^{8}Li energy within the gas cell was $E_{lab}(^8Li)$ = 3.75 MeV or E_{cm} = 1.25 MeV. The neutron detection was carried out with a 4π detector consisting of 12 ^{3}He -filled proportional counters embedded in a polyethylene moderator; the detector was provided by Caltech [6]. The moderator has a cubic form, 40 cm on a side, with an 11 cm x 11 cm channel through the center, for insertion of the ^{4}He gas cell and of the beam pipe. The neutrons created in the gas cell were moderated in the polyethylene material. Surrounding the 40 cm cube of polyethylene was a 4π layer of Cd shielding (0.6 mm thick), which in turn was surrounded by a 4π passive layer of polyethylene and borated paraffin, approximately 10 cm thick. This passive shielding served to absorb externally created neutrons. The 12 ^{3}He proportional counters were positioned about the beam pipe channel in a circle of about 12 cm radius.

DATA TAKING AND ANALYSIS

The firts run of the experiment allowed for accumulating sufficient statistics to perfom a preliminary data analysis.

The neutron cosmic background rate in the detector was observed to be 2 events/min, while the beam induced background was about 12 events/min arising predominantly from the ^{7}Li primary beam interacting with the vacuum pipe walls of the 66 magnet. Using a calibrated ^{252}Cf source, the neutron detection efficiency was found to be $\eta = 0.20 \pm 0.01$ (for a mean neutron energy of $E_n = 2.35$ MeV) consistent with previous work [6]. Since the neutrons emitted in ^{4}He(^{8}Li,n)^{11}B at $E_{cm} = 1.25$ MeV involve energies up to about 8 MeV, we have adopted a mean value of $\eta = 0.19 \pm 0.03$ according to Monte Carlo calculations [6]. The neutrons were moderated and absorbed in the detector with an observed "die-away" time of $\tau = \lambda^{-1} = 86 \pm 2$ μs. A gate set on the ^{8}Li events in the TOF spectrum started a time-to-amplitude converter (TAC), whereby the stop signals of the TAC (selected time range = 1 ms) were provided by the neutron detector (i.e. the events from the logic *or* of the 12 ^{3}He proportional counters). If neutrons were correlated with the ^{8}Li beam, one expected a coincidence time distribution dN(t)/dt declining exponentially with time (or rising linearly with exp(-λt)) according to

$$dN(t)/dt = N_{BG} + N_o \lambda \exp(-\lambda t), \qquad (1)$$

where N_{BG} is the time-independent background contribution and N_o is the total number of time-correlated events.

PRELIMINARY RESULTS AND FUTURE DEVELOPMENTS

Fig. 2 illustrates the results of such correlations, where data are shown versus the corresponding mean values of $\lambda \exp(-\lambda t)$. The result clearly shows the trend for a neutron yield component that is correlated with the interaction of the ^{8}Li beam with the gas cell.

We cannot give yet a value for the cross section of the ^{8}Li$(\alpha,n)^{11}$B process because it is not possible to disentangle a possible, though unlikely, contribution to the neutron yield due to the interaction of the ^{8}Li beam with other materials than the ^{4}He gas in the target, e.g. the nickel window. A measure of these contributions is planned.[1]

It should be noted that the present work is the first experiment that uses a radioactive ion beam experiment carried out at INFN-LNS. Within the future EXCYT project at INFN-LNS [7], a ^{8}Li ion beam will be available in 2004 with an intensity several orders of magnitude higher than that obtained in the present experiment. With such a beam current, data with high precision will be obtained with the present setup over the full astrophysical energy range of ^{8}Li$(\alpha,n)^{11}$B.

---

[1] The measurement was actually carried out after the present Conference and results will be submitted for publication soon.

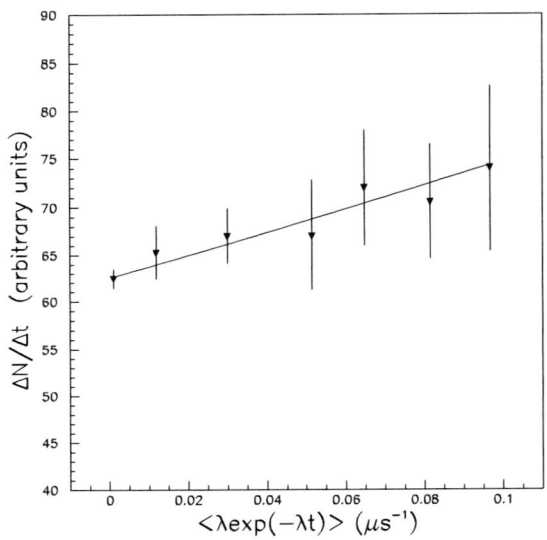

FIGURE 2. TAC-spectra (time range = 1 ms) for coincidence events between incident projectiles and neutrons are shown as a function of $<\lambda\exp(-\lambda t)>$

REFERENCES

1. T. Paradellis, S. Kossionides, G. Doukellis, X. Aslanoglou, P. Assimakopoulos, A. Pakou, C. Rolfs, and K. Langanke, *Z. Phys. A* **337** (1990) 211
2. R. N. Boyd, I. Tanihata, N. Inabe, T. Kubo, T. Nakagawa, T. Suzuki, M. Yonokura, X. X. Bai, K. Kimura, s. Kubono, S. Shimoura, H. S. Xu, and D. Hirata, *Phys. Rev. Lett.* **68**,(1992) 1283
3. X. Gu, R. N. Boyd, M. M. Farrell, J. D. Kalen, C. A. Mitchell, J. J. Kolata, M. Belbot, K. Lamkin, K. Ashktorab, F. D. Becchetti, J. Brown, D. Roberts, K. Kimura, I. Tanihata, K. Yoshida, and M. S. Islam, *Phys. Lett. B***343**, (1995) 31
4. R. N. Boyd, T. Paradellis, C. Rolfs, *Comments Nucl. Part. Phys.***22** (1996) 47
5. Y. Mizoi, T. Fukuda, Y. Matsuyama, T. Miyachi, and H. Miyatake, N.Aoi, N. Fukuda, M. Notani, Y. X. Watanabe, and K. Yoneda, M. Ishihara, H. Sakurai, Y. Watanabe, and A. Yoshida, *Phys. Rev C* **62** 065801
6. P.R.Wrean: Thesis, California Institute of Technology (1998)
7. INFN - LNS Activity Report 2002, pages 147-151

Single-Particle Structure of Neutron-Rich Nuclei

J.A. Cizewski, K.L. Jones J.S. Thomas*, D.W. Bardayan, J.C. Blackmon,
C.J. Gross, J.F. Liang, D. Shapira, M.S. Smith, D.W. Stracener[†],
R.L. Kozub, C.D. Nesaraja**, U. Greife, R.J. Livesay[‡] and Z. Ma[§]

Rutgers University, New Brunswick, NJ 08903 USA
[†]*Oak Ridge National Laboratory, Oak Ridge, TN 37831 USA*
**Tennessee Technological University, Cookeville, TN 38505 USA*
[‡]*Colorado School of Mines, Golden, CO 80401 USA*
[§]*University of Tennessee, Knoxville, TN 37996 USA*

Abstract. The d(^{82}Ge,p) reaction has been measured in inverse kinematics at the Holifield Radioactive Ion Beam facility, enabling a study of the evolution of single-particle structure above the N=50 shell gap for neutron-rich nuclei.

INTRODUCTION

An open question in understanding the structure of atomic nuclei is the evolution of nuclear shell structure away from the valley of stability, and in particular on the neutron-rich side as the drip line is approached. In very neutron-rich nuclei, as the nuclear surface becomes more diffuse, the shell gaps are predicted [1] to become quenched. These neutron-rich nuclei are also on the path of r-process nucleosynthesis, where the observed abundances of elements again suggest a quenching of the shell structure, and neutron-capture reactions could involve both direct and compound processes. To answer both the structure and reaction questions, we have begun a program to measure neutron-transfer reactions with radioactive ion beams.

EXPERIMENTAL DETAILS

The present measurements were conducted at the Holifield Radioactive Ion Beam Facility (HRIBF) at Oak Ridge National Laboratory. The neutron-rich beams are produced by bombarding a uranium carbide matrix with protons from the ORIC cyclotron, extracting negative ions after mass selection, and injecting the species into the 25-MV Tandem electrostatic accelerator. By extracting sulphides, elementally enhanced beams of Sn and Ge neutron-rich isotopes have been accelerated to energies above 4 MeV/nucleon.

The initial measurements of a (d,p) reaction with neutron-rich beams were made with a mixed beam of ^{82}Se/^{82}Ge accelerated to 4 MeV/nucleon. The beams impinged on a $\sim$400μg/cm^2 CD$_2$ target. The heavy recoils were detected in a segmented ion chamber at zero degrees, to enable elemental separation, as well as to monitor beam current; ΔZ=1 separation was readily obtained. Typical beam intensities were 7x10^4

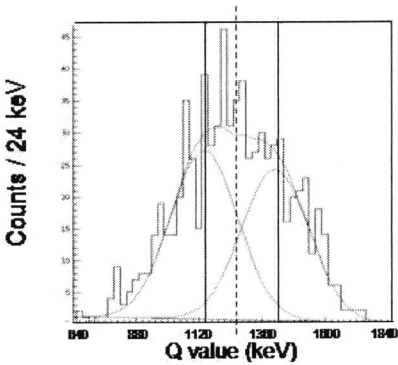

FIGURE 1. Partial Q-value spectrum for the d(^{82}Ge,p) reaction at 4 MeV/nucleon.

particles/s, with the stable ^{82}Se contaminant about 6 times more intense than ^{82}Ge, with minimal ^{82}As. The reaction protons were detected at back angles with the SIDAR array [2] of silicon strip detectors in a lampshade geometry, subtending 105-150 degrees in the laboratory. A TAC signal was used to help ensure good coincidences between light reaction products and the element of interest. The stable ^{82}Se contaminant in the beam enabled good energy and angle calibrations, since the ^{82}Se(d,p) reaction had been previously measured [3]. A portion of the Q-value spectrum, summed over all angles, for the d(^{82}Ge,p) reaction in the laboratory is displayed in Fig. 1.

From the ^{82}Se contaminant measurements, an empirical energy resolution of about 310 keV FWHM was determined. Therefore, the peak at highest Q-value in ^{83}Ge, displayed in Fig. 1, with a width of about 460 keV, is an unresolved doublet. By separating this doublet, the ground-state Q-value has been determined to be $\sim$1.47 MeV, and the first excited state is at $\sim$270 keV in energy. The preliminary analysis of the angular distributions for the ground and first excited states in ^{83}Ge is consistent with ℓ=2 transfer, presumeably $d_{5/2}$, to the ground state and a first excited 1/2$^+$ state at 270 keV populated with ℓ=0 transfer.

SYSTEMATICS OF N=51 ISOTONES

In Figure 2 the systematics of the N=51 isotones is displayed. The isotones near stability, ^{91}Zr and ^{89}Sr, have 5/2$^+$ ground states and first excited, 1/2$^+$ states above 1 MeV in energy. For Z<38, the 1/2$^+$ state comes down in energy, as low as 270 keV in ^{83}Ge. The next excited group in ^{83}Ge observed in the present (d,p) experiment is above 1 MeV in energy. The single-particle assignments in ^{85}Se are based on systematics; the d(^{84}Se,p) reaction will be measured in the near future to confirm these assignments.

It is well known that the order of neutron single-particle orbitals changes as a function of proton number. The usual explanation for this effect is the attractive monopole pairing

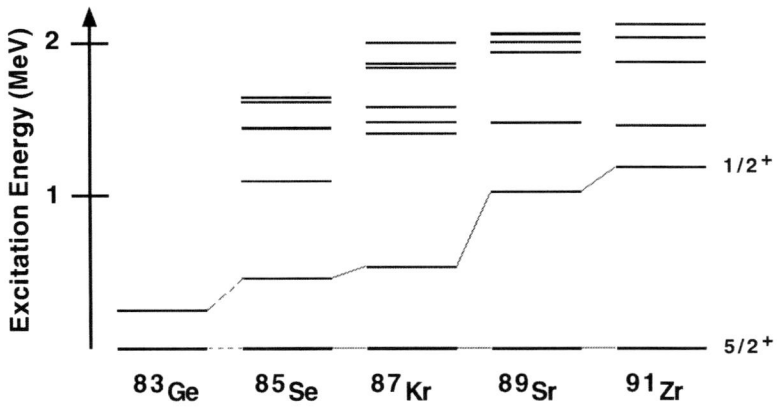

FIGURE 2. Systematics of N=51 isotones. Taken from [4, 5] and present results.

interaction between protons and neutrons in spin-orbit partners [6]. This explanation is not obviously the solution for the N=51 isotones summarized in Fig. 2, because the protons are in the $1f_{5/2}, 2p_{3/2}$ orbitals, while the valence neutrons are in the $2d_{5/2}$ and $3s_{1/2}$ orbitals. However, the more modest proton-neutron interaction between $1f_{5/2}$ protons and $2d_{5/2}$ neutrons could be lowering the $2d_{5/2}$ excitations, with respect to the $3s_{1/2}$ orbital, as the stable ^{89}Sr and ^{91}Zr are approached. Alternatively, the dramatic lowering of the $3s_{1/2}$ orbital in the neutron-rich N=51 isotones could be evidence for an enhanced diffuseness of the nuclear surface, which preferentially lowers $s_{1/2}$ orbitals [1]. To resolve the interpretation of the systematics of the single-particle orbitals in N=51 isotones will require theoretical shell model calculations, beyond the scope of the present experimental study.

FUTURE DIRECTIONS

The present work is the first study of a neutron-transfer reaction with a heavy neutron-rich beam, ^{82}Ge. However, a plethora of unstable neutron-rich species are available at HRIBF. In addition to completing the study of N=51 isotones with the d(^{84}Se,p) measurement, we plan to measure the d(^{132}Sn,p) reaction. Enriched ^{132}Sn beams with intensities of up to 10^5/s at 5 MeV/nucleon have been obtained at HRIBF. To enable such measurements in the near term, currently available, thin, position-sensitive silicon strip ΔE detectors and thick, pad E detectors will be used and mounted near 90 degrees in the laboratory. In the longer term, a barrel-shaped array of silicon ΔE-E telescopes is being developed to facilitate more detailed studies of the single-particle structure of neutron-rich A$\sim$130 nuclei.

ACKNOWLEDGMENTS

This work was supported in part by the National Science Foundation under contract number NSF-PHY-00-98800 and U.S. Department of Energy under contract numbers DE-FC03-03NA00143 (Rutgers), DE-AC05-000R22725 (ORNL), DE-FG02-96-ER40955 (TTU), DE-FG03-93ER40789 (Mines).

REFERENCES

[1]. J. Dobaczewski et al., Phys. Rev. C **53**, 2809 (1996).

[2]. D. W. Bardayan, et al.,Phys. Rev. C **62**, 055804 (2000).

[3]. L.A.Montestruque, M.C.Cobian-Rozak, G.Szaloky, J.D.Zumbro, S.E.Darden, Nucl.Phys. **A305**, 29 (1978).

[4]. Evaluated Nuclear Structure Data Files, http://www.nndc.bnl.gov.

[5]. J. P. Omtvedt, B. Fogelberg, and P. Hoff, Z. Phys. A **339**, 349 (1991).

[6.] P. Federman and S. Pittel, Phys. Lett. **69B**, 385 and **77B**, 29 (1977).

A Modular and Compact Multidetector System Based on Monolithic Telescopes

P. Figuera[a], F. Amorini[a,c], G. Cardella[a], A. Di Pietro[a], G. Fallica[b],
A. Musumarra[a,c], M. Papa[a], G. Pappalardo[a,c], F. Rizzo[a,c], W. Tian[a],
S. Tudisco[a,c], G. Valvo[b].

[a] *INFN-Laboratori Nazionali del Sud e Sezione di Catania, Catania Italy*
[b] *ST-Microelectronics, Catania, Italy*
[c] *Universita' di Catania, Catania, Italy*

Abstract. The characteristics of a new multidetector based on the use of Monolithic Silicon Telescopes are presented. Using suitable ion implantation techniques, the ΔE and residual energy stages of the telescopes have been integrated on a single Si chip, obtaining a typical thickness for the ΔE stage of the order of 2μm.

INTRODUCTION

The realization of compact detection systems with low identification threshold is one of the technological needs of heavy ion nuclear physics at low and intermediate energy. For those applications where compact detection systems with charged particle identification thresholds lower than 1 MeV/nucleon are requested, the conventional ΔE-E technique is difficult to apply and may require the use of expensive and fragile detectors. To overcome these difficulties, in the last years we started to work on the development of some prototypes of monolithic telescopes, where the ΔE and residual energy stages are integrated on a single silicon chip [1-3]. Using ion implantation techniques, such detectors have been built according to the scheme reported in figure 1.

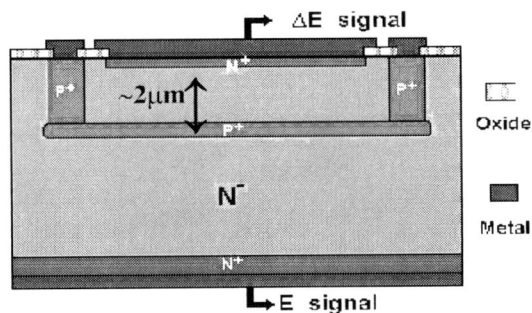

FIGURE 1. Scheme of the monolithic silicon telescope. See text for details.

Basically, the P^+ region (see figure) has been obtained using Boron implantation on an N^- bulk 500 μm thick and having a resistivity of 5000 Ω cm. Such P^+ region acts as a common ground electrode both for the ΔE and residual energy stages. In this way a ΔE thickness of the order of 2 μm (including dead layers) can easily be obtained, allowing for typical charge identification thresholds of the order of 300÷400 keV/nucleon in the mass region around Nitrogen. As an example, in figure 2 we show a ΔE versus residual energy spectrum obtained with the first prototype of the detector which had a size of only 4x4 mm^2.

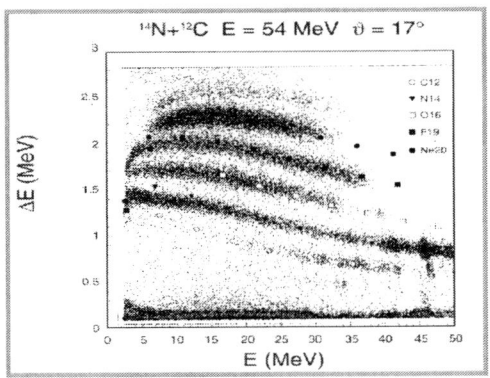

FIGURE 2. ΔE-residual energy spectrum measured in the reaction $^{14}N+^{12}C$ at E_{Lab}=54 MeV and ϑ_{Lab}= 17°.

The main limit of the first prototype was its small surface which would have required a large number of detectors to cover a large solid angle. A larger area detector of 20x20 mm^2, having similar characteristics as the first prototype, was also developed [3]. However, its area was not suitable to built a very compact detection system keeping a good angular resolution. Therefore, as we will discuss in the next section, we developed a more flexible device which allowed to built a compact and reliable multidetector which was already used in different experiments.

THE STRIP DETECTOR ARRAY

A good solution allowing to obtain a detector larger (15x4 mm^2) than the first prototype, working with standard commercial electronics and keeping a good position resolution, was found developing a device having 5 indipendent ΔE strips (3x4 mm^2) implanted on a common residual energy stage. Two of such devices mounted on a common ceramic package form a single detection module of our MONTE (MONolithic TElescope) array. Signals from the detectors are fed into compact preamplifier boards working under vacuum. The modular structure of the multidetector allows to mount it in very different configurations according to the experimental needs. In the first experiment we performed, 12 modules were mounted inside half a sphere (having a radius of only 5 cm) which was surrounded by BaF$_2$ crystals.

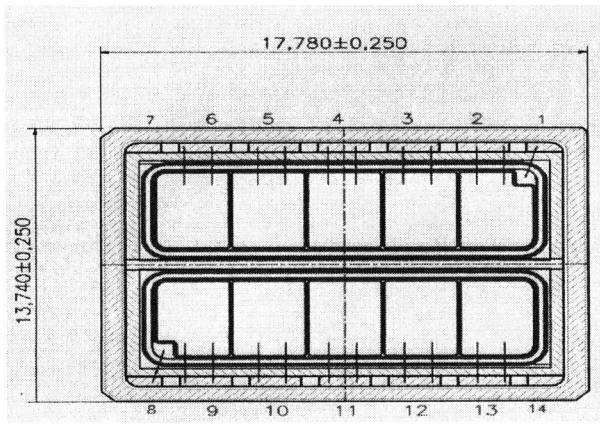

FIGURE 3. Scheme of the front view of the single detection module. Two detectors, having 5 independent ΔE strips implanted on a common ER stage, are mounted on the same ceramic package.

FIGURE 4. Picture of one of the geometrical configurations used for the MONTE array. Up to 24 detection modules can be mounted inside half a sphere having 5 cm radius. To estimate the dimensions note the 1 Euro coin in the picture.

In the second experiment we performed, 14 modules were mounted in plane covering the angular range 3^0-45^0 on both sides of the beam. The monolithic telescope setup, coupled with an array of conventional large area Si strip detectors (see figure 6), was used to study reactions induced by a radioactive ^{13}N beam at 30 and 45 MeV on a ^{9}Be target.

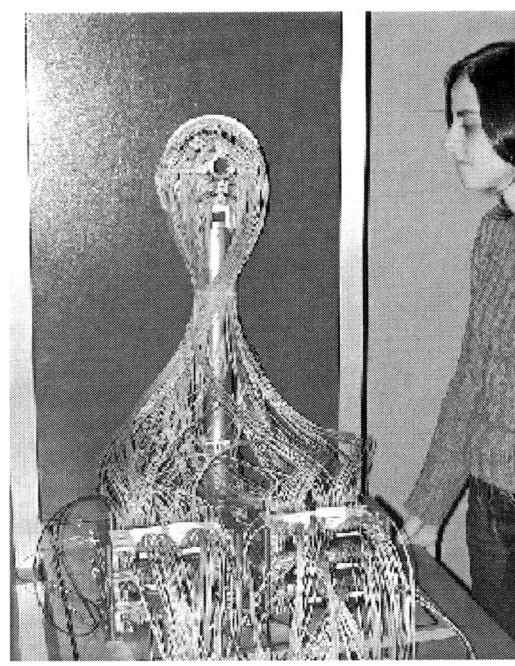

FIGURE 5. Picture of the setup shown in figure 4 seen from the back. One can see the cables coming out from the spherical support going into the preamplifier boards arranged in two groups of 3 boards each.

FIGURE 6. Picture of the MONTE setup mounted in plane coupled with an array of large area Si strip detectors.

A typical experimental problem in studying reactions induced by radioactive beams, is the presence of an intense β background due to decay of beam particles. The βs in fact, after entering inside a silicon detector, can be scattered at very large angles. Therefore they can travel for "long" distances inside the detectors depositing energies which typically can go up to 1 or 2 MeV. This is of course a real problem in those experiments were the detection of low energy particles is important. As an example, in Figure 7, we show an "on line" energy versus time spectrum collected by one of our telescopes in the reaction ^{13}N + ^{9}Be at 45 MeV. The time has been measured with respect to the cyclotron RF. One can see heavy ions, protons and α particles separated in time of flight and, at low energy, the β background which does not have a time correlation with the cyclotron RF.

Due to the extremely small thickness of the ΔE stage, the βs leave almost zero energy in the ΔE. Therefore their background is easily removed in ΔE versus residual energy spectra collected by using our MONTE array. This can be very useful if one is interested in detecting low energy particles in radioactive beam induced reactions.

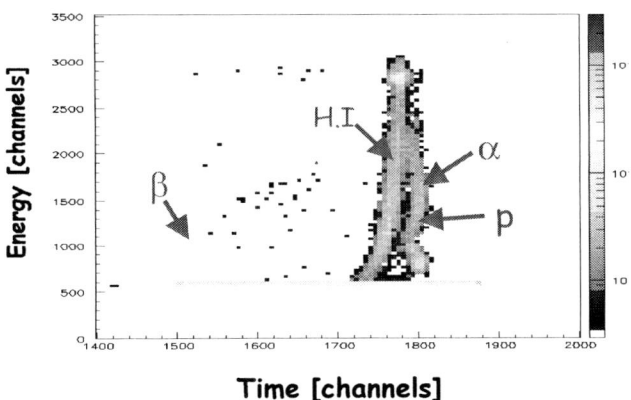

FIGURE 7. On line energy versus time spectrum measured by one of our detection modules in the collision ^{13}N+^{9}Be at 45 MeV. The β background, which is not correlated in time, is clearly visible.

REFERENCES

1. G.Cardella et al.: Nucl. Instr. and Meth. A378,262,(1996)
2. A.Musumarra et al.: Nucl. Instr. and Meth A409,414,(1998)
3. S.Tudisco et al.: Nucl.Instr. and Meth. A426,436,(1999)

Octupole Correlations In High-K States In ^{240}Pu

L. Amon[*], S. M. Mullins[¶], B. Akkus[*], M. N. Erduran[*]

[*] *Istanbul University, Department of Physics, Division of Nuclear Physics,*
34459 Vezneciler- Istanbul, TURKEY
[¶] *iThemba LABS , P.O. Box 722, Somerset West 7129 , South Africa*

Abstract. The aim was to search for reflection asymmetric high-K states in ^{240}Pu by measuring coincidences across the $T_{1/2}$ = 165 ns, K^{π} = 5$^-$ two-quasiproton isomer. The measurements were undertaken with the AFRODITE gamma-ray spectrometer when a pulsed beam of 30MeV α particles was directed onto a ^{238}U target. No evidence was found for the known γ rays that decay from the isomer into the ground state band.

INTRODUCTION

Relatively few nuclei are spherical in their ground state and a variety of shapes can be observed, often in the same nucleus. Most deformed nuclei possess a reflection-symmetric shape, which can be schematically subdivided into prolate, oblate and triaxial, depending on the relative axis values of the ellipsoid (Fig. 1).

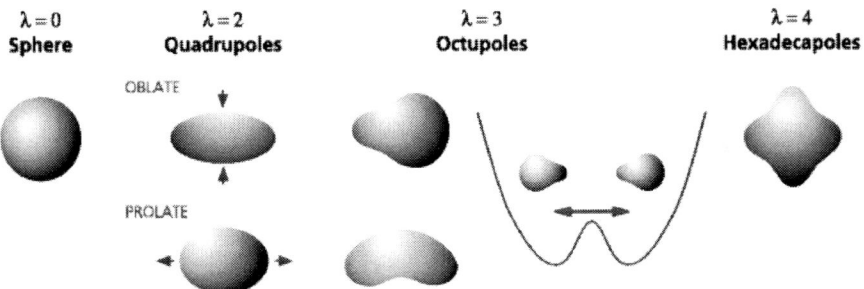

FIGURE 1. Schematic representation of some nuclear shapes [1] (originally published in Europhysics News).

Near closed shells spherical shapes prevail, while between closed shells the larger number of valence nucleons in orbits with large single particle angular momentum leads to nuclei with large deformations. The deformed shape breaks rotational invariance in the body-fixed frame of the nucleus, so that in general the intrinsic states are characterized by a mixture of j values, but the parity remains unmixed.

In the actinide region, rotational bands are observed that are made up of states of alternating positive and negative parity, which are linked by strong E1 transitions. In these cases, the nucleus is characterized by an octupole deformation so that it no longer has reflection symmetry, analogous to the shape of a pear. Octupole correlations are strongest when pairs of orbitals with $\Delta j = \Delta l = 3\hbar$ are near the Fermi surface [2], and it is noteworthy that the same sets of pairs, e.g., $j_{15/2} \otimes g_{9/2}$ neutrons and $i_{13/2} \otimes f_{7/2}$ protons, are at the Fermi surface in the heavy U-Pu-Cm actinides. The energy displacement between the positive and negative parity states can be used as an empirical measure of the strength of the octupole correlations [3].

K=5⁻ ISOMERIC STATE IN ^{240}Pu

K is the projection of the total angular momentum (or "spin"), usually called J, onto the symmetry axis of an axially symmetric deformed nucleus. That means the nucleus is stretched out (or squashed in) along one direction, and the symmetry axis corresponds to that direction. In the case of ^{240}Pu and in many deformed nuclei, the nucleus is stretched along the symmetry axis, and this shape is called "prolate". A consequence of this stretched-out prolate shape is that it is quantum-mechanically allowed for the nucleus to rotate collectively about an axis that is perpendicular to the symmetry axis.

Hence, for ^{240}Pu, there is rotational band built on the $J^\pi = 0^+$ ground state with spins that progress in units of $2\hbar$. But for each of these states, the total angular momentum vector still points up perpendicular to the symmetry axis, so that its projection onto this axis is zero, that is K = 0. This is a consequence of the nuclear pairing force, which keeps each proton paired off within another one, so that the summed angular momentum of each pair is zero, likewise for the neutrons. It is possible to form intrinsic excited states by "breaking" a pair, such that one particle occupies an orbital without a partner orbiting in the opposite direction.

Such is the case for K = 5 state in ^{240}Pu, where now a pair of protons is broken, and their angular momentum vectors instead of summing to K = 0, now sum together such that K = 5 i.e. the projection of summed angular momenta of the two unpaired protons onto the symmetry axis is $5\hbar$. Hence, for K = 5 state decay back to the K = 0 states based on the ground state, K has to change by $5\hbar$, whereas the E1 photon by which the decay takes place, only has $1\hbar$ of angular momentum. Hence, the decay is "K-forbidden". Since, however, one or both of the states does not correspond to purely K = 0 or K = 5, rather than being strictly forbidden, the decay is inhibited, such that the half-life is 165 ns (Fig. 2).

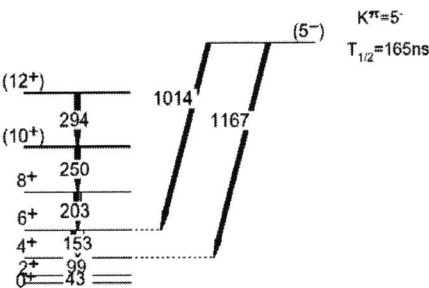

FIGURE 2. Low-spin level scheme of ^{240}Pu. The decays of the isomeric state populate the levels of the rotational band built on the ground state. The E1 transitions feed into the ground state band in which the E2 transitions are heavily electron-converted. This led to strong Pu X-rays that were measured in the LEPS detectors (Fig. 3).

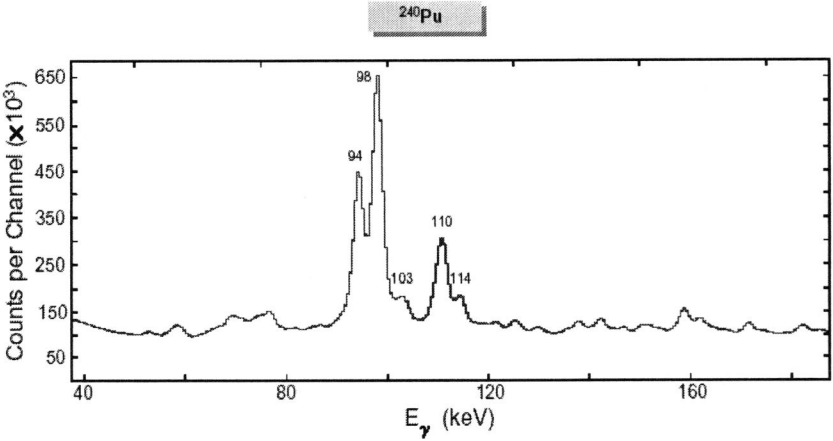

FIGURE 3. Total-projection LEPS spectrum obtained from the CLOVER-LEPS matrix. The dominant X rays are, $K_{\alpha 1}$ (103 keV) of ^{240}Pu and $K_{\alpha 1}$ (98 keV), $K_{\alpha 2}$ (94 keV) of ^{238}U.

EXPERIMENTAL SETUP

The measurements were undertaken with the AFRODITE gamma-ray spectrometer at the iThemba LABS located in near Cape Town, South Africa. In this configuration, AFRODITE consisted of seven clover detectors, each in BGO Compton suppression shield, and eight Low Energy Photon Spectrometers (LEPS). Beams of ^{4}He ions were delivered at an energy of 30 MeV from the Separated Sector Cyclotron (SSC) accelerator, and were directed onto a target which consisted of a ~300mg/cm^2 foil of ^{238}U (99.7 % enrichment). States in ^{240}Pu were populated by evaporation of two neutrons from the ^{242}Pu* compound system. The recoiling ions stopped in the focus of the HPGe detectors so that the measurement of "early-delayed" coincidences across the $T_{1/2}$ = 165 ns isomer was possible. Every fifth beam pulse was selected before injection, so that the time interval between successive beam pulses on target was ~300 ns. Events were written to tape when three or more of the fifteen HPGe detectors fired in coincidence with a resolving time of ~600ns. Efficiency and energy calibrations were undertaken with standard ^{152}Eu and ^{133}Ba sources.

DATA ANALYSIS & RESULTS

The raw experimental data were reduced into various E_γ - E_γ coincidence matrices with the Liverpool MTSort software. Different time conditions were imposed in order to enhance the sensitivity to the decay from the $K^\pi = 5^-$ isomer and hence to search for "early" transitions selected by the delayed decays (1014 and 1167keV γ rays) from the isomer. Analysis of the matrices was performed with the programs of the Radware package [4].

The vast majority of the events were generated by the decay of fission fragments which arose from the scission of the ^{242}Pu* compound nucleus. Any hope of finding the states of interest in ^{240}Pu hinged on being able to observe, and hence gate on, the gamma decays from the isomer. Unfortunately there was no evidence for either of the 1014 or the 1167 keV transitions in the delayed spectrum. This may suggest that the isomer was not populated or that the experimental setup did not have sufficient sensitivity. The latter possibility was investigated by a short measurement on ^{178}Hf *via* the ^{176}Yb(α,2n) reaction, in which the band associated with the $T_{1/2}$ = 78ns, $K^\pi = 6^+$ isomer [5] was observed with the same experimental setup and early-delayed analysis procedure. It is worth noting that the band could not be observed without the imposition of the early-delayed condition in the data reduction.

CONCLUSION

The AFRODITE spectrometer was employed to search for states above the $K^\pi = 5^-$ $T_{1/2} = 165$ns isomer in ^{240}Pu via early-delayed coincidence techniques. Although the experimental setup and beam-pulsing conditions were suitable for the measurement, no evidence was observed for the decay of the isomer. This could be because the isomer was not sufficiently strongly populated, since the spectra were dominated by gamma-ray lines from fission fragments. The background may have been reduced if a thinner target had been used, which would have allowed some of the fission fragments to escape while the stopping the ^{240}Pu evaporation residues within the focus of the array. A diagnostic measurement on the well-known $T_{1/2} = 78$ns, $K^\pi = 6^+$ isomer in ^{178}Hf showed that the experimental setup and data analysis procedures were sound.

ACKNOWLEDGMENTS

This project was a collaborative venture between iThemba LABS & Istanbul University. The contributions of M.Benatar, E.Gueorgiueva, P.Kwinana, J.J.Lawrie, G.K.Mabala, P.M.Maine, S.Mukherjee, N.J.Ncapayi, R.T.Newman, F.D.Smit and P.Vymers to the measurements are gratefully acknowledged. The operations crew of iThemba LABS are thanked for their delivery of the beam.

REFERENCES

1. R. Lucas, *Europhysics News*, **32**, 1 (2001).
2. G. Hackman et al., *Phys. Rev. C*, **57**, 3 (1998).
3. I. Wiedenhöver et al.., *Phys. Rev. Lett.*. **83**, 11 (1999).
4. D.C. Radford, *Nucl. Instrum. Methods Phys. Res.*, Sect. A **361**, 306 (1995).
5. T.L. Khoo and G. Lovhioden, *Phys. Lett. B* **67**, 271 (1977).

Collective and Single-Particle Properties of Neutron-Rich Nuclei Investigated via Reactions of Relativistic Radioactive Beams

Thomas Aumann

*Gesellschaft für Schwerionenforschung (GSI),
Planckstraße 1, 64291 Darmstadt, Germany*

Abstract. Measurements of the dipole continuum response of nuclei provide information on collective as well as single-particle properties. The rapidly varying electromagnetic field of a high-Z target experienced by a fast moving projectile with several hundred MeV/u kinetic energy causes dipole transitions into the continuum up to excitation energies of the giant dipole resonance. The extraction of differential cross sections with respect to excitation energy, which are directly linked to the dipole strength functions, can be accomplished by an exclusive measurement of the decay. This method was applied in a series of experiments at GSI utilizing fast secondary beams produced via fragmentation aiming at an investigation how the dipole response of nuclei evolves as a function of increasing isospin. Results of the experimental programme, which has concentrated so far on light neutron-rich nuclei ranging from helium to oxygen isotopes, are discussed. Much in contrast to stable nuclei, low-lying dipole excitations well below the giant dipole resonance region have been observed as a general phenomenon for these neutron-proton asymmetric nuclei. A quantitative analysis of low-lying threshold strength for loosely bound nuclei indicates that the characteristics of the dipole strength is directly related to the ground-state single-particle structure of the valence nucleon in the projectile. Finally, a brief outlook on future perspectives is given.

INTRODUCTION

Nuclear and electromagnetically induced reactions of radioactive beams studied in inverse kinematics provide a versatile tool for the investigation of nuclear-structure properties of short-lived nuclei far away from the valley of beta-stability. The relatively high energies (ranging from about 50 MeV/u to 1 GeV/u) of secondary beams produced via fragmentation are advantageous both from an experimental point of view as well as from theoretical considerations. The high beam energies result in short interaction times and small scattering angles, which allow the use of certain approximations and thus a quantitative description of the underlying reaction mechanisms. Experimental merits are the possibility of using relatively thick targets (in the order of g/cm^2) and kinematical forward focusing, which makes full-acceptance measurements feasible with moderately sized detectors. Thus nuclear-structure investigations of very exotic nuclei at the drip lines are possible even if such beams are produced with very low rates in the order of a few ions per second.

Here, we focus on dipole excitations to the continuum studied via electromagnetic-scattering experiments. At beam energies of several hundred MeV/u, the rapidly varying electromagnetic field of a high-Z target experienced by the fast moving projectile, generates dipole transitions with relatively large cross sections up to excitation energies of about 25 MeV, thus giving access to the dipole response of exotic nuclei. Theory predicts considerable changes of the excitation spectra for neutron-proton asymmetric nuclei in comparison to stable nuclei. For neutron-rich nuclei, for instance, different types of calculations predict pronounced effects, in particular a redistribution of the strength towards lower excitation energies well below the giant resonance region [1, 2, 3, 4, 5, 6]. The predicted strength functions depend thereby strongly on the effective forces used in the calculations [3] and thus, in turn, the measurement of the multipole response of exotic nuclei can yield information on the isospin dependent part of the in-medium nucleon-nucleon interaction.

The existence of a new collective dipole mode, the excitation of the more weakly bound valence neutrons against the core, was proposed [7, 8] for neutron-rich nuclei. Such a dipole vibration of the neutron-skin or halo against the core is predicted to occur at lower frequencies than the usual giant dipole resonance (GDR) due to a smaller restoring force [8]. A peak-like concentration of dipole strength below the GDR and involving many particle-hole states was found recently in theoretical calculations, e.g., using the relativistic mean field approach by Vretenar *et al.* [5], and by using the quasi-particle random-phase approximation (Sarchi *et al.* [9]). Such a collective 'soft mode' or 'Pygmy resonance' was also discussed to occur in neutron-rich stable nuclei. The cumulated dipole strength below the neutron threshold observed recently in real-photon scattering experiments on ^{208}Pb at the S-DALINAC in Darmstadt [10], was discussed in the context of a Pygmy dipole resonance excitation. A rather small exhaustion of the energy-weighted dipole sum rule, however, was observed. The effect is expected to be much more pronounced in unstable nuclei with large neutron excess.

So far, experimental information on the multipole response of unstable nuclei is rather limited. Low-lying dipole strength was observed in particular for nuclei exhibiting a halo structure, see, e.g., the $E1$ strength distributions measured for the two-neutron halo nuclei ^{6}He [11] and ^{11}Li [12], and for the one-neutron halo nuclei ^{11}Be [13] and ^{19}C [14]. The question of the resonant character of this low-lying strength is to our understanding still open for the two-neutron halo nuclei, while for the one-neutron halo nuclei the strength can be attributed to non-resonant dipole transitions to the continuum and is solely related to the single-particle properties of the weekly bound neutron. Such non-resonant transitions, however, are found to be a promising spectroscopic tool, as will be discussed in the present article.

EXPERIMENTAL TECHNIQUES

The experimental method applied by the LAND collaboration at GSI consists of producing high-energy radioactive beams and a kinematical measurement of breakup products produced in secondary targets. The excitation energy prior to decay is reconstructed by utilizing the invariant-mass method. The measurement is exclusive or kinematically complete in the sense, that all projectile-like decay products are detected, i.e. reaction products with velocities close to the beam velocity. Target-like reaction products are not measured (with the exception of γ-rays).

The experimental results exemplified in the next section utilized radioactive beams, which were produced by fragmentation of a primary ^{40}Ar beam delivered by the synchrotron SIS at GSI, Darmstadt. Fragment beams were selected by the Fragment Separator FRS according to their magnetic rigidity only, thus mixed secondary beams containing isotopes with similar mass-over-charge ratio were transported to the experimental area. The incident projectiles, however, were uniquely identified on an event-by-event basis by utilizing energy-loss and time-of-flight measurements.

In a similar manner, the fragments produced in the reaction target are identified. Here, the magnetic rigidity is determined from three position measurements defining the trajectories of the charged projectile residues in the magnetic field of a large-gap dipole magnet placed behind the target. Additional energy-loss and time-of-flight measurements allow unique identification of the outgoing fragments and determination of their momenta.

Neutrons emitted from the excited projectile or excited projectile-like fragments are kinematically focused in the forward direction and detected with high efficiency (90%) in the LAND neutron detector. The momenta of the neutrons are determined from the time-of-flight and position information. The angular range for fragments and neutrons covered by the detectors corresponds to a 4π measurement of the breakup in the rest frame of the projectile for fragment-neutron relative energies up to 5.5 MeV (at 500 MeV/u beam energy).

At the high beam energies used, the γ-rays need to be detected with good angular resolution in order to minimize Doppler broadening effects. Two detectors are used alternatively: the Crystal Ball spectrometer, which consists of 160 NaI detectors covering almost the full solid angle, or, a CsI array consisting of 144 submodules. The latter covering only the forward hemisphere, but with better angular resolution. Still, the resolution is limited by the Doppler broadening.

The excitation energy prior to decay is obtained by reconstructing the invariant mass combining the measurements described above. The resolution in excitation energy depends on the relative fragment-neutron kinetic energy and the resolution for measuring the gamma-sum energy (in the projectile rest frame) in case of the population of excited states. It changes from about 200 keV at the threshold to a few MeV in the region of the giant dipole resonance (at excitation energies around 15 MeV). In order to extract the electromagnetic excitation cross section from the measurement with the lead target, the nuclear contribution is determined from a measurement with a carbon target and scaled accordingly before subtraction.

THE DIPOLE RESPONSE OF LIGHT NEUTRON-RICH NUCLEI

Low-lying dipole modes in oxygen nuclei

The evolution of the dipole-strength distribution as a function of neutron-to-proton asymmetry was investigated systematically for the oxygen isotope chain ranging from a neutron excess of $N-Z=1$ up to $N-Z=7$ [15]. ^{16}O is a strongly bound doubly magic nucleus with a neutron separation energy $S_n = 16$ MeV, while, for the heavier isotopes,

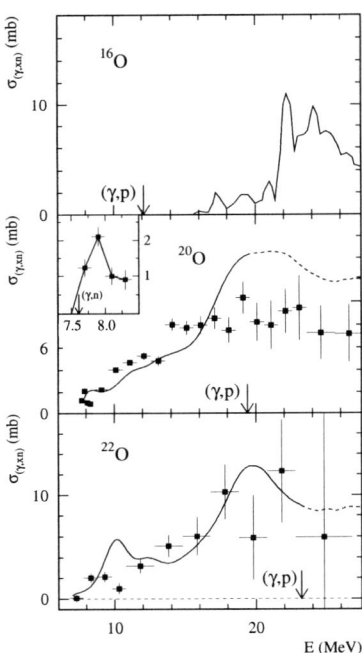

FIGURE 1. Photoneutron cross sections $\sigma_{(\gamma,xn)}$ for the unstable isotopes ^{20,22}O [15] and for ^{16}O [16] (upper panel). From Ref. [15].

the separation energy decreases to $S_n \approx 7$ MeV for the even isotopes with $A = 18$ to 22, and $S_n \approx 4$ MeV for the odd isotopes. Thus, one may expect a decoupling of the valence neutrons from the inert ^{16}O core and the appearance of a collective soft-dipole excitation.

The experimental results [15] for the unstable isotopes ^{20,22}O are shown in Figure 1. The photoneutron cross sections $\sigma_{(\gamma,xn)}$ for $x = 1,2,3$ were extracted from the measured electromagnetic excitation cross section by applying semiclassical calculations [17]. For comparison, the photoneutron cross section for the doubly magic nucleus ^{16}O [16] is also shown in the upper panel. Evidently, the dipole response changes significantly due to the presence of the valence neutrons. The strength appears to be strongly fragmented with a considerable fraction of the dipole absorption cross section located clearly below the GDR energy region. The integrated strength up to 15 MeV excitation energy amounts to about 10% of the classical energy-weighted dipole (TRK) sum rule for ^{20,22}O, while no dipole strength below 15 MeV is observed for the $N = Z$ nucleus ^{16}O. For ^{20}O, cumulation of dipole strength was also observed below the neutron threshold in the energy range from 5 to 7 MeV as extracted from a virtual-photon scattering experiment by Tryggestad *et al.* [18]. The integrated value of about 0.1 $e^2\text{fm}^2$ [18], however, corresponds to much less than 1% of the TRK sum rule only. The data are compared to a large-scale shell-model calculation [4] (solid curve in the lower two panels in Figure 1). Qualitatively, the shell-model calculations reproduce the experimental observation of a redistribution of the $E1$ strength towards excitation energies below the GDR for the neutron-rich isotopes.

The question to which extent the observed low-lying dipole strength involves coherent excitations was studied theoretically by Vretenar et al. [5] within the relativistic mean field approach. According to this calculation, the low-lying strength in the oxygen isotopes is mainly related to single neutron particle-hole excitations, while for heavier neutron-rich nuclei, e.g., for the tin isotopes, a resonance-like structure at low excitation energy resulting from a coherent superposition of particle-hole excitations is predicted. Similarly, Colò and Bortignon [6] found in their QRPA plus phonon coupling calculation only a small number of components in the wave functions of the low-lying structures in the oxygen isotopes. The low-lying dipole strength observed in neutron-rich oxygen nuclei may thus not be attributed to a collective soft dipole ('pygmy') mode.

Coulomb breakup and spectroscopy

The large cross sections observed for the electromagnetic dissociation of halo nuclei can be explained by non-resonant transitions to the continuum due to a large overlap between the tail of the neutron wave function and the continuum wave function with corresponding long wavelengths (direct-breakup model). Due to the smaller effective charge for higher multipolarities [19], the cross section is dominated by dipole excitations. The differential cross section for Coulomb breakup can thus be written as a product of the number $N_{E1}(E^*)$ of equivalent dipole photons with energy E^* associated to the rapidly varying Coulomb field of the target, and the square of the dipole matrix elements [20]:

$$\frac{d\sigma}{dE^*}(I_c^\pi) = (\frac{16\pi^3}{9\hbar c})N_{E1}(E^*)\sum_{nlj}C^2S(I_c^\pi,nlj) \times \sum_m |\langle q|(Ze/A)rY_m^1|\psi_{nlj}(\mathbf{r})\rangle|^2. \quad (1)$$

The Coulomb breakup cross sections are calculated for individual ground-state single-particle configurations of the neutron with a relative-motion wave function $\psi_{nlj}(\mathbf{r})$, with r being the relative distance between the core and the valence neutron, and corresponding core state (I_c^π). The final state with relative momentum $\mathbf{q}$ is represented either by plane or distorted waves describing the neutron in the continuum. As one can see from the above equation, the $E1$-strength distribution is very sensitive to the spatial extension of the single-particle wave function. A pronounced halo-like tail of the wave function leads to large transition probabilities at very low relative energies. The shape of the strength distribution is thus also sensitive to the angular momentum of the valence neutron. This characteristic behavior can be utilized to probe the ground-state single-particle structure by measuring the differential electromagnetic dissociation cross section: the l value of the valence nucleon is determined by the shape of the distribution, the core state I_c^π to which the nucleon is coupled can be identified by the characteristic γ-decay (in case of excited states), while the associated spectroscopic factors $C^2S(I_c^\pi,nlj)$ can be deduced from the absolute cross sections.

As an example, the results obtained for the halo nucleus ^{11}Be are shown in Figure 2 [21]. Results for carbon isotopes are published in Ref. [22]. The γ-ray sum-energy spectrum as recorded in coincidence with ^{10}Be fragments and a neutron (left panel)

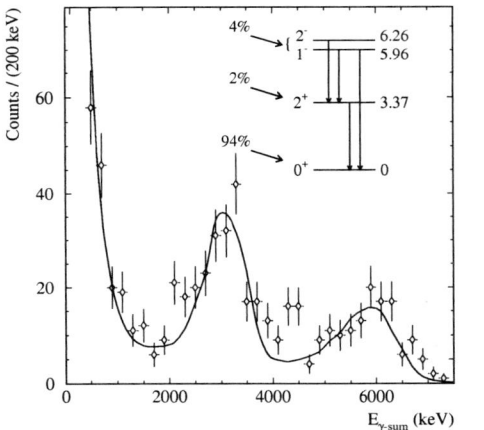
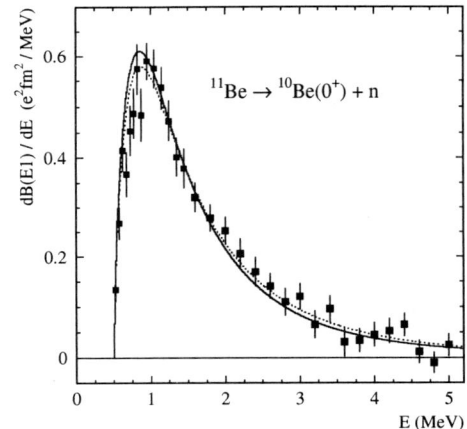

FIGURE 2. Left panel: Doppler corrected gamma-sum energy spectrum measured in coincidence with a ^{10}Be fragment and a neutron after inelastic scattering of ^{11}Be on a lead target. Right panel: Dipole-strength distribution for ground-state transitions. Taken from Ref. [21].

reveals contributions to the breakup cross section involving the 2^+ first excited state at 3.37 MeV but also higher lying states of ^{10}Be at around 6 MeV excitation energy. The latter result from removal of more deeply bound core neutrons from the p shell populating 1^- and 2^- states. The dominant part of the cross section yields the ^{10}Be fragment in its ground state. The resulting experimental dipole-strength function related to ground-state transitions is shown in the right panel of Figure 2 (symbols). The result of the calculation with the direct-breakup model (equation 1) is displayed by the dotted and solid curves for the plane-wave and distorted-wave calculations, respectively, resulting in a spectroscopic factor of 0.61(5) for the halo neutron in the $2s_{1/2}$ orbital coupled to the 0^+ ground state of the ^{10}Be core. From the integrated dipole strength, a corresponding root-mean-square radius for the s neutron halo density distribution of 5.7(4) fm was deduced [21]. The spectroscopic factor derived from the Coulomb breakup reaction is in good agreement with the one obtained from the nuclear one-nucleon removal cross section [23] measured at lower incident energy at MSU, as well as with the one deduced from a transfer reaction measured at GANIL [24]. We thus conclude that non-resonant Coulomb dissociation provides quantitative spectroscopic information on the single-particle structure of exotic nuclei. The enormous sensitivity of the cross sections to the tail of the wave function makes this method particularly well suited and efficient for the study of loosely bound systems.

FUTURE PERSPECTIVES

Experimental investigations concerning the multipole response of exotic nuclei were so far mainly limited to dipole excitations. In principle, $E2$ and $M1$ excitations to the continuum can be studied as well using electromagnetic excitation. The cross sections,

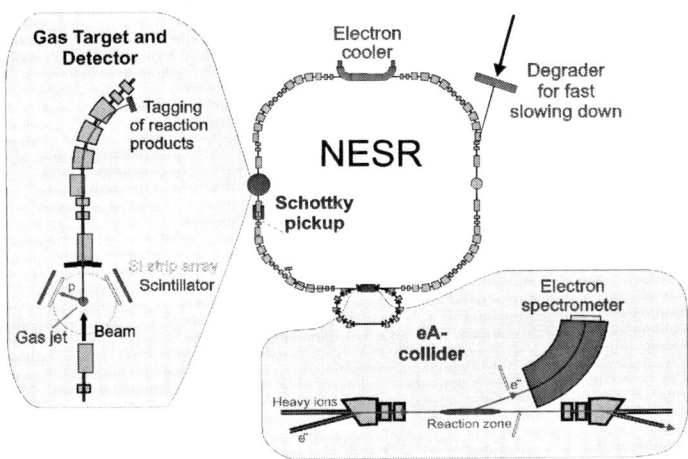

FIGURE 3. Schematic drawing of the storage ring NESR as planned for the new radioactive-beam facility at GSI [26, 27]. The lower insert displays the reaction zone of the electron heavy-ion (eA) collider and the electron spectrometer. The left insert shows schematically the setup for scattering experiments using gas-jet targets.

however, are much smaller and high statistical accuracy and good resolution in scattering angle are required in order to extract those excitation spectra from the dipole-dominated cross sections. Therefore, other reaction mechanisms are called for. The use of charge-exchange reactions in inverse kinematics seems to be a promising tool to study spin-dipole excitations in exotic nuclei, which are related to the neutron-skin thickness [25].

Electron-scattering experiments with exotic nuclei are still prohibited at present, however, a concept for an electron heavy-ion collider is proposed to be installed at the new 'International Accelerator Facility for Beams of Ions and Antiprotons' at GSI [26]. At this facility, high-energy radioactive beams produced by uranium fission will be available with intensities exceeding those of today by about three to four orders of magnitude. This is realized by i) faster cycling synchrotrons, ii) accelerating a lower charge state, iii) a large-acceptance separator optimized for the transmission of fission fragments, and iv) by an efficient injection into storage rings. The principle layout of the storage ring NESR is sketched in Figure 3. It is proposed to intersect the stored radioactive beam with an contra-propagating electron beam and to analyze the scattered electrons with good momentum resolution by a magnetic spectrometer with large acceptance (see lower part of Figure 3). Although the main purpose of this experiment will be elastic scattering with the aim of measuring charge distributions of exotic nuclei, the estimated luminosities (see Ref. [26], page 164) will make dipole and quadrupole giant resonance studies feasible.

The storage ring in conjunction with an internal gas target opens additional perspectives concerning inelastic hadron scattering. The giant monopole resonance, for instance, is usually studied utilizing alpha scattering at zero degrees. In inverse kinematics, this corresponds to alpha particles recoiling close to 90 degrees in the laboratory frame with very small kinetic energy. Thus thin targets are required in order to detect the alpha particles, which can be accomplished by a helium-gas-jet target in the storage ring. The drawback of a thin target can thereby be compensated in a storage ring by the revolution frequency of the ions of about 1 MHz resulting in luminosities (see Ref. [26]) making GMR studies possible with radioactive nuclei. One experimental challenge here is the precise measurement of the angle of the low-energy (~ 1 MeV) recoiling alpha particles. Since giant resonances decay by particle emission/evaporation, the magnetic rigidity of the resulting $(A - x)$ fragments will be different from the beam rigidity allowing a detection of fragments in the first dipole stage after the target (see insert in Figure 3). The same applies for the electron scattering experiments as well as the charge-exchange reactions mentioned above, resulting in a significant suppression of background as compared to the measurements in usual kinematics as performed with stable nuclei.

The experiments discussed in this article are the result of a collaborative effort. The author wishes to thank all members of the LAND/FRS collaboration: P. Adrich, K. Boretzky, D. Cortina, U.D. Pramanik, Th.W. Elze, H. Emling, H. Geissel, A. Grünschloß, M. Hellström, K.L. Jones, L.H. Khiem, J.V. Kratz, R. Kulessa, Y. Leifels, A. Leistenschneider, E. Lubkiewicz, G. Münzenberg, C. Nociforo, R. Palit, P. Reiter, H. Simon, K. Sümmerer, and W. Walús.

REFERENCES

1. I. Hamamoto and H. Sagawa, Phys. Rev. C **53** (1996) R1492.
2. F. Ghielmetti, G. Colò, P.F. Bortignon, and R.A. Broglia, Phys. Rev. C **54** (1996) R2143.
3. P.G. Reinhard, Nucl. Phys. A **649** (1999) 305c.
4. H. Sagawa and T. Suzuki, Phys. Rev. C **59** (1999) 3116.
5. D. Vretenar, N. Paar, P. Ring, and G.A. Lalazissis, Nucl. Phys. A **692** (2001) 496.
6. G. Colò and P.F. Bortignon, Nucl. Phys. A **696** (2001) 427.
7. Y. Suzuki, K. Ikeda, and H. Salto, Prog. Theor. Phys. **83** (1990) 180.
8. J. Chambers, E. Zaremba, J.P. Adams, and B. Castel, Phys. Rev. C **50** (1994) R2671.
9. D. Sarchi, G. Colò, and P.F. Bortignon, COMEX1, la Sorbonne, France, 2003, book of abstracts, p73.
10. N. Ryezayeva et al., Phys. Rev. Lett. **89** (2001) 272502.
11. T. Aumann et al., Phys. Rev. C **59** (1999) 1252.
12. M. Zinser et al., Nucl. Phys. A **619** (1997) 151.
13. T. Nakamura et al., Phys. Lett. B **331** (1994) 296.
14. T. Nakamura et al., Phys. Rev. Lett. **83** (1999) 1112.
15. A. Leistenschneider et al., Phys. Rev. Lett. **86** (2001) 5442.
16. E.G. Fuller, Phys. Rep. **127** (1985) 187; B.L. Berman, At. Data Nucl. Data Tables **15** (1975) 319.
17. C.A. Bertulani and G. Baur, Phys. Rep. **163** (1988) 299.
18. E. Tryggestad et al., Phys. Lett. B **541** (2002) 52.
19. S. Typel and G. Baur, Phys. Rev. C **64** (2001) 024601.
20. G. Baur and C.A. Bertulani, Nucl. Phys. A **480** (1988) 615.
21. R. Palit et al., Phys. Rev. C **68** (2003) 034318.
22. U. Datta Pramanik et al., Phys. Lett. B **551** (2003) 63.
23. T. Aumann et al., Phys. Rev. Lett. **84** (2000) 35.
24. J.S. Winfield et al. Nucl. Phys. A **683** (2001) 48.
25. A. Krasznahorkay et al., Phys. Rev. Lett. **82** (1999) 3216.
26. *An International Accelerator Facility for Beams of Ions and Antiprotons*, Conceptual Design Report, Publisher GSI (2001), http://www.gsi.de/GSI-Future/cdr/.
27. H. Geissel et al., Nucl. Instr. and Meth. in Phys. Res. B **204** (2003) 71.

The Giant Dipole Resonance at very high and very low excitation energies

F. Camera

Dipartimento di Fisica Universitá di Milano and INFN Sez of Milano via Celoria 16, 20133 Milano, Italy

Abstract. The spectra of high-energy γ rays emitted in the decay of the Giant Dipole Resonance (GDR) built on highly excited Ce isotopes (T > 2 MeV) and in warm ^{179}Au (T ≈ 0.7 MeV) nuclei have been measured to investigate the damping mechanisms of the GDR. The data on ^{179}Au have been taken at the Argonne National Laboratory using the LEPPEX array coupled to the FMA spectrometer. Using the symmetric fusion reaction ^{90}Zr (E_{beam} = 352 MeV) + ^{89}Yb it has been possible to obtain high energy gamma rays spectra associated with the evaporation of 0, 1 and 2 nucleons. The radiative fusion data suggest statistical emission from the compound nucleus. The analysis of the high-energy gamma ray spectra associated with the different evaporation channels at the present temperature of 0.7 MeV and spin range 15–20 ℏ show a fairly narrow GDR width of 5 MeV. This value is smaller than what would be expected in a nucleus where shell effects do not play a role. The data on highly excited Cerium has been taken at Legnaro INFN laboratories using the HECTOR array coupled to GARFIELD setup allowing the measurement of the heavy reaction residues and charged particles in coincidence with high energy γ-rays. Two reactions have been used to produce excited ^{132}Ce nuclei, a symmetric one (^{64}Ni + ^{68}Zn at beam energies of 300, 400 and 500 MeV) and an asymmetric one (^{16}O + ^{116}Sn at beam energies of 130, 250 MeV). The first preliminary data concerning GDR γ decay spectra will be presented.

INTRODUCTION

In these last years the experimental activity on the measurement of the GDR γ-decay in excited nuclei has investigated both the excitation energy and spin dependence of the GDR properties (see [1] for a review). Such measurements indicate that the FWHM of the resonance becomes broader as a function of both angular momentum and excitation energy. The centrifugal forces induced by high rotational frequencies deform the excited nucleus and, as the GDR couples to the quadrupole nuclear deformation, split the GDR components therefore increasing its FWHM. The temperature, instead, controls thermal shape fluctuations [2] that give rise to a dipole frequency distribution and, consequently, to a broadening of the GDR width. Beyond the bombarding energy at which the maximum angular momentum that the nucleus can sustain with no fissioning saturation, it has been observed that the GDR width increases more slowly. In fact, at T=2-3 MeV, only the temperature controls the thermal fluctuation effects due to the saturation of the spin effects and, consequently, also the width of the GDR seems to saturate. The comparison between experimental data and theoretical calculations has shown that the Collisional Damping, the mechanism responsible of GDR width at zero temperature, is rather independent of spin and temperature (at least up to T ≈ 2 MeV) [1,3]. It should be stressed that the great majority of GDR measurements have been done for a nuclear temperature interval between 1.2 to 2 MeV, region for which shell effects have already vanished, thermal fluctuations are fully established and the cooling mechanism of the composite system mainly follows the statistical model. Such scenario changes drastically outside such temperature interval. In fact, below 1.2 MeV, shells and pairing play an important role and might strongly affect GDR damping mechanisms, while for temperature higher than 2 MeV pre-equilibrium emission starts to play a role in the cooling process before thermalization and consequently, must be measured and taken into account in the interpretation of the data.

The very few experimental data in the temperature region T < 1.2 MeV have been obtained in experiments measuring high-energy γ-rays spectra from the decay of nuclei formed in fusion reactions with very light projectiles [4] or in inelastic scattering [5,6]. The results shows a GDR width similar to its T=0 value. Presently, at such low temperatures, the current theory of thermal fluctuations does not seem to be able to reproduce neither the experimental GDR width value nor its temperature dependence with exception of the nucleus ^{208}Pb. [6,7].

At high excitation energies a very recent measurement on tin isotopes has shown the presence of an unexpectedly strong pre-equilibrium component [8]. Its magnitude is so relevant that the excitation energy of the compound nucleus is reduced by about 20%. Based on that measurement together with a reanalysis of the existing results in Sn and nearby compound nuclei, it was proposed that the GDR width keeps increasing up to temperature of 3.2 MeV [8] instead saturating as previously observed. It should be stressed that all the conclusions of such work have been mainly deduced from inclusive measurements and in any case without coincidences with the evaporation residue, namely without a clean selection of the fusion channel.

In the present paper we present the results of two experiments where the properties of the GDR at very high and very low excitation energy have been measured. In the first experiment high energy γ rays up to 14 MeV emitted in the radiative fusion reaction ^{90}Zr+ ^{89}Y at E_{beam}= 352 MeV have been measured [9]. Such measurement provides the first investigation in the temperature region T=0.5-1 MeV at finite spin I =15-20 ℏ without the uncertainties of the previous measurements [4-6] because the excitation energy of the compound and the fusion channel are experimentally well defined. In the second set of experiments the hot rotating ^{132}Ce nuclei have been formed using two different entrance channels: ^{64}Ni+^{68}Zn (at E_{beam}=300, 400 and 500 MeV) and ^{16}O+^{116}Sn (at E_{beam}=250, 130 MeV). In such measurements high energy γ rays, residues and light charged particles have been measured to investigate, with exclusive data, the role of the pre-equilibrium emission in the cooling of the compound and in the decay of the GDR. In fact, the ^{64}Ni+^{68}Zn reaction at E_{beam}=500 and 300 MeV leads to the same excitation energy of the compound nucleus formed in the ^{16}O+^{116}Sn reaction.

THE GDR IN WARM 179AU NUCLEI

As previously discussed, the very few experimental data in the temperature region T < 1.2 MeV have been obtained in experiments measuring high-energy γ-rays spectra from the decay of nuclei formed in fusion reactions with very light projectiles [4] or in inelastic scattering [5,6]. An alternative way to populate medium-mass nuclei at rather low excitation energy and finite spin is to use the cold fusion in very symmetric heavy ion reactions. In fact, because of the Q-value of the reaction, the compound system is produced with excitation energies ranging between 10 to 40 MeV. However, in the present case, the radiative capture, namely the compound that does not emit any particle, has an unexpectedly large cross section (several tenths of μbarn) [10].

The first spectroscopic investigations in of the radiative fusion in such systems were based on measured γ-ray spectra which extend up to E_γ = 6-7 MeV [11]. Those experiments gave only some indications on the nature of the emission of high-energy γ-rays, which seemed to result not from of a direct process but instead from an equilibrated compound nucleus. It was not possible to deduce information about the GDR line-shape.

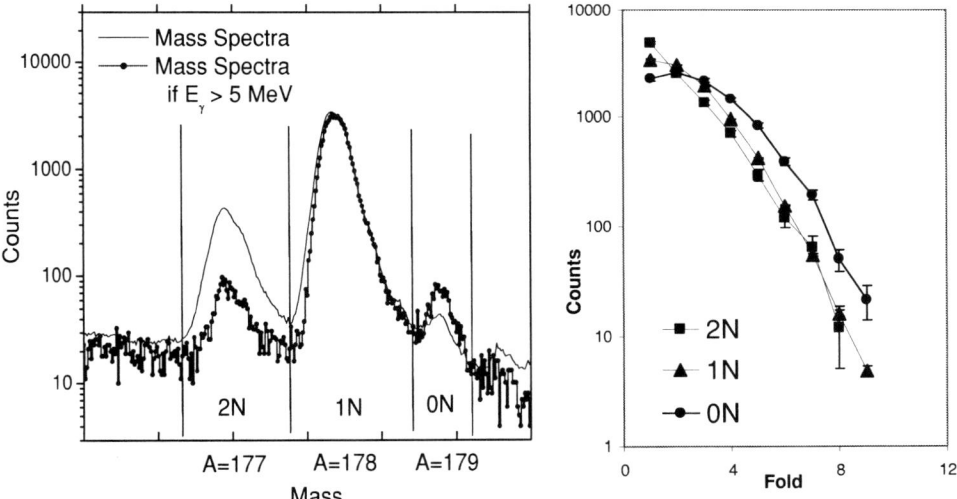

FIGURE 1. Left panel: Mass distributions of residue nuclei with (circles) and without (continuous line) the requirement of a coincidence with high-energy γ-rays (E_γ >5 MeV). The spectra have been normalized to the counts of the A=178 peak. Right panel: The fold distributions of low-energy γ-rays measured in the reaction ^{90}Zr + ^{89}Y at E_{beam}=352 MeV associated with high-energy γ-rays and the 0-, 1- and 2-particle emission channels (denoted in the legend, respectively, 0N, 1N and 2N). The three distributions are normalized at low folds. Note that the 0N fold distribution has larger yields at higher folds.

The experiment was performed at Argonne National Laboratory using a ^{90}Zr beam, provided by the ATLAS accelerator, impinging on an isotopically pure target of ^{89}Y, 400 μg/cm^2 thick. The isotopic purity of the target is guaranteed because ^{89}Y is the only stable isotope of Yttrium. The chosen bombarding energy, 352 MeV (346 MeV in the middle of the target), corresponds to the formation of the compound nucleus ^{179}Au, at an excitation energy E* ~ 20 MeV. The recoiling products pass through the Argonne Mass Analyzer (FMA) and subsequently are implanted into a 40x40 strips, double-sided silicon strip detector (DSSD). This DSSD was used not only to detect the implantation of the residues and determine their time of arrival with respect to the prompt γ-rays detected at the target, but also to measure the subsequent α decay(s) of the implanted ions. The high energy γ rays emitted by the reactions were measured with the ORNL/MSU/TAMU array of BaF$_2$ each having a hexagonal shape (with internal radius of 3.25 cm) and 20 cm deep. These detectors were grouped in 4 packs each made of 37 crystals. They were positioned at 90° and 149° with respect to the beam direction at a distance 39 cm from the target. In addition, low-energy γ-rays were also detected by a BGO spin/sum-energy array which was used primarily as a multiplicity filter with an efficiency of ~ 30% and cross talk probability of ~ 12%. The detectors of the multiplicity filter formed two honeycomb structures, one above and one below the reaction plane.

The left part of Figure 1 shows the spectra of the mass of the heavy recoiling nuclei detected at the focal plane of the FMA. The strongest peak in the middle corresponds to mass A=178 and the peak on the right correspond to mass A=179, namely, the radiative fusion channel, where the entire excitation energy is dissipated through γ emission. As the figure illustrates, when a high-energy γ-ray is detected, the population cross-section of the 0-particle channel increases while that for 2-nucleon emission decreases. This behavior is expected from simple phase-space considerations, as there is not enough energy for the emission of two particles and a high-energy γ-ray. In the right panel of figure 1 the low energy γ-ray fold spectra measured in the BGO multiplicity filter in coincidence with the different recoils measured in the FMA are shown.

The data indicate that the radiative fusion channel is associated to a higher gamma multiplicity and this finding is consistent with the statistical nature of the radiative fusion channel according to which the emission of high-energy γ-rays is more probable at high spin where particle emission is inhibited because of the low internal energy.

The measured high-energy γ-ray spectra in coincidence with the residues measured in the focal plane of the FMA have been analysed within the framework of the statistical model decay of the compound ^{179}Au nucleus. Because of the channel selection, a Monte Carlo approach was employed [11]. The calculations were performed assuming a level density parameter a = A/8 MeV^{-1} and 100% of the EWSR strength of the GDR. In addition, for the yrast line the parabolic parameterization deduced from the measured yrast line of Ref. [12] was used.

A grid of calculations was performed where the quadrupole deformation parameters (β,γ) of the compound nucleus and the intrinsic damping width $\Gamma_0^{\downarrow}$ were varied. At each value of the deformation parameters (β,γ) the values of the GDR centroids and widths were deduced using the Hill Wheeler parameterization [13,14] with the spherical value $E_0 = 14.2$. The relation used between the total width and the intrinsic width $\Gamma_0^{\downarrow}$ is $\Gamma = \Gamma_0^{\downarrow} (E_{GDR}/E_0)^{1.6}$ [1].

The calculations have been folded with the response function of the BaF$_2$ detector wall (calculated with the GEANT libraries) and then normalized to minimize the χ^2_{corr} between 5-9 MeV. In the calculation of χ^2_{corr} the exponential behaviour of the spectrum has been weighted accordingly [3]. For the best fit of the quadrupole deformation parameters (β,γ) and of the intrinsic width $\Gamma_0^\downarrow$ the χ^2_{corr} was calculated in the interval of E$_\gamma$ between 9 and 13 MeV.

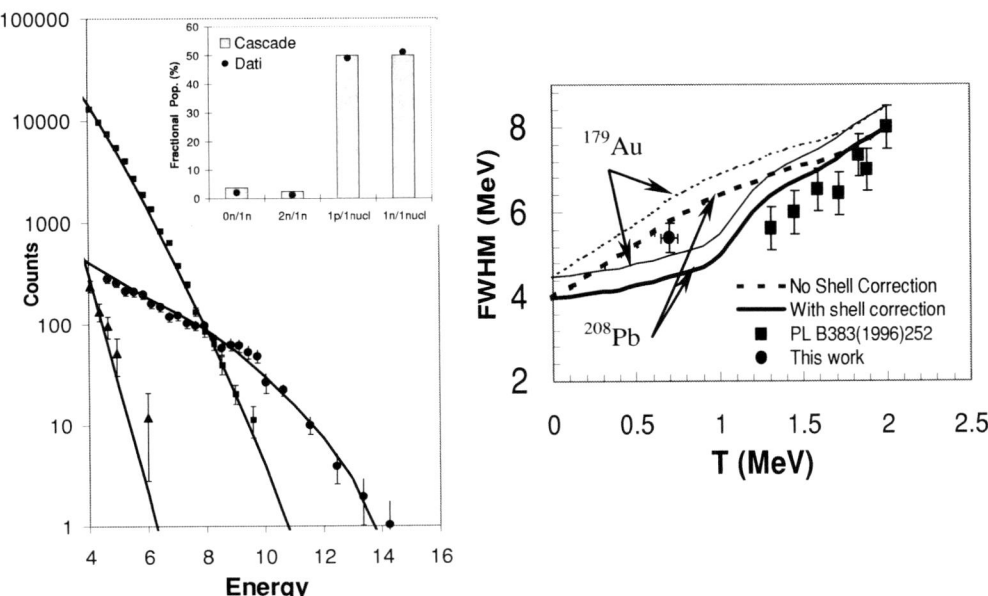

FIGURE 2. Left panel: The high-energy γ-ray spectra associated with the 0-, 1- and 2-nucleon emission channels (filled points, squares and triangles, respectively) are shown in comparison with statistical model calculations as described in the text. In the inset the measured fractional populations of the different residual nuclei are shown in comparison with statistical model predictions. Right Panel: The FWHM of the GDR (filled point) extracted from the fitted GDR lineshape deduced from the analysis of the high-energy γ-ray spectrum of the radiative fusion channel from the reaction ^{90}Zr + ^{89}Y at E$_{beam}$=352 MeV. Also shown are the measured widths for ^{208}Pb [5] and calculations from the thermal fluctuation model for ^{208}Pb [13] with shell correction (thick continuous line) and without (thick dashed line). The upper lines (thin continuous and dashed) correspond to a rigid shift of 0.5 MeV to take into account the different value of $\Gamma_0^\downarrow$ at T=0 for the two nuclei.

The angular momentum distribution of the compound nuclei which is needed for the statistical model calculations has been obtained comparing the coincidence-fold distribution measured with the multiplicity filter associated with all detected evaporation residues with the calculations obtained by folding a triangular smoothed spin distribution of the compound nuclei with the calculated response function of the multiplicity filter. The calculations were made for different assumptions of the value of the maximum angular momentum assuming a diffuseness d=2. The best fitting value $L_{max} = 16\ \hbar$ corresponds approximately to the value predicted for this reaction by the grazing model [15].

The high-energy γ-ray spectra measured in the BaF$_2$ detectors associated to the three different recoil masses measured in the FMA are displayed in the left panel of figure 2. The three high-energy γ-ray spectra corresponding to 0-, 1- and 2-nucleon emission are shown with filled points, square and triangles, respectively. In the same plot the best fitting statistical model calculations are shown. The obtained nuclear deformation is β= 0.1 while the intrinsic width is $\Gamma_0^\downarrow = 5$ MeV. In the inset of the left part of the figure the relative yields of the measured residual cross sections is shown in comparison with the statistical model prediction. The fact that the statistical model calculations simultaneously reproduce the different decay channels and the residual nuclei population provides a stringent consistency check and therefore strengthens the conclusion about the GDR width deduced from the radiative fusion γ–ray spectrum.

The measured total GDR width for the nucleus ^{179}Au as deduced from the present analysis is $\Gamma_0^\downarrow = 5 \pm 0.35$ MeV. This value is not very different from the zero temperature value measured in the same mass region ($\Gamma = 4.5$-4.8 MeV) and it corresponds to an average temperature of 0.7 MeV as deduced from the relation $T = [(E^*-E_{GDR} - E_{rot})/a]^{1/2}$, where $a = A/8$. It is interesting to compare the present result for the nucleus ^{179}Au with those obtained using inelastic scattering for ^{208}Pb [5]. As shown in the right panel of figure 2, in the case of the Pb nucleus the experimental data have been compared with two predictions based on the thermal fluctuation model [13], with and without the inclusion of shell corrections. One can make a simple rescaling of these calculations for ^{197}Au (as $\Gamma_0^\downarrow$(T=0) for ^{197}Au should be at least 0.5 MeV larger that that of ^{208}Pb) as a guideline for the interpretation of the results. In fact, shell effects in ^{197}Au could be smaller than in the closed shell ^{208}Pb nucleus.

GDR IN HOT 132CE NUCLEI

As previously discussed the problem of the saturation of the GDR width has been recently addressed in a work [8] using the ^{18}O+^{100}Mo = ^{118}Sn* reaction at E$_{beam}$ ranging from 122 to 214 MeV. Such experiments were inclusive and the studied systems were formed with quite asymmetric reactions. To understand better the role of the pre-equilibrium emission in the cooling process of the compound nuclei we have measured for different entrance channels and excitation energy the γ decay of the GDR in the ^{132}Ce nucleus. This nucleus is spherical like the Sn and, from the point of view of the thermal shape fluctuations, is expected to have a similar behaviour of the Sn isotopes but with somewhat reduced spin effects because of the larger moment of inertia.

In the experiment the high-energy gamma rays were measured by 8 large BaF$_2$ detectors of the HECTOR [16] apparatus, the residues were measured by a PSPPACs system covering the angles between 4° and 14° while the light charged particles were measured using one section of the GARFIELD detector array [17].

The BaF$_2$ detectors of the HECTOR array consist of 8 large cylindrical tapered scintillators 17.5 cm long and 14.5 cm large at the basis. These detectors have been placed inside the scattering chamber of GARFIELD, namely in vacuum. All the voltage dividers, to avoid overheating, have been placed externally. The calibration was made using ^{60}Co and ^{88}Y sources for low energy γ rays and the 15.1 MeV monochromatic emission produced in the reaction ^{11}B (19 MeV) + D = ^{13}C*.

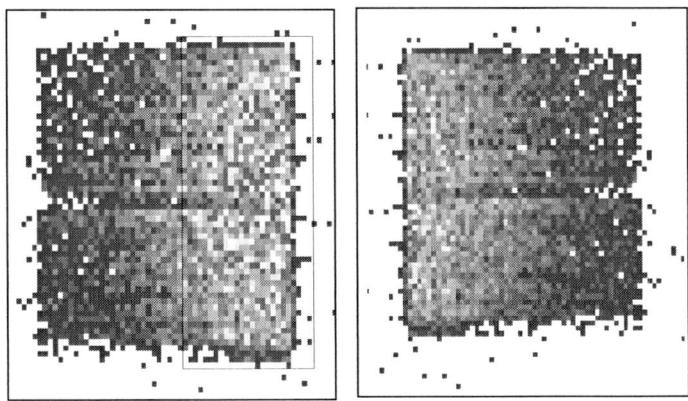

FIGURE 3. The PSPPAC hit pattern for the fusion residues measured in the reaction ^{64}Ni+^{68}Zn at E_{beam}= 400 MeV

The GARFIELD array [1] is composed by two large drift chambers filled with CF. Inside the chambers a micro-strip structure composed of 180 trapezoidal pads is placed. Each such pad provides the ΔE signals of the incoming light charged particles. In the same gas volume there are 180 CsI(Tl) which provide the measurement of the residual energy [2]. By the combined use of the E - ΔE and time signals it has been possible to detect and to identify the light charged particles emitted between 30° to 85°. The PSPPAC system consists of two position sensitive PPACs and a foil between them. The thickness of the foil has been chosen to stop only the fusion like particles and let the scattered beam or the project-like particles to pass through. Consequently, the anticoincidence between the two PPACs together with the measurement of the time of flight cleanly selects only the fusion residues. The left panel of figure 3 shows the hit pattern of the fusion residues measured in one of the PSPPAC system.

The hot and rotating ^{132}Ce nucleus was formed using two different entrance channels, one symmetric (^{64}Ni + ^{68}Zn at E_{beam}=300, 400 and 500 MeV) and the other very asymmetric (^{16}O + ^{116}Sn at E_{beam}=250 and 130 MeV), in order to investigate the role played by the pre-equilibrium emission. It is important to note that the ^{64}Ni+^{68}Zn reaction at E_{beam}=500 and 300 MeV leads to the same excitation energy of the compound nucleus formed in the ^{16}O + ^{116}Sn reaction at 250 and 130 MeV. In figure 4 the high-energy γ ray spectra measured in the reaction ^{64}Ni+^{68}Zn at E_{beam}= 400 MeV with and without the residues selection by the PSPPAC are shown. In the left panel the spectra are not normalized while in the right panel the spectra are normalized in the high energy part. It is clear that the gate on fusion residues, besides reducing the overall counting statistic of an order of magnitude, strongly affects the low energy part of the spectra. In fact, non fusion reactions contribute mainly to the low energy part of the spectra (E < 10 MeV). In the same plot a very preliminary statistical model calculation is displayed. In the calculation a GDR centroid of 14.5 MeV and a width of 11 MeV and 80% of the EWSR have been used.

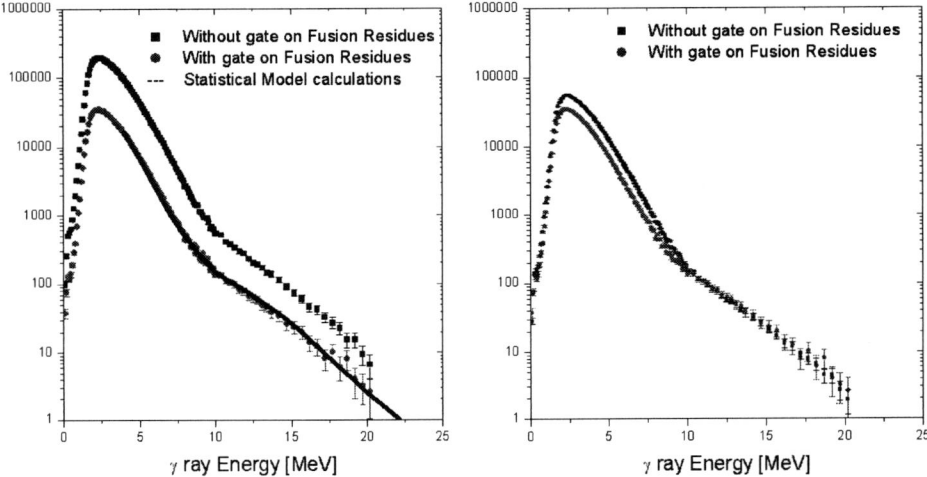

FIGURE 4. The high energy gamma rays spectra measured in the reaction ^{64}Ni+^{68}Zn at E_{beam}= 400 MeV. In both panels, the spectra measured without (with) a condition on the fusion residues is show with filled squares (filled points). In the left panel the data are shown without any kind of normalization with on the right the data has been normalized in the high energy part. The prediction from statistical model calculation (see text) is shown in the left plot with a continuous line.

CONCLUSIONS

The GDR γ decay of the warm ^{179}Au nucleus (T ≈ 0.7 MeV) and of the hot (T > 2 MeV) ^{132}Ce nucleus have been measured. In the first case the spectral shape of the radiative fusion channel is characterized by an exponential slope that is much less steeper than those of spectra measured at higher excitation energies. The statistical model analysis of the data indicates that the radiative fusion is not a direct process and that it can be described as a decay channel of the compound nucleus formed in the fusion process. In addition, the parameters of the GDR line shape used to reproduce the data are consistent with a nucleus with a deformation similar to that deduced from the yrast line. Altogether the present result suggests the persistence of shell effects in the region of temperature T = 0.5-1 MeV and calls for detailed calculations for this nucleus.

The data presented on ^{132}Ce nucleus are still in a preliminary form. Some preliminary results show that the selection of residues, although very expensive in terms of the counting statistics, is very critical at such high excitation energies. The preliminary statistical model calculations show, as expected, large values of the FWHM of the GDR lineshape.

ACKNOWLEDGMENTS

This work is based on two different experiments. The first has been carried out in Argonne National Laboratory in collaboration with B.B. Back, A .Bracco, M.P. Carpenter, F. Della Vedova, I. Dioszegi, K. Eisenman, R.V.F. Janssens, M. Halbert, A. M. Heinz, P. Heckman, D. Hoffman, D. Jenkins, T.L. Khoo, F.G. Kondev, T. Lauritsen, S. Leoni, C.J. Lister, A. Lopez-Martens, S. Mantovani, B. Mcclintock, B. Million, S. Mitsuoka, E.F. Moore, V. Nanal, M. Pignanelli, J. Seitz, D. Seweryniak, R. Siemssen, M. Thoennessen, R.J. Van Swol, R. Varner and O. Wieland. The second has been carried out in Legnaro in collaboration with U. Abbondanno, S. Barlini, A. Bracco, S. Brambilla, G. Benzoni, M. Brekietz, M. Bruno, G. Casini, M. Chiari, M. D'Agostino, A. Lanchais, S. Leoni, E. Geraci, A. Giussani, F. Gramegna, M. Kmiecik, A. Maj, G.V. Margagliotti, P.F. Mastinu, P.M. Milazzo, B. Million, A. Moroni, A. Nannini, A. Ordine, R. Sacchi, G. Vannini, L. Vannucci and O. Wieland. I thank all these collegue for their efforts

REFERENCES

1. P.F. Bortignon, A. Bracco and R.A. Broglia, Harwood Academic Publishers, Amsterdam (1998), volume of Contemporary Concepts in Physics.
2. W.E. Ormand et al. Phys. Rev. Lett. 77, 607(1996); Nucl. Phys. A614, 217(1997); A618, 20(1997).
3. A. Bracco et al. Phys. Rev. Lett. 74, 3748(1995).
4. M. Kicinska-Habior et al. Phys. Rev. C36, 612(1987).
5. E. Ramakrishnan *et al.*, Phys. Lett. B383, 252(1996).
6. P. Heckman et al. proceeding of the Conference on Frontiers of Nuclear Structure (July 29th - August 2nd, 2002) edited by P.Fallon in print by AIP.
7. D. Kusnezov, Y.Alhassyd and K.Snover Phys. Rev. Lett. 81, 542(1998).
8. M.P. Kelly et al., Phys. Rev. Lett. **82**(1999)3404.
9. F. Camera et al. Physivs Letter B 560(2003)155.
10. J.G. Keller *et al*, Nucl. Phys. A452, 173(1986).
11. K.-H. Schmidt *et al.*, Phys. Lett. B168, 39(1986).
12. W. Mueller et al. To be published
13. W.E. Ormand, Nucl. Phys. A649, 145c(1999) and references therein.
14. Y. Alhassid, Nucl. Phys. A649, 107c(1999) and references therein.
15. G. Pollarolo et al. Phys. Rev. C62, 054611(2000).
16. A. Maj et al. Nucl. Phys. A571(1994)185
17. F. Gramegna et al., *Nucl. Inst. Meth.* **A389**(1997)474

A Study of the Jacobi Shape Transition in Light, Fast Rotating Nuclei with the EUROBALL IV, HECTOR and EUCLIDES Arrays

A. Maj[1], M. Kmiecik[1], M. Brekiesz[1], J. Grębosz[1], W. Męczyński[1],
J. Styczeń[1], M. Ziębliński[1], K. Zuber[1], A. Bracco[2], F. Camera[2],
G. Benzoni[2], B. Million[2], N. Blasi[2], S. Brambilla[2], S. Leoni[2],
M. Pignanelli[2], O. Wieland[2], A. Airoldi[2], B. Herskind[3], P. Bednarczyk[1,4,11],
D. Curien[4], E. Farnea[5], G. de Angelis[6], D.R. Napoli[6], J. Nyberg[7],
M. Kicińska-Habior[8], C.M. Petrache[9], D. Petrache[9], N. Dubray[4], J. Dudek[4],
K. Pomorski[10]

[1] *The Henryk Niewodniczański Institute of Nuclear Physics, Polish Academy of Sciences, ul. Radzikowskiego 152, PL-31342 Kraków, Poland*
[2] *Dipartimento di Fisica and INFN sez. Milano, I-20133 Milano, Italy*
[3] *The Niels Bohr Insitute, Blegdamsvej 17, DK-2100 Copenhagen, Denmark*
[4] *Institut de Recherches Subatomiques, 23 rue du Loess, BP28, F-67037 Strasbourg, France*
[5] *INFN sez. Padova, I-35131 Padova, Italy*
[6] *INFN - Laboratori Nazionali di Legnaro, I-35020 Legnaro (PD), Italy*
[7] *Department of Radiation Sciences, Uppsala University, SE-75121 Uppsala, Sweden*
[8] *Institute of Experimental Physics, Warsaw University, PL-00681 Warsaw, Poland*
[9] *Dipartimento di Fisica, Universita di Camerino, I-62032 Camerino (MC), Italy*
[10] *Katedra Fizyki Teoretycznej, Uniwersytet Marii Curie-Skłodowskiej, PL-20031 Lublin, Poland*
[11] *GSI Darmstadt, Planckstr. 1, D-64291 Darmstadt, Germany*

Abstract. The high-energy and discrete γ-ray spectra, as well as the charged particle angular distribution have been measured in the reaction 105 MeV $^{18}O+^{28}Si$ using the EUROBALL IV, HECTOR and EUCLIDES arrays in order to investigate the predicted Jacobi shape transition in light nuclei. A comparison of the GDR line shape data with the predictions of the thermal shape fluctuation model, based on the most recent rotating liquid drop LSD calculations, shows evidence for such Jacobi shape transition in hot, rapidly rotating ^{46}Ti. The found narrow low-energy component in the GDR line shape is interpreted as the consequence both of the elongated shape and of the Coriolis effect.

INTRODUCTION

The Jacobi shape transition, an abrupt change of nuclear shape from an oblate ellipsoid non-collectively rotating around its symmetry axis to an elongated prolate or triaxial shape rotating collectively around the shortest axis, has been predicted to appear in many nuclei at angular momenta close to the fission limit [1,2]. A recently developed LSD (Lublin-Strasbourg Drop) model [3,4] has been used to calculate the Jacobi transition mechanism in ^{46}Ti nucleus. This model is based on the standard liquid drop approach with the newest, and so far the best, parametrization of the fission barrier height for all known nuclei. The model neglects the shell effects, but can be used for higher temperatures ($T > 1$ MeV), where the influence of the shell effect is small, e.g. for the description of compound nuclei produced in heavy-ion reactions. The results of the LSD calculation for ^{46}Ti are illustrated in Fig. 1, where the quadrupole deformation parameter β_2, as well as the corresponding ratio of the long to short semi axis, are shown for different spin values. As can be seen, the equilibrium shape of the nucleus (in this liquid drop approximation) is spherical at I = 0 and nearly spherical for $I < 10$ $\hbar$, becomes oblate at larger spin value and the size of the oblate deformation becomes larger when the spin increases. At around $I = 28$ $\hbar$ the Jacobi shape transition sets in: the nucleus becomes unstable towards triaxial and then (for $I > 34$ $\hbar$) it follows prolate shape configurations, with rapidly increasing size of the deformation up to the fission limit (around $I = 40$ $\hbar$).

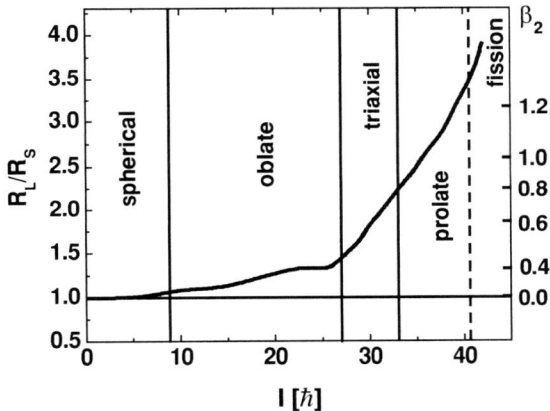

FIGURE 1. Long-to-short axis ratio and the β_2 deformation parameter for the equilibrium shape of ^{46}Ti as a function of spin, obtained from the LSD model calculations.

Signatures of the presence of elongated shapes from the Jacobi regime should be visible, among others in: the γ-decay of the Giant Dipole Resonance (GDR), the giant back-bend of the E2 γ-transition energies [5] and the angular distribution of the emitted charged particles.

So far, the question of Jacobi transition in light nuclei has been addressed by studying the GDR γ-decay from compound nuclei ^{45}Sc [6] and ^{46}Ti [7,8] in rather inclusive experiments. In both cases the data suggest the presence of large deformations that could be related to the Jacobi transition. In medium-mass nuclei the possible giant backbend of the quasi-continuous E2-radiation due to the Jacobi transition has been recently investigated [5]. In all these studies, however, an unambiguous experimental evidence for the Jacobi shape transition in nuclei has not been found.

In this paper we present new results from a highly selective experiment for the ^{46}Ti compound nucleus. They concern the measurement of high energy γ-rays and charge particle angular distributions with very stringent conditions selecting the highest part of the spin distribution, namely by gating on well-known discrete lines of the evaporation residual nuclei and using the multiplicity filter to select the spin range. Additionally, the discussion of these results shows the importance of the Coriolis splitting of the GDR components at the high rotational frequencies that can be reached only in light nuclei.

THE EXPERIMENT AND RESULTS

The experiment was performed at the VIVITRON accelerator of the IReS Laboratory of Strasbourg (France), using the EUROBALL phase IV Ge-array coupled to the HECTOR array [9] and the charged particle detector EUCLIDES. The ^{46}Ti compound nucleus was populated in the ^{18}O + ^{28}Si reaction at $E_B = 105$ MeV. The excitation energy of the ^{46}Ti was 86 MeV and the maximum angular momentum $l_{max} \approx 34\ \hbar$. The configuration of the EUROBALL array for this experiment was 26 germanium clover and 15 cluster detectors (all with the BGO Compton suppression shields), and the Inner-ball consisting of 83 BGO crystals, which together with the germanium detectors resulted in ca. 65% efficiency for the multiplicity determination. The 8 large volume BaF$_2$ detectors of the HECTOR were placed in the forward hemisphere, together with 4 small BaF$_2$ detectors to provide a good time reference signal. The charged particles array EUCLIDES consisted of 40 silicon telescope detectors, arranged in 6 rings, covering approximately 90% of the solid angle. The trigger condition was such that events having at least two Ge signals (BGO-shield suppressed) and one high-energy γ-ray were accepted. A total number of 10^8 events was collected, in which the γ-ray energy in the BaF$_2$ detector was $E_\gamma > 4$ MeV.

To obtain the clean high-energy γ-ray spectrum, associated with the high spin part of the ^{46}Ti compound nucleus, the gates were set on known, well resolved low energy γ-ray transitions of ^{42}Ca measured in the Ge-detectors of the EUROBALL array. This particular residual nucleus was expected to be populated only in the 2p2n- or α-decay from the highest spin region of the compound nucleus and, therefore, by gating on it we expected to obtain data corresponding to the highest spins and free from fission contributions. One should note that the low spin transitions in ^{42}Ca, used as the gates, were found (see [10] and references therein) to be strongly fed from the highly deformed bands in the same nucleus. The left panel of Figure 2 shows the measured high-energy γ-ray spectrum obtained with such condition.

In order to extract the information on the GDR parameters which contribute to the shape of the measured high-energy γ-ray spectra, a chi-square fit using as the theoretical function the calculated spectrum from the statistical model code CASCADE [11] is normally used. When, however, the experimental spectrum is obtained with specific conditions which modify the available phase-space for the decay, the Monte-Carlo implementation of this model has to be applied, in which exactly the same conditions can be modeled. We used the Monte Carlo version of the CASCADE code, previously employed for the analysis of the heavier mass data [12,13], adapted for this mass region by choosing the most appropriate values of the statistical model parameters (e.g. level density was assumed to be in accordance with the Reisdorf prescription [14,15], the yrast line was taken from the experiment and extrapolated by the liquid drop values).

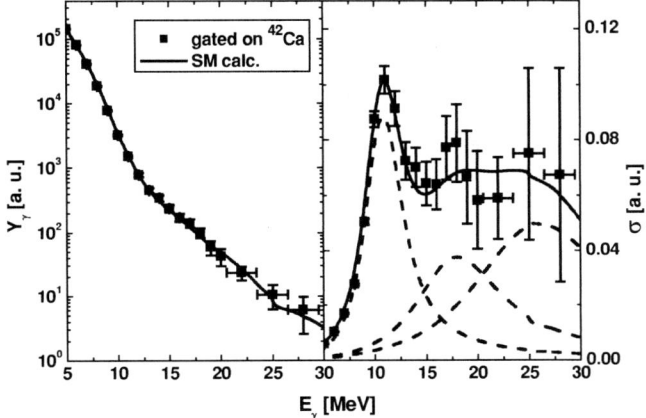

FIGURE 2. Left panel: The high-energy γ-ray spectrum gated by the ^{42}Ca transitions and by high fold region, in comparison with the best fitting statistical model calculations (full drawn line) assuming a 3-Lorentzian GDR line shape with energies: E_{GDR} = 10.8(1), 18(1) and 26(2) MeV; widths: Γ_{GDR} = 4.0(5), 10(1) and 16(3) MeV; and strengths: σ_{GDR} = 0.35(5), 0.38(10) and 0.7(2). Right panel: The deduced experimental GDR strength function (full drawn line) together with best fitting 3-Lorentzian function and its individual components.

In the left panel of the Fig. 2, the best fit to the experimental data is shown with the full line. It corresponds to a GDR strength function consisting of three components, each described by a Lorentzian, at E_{GDR} = 10.8, 18 and 26 MeV, exhausting approximately the entire EWSR strength. The quality of the fit can be judged more clearly by inspecting the right panel of Fig. 2, where the GDR line shape, i.e. the extracted absorption cross-section using the method described in e.g. [6], is shown. One feature of the obtained GDR line shape is, similarly as found in the less exclusive data [6-8], a broad high-energy component centered at around 25 MeV. But additionally, a narrow low-energy component at 10.5 MeV is clearly visible. It should be stressed, that the 3-Lorentzian function, which fits the best the GDR line-shape, does not mean that the nucleus has a well-defined triaxial *equilibrium* shape. It corresponds to the *effective* nuclear shape probed by the GDR oscillations, thus can have contributions from many different shapes. This fit shows that the average GDR line shape has to be approximated with at least 3 components.

Since one expects that a transition of the nuclear shape from oblate to very elongated prolate might manifest itself by a change of the angular distribution of emitted charged particles, we have measured the angular distributions of protons as a functions of measured γ-folds. To neglect the differences of the detector's efficiencies (due to the different absorber thickness, emission from moving frame of reference, etc.), the angular distribution functions measured in the laboratory frame for different fold windows are normalized to the one measured at fold region 5-10, and converted to the center of mass reference assuming that the kinetic energy of protons is, in average, the same for each fold window. Figure 3 shows such relative angular distributions for different fold regions, obtained under the condition that the low energy transitions in ^{42}Ca are simultaneously measured in EUROBALL Ge-array (exactly as the condition for the high-energy γ-spectrum shown in Fig. 2). As can be seen, the overall magnitude of the relative angular distribution is rather small. Nevertheless, despite the error bars, for the highest fold window which corresponds to the angular momenta close to the $l_{max} \approx 34\ \hbar$, a minimum at around 90° and maxima around 50° and 130° become to be visible.

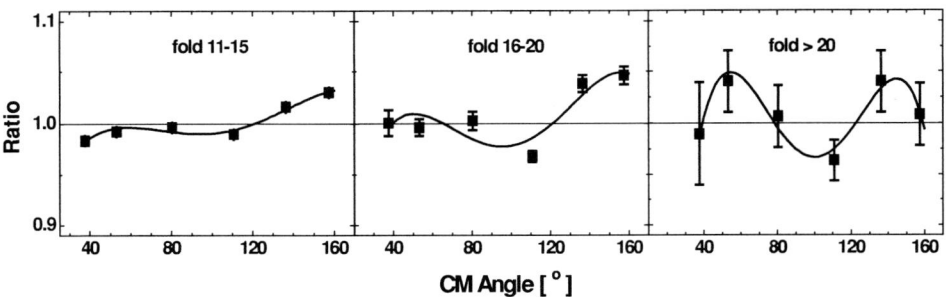

FIGURE 3. Ratio of the angular distribution of protons for different fold windows to the one measured at lower folds (the lines are to guide the eye).

DISCUSSION

In order to interpret the obtained GDR line shape (see Fig.2, right panel), we used the same approach as has been adopted in many studies concerning the GDR in hot and rotating nuclei [9, 16-19]. This is based on the *thermal shape fluctuation model* (see Ref. [18,19] and references therein) which assumes that the *average* GDR line shape (for given excitation energy and spin) is the weighted sum of individual (i.e. at given deformation) GDR line shapes:

$$f^{av}_{GDR}(E^*,I) = \iint f_{GDR}[\beta,\gamma;\omega_{rot}(I,\beta,\gamma)] \exp[\frac{-F(T,I;\beta,\gamma)}{T(E^*,I;\beta,\gamma)}]\beta^4|\sin(3\gamma)|d\beta d\gamma \quad (1)$$

where f_{GDR} is GDR line shape at a given deformation and triaxiality value, ω_{rot} is rotational frequency of the nucleus, and the integral covers the 0° < γ < 60° sextant of the β-γ plane. The weighting factors (that can be interpreted as the probability of finding the nucleus at a given deformation value) are given by the Boltzmann factor exp(-F/T), where F = U-TS is the free energy, T - temperature, U – thermal energy, and S – entropy. In our work the free energy F has been estimated by using the realistic Woods-Saxon spectra at the deformations considered in the calculations to obtain T and S; and by the macroscopic deformation-dependent LSD energy to calculate U [20].

As far as the individual GDR line shape $f_{GDR}(β,γ)$ is concerned, one uses very often a *3-Lorenztian parametrization*, with the centroid energies of the three components given by the Hill-Wheeler formula as

$$E_k = E_0 \exp\left[-\sqrt{\frac{5}{4\pi}}\beta \cos(\gamma - \frac{2\pi}{3}k)\right] \quad (2)$$

with $k = 1,2,3$ denoting the major axis of the nucleus and E_0 – the centroid energy for the spherical nucleus.

In general, however, f_{GDR} could also depend on the rotational frequency. Namely one should take into account an additional energy splitting of those two components which are perpendicular to the spin. This splitting is due to the *Coriolis effect* [21-23], and appears only if the shape is prolate or triaxial. The size of this splitting is approximately $2\hbar\omega_{rot}$, what is negligible for the medium or heavy masses, or for moderate spins. For such light nuclei, as ^{46}Ti, at $I \approx 30~\hbar$ the rotational frequency becomes $\hbar\omega_{rot} \approx 2.8 - 3$ MeV which gives an enormous splitting of ca. 6 MeV, and definitely cannot be neglected. Therefore in our approach we included the Coriolis splitting in the calculation of the f_{GDR} for given spin and deformation value using the rotating harmonic oscillator model of Neergård [21]. Thus for each deformation point the GDR line shape consisted of *5-Lorentzian parametrization*.

The Boltzmann factor distributions obtained in our approach for two spin regions are plotted in Fig. 4. The maximum of the distribution for $I = 24~\hbar$ is centered on the oblate shape ($\beta_2 \approx 0.3$-0.4), while the distribution for the $28~\hbar < I < 34~\hbar$ shows importance of elongated triaxial shapes ($\beta_2 \approx 0.6$-0.7). The width of these probability distributions reflects the extent of the thermal shape fluctuations on the β-γ plane.

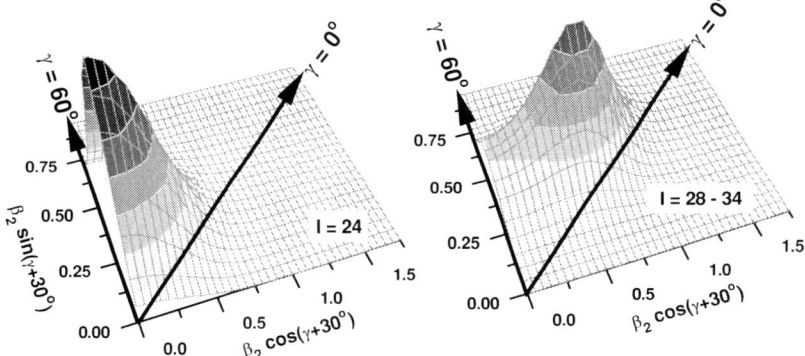

FIGURE 4. Boltzmann factor distributions for ^{46}Ti, obtained from the LSD model for two spin regions

These probability distributions were used to calculate the expected GDR line shape. For the comparison with the experimental data at the highest spins populated in the reaction, we selected for the calculation of Eq. 1 a spin interval 28-34 $\hbar$ (which corresponds to the Jacobi shape transition regime, as shown in the right-hand side of Fig. 4). The resulting averaged GDR line shape is displayed with the full line in Fig. 5 (left-hand side) together with the experimental data. A remarkable good agreement between the theoretical predictions and the present experimental results speaks very much in favor of the presence of the Jacobi transition. For comparison, the calculated averaged GDR line shape for $I = 24~\hbar$, i.e. in the oblate regime (left-hand side of Fig. 4), is shown with dashed line in the same figure. More evidence for the Jacobi transition is given by the fact that the peak around 10 MeV can originate only due to the Coriolis splitting (and shifting down part of the GDR strength), which is present only when the nucleus is either triaxial or prolate (see discussion in Ref. [24]). To demonstrate the importance of the Coriolis effect, Fig. 5 (right-hand side) shows the same calculations of the average GDR line shape, but neglecting the Coriolis splitting. As can be seen, here the low-energy component (for the I = 28-34 $\hbar$ case) has higher energy than the experimental one, and also the entire GDR line shape does not reproduce the experimental data. The predictions for the oblate regime (I = 24 $\hbar$) are very weakly sensitive to the Coriolis effect.

Still further signature for the Jacobi transition may come from the presence of two other broad components in the strength function (Fig. 5, left-hand side) at higher energies both in the experiment and calculations. One component, at around 17 MeV, is clearly seen in the present data, and the other which is very broad (at 20-30 MeV) can also be identified despite of the very limited statistics at these high energies in the present very exclusive spectrum.

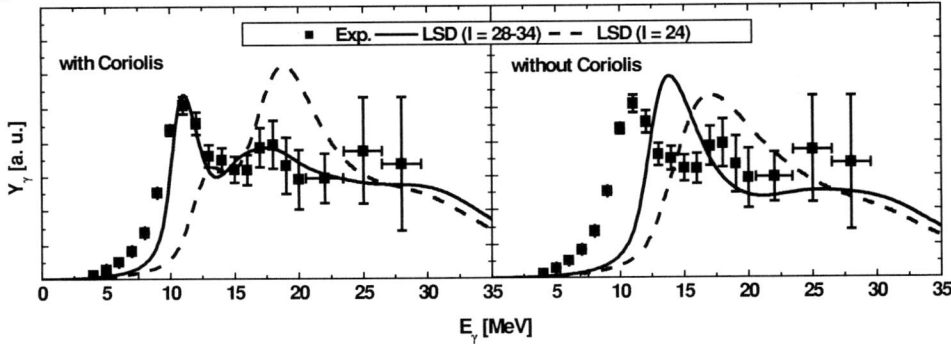

FIGURE 5. Left-panel: The full drawn line shows the theoretical prediction (at $<T> \approx 2$ MeV and in the spin region 28÷34 $\hbar$) of the GDR line shape in ^{46}Ti obtained from the thermal shape fluctuation model (including the Coriolis splitting of the GDR components) based on free energies from the LSD model calculations. The dashed line shows similar prediction for $I = 24$ $\hbar$. The filled squares are the experimental data shown also in the right panel of Figure 2. Right panel: the same, as in the left panel, but in the calculations the Coriolis splitting was not taken into account.

As far as the angular distribution of protons emitted from the very elongated and rotating nucleus is concerned, the following features are expected. Classically one would expect to have an enhancement of the emission from the "tips" of the nucleus, since the Coulomb barrier is here the smallest. However, this is quantum-mechanically valid only for protons having the single-particle angular momentum $l = 0$, while for higher l-values the centrifugal barrier makes such emission impossible [25]. Therefore, in this very simple picture, one expects emission probability to be the highest at ca. 45° cone around the "tips" of the nucleus. Converting this picture in to the angular distribution relative to the beam axis (i.e. averaging over all possible spin directions), one is expecting to have a P_4-like angular dependence - a minimum at 90°, and maxima at 45° and 135° degrees. Such a behavior seems to be visible in Figure 3 (right panel), where the relative angular distribution of protons in CM-frame for the highest fold region is shown; although the effect is small and error bars are large.

SUMMARY

The present work on the GDR γ-decay in the hot rotating nucleus, ^{46}Ti, shows the evidence for the expected Jacobi shape transition. It is based on the observation of two particular features in the measured GDR line shape. The first, already observed previously [6-8], is the presence of a high-energy component related to large deformations. The second feature, identified for the first time in the present experiment, is the appearance of a GDR component at 10 MeV (in region where the statistics is very high). This is interpreted as due to the Coriolis splitting of the lowest vibrational frequency, which corresponds to the dipole vibration along the longest axis of the well deformed prolate or triaxial shape, and consequently shifting down a part of the strength. An additional finding, which also might point to the presence of very elongated shapes at the highest spins, is the behavior of the relative angular distribution of emitted protons, where a weak minimum at 90° emerges when gated by the highest spins.

ACKNOWLEDGMENTS

This work was supported by the Polish State Committee for Scientific Research (KBN Grant No. 2 P03B 118 22), the European Commission contract EUROVIV, the Danish and Swedish Research Councils, and the Italian INFN.

REFERENCES

1. Beringer, R., and Knox, W.J., *Phys. Rev.* **121**, 1195 (1961).
2. Myers, W.D., and Świątecki, W.J., *Acta. Phys. Pol.* **B32**, 1033 (2001).
3. Pomorski, K., and Dudek, J., *Phys. Rev.* **C67**, 044316 (2003).
4. Dudek, J., and Pomorski, K., *Eur. Phys. J.* **A**, in press; nucl-th/0205011.
5. Ward, D., et al., *Phys. Rev.* **C66**, 024317-1 (2002).
6. Kicińska-Habior, M., et al., *Phys. Lett.* **B308**, 225 (1993).
7. Maj, A., et al., *Nucl. Phys.* **A687**, 192 (2001).
8. Maj, A., et al., *Acta. Phys. Pol.* **B32**, 243 (2001).
9. Maj, A., et al., *Nucl. Phys.* **A571**, 185 (1994).
10. Lach, M., et al., *Eur. Phys. J.* **A16**, 309 (2003).
11. Pühlhofer, F., *Nucl. Phys.* **A280**, 267 (1977).
12. Camera, F., et al., *Phys. Rev.* **C60**, 014306 (1999).
13. Kmiecik, M., et al., *Eur. Phys. J.* **A12**, 5 (2001).
14. Reisdorf, W., *Z.Phys.* **A300,** 227 (1981).
15. Kicińska-Habior. M., et al., *Phys. Rev.* **C36**, 612 (1987).
16. Snover, K.A., *Annu. Rev. Nucl. Part. Sci.* **36**, 545 (1986).
17. Gaardhøje, J.J., *Annu. Rev. Nucl. Part. Sci.* **42**, 483 (1992).
18. Bortignon, P.F., Bracco, A., and Broglia, R.A., *Giant Resonances: Nuclear Structure at Finite Temperature*, New York: Gordon Breach, 1998.
19. Ormand, W.E., Bortignon, P.F., and Broglia, R.A., *Nucl. Phys.* **A618**, 20 (1997).
20. Dubray, N., et al., to be published.
21. Neergård, K., *Phys. Lett.* **110B**, 7 (1982).
22. Gallardo, M., et al., *Nucl. Phys.* **A443**, 415 (1985).
23. Gaardhøje, J.J., et al., *Acta Phys. Pol.* **B24**, 139 (1993) ; Døssing, T., private communication.
24. Maj, A., *Nucl. Phys.* **A**, in press; nucl-ex/0309018.
25. Pomorski, K., et al., *Nucl. Phys.* **A605.** 87 (1996).

Spectroscopy of light exotic nuclei using nuclear break-up

D. Cortina-Gil*, J. Fernandez-Vazquez*, T. Aumann[†], T. Baumann**,
J. Benlliure[‡], M.J.G. Borge[§], L.V. Chulkov[†], U. Datta Pramanik[¶],
C. Forssén[‖], L. M. Fraile[§], H. Geissel[†], J. Gerl[†], F. Hammache[††],
K. Itahashi[‡‡], R. Janik[§§], B. Jonson[‖], S. Mandal[†], K. Markenroth[‖],
M. Meister[‖], M. Mocko[§§], G. Münzenberg[†], T. Ohtsubo[¶¶], A. Ozawa[‡‡],
Y. Prezado[§], V. Pribora***, K. Riisager[†††], H. Scheit[‡‡‡], R. Schneider[§§§],
G. Schrieder[¶¶¶], H. Simon[†], B. Sitar[§§], A. Stolz**, P. Strmen[§§],
K. Sümmerer[†], I. Szarka[§§] and H. Weick[†]

*Universidad de Santiago de Compostela, E–15706 Santiago de Compostela, Spain
[†]Gesellschaft für Schwerionenforschung (GSI), D–64291 Darmstadt, Germany
**NSCL, Michigan State University, East Lansing, MI–48824, USA
[‡]Universidad de Santiago de Compostela, E–15706 Santiago de Compostela, Spain
[§]Instituto de Estructura de la Materia, CSIC, E–28006 Madrid, Spain
[¶]Saha Institute of Nuclear Physics I-700064 Kolkata,India
[‖]Chalmers Tekniska Högskola och Göteborgs Universitet, SE–412 96 Göteborg, Sweden
[††]Institut de Physique Nucléaire, IN2P3-CNRS, F-91406 Orsay Cedex, France
[‡‡]RIKEN,2-1 Hirosawa Wako, Saitama 3051-01,Japan
[§§]Faculty of Mathematics and Physics, Comenius University, 84215 Bratislava, Slovakia
[¶¶]Department of Physics, Niigata University, 950-2181, Japan
***Kurchatov Institute, RU–123182 Moscow, Russia
[†††]Institut for Fysik og Astronomi, Aarhus Universitet, DK–8000 Aarhus C, Denmark
[‡‡‡] Max-Planck Institut für Kernphysik, D–69117 Heidelberg, Germany
[§§§]Physik-Deptartment E12, Technische Universität, München, D–85748 Garching, Germany
[¶¶¶]Institut für Kernphysik, Technische Universität, D–64289 Darmstadt, Germany

Abstract.
One-nucleon removal reactions at relativistic energies have been used as spectroscopic tool to obtain information about the ground state properties of several neutron-rich isotopes in the sd-shell. Using the FRS at GSI, the longitudinal momentum distributions of the emerging fragments after one-nucleon removal cross-sections were measured. The relative contributions of the remaining fragments in their ground and excited states have been determined from measurements of γ rays in coincidence with the longitudinal momentum distributions. The interesting case of ^{23}O will be directly address. The interpretation of our experimental results in the framework of a simple theoretical model will favours a spin and parity assignment of $1/2^+$ for the ^{23}O ground state in agreement with shell model predictions.

Neutron-rich oxygen isotopes near the neutron dripline present very exciting issues. It is today well stablished the last bound oxygen isotope is ^{24}O [1, 2, 3]. This experimental fact reinforce the idea of the $N = 16$ magic number replacing the $N = 20$ gap for the heavy dripline nuclei [4]. ^{23}O is a key nucleus to understand the structure of light neutron -rich isotopes. Consequently, it has been subject of interest and several experiments have been dedicated to its study in the last years gaining a new interest because the interpretation of different inclusive experimental results [5, 6] yield for different spin and parity assignment for its ground state. A first experiment performed at GANIL [5] measuring the inclusive longitudinal momentum distributions of ^{22}O fragments emerging after breakup of ^{23}O in a C target interpreted the results in the framework of the $p-sd$ shell-model. Almost at the same time, another experiment at GSI [7] revealed a significantly larger interaction cross-section for this nucleus. This fact was associated with a possible neutron halo structure of the ^{23}O ground state and with the existence of a sub-shell closure at $N=14$ in the oxygen chain. A more recent experiment performed at RIKEN [6], in which the inclusive one- and two-neutron breakup were investigated suggested evidence of structural changes of the ^{22}O core in ^{23}O ground state and proposed $I^+=5/2^+$ as ground state spin and parity of ^{23}O which is different from the shell-model predictions. The interpretation of these results is however controversial and was refuted in [8]. For better understanding of this problem gamma coincidence data are crucial. We have therefore performed an experiment at GSI to distinguish between the ^{22}O $g.s$ contribution to the ^{23}O wave function from any other contribution of ^{22}O excited states.

The experiment was performed at the FRagment Separator(FRS) [9] at GSI. It was dedicated to the study of the nuclear structure evolution when approaching the dripline using the one-neutron breakup of light neutron-rich secondary beams. The secondary beams were produced by nuclear fragmentation of relativistic ^{40}Ar at 1 GeV/nucleon, on a carbon target. A complete description of the experimental technique used can be found in [10, 11].

The obtained results for the inclusive breakup fragment longitudinal momentum distribution widths after correction of the intrinsic momentum resolution are shown in figure 1 for all the secondary beams studied in the experiment. The longitudinal momentum distribution have been obtained following the procedure described in [10, 11]. They show a quite constant value (~ 200 MeV/c) and a sudden decrease for ^{22}O and ^{23}F fragments corresponding with the filling of the $2s_{1/2}$ shell. The intrinsic momentum resolution was experimentally evaluated for each secondary beam and amounts 19 ± 1 MeV/c (FWHM) for ^{23}O.

The corresponding one-neutron removal cross-sections have been extracted counting the incident projectiles and the fragments in front and behind the breakup target which provide a very direct method of deducing the cross-section. This ratio was corrected by the experimental transmission evaluated with the ion optics code MOCADI [12] after adjustment of the simulated fragment longitudinal momentum width to the measured. We present in table 1 the results and associated errors obtained in this experiment for oxygen isotopes. The error includes statistical and transmission errors. The observed experimental trend shows a smooth increase in the one-neutron removal cross-section. We can observe that in contrast with the results obtained in [6, 7] the one-neutron removal cross-section of ^{23}O is not significantly big to assign a neutron halo character to this nucleus.

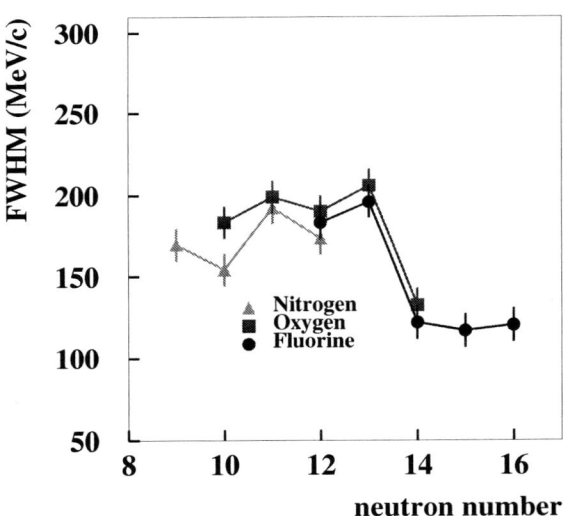

FIGURE 1. FWHM's of longitudinal momentum distribution for different isotopes of Nitrogen (triangles), Oxygen (squares) and Fluorine (circles) after one-neutron removal on a carbon target.

TABLE 1. Results for the inclusive one-neutron removal cross section after one-neutron removal of $^{19-23}$O at 939 MeV/u on a carbon target.

Fragment	σ_{-1n}(mb)	Fragment	σ_{-1n}(mb)
^{18}O	56 ± 10	^{21}O	70 ± 10
^{19}O	56 ± 9	^{22}O	85 ± 15
^{20}O	72 ± 9		

We will analyse in detail the singular case of ^{23}O. The inclusive longitudinal momentum distribution (p_{long}), of ^{22}O fragments after one-neutron removal from ^{23}O obtained in this experiment is shown on the left side of figure 2 where the solid line corresponds to a fit to the experimental data to extract the width of the momentum distribution. The FWHM for this inclusive measurement is found to be 134 ± 10 MeV/c. A minor correction for the intrinsic momentum resolution gives a final width of 133 ± 10 MeV/c.

The γ-rays emitted during de-excitation of ^{22}O were recorded with NaI detectors. The analysis of the γ-ray spectrum reveals three γ energies at 1.3, 2.6, and 3.1 MeV. This spectrum has been interpreted based on previous experimental data and shell model calculations [8, 13, 14]. Assuming that all the ^{22}O excited levels decay through the first excited state at 3.1 MeV, we would use this peak to gate the longitudinal momentum distribution in order to obtain the exclusive distribution that are show on the right side of figure 2.The FWHM for these exclusive measurement are found to be 126 ± 20 MeV/c and 236 ± 20 MeV/c for ^{22}O in its ground and any excited state respectively. The

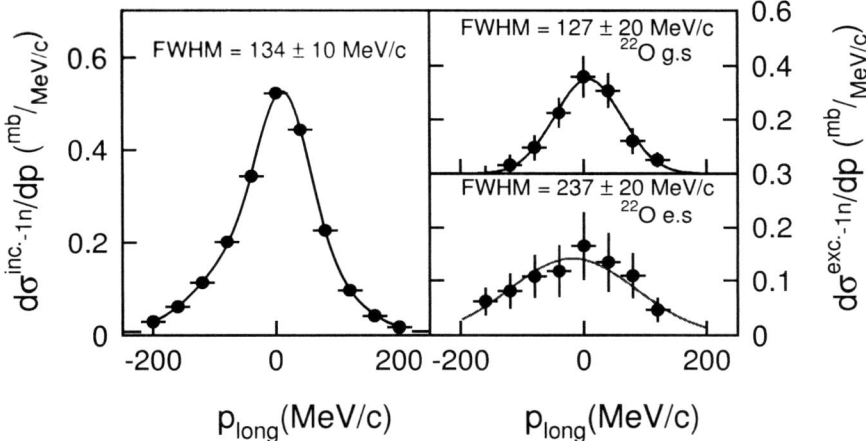

FIGURE 2. Left: Inclusive longitudinal momentum distributions (p_{long}) for ^{22}O fragments after one-neutron removal from ^{23}O. Right top: Exclusive longitudinal momentum for ^{22}O in its ground state. Right bottom: Exclusive longitudinal momentum distribution for ^{22}O in any excited state.

experimental procedure to analyse the coincidence spectrum has already been explained in previous papers [10, 11]. The experimental exclusive momentum distribution allows the extraction of the corresponding cross-section that amounts for 50±12 mb and 35±9 for the ^{22}O ground state and excited states contributions.

The experimental momentum distribution for the one-neutron removal channel leaving the ^{22}O core in its ground state is compared in figure 3 to theoretical momentum distributions calculated in an Eikonal model for the knockout process [15]. Two calculations are shown for angular momenta $l = 0$ and $l = 2$. Clearly, the distribution assuming a $2s_{1/2}$ neutron coupled to the ^{22}O(0^+) core is in much better agreement with the data. We can thus conclude that the ground-state spin of ^{23}O is $I^\pi = 1/2^+$. The experimental distribution is, however, slightly wider than the prediction for $l = 0$

Neutron-knockout cross sections were calculated for the configuration $(d_{5/6})^6 (s_{1/2})^1$. For the neutron knockout from the $1s$-shell the calculated cross section is equal to 51 mb and thus in agreement with the experimental value of 50 ± 12 mb. This result confirms the large spectroscopic factor for the s-neutron ($C^2S = 0.8$) obtained by Brown et al [8]. However, a quite significant discrepancy has been found when compared the neutron-knockout cross section contributions involving the 2^+ and 3^+ state and other contributions from deeper p shells ($1^-, 0^-$ state) with the experimentally deduced cross-section. The confirmation of these experimental results would suggest a failure in the theoretical description used either in the nuclear structure or the reaction mechanism part.

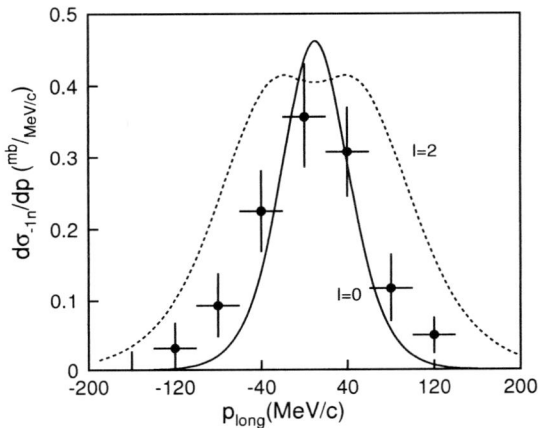

FIGURE 3. Ground state exclusive momentum distribution for ^{22}O fragments after one-neutron knock-out reaction from ^{23}O compared with calculations assuming $l = 0$ and $l = 2$

In conclusion we have measured for the first time the ^{22}O distribution after one-neutron knock-out of ^{23}O in coincidence with the ^{22}O γ de-excitation thereby demonstrating that the ground-state spin of ^{23}O is $I^\pi = 1/2^+$. This result provides a clear solution to the discrepancy of the ground-state spin and parity assignment of ^{23}O.

REFERENCES

1. O.Tarasov et al., Phys. Lett **B409** (1997) 64.
2. A. Ozawa et al., Nucl. Phys. **A673** (2000) 411.
3. H. Sakurai et al., Phys. Lett. **B448** (1999) 180.
4. R. Kanungo et al., Phys. Lett. **B 528**, 58 (2002)
5. E. Sauvan et al., Phys. Lett. **B 491**, 1 (2000)
6. R. Kanungo et al., Phys. Rev. Lett. **88**, 142502 (2002)
7. A. Ozawa et al, Nucl. Phys. **A693** (2001) 32.
8. A. Brown, G. Hansen and J. Tostevin., Phys. Rev. Lett **90** (2003) 159201-1.
9. H. Geissel et al., Nucl. Inst. and Meth.**B70** (1992) 286.
10. D. Cortina-Gil et al., Phys. Lett. **B 529**, 36 (2002)
11. D. Cortina-Gil et al., Nucl. Phys. **A 720**, 3 (2003)
12. N. Iwasa et al., Nucl. Inst. and Meth. **B 126** (1997) 284.
13. P.G. Thirolf et al., Phys. Lett. B **485** (2000) 16.
14. O. Sorlin et al., proceedings this conference and M. Stanoiu et al., submitted to Phys. Rev. C.
15. P.G. Hansen, Phys. Rev. Lett. **77** (1996) 1017.

Protein folding and non–conventional drug design: a primer for nuclear structure physicists

R. A. Broglia*, G. Tiana† and D. Provasi†

*Dipartimento di Fisica, Università di Milano, Via Celoria 16, I-20133 Milano, Italy; INFN, Sezione di Milano, Via Celoria 16, I-20133 Milano, Italy; The Niels Bohr Institute, University of Copenhagen, 2100 Copenhagen, Denmark.
†Dipartimento di Fisica, Università di Milano, Via Celoria 16, I-20133 Milano, Italy; INFN, Sezione di Milano, Via Celoria 16, I-20133 Milano, Italy

Abstract. Some of the paradigms emerging from the study of the phenomena of phase transitions in finite many–body systems, like e.g. the atomic nucleus can be used at profit to solve the protein folding problem within the framework of simple (although not oversimplified) models. From this solution a paradigm emerges for the design of non–conventional drugs, which inhibit enzymatic action without inducing resistance (mutations). The application of these concepts to the design of an inhibitor to the HIV–protease central in the life cycle of the HIV virus is discussed.

1. INTRODUCTION

Proteins are linear sequences of twenty different types of amino acids. They are produced by the ribosomes of cells following the blueprint of the genetic information. They fold into a unique three–dimensional conformation (native conformation) biologically active, in typical times of $1\mu s - 1s$. It is well established that the folding of proteins is completely determined by the sequence of its amino acids. However, one does not yet know how to extract, from the knowledge of this linear sequence, the three–dimensional native conformation of the proteins. This is the protein folding problem.

In trying to solve this problem simplified, although not oversimplified, lattice models were developed (Fig. 1) and the so called inverse folding problem was formulated [1]. Within this scenario, the protein folding problem is turned upside down into the quest for the sequences of amino acids which fold on short call into a selected native conformation. This problem has, at least for small, single domain proteins, a simple solution: good folder sequences are characterized by a large gap $\delta (\equiv E_n - E_c)$ (compared to the standard deviation σ of the contact energies between the amino acids) between the energy E_n of the sequence in the native conformation and the lowest energy E_c of the conformations structurally dissimilar[1] to the native conformation [2, 3, 4, 5, 6], E_c being a quantity which is solely determined by the composition of the protein, i.e. by the (experimental) relative presence of the different types of amino acids. In other words, good folders are associated with a normalized gap $\tilde{\xi} = \delta/\sigma \gg 1$, quantity closely related to

---

[1] That is, conformations for which the value q of the relative native contact is ≤ 0.6, $q = 1$ indicating the native structure.

the z-score [7]. Furthermore, starting from a designed sequence which displays a large gap, all mutated sequences which preserve (to some extent) the gap fold into the native conformation [8]. Good folders are obtained by minimizing the energy of the chain in the native conformation with respect to amino acid sequence for fixed composition.

Within the lattice model, the configurational energy of a chain of N monomers is given by

$$E = \frac{1}{2} \sum_{i,j}^{N} U_{m(i),m(j)} \Delta(|\vec{r}_i - \vec{r}_j|), \qquad (1)$$

where $U_{m(i),m(j)}$ is the effective interaction potential between monomers $m(i)$ and $m(j)$, $\vec{r}_i$ and $\vec{r}_j$ denote their lattice positions and $\Delta(x)$ is the contact function. In Eq. (1) the pairwise interaction is different from zero when i and j occupy nearest–neighbour sites, i.e., $\Delta(a) = 1$ and $\Delta(na) = 0$ for $n \geq 2$, where a indicates the step length of the lattice. In addition to these interactions, it is assumed that on–site repulsive forces prevent two amino acids to occupy the same site simultaneously, so that $\Delta(0) = \infty$ (excluded volume ansatz). The folding of the chain is simulated by Monte Carlo (MC) methods. We shall consider throughout a 20–letters representation of protein sequence where U is a 20×20 matrix. A possible realization of this matrix is given in ref. [9] (Table VI), where it was derived from frequencies of contacts between different amino acids in protein structures. The model we study here is a generic heteropolymer model which has been shown to reproduce important generic features of protein folding thermodynamics and kinetics, independent on the particular potential chosen [6, 10]. This is achieved by using the same potential to design sequences and to simulate folding[2]. However, in using such an approach, one should keep in mind that the labelling of amino acids (spherical beads all of the same size and with no side chain) is generic too and may be no obvious relation between those labels and labels for real amino acids.

We shall now discuss some of the results of a Monte Carlo simulation study of the dynamics of a 36–monomers chain characterized by a polymer sequence, denoted S_{36} (cf. Fig. 1(b)), designed by minimizing the energy in the target (native) conformation shown in Fig. 1(a). In the units we are considering ($RT_{room} = 0.6 kcal/mol$), the energy of S_{36} is $E_{nat} = -16.5$, while $E_c = -14$. Although this is not the sequence of lowest–energy, in particular the sequence displayed in Fig. 1(c) has an energy in the native conformation of -17.13, S_{36} has a sufficiently low energy and a large value of ξ ($= 8.33$) so that it can encode for a protein.

Monte Carlo simulations of folding performed on S_{36} at $T = 0.20$ (in our units) and using a standard algorithm described extensively in the literature [11, 12], in which, at each MC step [13], a monomer is picked up at random and end and crankshaft moves as well as corner flips are considered, indicate that this designed chain folds in a rather short "time" of $8 \cdot 10^6$ MC steps, and that at $T = 0.28$ the folding time is even shorter, $6.5 \cdot 10^5$ MC steps. The fractional population of the native state corresponding to these two temperatures is 91% and 10%, respectively, to be compared with a population of

---

[2] This point is of crucial importance to generalize model studies to realistic situations, in that if one would like to carry out a realistic calculation of a real protein, one would need to known (calculate) the interaction between all the amino acids in a full atom calculation.

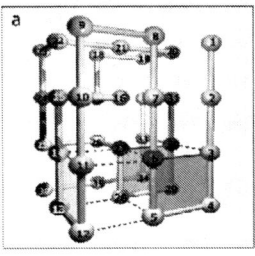

b SQKWLERGATRIADGDLPVNGTYFSCKIMENVHPLA

c YPDLTKWHAMEAGKIRFSVPDACLNGEGIRQVTLSN

FIGURE 1. (a) The conformation of the 36-mer chosen as the native state in the design procedure. Each amino-acid residue is represented as a bead occupying a lattice site. The design tends to place the most strongly interacting amino acids in the interior of the protein where they can form most contacts. The strongest interactions are between groups D, E and K (cf. (b) and Fig. 1), the last one being buried deep in the protein (amino acid in site 27). Amino acids occupying "hot" sites (sites 6, 27, 30) have been represented by deep grey beads, those occupying "warm" sites (sites 3, 5, 11, 14, 16 and 28) by grey beads and those occupying cold sites by light grey beads. The local elementary structures (LES) formed by the amino acid sequences $S_4^1 \equiv (3,4,5,6)$, $S_4^2 \equiv (27,28,29,30)$ and $S_4^3 \equiv (11,12,13,14)$ and stabilized by the contacts 3–6, 27–30 and 11–14 (drawn by continuous lines) are explicitely shown by shadowed areas. The contacts between the LES are shown by dashed lines. (b) Designed amino acid sequence S_{36}. (c) Designed sequence S'_{36} corresponding to $E_n = -17.13$.

0.5 and of 10^{-5} for the heteropolymer folding temperatures of $T = 0.25$ (temperature at which the probability for folding as well as for unfolding is $1/2$) and $T = 0.40$ (temperature at which bonds break essentially as fast as they are formed due to thermal fluctuations) respectively. All the calculations discussed below were carried out (unless otherwise stated) at the temperature $T = 0.28$, optimal from the point of view of allowing for the accumulation of representative samples of the different simulations, and at the same time leading to a consistent population of the native conformation.

In Fig. 2 we display the distribution of the similarity parameter (relative native contacts q, i.e. $q = 0$ unfolded, $q = 1$ folded) associated with these simulations at $T = 0.25$. The distribution displays two peaks associated with the denaturated (small q) and the native (large q) conformations of the designed S_{36} protein. These resutls indicate that lattice models, although being a quite simplified representation of proteins, are able to account for some important features of these polypeptides, in particular the all–or–none (phase) transition between the denaturated (unfolded) and native conformations experimentally observed (cf. Fig. 3).

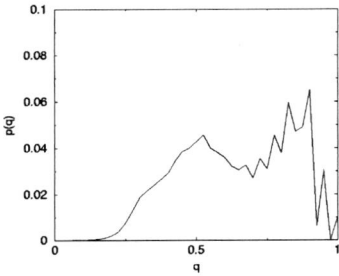

FIGURE 2. The equilibrium distribution of the order parameter q at the folding temperature $T = 0.26$.

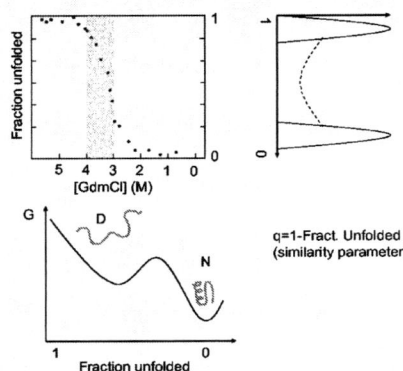

FIGURE 3. Phase transition in proteins. GdmCl changes the pH of the solvent and acts as a denaturant (after Fig. 17.2 of ref. [24]).

2. PHASE TRANSITION

In a phase transition taking place in an infinite system none of the particles involved play a special role (cf. e.g. Fig. 4(a)), while in the case of finite systems phase transitions are controlled by few quantal states (Fig. 4(b)). In other words, while in the description of phase transitions taking place in finite systems one does not need to make use of the singular function used to describe infinite systems, one can describe the phase transition in terms of individual quantal states (Fig. 4(c)).

2.1. Hot orbitals (and Cooper pairs)

The specific tools to probe pairing correlations is two–particle transfer reactions (cf. e.g. [14] and refs. therein). Much work has been carried out through (t,p) and (p,t) reactions. In this case, the relative motion of the two neutrons in the tritium has a

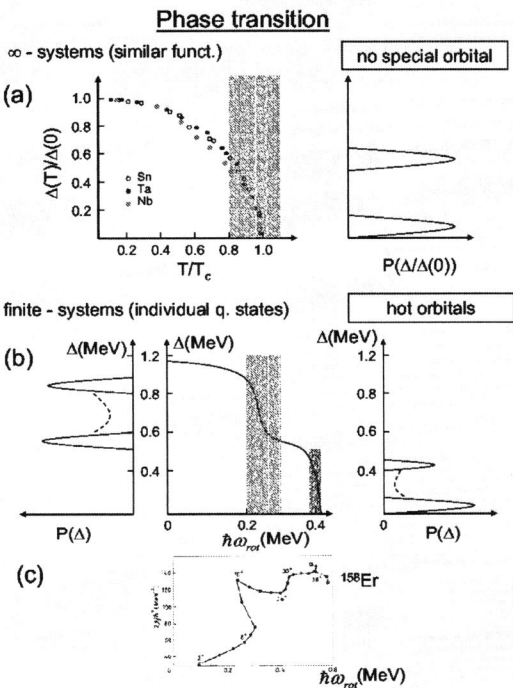

FIGURE 4. Pairing phase transitions in: (a) infinite systems (cf. p. 345 of ref. [28]), (b) and (c) finite systems (cf. [29, 30]).

large 0s–component. Consequently, pure two–particle configuration $j^2(0)$ with a large 0s–component will indicate which are the hot pairing orbitals j, the corresponding configurations $j^2(0)$ being hot configurations.

From the relation (gap equation)

$$\Delta = G \sum_v \frac{2j_v+1}{2} U_{j_v} V_{j_v}, \qquad (2)$$

one can state that orbitals with high angular momentum and lying close to the Fermi energy are central to the stability of pairing correlations in nuclei. It is then not surprising that the pairing phase transition taking place in deformed, rapidly rotating nuclei is mainly controlled by the breaking of the Cooper pairs built on these orbitals (hot Cooper pairs, like e.g. $i^2_{13/2}(v)$ and $h^2_{11/2}(\pi)$) and leading to the so called back bending phenomena (cf. Fig. 4(c)). It is well known that the blocking of these hot orbitals, and thus of the possibility of forming the corresponding hot Cooper pairs completely modifies the behaviour of the system as a function of angular momentum, as well as the properties of its ground state. In fact, and in keeping with Eq. (2), blocking the hot orbitals the gap equation admits as only solution $\Delta = 0$, or eventually a rather small value.

2.2. Hot amino acids (and local elementary structures (LES))

In order to study the way in which each amino acid (site) contributes to the stability of the native conformation of a designed protein, e.g., the S_{36} sequence, point mutations were introduced by replacing a single amino acid by an amino acid of different type. There thus exist 19 such possible substitutions for each of the 36 amino acids of S_{36}. Studying the evolution of the resulting sequences in conformational space through Monte Carlo simulations, one arrives to the complete characterization of the 36 sites displayed in Fig. 1. It is found that 27 sites can be considered "cold" sites (light grey beads), 6 "warm" sites (grey beads, #3,5,11,14,16,28) and only three hot sites (dark grey beads, #6,27,30). Thus, about 75% of the heteropolymer chain admits single error transcription in the designed (correct) amino acid sequence yielding altered chains which still fold to the native structure. In fact, introducing multiple mutations in the cold sites (by swapping amino acids of this type in order to conserve composition), the resulting sequences ($\approx 10^{30}$, all preserving to some extent the gap $\delta = E_n - E_c$) still fold to the same native structure [15]. Mutations in the yellow sites only delay the folding of the protein but not, as a rule, the ability the protein has to fold to the native structure or to a structures very similar to it. Mutations in the hot sites of the chain, which amount to $\approx 8\% - 10\%$ of all the amino acids lead to protein denaturation (misfolding), that is block the (unfolded) denaturated→native ($D \to N$) phase transition (cf. Fig. 3), in a similar way in which odd nucleons in the $i^2_{13/2}(\nu)$ and $h^2_{11/2}(\pi)$ orbitals block the normal→superfluid pairing phase transition in rotating, superfluid nuclei (like, e.g., ^{158}Er).

As stated above, mutations in the other sites lead to sequences which still fold to the native conformations and thus qualify as good folders. The resulting families of analogous proteins (proteins folding to the same native structure but not having a common ancestor) display in common essentially the few amino acids which occupy the hot sites. Examples of these families are well known experimentally, an example being provided by the immunoglobulin family [16].

The hot amino acids not only determine the stability of the protein but also the hierarchy of native contacts formation through which the protein, starting from an elongated phase reaches the native conformation (cf. Fig. 5): 1) formation, almost instantaneously of few local elementary structures (LES, i.e. hidden intermediates corresponding to incipient α–helices and β–sheets, the secondary structures of proteins) stabilized by the interaction between the hot amino acids, b) formation of the minimum set of native contacts which brings the system over the major free energy barrier of the whole folding process resulting from the docking of the LES (i.e., formation of the post–critical folding nucleus (FN)), c) relaxation of the remaining amino acids onto the native structure

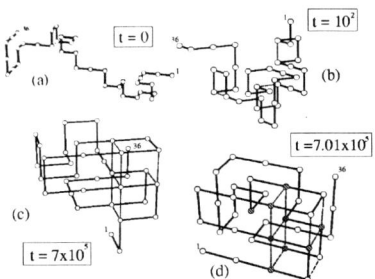

FIGURE 5. Snapshots of the folding of the sequence S'_{36}), whose energy in the native conformation is $E_n = -17.13$. Starting from a random conformation (a), the system forms after $\approx 10^2$ MC steps partially folded intermediates (b), involving three sets of four amino acids (3–6, 11–14, 17–30), whose stability is provided by the bonding indicated by dotted lines. When the partially folded intermediates come together to form the folding core (indicated by dotted and dashed lines) after $7 \cdot 10^5$ MC steps (c), the system folds to the native conformation after only 10^3 MC steps (d). The amino acids participating in the bonding of the partially folded intermediates (dotted lines) are among some of the most strongly interacting amino acids, which occupy, in the native conformation (d), "hot" and "warm" sites indicated by dark– and light–grey beads, respectively. The monomers number 36 of the sequence S'_{36} are indicated for each conformation.

shortly after the formation of the FN giving rise to a unique system with an energy below E_c [17, 6]. Summing up, the folding of proteins is controlled by the corresponding hot amino acids through the LES, ultimate building blocks of this molecular LEGO [18, 19]. In other words, the simple, most important feature common to all designed sequences folding to the same native structure is the presence of few, highly conserved, strongly interacting, hot, amino acids which stabilize the LES and which are buried inside the folding nucleus of the protein in its native conformation.

With the help of the results discussed above, one has been able to develop a strategy which allows to predict the three–dimensional native conformation of a model protein from its amino acid sequence, that is to solve the protein folding problem (within lattice models) provided the contact energies acting among the amino acids are known [20]. This strategy, called the three step strategy (3SS), is the first *bona fide* solution of the protein folding problem, as the solution of the inverse folding problem requires the knowledge of the native conformation.

The generalization of the 3SS to the case of real proteins implies, in principle, the detailed knowledge of only the interaction between the hot amino acids of the protein under study. Such an interaction can be obtained by calculating *ab initio* the corresponding contact energies between pairs of amino acids (cf. e.g. [21, 22]). It is an open question whether three–body or many–body effects (i.e., simultaneous interaction of more than two amino acids) are important within this context (additivity condition).

In any case, we shall not dwell any more on this (central) subject of the program of our group concerning the protein folding problem, but go directly to the subject of drug design, a subject which is providing important information also concerning the validity of the 3SS in general, and on the role LES play in the folding process in particular.

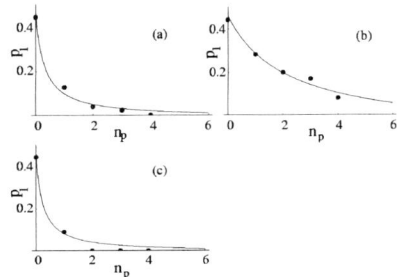

FIGURE 6. Stability p_1 (for T=0.24) of the native structure of S36 (Fig. 1(a)) as a function of the number n_p of p–LES of kind 3'–6' (a), 11'–14' (b) and 27'–30' (c) present in the cubic cell (solid dots). The results displayed by the continuous curve was determined making use of Eq. (2) of ref. [23]

3. DRUG DESIGN

LES elementary structures are also at the basis of a protocol for non–conventional drug design recently proposed by us [23]. Conventional drugs perform their activity either by activating or by inhibiting some target component of the cell. In particular, many inhibitory drugs bind to an enzyme and deplete its function by preventing the binding of the substrate, that is the molecule which has to be acted upon by the enzyme. This is done by either capping the active site of the enzyme (competitive inhibition) or, binding to some other part of the enzyme, by provoking structural changes which make the enzyme unfit to bind the substrate (allosteric inhibition). The two main features that inhibitory drugs must display are efficiency and specificity. In other words, it is not sufficient that the drug binds to the substrate and reduces efficiently its activity. It is also important that it does not interfere with other cellular processes, binding only to the protein it was designed for. These features are usually accomplished designing drugs which mimick the molecular properties of the natural substrate. In fact, the pair enzyme/substrate have undergone milions of years of evolution in order to display the required features. Consequently, the more similar the drug is to the substrate, the lower is the probability that it interferes with other cellular processes. Something that this kind of inhibitory drugs are not able to do is to avoid the development of resistance, a phenomenon which is typically related to viral protein targets. Under the selective pressure of the drug, the target is often able to either mutate the amino acids at the active site or at sites controlling its conformation in such a way that the activity of the enzyme is essentially retained, while the drug is no longer able to bind to it.In keeping with this result and with the central role played by LES in the folding process of proteins, we suggest the use of short peptides with the same sequence as LES (p–LES) as non–conventional drugs which interfere with the folding mechanism of the target protein, destabilizing it and making it prone to proteolysis [3]. These drugs are efficient, specific

---

[3] Process in which a protein undergoes cleavage by an enzyme which cuts a covalent bond (backbone bond) with accompanying addition of water, –H being added to one product of the cleavage and –OH to the other.

and do not suffer from the upraise of resistance.

In fact, the very reason why LES make single domain proteins fold fast confers p–LES the required features to act as effective drugs, that is, efficiency and specificity. They are efficient because they bind as strongly as LES do. Since LES are responsible for the stability of the protein, their stabilization energy must be of the order of several times kT. These peptides are also as specific as LES are. In fact LES have evolved over millions of years so as to prevent the upraise of metastable states and to avoid aggregation, aside from securing that the protein to fold fast. In Fig. 6 we display results of model calculations which provide a quantitative assessment of the validity of these expectations.

The possibility of developing non–conventional drugs for actual situations is tantamount to being able to determine the LES for a given protein. This can be done either experimentally (e.g. through molecular engineering [24]), or extending the algorithm discussed in ref. [20] making use of a realistic force field. The resulting peptides can be used either directly as drugs, or as templates to build mimetic molecules, which eventually do not display side effects connected with digestion or allergies. A feature which makes, in principle, these drugs quite promising as compared to conventional ones is to be found in the fact that the target protein cannot evolve through mutations to escape the drug, as happens in particular in the case of viral proteins in response to conventional drugs, because the mutation of residues in the LES would, anyway, lead to protein denaturation.

An important example of drug-resistance is connected with AIDS. In this case, one of the main target proteins, HIV-protease, a dimer formed out of two identical chains each containing 99 residues and folding according to the LES paradigm discussed above (cf. e.g. [25]), is able to mutate its active site so as to avoid the effects of drug action within a period of time of 6-8 months from the beginning of the therapy [26, 27].

Making use of empirically determined potentials to describe the interaction among amino acids a Molecular Dynamics calculation was carried out starting from the native conformation of the protease, and followed it through 100 ps. Although this time is negligible compared to the folding time of the dimer (of the order of milliseconds), it is still sufficient to provide information concerning the interaction among the amino acids. In particular, it provides information concerning the 151 native contacts of each of the monomers. Although this is not an *ab initio* interaction, one can use it within the framework of the 3SS to determine the LES and the FN, aside from the hot sites.

The results indicate the presence of two closed LES, each composed of $\approx 8 - 10$ amino acids, and one open containing 4–5 amino acids, all containing some strongly interacting highly hydrophobic (non–polar) amino acids. These LES are found to be in highly protected regions of the protein, and the few drug induced mutations which affect them are all conservative (i.e. highly hydrophobic amino acids are replaced by dito).

Based on these results we have chosen one of the (closed) LES as the most likely HIV–1–Pr inhibitor (drug). This peptide was ordered to a firm in the Milano area (Primm spa), while the enzyme itself (together with the substrate) was obtained from the National Institute of Health (NIH, Bethesda, Maryland, USA). The experiment, carried out by our group at the Department of Biology of the University of Milano gave a clear signal of the inhibitory properties of the peptide.

Although these results are only preliminar and much (experimental and theoretical) work has to be carried out to determine the inhibition constants of the designed drug as well as its ability to be active also in the presence of mutations (let alone the eventual medical implications), it vindicates the central role played by LES (and thus hot amino acids) in the D→N phase transition of real proteins, and thus the fertility of the nuclear structure paradigms which are at the basis of these concepts.

REFERENCES

1. E. I. Shakhnovich, Phys. Rev. Lett. **72** (1994) 3907
2. R. Goldstein, Z. Luthey–Schulten and P. Wolynes, *Optimal folding codes from spin–glass theory*, Proc. Natl. Acad. Sci. USA **89** (1992) 4918–4922
3. A. Sali, E. I. Shakhnovich and M. Karplus, *Kinetics of proteinfolding: a lattice model study for the requirements for folding to the native state*, J. Mol. Biol. **235** (1994) 1614–1636
4. E. I. Shakhnovich and A. Gutin, *A novel approach to the designof stable proteins*, Protein Eng **6** (1993) 793–800
5. J. Bryngelson, J. N. Onuchic, N. D. Socci and P. Wolynes, Proteins:Struct. Funct. and Gen. **21** (1995) 167–195
6. E. I. Shakhnovich, Curr. Opin. Struct. Biol. **7** (1997) 29–40
7. J. Bowie, R. Luthey–Shulten and D. Eisemberg, *A method to identify protein sequences that fold into a known three–dimensional structure*, Science **253** (1991) 164–169
8. G. Tiana, R. A. Broglia H. E. Roman, E. Vigezzi and E. I. Shakhnovich, *Folding and misfolding of designed proteinlike chains with mutations*, J. Chem. Phys. **108** (1998) 757–761
9. S. Miyazawa and R. Jernigan, Macromolecules **18** (1985) 534
10. E. I. Shakhnovich, Folding and Design **1** (1996) R50
11. P. H. Verdier, J. Chem. Phys **59** (1973) 6119
12. H. J. Hilhorst and J. M. Deustch, J. Chem. Phys **63** (1975) 5153
13. N. Metropolis, A. W. Rosenbluth, M. N. Rosenbluth, A. H. Teller and E. Teller, J. Chem. Phys. **21**, 1087 (1953)
14. R. A. Broglia, O. Hansen and C. Riedel, Adv. in Nucl. Phys. (1973) Vol. 6, 287–457
15. R. A. Broglia, G. Tiana, H. E. Roman, E. Vigezzi and E. I. Shakhnovich, Phys. Rev. Lett. **82** (1999) 4727
16. T. E. Creighton, *Proteins*, W. H. Freeman and Co., Newy York (1994)
17. R. A. Broglia and G. Tiana, J. Chem. Phys. **114** (2001) 7267
18. G. Tiana and R. A. Broglia, J. Chem. Phys. **114** (2001) 2503
19. R.A. Broglia, G. Tiana, S. Pasquali, H. E. Roman, E. Vigezzi, Proc. Natl. Acad. Sci. USA, **95** (1998) 12930
20. R. A. Broglia and G. Tiana, Proteins **45** (2001) 421
21. D. Provasi, Ph. D. Thesis (2002) University of Milano (unpublished)
22. R. A. Broglia, G. Coló, G. Onida, H. E. Roman, Springer Verlag (in press)
23. R. A. Broglia, G. Tiana and R. Berera, J. Chem. Phys. **118** (2003) 4754
24. A. Fersht (1999) *Structure and mechanism in protein science*, W. H. Freeman, New York (1999)
25. G. Tiana and R. A. Broglia, Proteins **49** (2002) 82
26. A. Wodawer and J. Vondrasek, Annu. Rev. Biophy. Struct. **27** (1998) 249
27. A. G. Tomaselli and R. L Henrikson, Biochim. Biopys. Acta **1477**, 189 (2000)
28. C. Kittel, *Introduction to solid state physics*, J. Wiley and sons, New York (1996)
29. J. Burdel et al., Phys Rev. Lett. **48** (1982) 530
30. Y. R. Shimitzu et al., Rev. Mod. Phys. **61** (1989) 131

Gamma-ray spectroscopy studies at GANIL: status and perspectives

G. de France

GANIL, BP55027 F-14076 CAEN cedex 5

Abstract. The SPIRAL facility started to deliver radioactive ion beams (RIB) in september 2001 and some experiments have been performed in particular using the EXOGAM array. These experiments will be briefly described and the first outputs will be shown. The in-beam performances of EXOGAM will also be discussed.

INTRODUCTION

The SPIRAL facility [1], producing RIB at GANIL has started to deliver beams for experiments in autumn 2001, a year ago. SPIRAL uses the primary stable beams delivered by the two separated sector cyclotrons of GANIL. This high energy, large intensity beam is then fragmented onto a very thick Carbon target producing a huge number of exotic species which are extracted from the target, selected by a low energy separator and reinjected into a new K=265 compact cyclotron (CIME). The very first beam produced and accelerated with CIME was ^{18}Ne produced by a primary beam of ^{20}Ne^{10+} at an energy of 95 A.MeV. With a primary beam intensity of 1.6 mA (300 W), the ^{18}Ne secondary beam at 7.2 MeV/A has been measured to be 10^6 pps as expected. Later on, several campaigns of RIB production have taken place. All together 8 experiments have used ^{18}Ne, ^{6,8}He and ^{74,76}Kr: two with LISE2000 (upgrade of the achromatic LISE3 spectrometer); three with SPEG and 4 with EXOGAM. I will concentrate on these latter ones.

In this talk I will very briefly give the main features of EXOGAM and describe the various experiments which have used SPIRAL beams and EXOGAM. I will finally draw some conclusions from these first experiments.

THE EXOGAM ARRAY

A detailed presentation of the EXOGAM array can be found in Ref. [2]. In this talk only the main characteristics of the spectrometer and some sub-systems of which it is composed will be given.

EXOGAM is an array which is dedicated to study the spectroscopy of exotic nuclei and in particular using the RIBs from the SPIRAL facility at GANIL. This has imposed severe constraints on the design specifications: large efficiency for low and medium gamma-ray multiplicity; good signal-to-noise ratio; coupling with many auxiliary devices to cope with the various experimental conditions. These requirements end up with

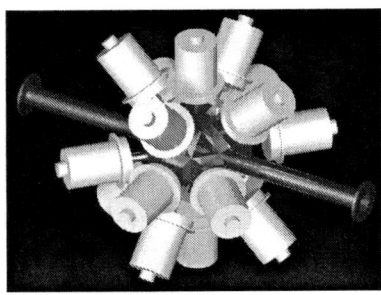

FIGURE 1. Scheme of the EXOGAM array in its 16 Clover configuration. The anti-Compton shields are not represented.

a design using 16 Compton suppressed large Clovers giving a photopeak efficiency of about 20% for a single gamma of $E_\gamma = 1.3$ MeV. The array (Fig. 1) has a versatile geometry which allows us to have various configurations: a compact one (conf. A) where the detectors are located at 11.4 cm from the target and a pulled-back configuration (conf. B) where the distance is 14.7 cm.

The four Ge crystals composing the Clover are 9 cm long and have a diameter of 6 cm before shaping. They are tapered over 3 cm with an angle of 22.5 degrees. Each diode is electrically segmented in 4 in order to enhance the Doppler correction capability of the detectors.

The shield design is based on 3 distinct layers: the first one is the back catcher in CsI; the second one, the side catcher, is made of BGO and surrounds the rear part of the clover while the tapered length of the Ge cans is not covered; the third layer, the side shield, is a long piece in BGO which covers the whole Clovers from 2 cm before the Ge front face to the backcatcher. In conf. A, the shield consists of the two first layers: the back- and the side-catcher. The Ge are positionned in such a way that their tapered faces are in contact. This optimize the efficiency. In particular at very low gamma ray multiplicity, it is possible to add the energy deposited in neighbouring clovers without loosing too much with pile up effects (the *inter-Clover addback*). The gain in efficiency is about 10 %. When the multiplicity is larger, leading in conf. A to large of pile-up probability, the detectors are pulled back in conf. B and the additional side shield can be mounted to significantly enhance the peak-to-total ratio. The calculated efficiencies and peak-to-total ratio are shown in table 1 for the various configurations. The measured value for the peak-to-total ratio in conf. B is 53%, about 10% less than the calculated values. This is most probably due to the approximate geometry used in the calculations.

The electronics is a VXI based electronics. A separate card instrument the inner (high resolution) contacts; the outer contacts (segments, with a lower resolution); and the shields. One particularity of the inner and outer contact cards is that they contain the electronics to perform some pulse shape analysis (flash ADCs and DSPs). With the 2D segmentation, this makes of EXOGAM an excellent device to understand more in detail the relations between pulse shapes and the interaction of gamma-ray with matter. This is therefore a particularly relevant way to implement, test and evaluate the gamma-ray tracking technique possibilities.

TABLE 1. Total photopeak efficiency and peak-to-total ratio calculated for EXOGAM. The cube configuration consists in 4 Clovers with full suppression at 6.8 cm from the target.

	Phot. efficiency (%)		Peak-to-total (%)	
	662 keV	1.3 MeV	662 keV	1.3 MeV
Conf. A	28	20	57	47
Conf. B	17	12	72	60
Gamma-Cube	15	10	72	60

EXOGAM will be used in coincidence with VAMOS Ref. [3], the GANIL large solid angle spectrometer designed for RIB which is installed in an other GANIL area. The coupling of the devices, is made by moving the whole EXOGAM array from one room to the other with a crane. To do that, EXOGAM is mounted on a single platform on which is installed the mechanics with the detectors; the electronics in their cooled racks; the autofill, bias, and high voltage systems for the detectors; etc. This allows to keep everything cabled between the array, the electronics and the various sub-systems.

GAMMA-RAY SPECTROSCOPY WITH RIBS

Structure of excited states in heavy nuclei populated in the ^{8}He+^{208}Pb fusion evaporation channel.

The first experiment using a RIB in EXOGAM has been run in april 2002 Ref. [4]. The beam was ^{8}He produced by the fragmentation of ^{13}C at 75 MeV on the Carbon target. The ^{8}He beam extracted from the target-source was then accelerated into CIME at an energy of 28 MeV to impinge a 30 mg.cm^{-2} thick ^{208}Pb target. The beam intensity for this isotope was between 3×10^4 to 3×10^5 pps. The effective measurement time was about 58 hours before the SPIRAL target-sources broke. The failure was due to mechanical constraints within the ensemble. Since then it has been successfully corrected. The main motivations for this experiment was to study the 'high' spin states in the ^{212}Po and ^{213}At nuclei and in particular the competition between the octupole degree of freedom and the multi-particle excitations. Furthermore, it was also planned in the same experiment to measure the lifetime of the short lived isomers in the picosecond region.

The setup for this experiment consisted in four Clovers in cube geometry: three EXOGAM ones with shield plus an EUROGAM size segmented Clover without shield. For the fast timing aspect of this run, 8 small BaF2 scintillators were installed as close as possible to the target without shadowing the germanium detectors. In this geometry, the efficiency has been estimated to be 3.4 % at E_γ = 662 keV. At 26 MeV the strongest populated channel is ^{212}Po after the evaporation of 4 neutrons. As expected from the low beam intensity, the spectra obtained from this run are dominated by:

- the room background radioactivity consisting of essentially the 1461 keV line from

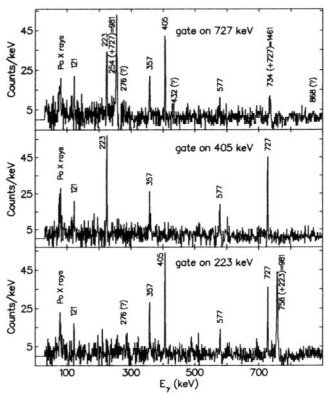

FIGURE 2. Coincidence spectra obtained for ^{212}Po by gating on the $2^+ \rightarrow 0^+$ transition (top); $4^+ \rightarrow 2^+$ transition (middle) $6^+ \rightarrow 4^+$ transition (bottom) (reprinted from [4], with permission from Elsevier).

the β-decay of ^{40}K and some decay chains from the natural Th and U isotopes. The endpoint of these decays lie in the mass region where the nuclei of interest are produced in the fusion-evaporation process.

- beam decay: ^{8}He ($T_{1/2}$ = 119 ms) β-decays to a 1^+ state in ^{8}Li ($T_{1/2}$ = 838 ms) with a probability of 86 %. The β particle is followed by a 981 keV gamma-ray to reach the ground state. In the remaining 14 % of the decays, ^{8}Li emits a neutron to give ^{7}Li whose first excited state decays via a 478 keV γ-ray. In the spectra, this transition is rather broad. This is due to the Doppler broadening caused by the neutron recoil.

Gamma-gamma spectra obtained by requiring the coincidence between two Clovers are very different from single γ-ray spectra. However the reduction observed for most of the background (i.e. room background and beam decay) transitions is not seen on others. This is for example the case of the 1461 keV line (which is strongly reduced) and the 981 or 478 keV γ-ray which are still very strong. To understand the difference one has to look at the details of the decay: the 1461 keV line is originating from an electron capture in ^{40}K leading to single gamma transition. The 981 keV (and also the 478 transition) follows the β^- decay of ^{8}Li and is therefore always accompany by an electron. This latter or the brehmsstrahlung induced by its slowing down can be detected in coincidence with the γ-ray observed in the Clovers. This explains why the reduction factor resulting from the γ-γ coincidence requirement is very different. The ^{212}Po nucleus has been previously studied by A.R Poletti et al (see [5]). The level scheme was established up to 2.885 MeV with a tentative 14^+ spin. The gamma spectrum obtained in our case for ^{212}Po after gating on the 223 keV line (6^+ to 4^+ transition) is shown in Fig. 2. With only 58 hours of beam time at this very low beam intensity (of the order of 10^5 particle per second), this spectrum show that high-resolution gamma-ray spectroscopy is still feasible.

The cross section for elastic scattering of the ^{8}He beam is extremely large as compared to any other processes. A very simple wrapping of lead foils (for a total of 9 mm thick) around the pipe gives a reduction of about 38 % of the 981 keV line. Even if the cross section decreases with scattering angle, the implantation of the scattered beam into the pipe e.g. is huge and must be properly shielded. An auxiliary detector must be used to go further. A first solution is to trig the acquisition system when a certain condition on the recoil products or the decaying particles is fulfilled. When this is not easy or possible, another solution is to use beam particle detectors. The two possibilities can of course be combined. With a cyclotron RF of 11 MHz and an intensity of 10^5 particles per second, we have rougly 1 particle every 100 cycles. Combining the RF (to have an optimum time reference) and a beam detector (to trig the acquisition) would certainly improve significantly data quality.

Study of very deformed ground-state in neutron-deficient light rare-earth nuclei populated using ^{76}Kr beam and inverse kinematics.

The second experiment using SPIRAL beams and EXOGAM was aiming at looking the very deformed ground-state in the light rare earth nuclei around ^{130}Sm (for a more detailed description of this run see [6]). These nuclei are calculated to have a ground-state deformation of $\beta \sim 0.40$ which is equivalent to the deformation measured for superdeformed bands in the Ce region. To produce these isotopes, we used ^{76}Kr on a ^{58}Ni target at 350 MeV. Six EXOGAM Clovers and 2 smaller ones were in the setup for this experiment which gave a photopeak efficiency of 11 % at 344 keV. The DIAMANT charged particle detector was used together with the Debrecen chessboard (both consisting in CsI detectors) to perform a better channel selection. The α and proton detection efficiency for this setup was approximately 70%. In this reaction, more than 10 channels were opened with a cross section larger than 10 mb.

^{76}Kr has a $T_{1/2}$ = 14.8h which build up into the target as well as in materials which is hitted by the scattered beam. Without any particle selection it is not possible to observe any γ-ray from fusion events. The particle tagging allows us to clearly identify the 4p, 3p, α2p and α3p channels leading respectively to ^{130}Nd, ^{131}Pm, ^{128}Nd and ^{127}Pr which are the most neutron-deficient known nuclei.

The spectra show that we have been able to observe the yrast band up to spin 18^+ in ^{130}Nd for example. Despite several diffculties, this run has nevertheless shown that the coupling of EXOGAM and DIAMANT is very efficient in selecting output channels. It was already known with stable beams induced fusion reaction that particle identification was really efficient and required when looking for neutron-deficient nuclei. For RIB this is even more critical and we observe that all the radioactivity γ-ray lines have disappeared in the α or proton gated spectra.

Shape coexistence studied by Coulomb excitation of ^{74,76}Kr beams.

Following the fusion-evaporation channel, another experiment using ^{74,76}Kr radioactive isotopes took place. The goal was to study the shape coexistence in the Kr isotopes.

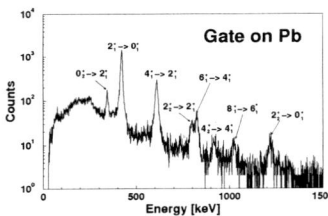

FIGURE 3. Doppler corrected γ-ray spectrum obtained from ^{76}Kr Coulomb excitation on a ^{208}Pb target [8].

This has been done using Coulomb excitation of the beam by various targets and at sub-Coulomb barrier incident energy. This avoid nuclear processes to take place and select only the electromagnetic part of the interaction. The charge distritution (Q_0) is then extracted from the differential cross section of the 2^+ state excitation probability. Several targets were used because low lying collective states or a change in B(E2) values can perturb the measurements for a single target. We used nearly the same gamma-ray setup as previously described (we had only 1 smaller detectors instead of 2 for the previous one). To detect the scattered beam, an annular Si strip detectors has been installed at forward angles. One of the EXOGAM trigger was conditionned by the particle detection in the Si detector. The method has the advantage to have a large reaction cross section which makes it possible to run with very low beam intensity (it was down to 10^4 pps for ^{74}Kr) and low gamma-ray multiplicity. With the annular and sector segmentation of the Si detector it was possible to optimize the Doppler correction (the recoil velocity was up to $\beta \sim 10\%$). Under these conditions, this experiment went very smoothly. In Fig. 3 is shown the online spectrum obtained with ^{76}Kr beams on the Ti target. As can be seen, with the Si detector in coincidence with EXOGAM as a trigger, we obtained nearly background free gamma-ray spectra.

Entrance channel effect on fusion cross-section using ^{6,8}He beams.

The last experiment we have made was aiming at investigating the entrance channel effects in the fusion process around the Coulomb barrier [10]. More precisely the idea was to look at the effect of the loosely bound neutrons in ^{6,8}He on the fusion with different isotopes to reach the same compound nucleus. The best target-projectile couples we have found was ^{6}He + ^{190}Os and ^{8}He+^{188}He giving ^{196}Pt*. As a reference, we have run the α+ ^{192}Os in Mumbai (India) using the TIFR-BARC pelletron tandem accelerator at various bombarding energies.The backing for the Osmium targets was ^{63,65}Cu and this gave also very interesting results as will be shown later.

The specific difficulties of these type of measurements with SPIRAL beams are two fold: 1) measuring an excitation function with a cyclotron and 2) precise determination of the absolute cross-section and in particular with low intensity RIBs. This has been done using several tools to be able to cross-check the informations: highly sensitive Faraday cup, plastic detectors and annular segmented Si strip detector. For this run the EXOGAM implementation consisted in 5 EXOGAM Clovers with their anti-Compton shield and 3 smaller Clovers. Knowing the array efficiency and the ^{8}He decay we estimated the beam intensity to be $i_{beam} \sim 8 \times 10^4$ pps with the newly designed target-source ensemble to produce this beam.

Sub-barrier fusion is possible because of quantum mechanics. It is well known that the one dimensional (1D) barrier penetration model is far too simplistic and that the intrinsic structure of the colliding nuclei must be considered when one hopes to reproduce data. It is also essential to take into account other channels like inelastic or transfer which are open at the same time. Because of its very special structure, it is interesting to use a halo nucleus as collision partner at energies close to the Coulomb barrier. Expectations from various models are really diverging: some predict an enhancement of sub-barrier fusion cross-section while others give rise to flux reduction. Related to that is the effect of the coupling to break up channel. This was part of the motivations for this experiment. The idea was to measure absolute cross section via the γ-decay of the residues produced in fusion. The total fusion cross-section being deduced from the summed partial cross-sections measured for each evaporation residue. The elastic channel was monitored by the annular silicon strip detector positionned at forward angles. The total intensity has been measured by a very sensitive Faraday cup (^{6}He) and a plastic detector (^{8}He). Typical spectra for the He+Cu (target backing) reaction are shown in Fig. 4. The RF gated singles γ-ray spectra obtained with a RIB are very clean. Gamma-rays from Cu isotopes also appear very clearly in coincidence with a charged particle in the Si detector. This corresponds to neutron transfer to the target. Statistical model calculations have been made with CASCADE which reproduce very nicely all the evaporation channels in the ^{4}He and ^{6}He + ^{65}Cu except a single one: ^{66}Cu in ^{6}He+^{65}Cu. CASCADE underestimate the cross section for this particular channel by more than one order of magnitude, pointing to evidence for a mechanism without compound nucleus formation. This is interpreted as a direct observation of neutron transfer followed by evaporation. Therefore, and for the first time, it is possible to distinguish between break-up and transfer.

CONCLUSIONS

The EXOGAM array and the SPIRAL facility have run a serie of experiments in the last months. During these runs several auxiliary devices have been coupled to EXOGAM. This has been shown to be absolutely vital to eliminate the background and to obtain sensible gamma-ray spectra. In the next coming years, the experimental programme will develop at GANIL and there is no doubt that many challenging experiments will be run and will give new resuts. At the end of this paper, I want to thank the spokespersons of the experiments which I have briefly described as well as the people who worked on the data: Ph. Walker, Zs. Podolyak, N. Redon, P. Nolan, O. Stezowski, W. Korten, E. Bouchez, A. Navin, J.M. Casandjian and the numerous collaborators who took part in

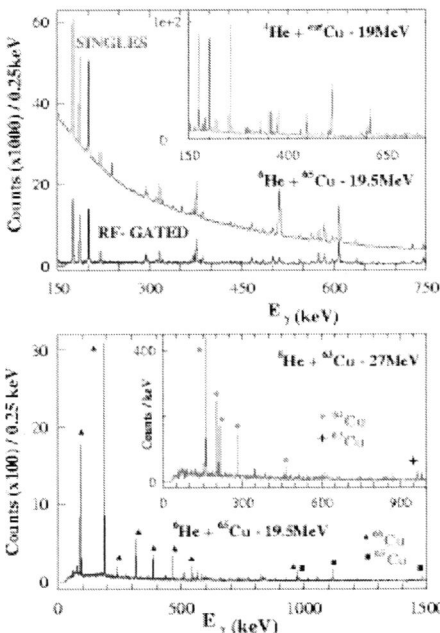

FIGURE 4. a) singles and RF gated γ-ray spectra for ^{4}He+^{nat}Cu and ^{6}He+^{65}Cu and ^{8}He+^{63}Cu. The bobarding energies are indicated. b) γ-ray spectra from ^{6}He+^{65}Cu and ^{8}He+^{63}Cu in coincidence with a charged particle in the Si detector. The main peaks are labeled.

one way or another to the experiments. My thanks are also going to the local engineers and technicians of GANIL who worked really hard and delivered very good beams in sometime difficult conditions and also took care of EXOGAM, in particular to J. Ropert and G. Voltolini.

REFERENCES

1. GANIL web site http://www.ganil.fr/spiral/
2. GANIL web site http://www.ganil.fr/exogam/
3. GANIL web site http://www.ganil.fr/vamos/
4. Zs. Podolyak *et al.* NIM A511, 354 (2003)
5. Poletti *et al.*
6. N. Redon *et al.* contribution to this proceedings.
7. O. Stezowski, private communication.
8. E. Bouchez *et al.* to be published.
9. V. Tripathi *et al.*, Phys. Rev. Lett. 88, 172701 (2002) and references therein.
10. A. Navin *et al.*, submitted.

Nuclear moments of nuclei in the neighborhood of the neutron drip line

H. Ueno*, K. Asahi[†*], H. Ogawa**, D. Kameda[†], H. Miyoshi[†],
A. Yoshimi*, H. Watanabe*, K. Shimada*, W. Sato[‡], K. Yoneda*, N. Imai*,
Y. Kobayashi*, M. Ishihara* and W.-D. Schmidt-Ott[§]

RIKEN, 2-1 Hirosawa, Wako, Saitama 351-0198, Japan
[†]*Department of Physics, Tokyo Institute of Technology, 2-12-1 Oh-okayama, Meguro, Tokyo 152-8551, Japan*
**Photonics Research Institute, AIST, Tsukuba, Ibaraki 305-8568, Japan*
[‡]*Department of Chemistry, Osaka University, Toyonaka, Osaka 560-0043, Japan*
[§]*II. Physikalisches Institut, Der Universität Göttingen, Bunsenstrasse 7-9, D-3400 Göttingen, Germany*

Abstract. Based on the β-NMR method with spin-polarized radioactive-isotope beams, we have been conducting a series of experiments at RIKEN for the measurement the magnetic moments and electric quadrupole moments of light unstable nuclei. From the systematic measurements of nuclear moments in the unstable nuclear region, the effects of the neutron excess on the structure have been discussed. In the μ-moment measurements of the nitrogen isotopes, the outward-directed deviation of $\mu(^{17}N)$ and $\mu(^{19}N)$ from the Schmidt value were observed. In the case of Q-moment, our experimental data show large discrepancy from the shell-model predictions if neutron and proton effective charges are assumed to be constant. By introducing an isospin dependence to the effective charges, the observed Q-moments of boron isotopes can be explained. We also applied the result of μ-moment measurements to the I^π assignment. Owing to the large difference of μ values among expected I^π candidates, $I^\pi(^{17}C)=3/2^+$ has been definitely assigned.

INTRODUCTION

Nuclear electromagnetic moments are one of the basic probes to obtain information on the nuclear structure. The availability of recently developed spin-polarized radioactive-isotope beams (RIBs) [1] offers us the opportunity of studying on the structures of nuclei far from the stability line, through the measurement of electromagnetic moments [2, 3]. In this method the spin-polarization is produced through the projectile-fragmentation reaction. It provides the following advantage in the measurement of nuclear moments. Firstly, the polarization can be obtained by simply selecting the emission angles and the momenta of fragments. Secondly, the size of the polarization is typically 1-5 %, which is large enough to accomplish a β-NMR measurement on quite a far-off-stability nucleus within several days. Thirdly, from the general feature of the fragmentation reaction, essentially any fragments with $Z \leq Z_{\text{projectile}}$ and $N \leq N_{\text{projectile}}$ can be produced and polarized irrespective of their chemical properties. Finally, because such fragments are emitted at velocities close to the beam velocities, they can be deeply implanted into the stopper material where any disturbance from the sample surface does not exist.

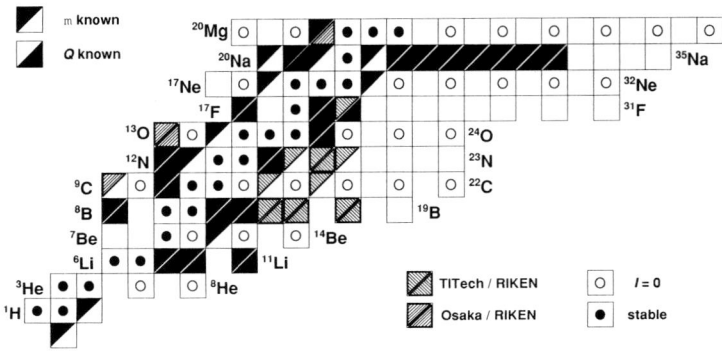

FIGURE 1. Nuclei whose magnetic dipole moment or electric quadrupole moment have been measured at RIKEN.

Figure 1 shows the nuclei whose magnetic moment (μ) or the electric quadrupole moment (Q) have been determined at RIKEN using spin-polarized RIBs. Up to now, eleven new μ moments and six new Q moments have been determined. These systematic measurements provide us opportunity for investigating the effect of neutron and proton excesses on the nuclear structure microscopically. In this work we report on the recent progress in the μ- and Q-moment measurements and analyses. To begin with, we briefly discuss how the spin polarization of RIBs is obtained in the projectile fragmentation process.

EXPERIMENTAL PROCEDURE

Spin-polarization in the projectile fragmentation

The fragmentation of a projectile nucleus in high-energy nucleus-nucleus collisions is well described by a model based on the assumption that a projectile fragment is a "spectator" portion of the projectile nucleus, with a "participant" portion being abraded off through the reaction. Mechanism of spin polarization in the projectile fragmentation reaction [1] is essentially related to the fact that a portion of the projectile to be removed through the fragmentation process (i.e. the participant) has non-vanishing orbital angular momentum due to the Fermi motion of nucleons.

Our earlier experiments have reveled that fragments can be spin-polarized as a function of outgoing momentum of the fragment. Figure 2 shows the polarization measured as a function of the momentum for ^{12}B (^{13}B) fragments from the collision of a ^{14}N (^{15}N) projectile with different targets and incident energies. Figures 2 (a) and (e) are the two extreme cases for the polarization behavior. Two remarkable features are noted: i) the signs of the polarization at the high and low momentum regions are opposite to each other, and ii) the polarization crosses zero at the momentum corresponding to the beam (or, projectile) velocity. These two features are in agreement with above discussion. It should also be noted that the signs of polarizations as a whole in (a) and (e) are opposite

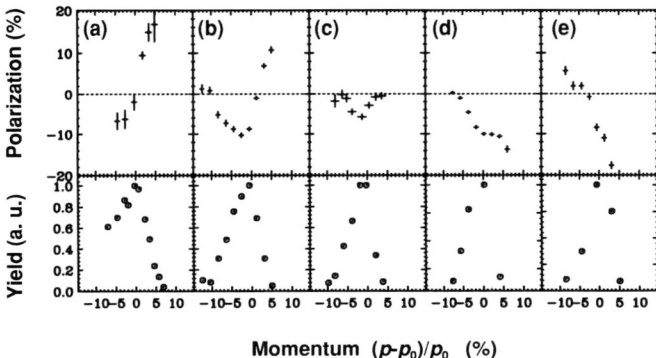

FIGURE 2. The behavior of the spin polarization in the projectile fragmentation reaction. The yield (upper) and spin polarization (lower) are shown as functions of momentum for two neutron removal reactions of (a) ^{14}N (39.4 AMeV) + ^{197}Au → ^{12}B(θ_L=5.0°), (b) ^{14}N (68.0 AMeV) + ^{197}Au → ^{13}B(θ_L=4.0°), (c) ^{14}N (109.6 AMeV) + ^{197}Au → ^{13}B(θ_L=2.0°), (d) ^{14}N (67.3 AMeV) + ^{93}Nb → ^{13}B(θ_L=2.5°), (e) ^{14}N (68.0 AMeV) + ^{27}Al → ^{13}B(θ_L=1.0°). The p and p_0 denote the momentum and the momentum corresponding to the beam velocity.

from each other. This feature is understood well within the above model of projectile fragmentation: The collision with a high-Z target such as ^{197}Au is expected to proceed through near-side trajectories because of the predominating role of the Coulomb force, while that with a low-Z target such as ^{27}Al conceivably proceeds through far-side trajectories because of an attractive nuclear force. The cases (b), (c), and (d) can be considered as manifestation of the competition between the near- and far-side trajectories. Details of the mechanism are discussed in Ref. [1].

Experimental setup

Figure 3 shows the arrangement of the RIKEN projectile-fragment separator RIPS [4] for the production of spin-polarized RIBs. In order to have RIBs spin-polarized, the emission angle and the outgoing momentum are selected for the fragments. The fragments emitted at finite angles are accepted to RIPS by the method shortly described. Also, an appropriate range of momentum is selected by a slit at the momentum-dispersive focal plane F1. To attain the $\theta_L \neq 0$ setting, which is essential for obtaining the non-vanishing spin polarization, a beam swinger installed upstream of the target is used. The isotope separation is provided through the combined analysis of the magnetic rigidity and momentum loss. Then polarized fragments are introduced to the NMR apparatus located at the final focus.

The NMR apparatus is also shown in Fig. 3. The spin-polarized fragments are implanted into the stopper located at the central part of the system. A static magnetic field is applied to the stopper in order to preserve spin polarization. Also, a radio-frequency oscillating field is applied perpendicular to the static field. β-Rays emitted from the implanted fragments are detected by using plastic scintillator telescopes located above

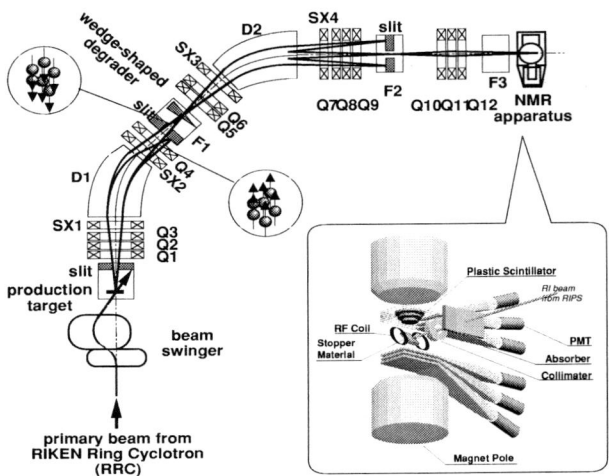

FIGURE 3. Arrangement of the RIKEN projectile fragment separator RIPS for the production of spin-polarized RIBs, and the schematic layout of the β-NMR apparatus.

and below the stopper. We employ the β-NMR method to determine the magnetic moment. In this method, the nuclear magnetic resonance is detected through a change in the up/down asymmetry of the β-ray angular distribution. The stopper material is chosen so as to provide long spin-relaxation time compared with the β-decay life-time. As the NMR technique, the adiabatic fast passage method is employed. Thus, the frequency of the oscillating field is swept over a certain region of frequency, and when it is swept across the Larmor frequency, the spin reversal takes place.

RESULT AND DISCUSSION

Magnetic moments of nitrogen isotopes

Figure 4(a) shows the obtained NMR spectrum of ^{17}N. The experimental magnetic moment of ^{17}N shows a substantial deviation from the Schmidt moment (μ_{Schmidt}) [3], as shown in Fig. 5. Most remarkably, the deviation is outward directed, while most of the experimental magnetic moments fall into the region between the Schmidt lines. There are several other examples showing such the outward deviations in the light-mass region. However, the observed $\mu(^{17}$N) value shows the largest deviation among them. Since the M1-type configuration mixing quenches μ-moment, which gives opposite-directed contribution, only the configurations in which two sd-hell neutrons coupling to form 2^+ is expected to shift the μ-moment outward from μ_{Schmidt}. The amplitude of the corresponding configuration is underestimated by the standard shell-model calculations [3].

The tendency for a neutron pair to couple to form $I^\pi = 2^+$ rather than $I^\pi = 0^+$, should be largely governed by the difference between the two-body matrix elements,

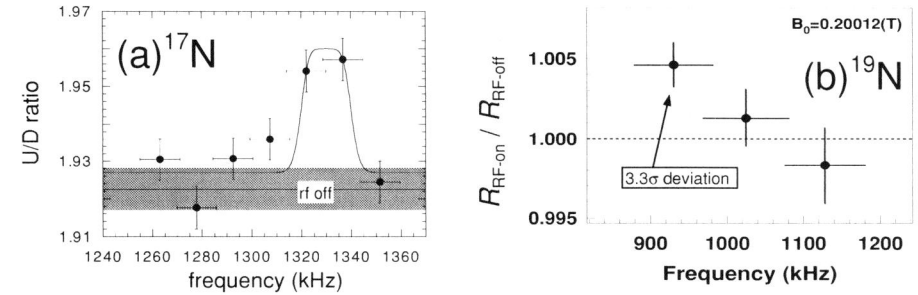

FIGURE 4. Obtained NMR spectra obtained for (a) ^{17}N and (b)^{19}N. In the panel (a) the up/down ratio observed as a function of the frequency of the oscillating field is plotted. In the panel (b) the vertical axis is expressed with double asymmetry. The data points deviating from that obtained without the oscillating field, denoted by rf-off, in the panel (a), and the unity in (b) indicate the spin alternation by AFP.

V_{01} in a channel with angular momentum $I=0$; $T=1$, and V_{21} in a channel with angular momentum $I=2$; $T=1$, for neutrons in the sd-orbit. Thus in shell models, the contribution of 2^+ coupling should increase when either the pairing energy of V_{01} is reduced or the energy V_{21} is increased. Here, it is interesting to note two points. For one thing, a signature for the diminished pairing energy is found in recent study of binding energies for the neutron-rich carbon isotopes [5]. The other is that, since a large part of pairing force between the d-orbit neutron is known to be mediated by the virtual excitation of the core [6], the reduction in the d-neutron pairing energy may suggest that weakening of the coupling between excess neutrons and a core should take place in nuclei far from stability. These observations point to the V_{01} reduction rather than the V_{21} enhancement, as an origin of the observed deviation. Thus we carried out shell-model calculation with V_{01} diminished by about 30 %. Then result reproduced very well the experimental $\mu(^{17}N)$ value. Similar result have been gained in the μ-moment measurements for boron isotopes [3].

From this point of view, μ-moment of ^{19}N was expected to show a larger deviation from the $\mu_{Schmidt}$ value because of the increased neutron-richness. Recently we have succeeded in the μ-moment measurement of ^{19}N whose NMR spectrum is shown in Fig. 4(b). The observed deviation of the $\mu(^{19}N)$ value, however, is found smaller than the $\mu(^{17}N)$ value. In order to clarify the situation, we needs further experimental and theoretical investigations.

Spin-parity assignment of the ^{17}C ground state

Experimentally, little is known concerning the structure of ^{17}C. Only one excited state has been found at $E_x=295$ keV. For the excited state as well as the ground state, spin-parities have not been assigned. Theoretically, shell-model calculations [7] predict the existing of three low-lying positive-parity states, $I^\pi=1/2^+$, $3/2^+$, and $5/2^+$, below $E_x=500$ keV. The ordering of them, however, changes from model to model depending on the choice of the effective interactions [8, 9, 10]. Although among these three

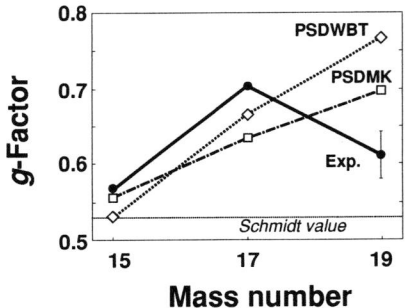

FIGURE 5. Comparison of the experimental g-factor for odd-mass nitrogen isotopes with shell model calculations. In this plot, the values g-factor are expressed with their absolute values.

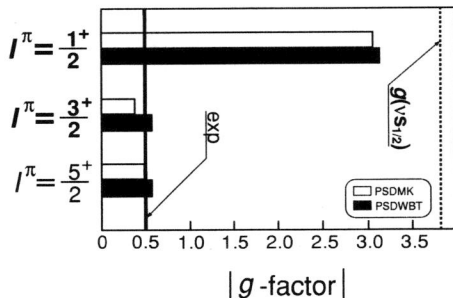

FIGURE 6. Comparison of the experimental g-factor of the ^{17}C ground state with shell model predictions with the PSDMK, and PSDWBT interactions.

candidates $I^{\pi}=5/2^+$ is predicted for the ground state in the naive shell model, this possibility is excluded by the β-decay properties of ^{17}C [8]: the ^{17}C ground state decays into the $1/2^+$ excited state of ^{17}N through the β-decay with a non-retarded ft value. In the remaining two candidates, $I^{\pi}=1/2^+$ and $3/2^+$, the latter has been proposed in the recent experiments based on the β-delayed neutron spectroscopy [11] and on the nuclear reactions using ^{17}C secondary beams [12]. If the spin-parity of the ^{17}C ground state is $I^{\pi}(^{17}C)=1/2^+$, the configuration in which the $I^{\pi}=0^+$ ^{16}C ground state couples with $\nu s_{1/2}$ will be dominant. In this situation, g-factor should be close to that of neutron, $g_n=-3.83$. If, on the other hand, $I^{\pi}(^{17}C)$ is $3/2^+$, the ground state would be dominated by two major configurations, one with three neutrons in the $d_{5/2}$ orbit to form $I^{\pi}=3/2^+$, and another with two $d_{5/2}$ neutrons (coupled to $I^{\pi}=2^+$) times one $s_{1/2}$ neutron. The corresponding g-factor should take a value between $g=-0.77$ and -0.15, the g-factors corresponding to the former and the latter configurations, respectively. Compared with these values, the $1/2^+$ value is 5 times or larger, as shown in Fig. 6. Owing to the large difference between the calculated g-values for the $I^{\pi}=1/2^+$ and $3/2^+$ cases, the spin-parity should be reliably assigned from the determined experimental value of $|g(^{17}C)|=0.5054(24)$.

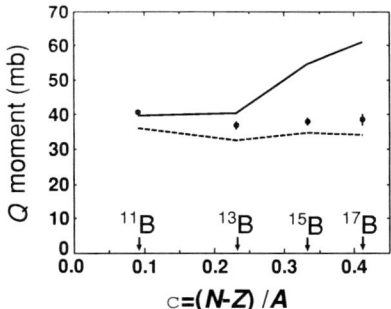

FIGURE 7. Comparison of the experimental Q-moments of odd-mass boron isotopes and shell model calculations with the standard values of constant effective charges $e_n=0.5$ and $e_p=1.3$ (solid lines) and those obtained taking into account the isospin dependence of the polarization charges (dashed lines).

Electric quadrupole moments of boron isotopes

Figure 7 shows the comparison of experimental Q-moments of boron isotopes with shell model predictions. Firstly, we compared the data with standard shell-mode calculations [9], where constant effective charges $e_n=0.5$ and $e_p=1.3$ are taken. These values are commonly used for the sd-shell nuclei. The steep increase of Q for $A=13$ to 17 is in sharp contrast with the experimental observation that the Q-moments are almost constant against A. In order to obtain agreement, we took an isospin dependence of the effective charges. According to Bohr and Mottelson, polarization charges, which are associated with the effective charges, have isospin dependence. The result of the calculations [13] shows good agreement using the same shell-model wave-functions, as shown in Fig. 7. This is the first observation of isospin dependence of the E2 effective charges.

Besides such the microscopic calculation, we also performed simple estimation with reduced neutron effective charges. We found mass dependence of Q-moment is also well reproduced if the neutron effective charge e_n^{eff} is reduced while (arbitrarily) retaining that for proton ($e_p^{\text{eff}}=1.5$). The e_n^{eff} values determined to reproduce the existing experimental values are plotted in Fig. 8. This result might indicate that the quenching of e_n^{eff} develops gradually toward zero as the N/Z ratio is increased.

SUMMARY

Based on the β-NMR method with spin-polarized RIBs, nuclear moments of unstable nuclei have been measured in the light-mass region. The obtained experimental nuclear moments were shown quite effective in discussing the effect of neutron excess on their structure. From the μ-moment measurement of the nitrogen isotopes, the nuclear structures were discussed. The observed deviation of $\mu(^{17}\text{N})$ from the Schmidt value was ascribed to an enhanced contribution of the $I^{\pi} = 2^+$ neutron configurations. This result was accounted for by assuming that the pairing energy for the sd neutrons to form $I^{\pi}=0^+$

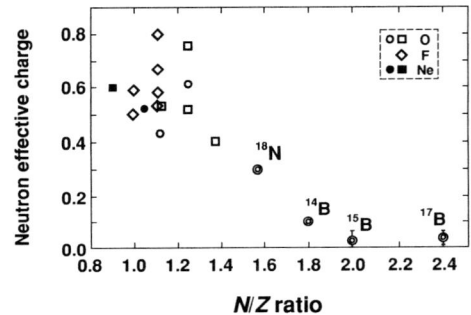

FIGURE 8. Empirical effective charges for neutrons in the *sd* orbits determined from the experimental static Q-moments (circles) and E2 transition probabilities (boxes and diamonds) based on shell model calculations.

is considerably reduced. A drastic change of situation may occur as the neutron excess is further increased, since the observed $\mu(^{19}\text{N})$ does not show such larger deviation. In the case of Q-moment, our experimental data show large discrepancy from the shell-model predictions. By taking the isospin dependence of the effective charges, the observed Q-moments of boron isotopes could be explained. Apart from this analysis, we also found that the Q-moments could be also reproduced by the reduced neutron effective charges. Both of these hypotheses are able to explain our experimental observations, and they still remain to be proved. We also showed another application of μ-moment measurements. From the obtained $\mu(^{17}\text{C})$ value, the discussion was made on the ground-state spin-parity of ^{17}C, in quite a different approach from those taken by the recently performed β-delayed neutron spectroscopy and nuclear-reaction studies. Among the candidates of the ^{17}C ground-state spin-parity, $I^{\pi}(^{17}\text{C})=3/2^{+}$ was definitely assigned from the measured g-factor.

REFERENCES

1. K. Asahi *et al.*, Phys. Lett. **B251** (1990) 488; H. Okuno *et al.*, Phys. Lett. **B335** (1994) 29.
2. H. Okuno *et al.*, Phys. Lett. **B354** (1995) 41; H. Izumi *et al.*, Phys. Lett. **B366** (1995) 51; M. Schäfer *et al.*, Phys. Rev. C **57** (1998) 2205; H. Ogawa *et al.*, Phys. Lett. **B451** (1999) 11.
3. H. Ueno *et al.*, Phys. Rev. C **53** (1996) 2142.
4. T. Kubo *et al.*, Nucl. Instr. Meth. **B70** (1992) 309.
5. D. Bazin *et al.*, Phys. Rev. Lett. **74** (1995) 3569.
6. T.T.S. Kuo and G.E. Brown, Nucl. Phys. **85** (1966) 40.
7. B.A. Brown A. Etchegoyen and W.D.M. Rae, OXBASH, MSU Cyclotron Laboratory Report No. 524, (1986).
8. E.K. Warburton and D.J. Millener, Phys. Rev. C **39** (1989) 1120.
9. E.K. Warburton and B.A. Brown, Phys. Rev. C **46** (1992) 923.
10. D.J. Millener and D. Kurath, Nucl. Phys. **A255** (1975) 315.
11. P.L. Reeder *et al.*, Phys. Rev. C **44** (1991) 1435; K.W. Scheller *et al.*, Nucl. Phys. **A582** (1995) 109.
12. T. Baumann *et al.*, Phys. Rev. Lett. **439** (1998) 256; D. Bazin *et al.*, Phys. Rev. C **57** (1998) 2156; E. Sauvan *et al.*, Phys. Lett. **B491** (2000) 1; V. Maddelena *et al.*, Phys. Rev. C **63** (2001) 024613.
13. H. Ogawa *et al.*, Phys. Rev. C **67** 064308 (2003).

RISING Status Report

M. Górska on behalf of the RISING collaboration

GSI, Darmstadt, Planckstr. 1, 64291 Darmstadt, Germany

Abstract. Unique opportunity for study of nuclear structure is offered by use of Euroball Cluster detectors at the GSI facility. Radioactive beams produced in fragmentation or fission of heavy ions at relativistic energies are combined with high resolution γ-ray spectroscopy. Challenging new possibilities to investigate exotic nuclei after Coulomb excitation, knock-out and secondary fragmentation reactions are given with beams of rare isotopes which can be produced even in high spin isomeric states. The first RISING campaign at GSI started on August the 5th 2003 with the commissioning run followed by three experiments scheduled for this year. The general description of the project and its physics is given. The up-to-date status of the project is discussed with respect to the tracking detectors of the FRS spectrometer as well as the Ge-array and ancillary detectors.

INTRODUCTION

The CLUSTER [1] detectors part of the EUROBALL array is presently at GSI where they are being used to explore exotic nuclei with radioactive beams. The SIS/FRS [2] facility provides secondary beams of unstable rare isotopes after fragmentation reactions or secondary fission of relativistic heavy ions with sufficient intensity for in-beam gamma-ray spectroscopy measurements. Following a workshop held on November 23 - 24, 2000 at GSI, the RISING (Rare Isotope Spectroscopic INvestigation at GSI) collaboration has been initiated to pursue this project. There are 39 institutions from all over Europe collaborating in this project.

Secondary beams can even be produced in high spin isomeric states. The beams can be used either at i) high energies (~100 MeV/u) for relativistic Coulomb excitation and fragmentation reactions, ii) slowed down to Coulomb barrier energies, enabling fusion, multiple Coulomb excitation and direct reactions or iii) stopped for decay studies. In three workshops held at University Göttingen, INFN Legnaro and GSI Darmstadt in 2001 the nuclear structure physics addressable in these domains has been discussed in some detail. The letter of intent summarizing the proposed physics and the intended experimental technique of the RISING project is available [3].

The proposed experiments for the first RISING campaign, which is the main topic of this paper, exploit unique beams at relativistic energies in the range from 100 MeV/u to 400 MeV/u [4]. The experiments, dependent on fragmentation products from heavy primary beams, are only feasible at GSI. The RISING spectrometer is employed not only for relativistic Coulomb excitation but also for pioneering high-resolution γ-spectroscopy experiments after secondary nucleon removal reactions and secondary fragmentation. New experimental methods for spectroscopy at relativistic energies are investigated in order to obtain important nuclear structure observables for these exotic nuclei. Several of the accepted experiments focus on new methods and techniques in order to determine magnetic moments and life times of short lived states, properties of isomeric states and spectroscopic factors at relativistic energies for the first time.

PHYSICS OF THE FAST BEAM CAMPAIGN

The motivations to explore nuclear structure of exotic nuclei at relativistic beam energies focus on a) shell structure of unstable doubly magic nuclei and their vicinity, b) symmetries along the N=Z line and mixed symmetry states, c) shapes and shape coexistence and d) collective modes and E1 strength distribution.

Shell Structure

Spectroscopic data on the single particle structure of instable doubly magic nuclei and their nearest neighbours are indispensable for theoretical description of the effective interactions in large-scale shell-model calculations. The studies around the N=Z doubly magic nuclei ^{56}Ni and ^{100}Sn provide an excellent probe for single-particle shell structure, proton-neutron interaction and the role of correlations, normally not treated in mean field approaches. The B(E2, $0^+ \rightarrow 2^+$) values in semi-magic Sn nuclei provide a sensitive test for changing (sub)shell structure, the E2 polarisability and the shape response of the magic core. For several reasons conventional techniques, employing (HI,xn) reactions, are very difficult or even impossible and the proposed Coulomb excitation measurements are the only way to obtain the information needed to investigate the evolution of shell structure close to the proton drip line.

Beyond the very neutron-rich shell closures from ^{34}Si to ^{132}Sn the possible disappearance of the familiar Woods-Saxon shell closures and their reappearance as harmonic magic numbers is predicted by several mean-field calculations. Moreover, ^{78}Ni and ^{132}Sn are located close to the astrophysical rapid neutron capture process path and indirect evidence for an altered shell structure and shell quenching of magic gaps at N=82 and N=126 is derived from recent r-process network calculations. Alternatively, observed new shell structure in light and medium-heavy nuclei can be qualitatively understood in terms of monopole shifts of selected nucleon orbitals, with deviating predictions for new shell gaps and spin-orbit splitting. So far the investigative methods have not been able to gather information in the region of neutron-rich Ni and Sn isotopes concerning the most significant interaction matrix elements, the magnetic moments and the spectroscopic factors, which are sensitive indicators of their structure.

Symmetries

The validity of the isospin symmetry for the strong interactions is a fundamental assumption in nuclear physics. The degree to which this symmetry holds as the proton drip line is approached, remains an open question. Experimentally, the symmetry shows as nearly identical spectra in pairs of mirror nuclei obtained by interchanging protons and neutrons. A slight breakdown of the symmetry arises from the Coulomb interaction, causing small differences in excitation energies between isobaric analog states. The Coulomb energy differences between excited analog states in mirror nuclei have been studied for proton rich (N=Z-2) isobars in the A=50 region revealing subtle nuclear structure effects as a function of spin and energy. RISING will facilitate experimental studies of medium mass mirror nuclei with larger values of isospin T≥3/2 in order to provide rigorous tests of large-scale pf-shell model calculations.

Within the framework of the proton-neutron version of the interacting boson model (IBM-2) the existence of low-lying valence shell excitations, which are not symmetric with respect to the proton-neutron degree of freedom, are predicted. These states are called mixed-symmetry states and are e.g. established in all stable N=52 isotones. Coulomb excitation experiments with radioactive beams are required to observe the evolution of these states for lighter N=52 isotones below Z=40.

Shapes

The phenomenon of shape coexistence is caused by various structure effects, which can be traced back to the polarisation by high spin intruder orbitals, which happens to occur primarily in exotic regions of the chart of nuclides. Along semi magic isotopic and isotonic chains, midway between shell closures (N=82 and N=126), 2p-2h and 4p-4h proton core excitations into high-spin orbitals cause coexistence of spherical, oblate and prolate shape as seen in the Pb (Z=82) isotopes. States observed experimentally have no firm assignment of the specific shape so far [5]. Aided by shell gap melting, even the ground state may become deformed as in the well known N=20 nucleus ^{32}Mg. High-K isomers, built by multi-quasi particle configurations, due to their influence on the amount of pairing and/or collectivity left, may give rise to shape changes. Most often deformed nuclei obey axial and reflection symmetry. However, in specific nuclear regions where the Fermi level approaches pairs of close lying, opposite-parity orbitals with $\Delta j=\Delta l=3$ reflection asymmetric shapes are predicted, e.g. the heavy Ba nuclei may show octupole deformed states.

Collective Excitations

Collective excitations such as the giant dipole resonance (GDR), built from collections of single-particle excitations are necessarily influenced by the nuclear shell structure. In exotic nuclei like $^{68-78}$Ni the proton-neutron asymmetry may give rise to changes in the shell structure. Theoretical calculations predict a significant change in the GDR strength distribution as one progress towards the doubly magic ^{78}Ni. The excitation function of the isovector GDR mode is expected to fragment substantially, favoring a redistribution of the strength towards lower excitation energies. Measurements of the GDR strength function provide access to the isospin dependent part of the in-medium nucleon-nucleon interaction and on dipole type vibrations of the excess neutrons. The predicted low-energy shift of the GDR strength was confirmed in neutron-rich oxygen isotopes by the LAND group at GSI by means of virtual photon absorption measurements [6]. The GDR experiment in ^{68}Ni will apply a complimentary method, virtual photon scattering, which relies on real projectile γ-ray emission following the virtual excitation. In order to observe discrete γ-transitions with high resolution and γ-decay from the GDR, the RISING array will be augmented by BaF scintillators.

EXPERIMENTAL TECHNIQUE

The γ-ray spectroscopy of nuclei from exotic beams is performed after in-flight isotope separation. The exotic beams are produced by fragmentation of a heavy stable primary beam or fission of a ^{238}U beam on a ^{9}Be or ^{208}Pb target, in front of the fragment separator, FRS.

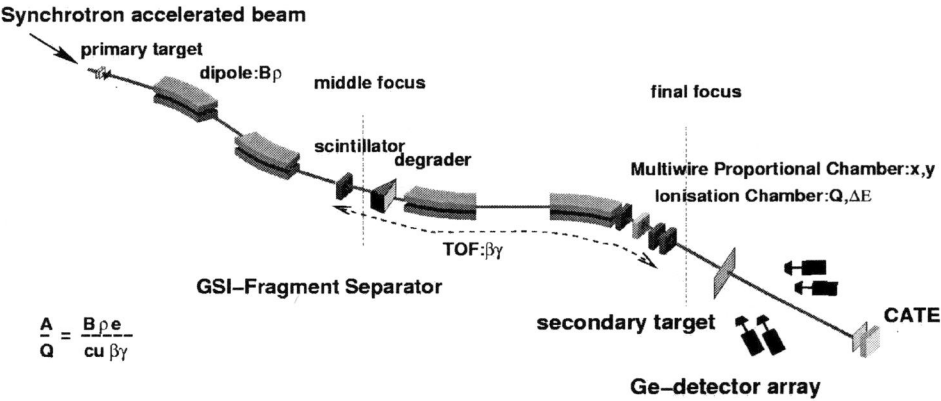

FIGURE 1. The RISING fast beam set-up at the FRS.

Particle identification and tracking

The FRS is be operated in a standard achromatic mode, which allows a separation of the species of interest by combining magnetic analysis with energy loss in matter. The transmission through the FRS is typically 20-70% (depending of a mass region) for fragmentation and 1-2% for fission. The separated ions are identified on an event-by-event basis with respect to mass and atomic number via combined time-of-flight, position tracking, and energy loss measurement. As shown in FIG.1 this is achieved with plastic detectors, multi-wire proportional chamber [7] and a MUSIC ionisation chamber [8].

After passing the identification set-up, the radioactive ions at relativistic energies are focussed onto a secondary target, positioned approximately 4m behind the last FRS magnet in the experimental area called S4. Massive slits reduce the amount of unwanted species reaching the secondary target. Behind the target the calorimeter

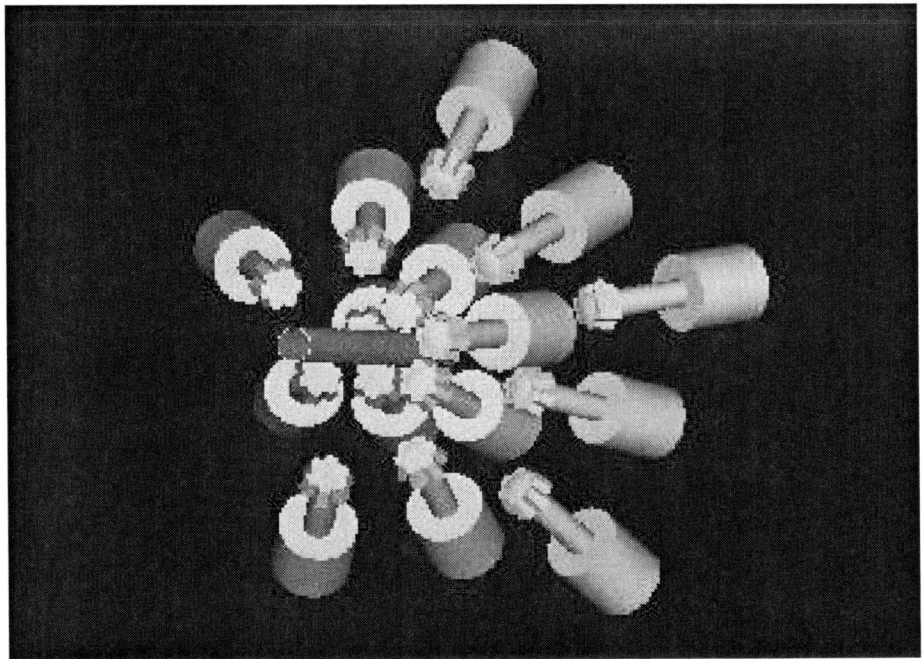

FIGURE 2. Configuration of 15 Cluster Ge-detectors without BGO shields for experiments with relativistic beams.

telescope, CATE, is used for channel selection. CATE consists of position sensitive Si ΔE detectors (lab. angular range $\pm 3°$) and CsI scintillators for total energy measurement. The goal is to measure mass and charge of the secondary fragments with the accuracy of 1%. It is assumed that the kinetic energy loss of a projectile-like nucleus due to nuclear excitation is negligible and thus the velocity behind the secondary target does not differ significantly from the velocity of those projectiles, which do not undergo nuclear interaction. In that case the mass of the projectile-like nucleus can be taken from the total energy without additional time-of-flight measurement. For heavy nuclei the mass resolution is not sufficient to discriminate single masses. However, due to the high resolution of the γ detectors and the expected low to medium γ multiplicities this is generally not required. The identification of a nucleus by gating on its characteristic γ line is of course always possible.

Gamma detection

For the excited fragments moving at a high velocity (v/c = 0.43 at a fragment energy of 100 MeV/u) the γ detectors have to be positioned at either forward or backward angles in order to minimize Doppler broadening. The distance to the target depends on the required energy resolution. The best possible configuration of the 15 Cluster detectors available for experiments with fast beams is displayed in FIG. 2.

Due to the Lorentz boost the main efficiency contribution comes from the first ring. This ring is positioned to achieve 1% resolution within the constraints of the beam pipe diameter of 16cm. One should note that the 5 Clusters in the 2nd and 3rd rings could be moved closer to the target position, for an increased efficiency but a lower overall resolution. Table 1 summarizes the positions of the detectors. If a γ-ray energy resolution of 1% is required the target distance of ring two (three) needs to be 112 cm (137 cm). In that case the total full energy efficiency at 1.3 MeV is 1.7% for γ rays emitted in flight (v/c=0.43).

If both rings are placed at a minimum distance of ~70cm then the configuration reaches a total efficiency of 2.9% while the average weighted energy resolution is only slightly increased to 1.24%.

TABLE 1. Position and distances of the Cluster detectors.

	Cluster detectors	Angle	Target distance
Ring 1	5	15.0°	680 mm
Ring 2	5	26.5°	680 – 1400 mm
Ring 3	5	34.0°	680 – 1400 mm

FIRST EXPERIMENTS

The first RISING campaign started on August the 5th with commissioning of the setup. This learning phase was followed in September and October by the performance of three of the 12 accepted RISING experiments. During each of the experiments the preliminary results were seen in the near line analysis stage thanks to highly advanced on(of)-line analysis software. Coulomb excitation of exotic nuclei, as well as knock-out and secondary fragmentation reactions, were successfully established.

REFERENCES

1. J. Eberth, Conf. on Physics from large gamma-ray Detector Arrays, Berkeley, LBL-35687 (1994).
2. H. Geissel et al., Nucl. Instrum. Meth. B70 (1992) 286
3. G. de Angelis et al., http://www-aix.gsi.de/~wolle/EB_at_GSI/main.html
4. P. Reiter et al., http://www-aix.gsi.de/~wolle/EB_at_GSI/main.html
5. A. Andreyev et al., Nature 405 (2000) 430
6. A. Leistenschneider et al., Phys. Rev. Lett. 86 (2001) 5442
7. H. Stelzer, Nucl. Instr. and Meth. A310 (1991) 103.
8. M. Pfützner et al., Nucl. Instr. and Meth. B 86 (1994) 213

Structure of proton-neutron multiplets around ^{132}Sn

A. Gargano, L. Coraggio, A. Covello, and N. Itaco

Dipartimento di Scienze Fisiche, Università di Napoli Federico II, and Istituto Nazionale di Fisica Nucleare, Complesso Universitario di Monte S. Angelo, Via Cintia, I-80126 Napoli, Italy

Abstract. We report on a study of nuclei around ^{132}Sn in terms of the shell model making use of an effective interaction derived from the CD-Bonn nucleon-nucleon potential. Attention is focused on the nuclei ^{132}Sb, ^{134}I, ^{134}Sb, and ^{136}I, which are a direct source of information on the particle-hole and particle-particle matrix elements of the proton-neutron effective interaction. It turns out that the core polarization gives a significant contribution to the effective interaction which makes it adequate to produce a satisfactory description of all considered nuclei. Some discrepancies for the particle-particle multiplets seem to evidence the need for a stronger core-polarization contribution. The existence of a remarkable similarity between the ^{132}Sn and ^{208}Pb regions is substantiated by our calculations.

1. INTRODUCTION

Nuclei at double shell closures are the most appropriate systems to bring to light the properties of the interaction between valence nucleons. In fact, low-energy states as well as some high-spin states in these nuclei can be described in terms of few shell-model configurations, thus providing a direct knowledge of the matrix elements of the two-body effective interaction.

In this context, nuclei in the close vicinity to ^{208}Pb, for which a large body of data is available, have been extensively investigated. This is not the case for nuclei in the neighborhood of ^{100}Sn and ^{132}Sn, which, lying far from the stability line, were until recently not accessible to spectroscopic studies. Thanks to substantial progress in the experimental techniques, new information on these nuclei is now emerging which makes them very attractive from the theoretical point of view.

During the last few years, we have studied [1, 2, 3] several nuclei around ^{100}Sn and ^{132}Sn within the framework of the shell-model employing effective interactions derived from modern nucleon-nucleon (NN) potentials. In this paper, we report on some results of our current work in the ^{132}Sn region. In particular, results of realistic shell-model calculations, in which the CD-Bonn free NN potential [4] is used, are presented for the odd-odd nuclei ^{132}Sb, ^{134}I, ^{134}Sb, and ^{136}I. In fact, the main motivation for this study is to acquire information on the effective interaction in the proton-neutron channel. To this end, special attention is focused on the proton particle-neutron hole and proton particle-neutron particle multiplets in ^{132}Sb and ^{134}Sb, as their behavior is directly related to the particle-hole and particle-particle matrix elements.

To our knowledge, a similar study [5] was performed only for the multiplets of the

TABLE 1. Proton single-particle and neutron single-hole and -particle energies (in MeV).

Proton-particle level	ε	Neutron-hole level	ε	Neutron-particle level	ε
$0g_{7/2}$	0	$1d_{3/2}$	0	$1f_{7/2}$	0
$1d_{5/2}$	0.962	$0h_{11/2}$	0.100	$2p_{3/2}$	0.854
$1d_{3/2}$	2.439	$2s_{1/2}$	0.332	$0h_{9/2}$	1.561
$0h_{11/2}$	2.793	$1d_{5/2}$	1.655	$2p_{1/2}$	1.656
$2s_{1/2}$	2.800	$0g_{7/2}$	2.434	$1f_{5/2}$	2.055
				$0i_{13/2}$	2.694

heavier particle-hole nucleus ^{208}Bi more than thirty years ago. A peculiar behavior [6] was evidenced by the experimental data, which was well reproduced by the shell-model calculation of Ref. [5], where particle-hole matrix elements deduced from the Hamada-Johston NN potential [7] were employed. Actually, the nucleus ^{208}Bi may be seen as the counterpart of ^{132}Sb in the lead region while ^{210}Bi corresponds to ^{134}Sb. This correspondence becomes more significant when considering the resemblance between the spectroscopy of the two regions. This point will be briefly discussed in Sec. 3.

In Sec. 2 we give a brief outline of the theoretical framework in which our shell-model calculations have been performed. The results are shown in Sec. 3, where they are also compared with the available experimental data. A summary of our conclusions is given in Sec. 4.

2. OUTLINE OF CALCULATIONS

In our calculations we assume that ^{132}Sn is a closed core and let the valence protons occupy the five levels $0g_{7/2}$, $1d_{5/2}$, $1d_{3/2}$, $2s_{1/2}$, and $0h_{11/2}$ of the 50-82 shell. These orbits are also occupied by the neutron holes, while for the valence neutrons the model space includes the six levels $1f_{7/2}$, $2p_{3/2}$, $0h_{9/2}$, $2p_{1/2}$, $1f_{5/2}$, and $0i_{13/2}$ of the 82-126 shell. Our adopted values for the single-particle and single-hole energies are reported in Table 1 and have been taken from the experimental spectra of ^{133}Sb [8], ^{131}Sn [9], and ^{133}Sn [10], respectively. Some single-particle levels, however, are still missing in the spectra of ^{133}Sb and ^{133}Sn. This is the case of the proton $2s_{1/2}$ and neutron $0i_{13/2}$ levels, whose energies have been taken from Refs. [11] and [12], respectively. As regards the neutron-hole energy $\varepsilon_{h_{11/2}}$ in ^{131}Sn, we have adopted the value of 0.100 MeV which has been recently suggested in Ref. [13]. This is slightly smaller than that reported in Ref. [9].

As already mentioned in the Introduction, in our shell-model calculations we have made use of a realistic effective interaction derived from the CD-Bonn free NN potential [4]. This potential, as all modern NN potentials, contains a strong repulsive core which prevents its direct use in nuclear structure calculations. To overcome this difficulty, we have employed a new approach [14] which provides an advantageous alternative to the

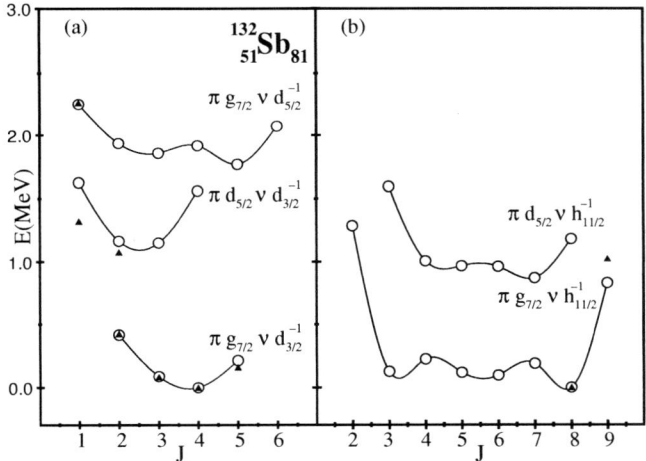

FIGURE 1. Proton particle-neutron hole multiplets in ^{132}Sb. The theoretical results are represented by open circles while the experimental data by solid triangles. The lines are drawn to connect the points.

use of the traditional Brueckner G matrix. It consists in renormalizing the short-range repulsion of the free NN potential by integrating out its high momentum components beyond a certain cutoff momentum Λ. The resulting potential, which we call $V_{\text{low}-k}$, is a smooth potential that preserves the low-energy physics of V_{NN} up to Λ and can be therefore used directly as input for the calculation of shell-model effective interactions. A detailed description of the derivation of $V_{\text{low}-k}$ can be found in [14], where various nuclei with few valence particles have been studied leading to the conclusion that $V_{\text{low}-k}$ can be profitably used in shell-model calculations. In Ref. [14] we have also given a criterion for the choice of Λ. According to this criterion, we have used here the value 2.1 fm^{-1}.

Once the $V_{\text{low}-k}$ is obtained, the calculation of the effective interaction V_{eff} is carried out within the framework of a folded-diagram method, including $V_{\text{low}-k}$ diagrams through second order. This method is described, for instance, in Ref. [1], where a list of relevant references can also be found. Here, we only mention that the proton-neutron matrix elements have been derived in the particle-hole and particle-particle representation for ^{132}Sb, ^{134}I and ^{134}Sb, ^{136}I, respectively.

3. RESULTS AND COMPARISON WITH EXPERIMENT

In this section, we show the results of our calculations for the proton-neutron multiplets around ^{132}Sn and compare them with the experimental data. The proton particle-neutron hole and the proton particle-neutron particle multiplets are presented separately in the two following subsections. In particular, the first one concerns ^{132}Sb and ^{134}I while the second is devoted to ^{134}Sb and ^{134}I. In both cases the behavior of the multiplets is analyzed in terms of the effective interaction and attention is focused on the effects of

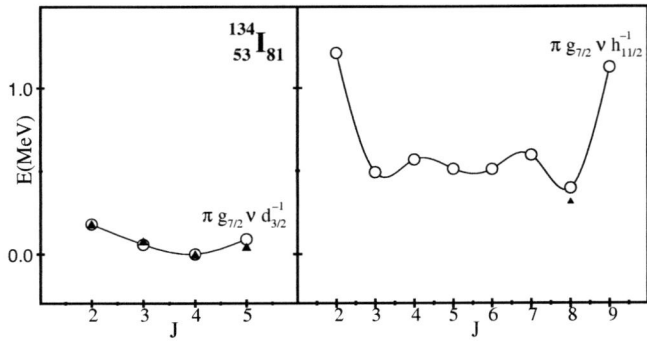

FIGURE 2. Same as Fig. 1, but for ^{134}I.

the core polarization. All the results present here have been obtained using the OXBASH shell-model code [15].

3.1. Particle-hole multiplets

The most appropriate system to study the particle-hole matrix elements of the effective interaction in the proton-neutron channel is ^{132}Sb, with one proton valence particle and one neutron valence hole. It should be mentioned that this nucleus has been the subject of our study of Ref. [1], to which we refer for details. However, for the sake of completeness, some calculated multiplets are shown in Fig. 1 and compared with the existing experimental data [16, 17, 18]. The wave functions corresponding to the calculated multiplets are of almost pure configuration, the percentage of the leading component being at least 85%. Note that the energies in Fig. 1(a) are relative to the 4^+ state while in Fig. 1(b) they are relative to the 8^- state. This is because the experimental spacing between these two levels is unknown. There are indications, however, that the 8^- state lies about 200 keV above the 4^+ ground state [16], in agreement with our calculation which predicts an excited 8^- state at 226 keV. From Fig. 1 we see a good agreement between the calculated and observed excitation energies.

A main feature of the calculated multiplets reported in Fig. 1 is that the states with the minimum and maximum J have the highest excitation energy, while the state with next to highest J is the lowest one. This is in agreement with the experimental pattern exhibited by the $\pi g_{7/2} \nu d_{3/2}^{-1}$ multiplet, whose members are all known. The other experimental multiplets, for which no more than two states have been identified, go in the same direction.

Further support to our predictions is given by the study [1] of the multiplet $\pi g_{7/2} \nu h_{11/2}^{-1}$ in ^{130}Sb. For this nucleus, the members of the calculated multiplet have been identified as those states dominated by the $\pi g_{7/2} \nu h_{11/2}^{-1}$ configuration with the two remaining neutron holes forming a zero-coupled pair. Along the same lines, we have investigated the nucleus ^{134}I having two additional protons with respect to

TABLE 2. Diagonal particle-hole matrix elements of V_{eff}, $V_{\text{low}-k}$, and V_{ph} for ^{132}Sb. See text for comments.

Proton-particle level	Neutron-hole level	J	$<V_{\text{eff}}>$	$<V_{\text{low}-k}>$	$<V_{\text{ph}}>$
$0g_{7/2}$	$1d_{3/2}$	2	0.508	0.558	0.122
		3	0.124	0.196	-0.021
		4	0.001	0.109	-0.082
		5	0.196	0.129	0.106
$1d_{5/2}$	$0h_{11/2}$	3	0.643	0.725	0.132
		4	0.158	0.220	-0.005
		5	0.109	0.199	-0.038
		6	0.114	0.136	0.000
		7	-0.012	0.085	-0.086
		8	0.308	0.174	0.165

^{132}Sb. In Fig. 2 two multiplets are reported. Their behavior is in agreement with the experimental data [9] and quite similar to that found for the two Sb isotopes.

The peculiar behavior of the particle-hole multiplets is directly related to the proton-neutron effective interaction since, as mentioned above, all the states in question are characterized by little configuration mixing. In this context, we have found that the core-polarization contribution represented by diagram (e) of Fig. 2 in Ref. [1] (the so-called bubble) plays a significant role. By way of illustration, the diagonal matrix elements of the effective interaction are reported in Table 2 for two particle-hole configurations. In this table, we also show the corresponding matrix elements of $V_{\text{low}-k}$ together with the contribution $<V_{\text{ph}}>$ from the above diagram. Here, we do not consider the contributions coming from either other second-order diagrams or folded diagrams, both of which we have found irrelevant for the present discussion. Actually, it turns out that the matrix elements of V_{ph}, although not very large in magnitude, are quite relevant in determining the pattern of the particle-hole multiplets. In fact, they are repulsive for states with minimum and maximum J and attractive for $J = J_{\text{max}} - 1$. It is worth noting that the repulsive contribution for the J_{max} state is essential to place it above the state with $J = J_{\text{min}} + 1$ in both configurations considered in Table 2.

To conclude this subsection, we would like to draw attention to the similarity between the behavior of the proton particle-neutron hole multiplets in the ^{132}Sn region and that of the observed multiplets in ^{208}Bi. As regards the latter, a shell-model study using the CD-Bonn NN potential will be the subject of a future publication. However, in the light of the previous discussion on the effective interaction, it is worth mentioning here that the particle-hole matrix elements for the ^{208}Pb region show the same features as those evidenced for ^{132}Sb. In particular, we have found that also in this case the core polarization plays a significant role.

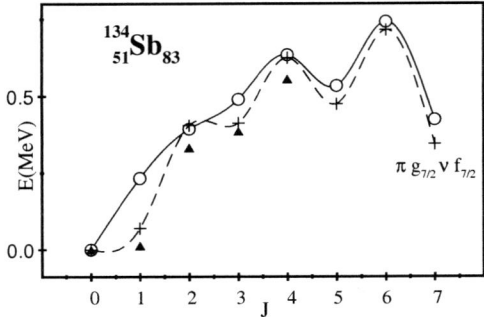

FIGURE 3. Proton-neutron multiplet $\pi g_{7/2} \nu f_{7/2}$ in ^{134}Sb. The theoretical results are represented by open circles and crosses (see text for comments) while the experimental data by solid triangles. The lines are drawn to connect the points.

3.2. Particle-particle multiplets

Let us start with the one-proton, one-neutron nucleus ^{134}Sb. In Fig. 3 the calculated $\pi g_{7/2} \nu f_{7/2}$ multiplet (open circles) is shown and compared with the experimental data [19]. The experimental ground-state has $J^\pi = 0^-$, as predicted by the Nordheim strong rule, but a 1^- state has been observed at 13 keV excitation energy. In this connection, it is worth mentioning that in ^{210}Bi, the analogous two-valence particle nucleus in the lead region, the 1^- state of the $\pi h_{9/2} \nu g_{9/2}$ configuration is the ground state, the 0^- state of the same configuration lying at about 50 keV excitation energy.

From Fig. 3, we see an overall good agreement between theory and experiment, the most significant discrepancy regarding just the position of the 1^- state, whose excitation energy is overestimated by about 200 keV. Note that for ^{210}Bi we predict for the ground state $J^\pi = 0^-$ with a 1^- state at about 150 keV excitation energy.

We now come to ^{136}I, which has two additional protons with respect to ^{134}Sb. In Table 3 we compare the three lowest calculated states (Calc. I) with the experimental ones [9]. These states are members of the $\pi g_{7/2} \nu f_{7/2}$ multiplet with the two remaining protons forming a zero-coupled pair. In this case, however, the admixture with configurations other than dominant one is rather large, so we do not consider it of interest to show the pattern of the whole multiplet. Our calculated spectrum is slightly compressed as compared to the experimental one, and the ground state is predicted to have $J^\pi = 2^-$ instead of 1^- as experimentally observed.

From a quantitative point of view, these discrepancies may be considered not very relevant. However, they have motivated us to examine in detail our effective interaction. We have therefore performed an analysis of the proton-neutron matrix elements. The diagonal matrix elements of V_{eff} for the $\pi g_{7/2} \nu f_{7/2}$ configuration are reported in Table 4 together with those of $V_{\text{low}-k}$ and V_{ph}. We see that the core-polarization contribution, unlike the particle-hole case, is repulsive for all states except $J^\pi = 1^-$, thus giving more binding to it than to the $J^\pi = 0^-$ state. This effect goes in the right direction, although, as is evident from Fig. 3 (open circles), not sufficient to bring the 1^- to the

TABLE 3. Experimental and calculated excitation energies (in MeV) for ^{136}I. See text for comments.

Expt.		Calc. I		Calc. II	
J^π	E	J^π	E	J^π	E
(1^-)	0.0	2^-	0.0	1^-	0.0
$(0^-,1^-,2^-)$	0.087	1^-	0.021	2^-	0.103
$0,1,2^-$	0.222	0^-	0.051	0^-	0.211

correct position. On these grounds, we have found it interesting to modify our effective interaction for both ^{134}Sb and ^{210}Bi by simply increasing the diagonal matrix elements of V_{ph} for the $\pi g_{7/2} \nu f_{7/2}$ and $\pi h_{9/2} \nu g_{9/2}$ configuration, respectively. We find that all the calculated energies go in the right direction, a factor of 2.6 being sufficient sufficient to give the correct spacing between the 0^- and 1^- state in both nuclei. The results of this calculation for ^{134}Sb are shown in Fig. 3, where they are represented by crosses. As for ^{136}I, the calculated spectrum obtained with the modified interaction is shown in Tab. 3 (Calc. II), where it is seen that for the ground state we now predict $J^\pi = 1^-$. Based on the good correspondence between theoretical and experimental energies, we propose the spin-parity assignment 2^- and 0^- for the first and second excited state, respectively.

TABLE 4. Diagonal particle-particle matrix elements of V_{eff}, V_{low-k}, and V_{ph} for ^{134}Sb. See text for comments.

Proton-particle level	Neutron-particle level	J	$<V_{eff}>$	$<V_{low-k}>$	$<V_{ph}>$
$0g_{7/2}$	$1f_{7/2}$	0	-0.627	-0.633	0.044
		1	-0.481	-0.389	-0.076
		2	-0.267	-0.361	0.094
		3	-0.204	-0.218	0.023
		4	-0.080	-0.171	0.094
		5	-0.169	-0.204	0.044
		6	0.006	-0.073	0.087
		7	-0.339	-0.359	0.012

4. CONCLUDING REMARKS

We have presented here some results of a shell-model study of neutron-rich nuclei close to doubly magic ^{132}Sn, focusing attention on the four odd-odd nuclei ^{132,134}Sb, and ^{134,136}I. In our calculations we have used a realistic effective interaction derived from the CD-Bonn NN potential.

The main aim of our work has been to study the effects of the neutron-proton interaction on the pattern of the particle-hole and particle-particle multiplets. In this context, we have found out that the core polarization plays a significant role. We have also found out that the main features of our neutron-proton effective interaction for the ^{132}Sn region are essentially the same as those of the effective interaction derived for the ^{208}Pb region.

ACKNOWLEDGMENTS

This work was supported in part by the Italian Ministero dell'Istruzione, dell'Università e della Ricerca (MIUR).

REFERENCES

1. Coraggio, L., Covello, A., Gargano, A., Itaco, N., and Kuo, T. T. S., *Phys Rev. C* **66**, 064311 (2002), and references therein.
2. Covello, A., Coraggio, L., Gargano, A., and Itaco, N., *Acta Phys. Pol. B* **34**, 2257 (2003).
3. Genevey, J., Pinston, J. A., Faust, H. R., Orlandi, R., Scherillo, A., Simpson, G. S., Tsekhanovich, I. S., Covello, A., Gargano, A., and Urban, W., *Phys. Rev. C* **67**, 054312 (2003).
4. Machleidt, R., *Phys. Rev. C* **63**, 024001 (2001).
5. Kuo, T. T. S., *Nucl. Phys.* **A112**, 325 (1968).
6. Moinester, M., Schiffer, J. P., and Alford, W. P., *Phys. Rev.* **179**, 984 (1969).
7. Hamada, T., and Johnston, I. D., *Nucl. Phys.* **34**, 382 (1962).
8. Sanchez-Vega, M., Fogelberg, B., Mach, H., Taylor, R. B. E., Lindroth, A., Blomqvist, J., Covello, A., and Gargano, A., *Phys. Rev. C* **60**, 024303 (1999).
9. Data extracted using the NNDC On-Line Data Service from the ENSDF database, file revised as of September 24, 2003.
10. Hoff, P., et al., *Phys. Rev. Lett.* **77**, 1020 (1996).
11. Andreozzi, F., Coraggio, L., Covello, A., Gargano, A., Kuo, T. T. S., and Porrino, A., *Phys Rev. C* **56**, R16 (1997).
12. Coraggio, L., Covello, A., Gargano, A., and Itaco, N., *Phys Rev. C* **65**, 051306(R) (2002).
13. Genevey, J., Pinston, J. A., Faust, H., Foin, C., Oberstedt, S., and Weiss, B., *Eur. Phys J. A* **7**, 463 (2000).
14. Bogner, S., Kuo, T. T. S., Coraggio, L., Covello, A., and Itaco, N., *Phys Rev. C* **65**, 051301 (2002).
15. Brown, B. A., Etchegoyen, A., and Rae, W. D. M., *The computer code OXBASH*, MSU-NSCL, Report No. 524.
16. Stone, C. A., Faller, S. H., and Walters, W. B., *Phys Rev. C* **39**, 1963 (1989).
17. Mach, H., Jerrestam, D., Fogelberg, D., Hellström, M., Omtvedt, J. P., Erokhina, K. I., and Isakov, V. I., *Phys Rev. C* **51**, 500 (1995).
18. Bhattacharyya, P., et al., *Phys Rev. C* **64**, 054312 (2001).
19. Korgul, A., et al., *Eur. Phys J. A* **15**, 181 (2002).

Chaotic Dynamics in Warm Rotating Nuclei

M. Matsuo[1], E. Vigezzi[2], S. Leoni[2], A. Bracco[2], G. Benzoni[2], T. Døssing[3], B. Herskind[3], G.B. Hagemann[3], A. Lopez-Martens[4], T.L. Khoo[5]

[1] *Graduate School of Science and Technology, Niigata University, Niigata 950-2181, Japan*
[2] *INFN sez. Milano, and Department of Physics, University of Milan, Milan, Italy*
[3] *Niels Bohr Institute, University of Copenhagen, Copenhagen Ø, Denmark*
[4] *CSNSM-IN2P3-CRNS, Orsay, France*
[5] *Argonne National Laboratory, Argonne, U.S.A.*

Abstract. Properties of thermally excited rotating nuclei are discussed by means of the cranked shell model. Focus is put on the violation of the K-quantum number in the rare-earth deformed nucleus ^{163}Er and on the peculiar features of rotational motion in superdeformed Hg nuclei, which appear to be precursory of ergodic rotational bands.

INTRODUCTION

Rapidly rotating nuclei formed by the fusion reaction and subsequent neutron cooling have a finite thermal excitation energy $U=1$-8 MeV in the initial and intermediate stages of the gamma-decay cascades leading towards the yrast line. Since the level density at given spin I increases exponentially as a function of U, the two-body interaction causes complex configuration mixing of many-particle many-hole excitations with increasing U. This leads to an order-to-chaos transition, which is predicted to occur for U equal to a few MeV. The chaotic motion of intrinsic many-particle many-hole excitations influences the collective rotational motion, resulting in different possible modes of rotation of warm deformed nuclei. One can probe the chaotic dynamics by means of rotational E2 gamma-rays emitted by the warm rotating nuclei. In this paper we give two illustrative examples: 1) violation of the K-quantum number which is connected to the chaotic configuration mixing, and 2) the peculiar features displayed by superdeformed Hg nuclei, which appear to be precursor of ergodic rotational bands.

VIOLATION OF THE K-QUANTUM NUMBER

The K-quantum number is an approximate quantum number associated with the axial symmetry of the nuclear shape[1]. Its conservation is well established as far as the discrete rotational bands with low thermal energy ($U<1$ MeV) observed near the yrast line are concerned. The situation will be different at higher thermal excitation energy since the chaotic dynamics is expected to enhance violation of the approximate quantum numbers [2-4]. An analogy is the parity violation in the compound states at thermal neutron energy $U=8$ MeV and at low spins[2]. One may thus expect that the extent to which the K-quantum number is violated may provide a probe to characterize the order-to-chaos transitions in warm rotating nuclei. At high excitation energy ($U>8$ MeV), the chaotic configuration mixing has been evaluated by means of the random matrix models[5], and it has been predicted to be so strong that the approximate K-quantum number is fully violated[6].

In order to study experimentally the K-quantum number in warm rotating nuclei, we analyze the quasi-continuum gamma-ray spectra obtained by gating on both the high-K and low-K rotational bands, which are observed in the rare-earth nucleus ^{163}Er[7,8]. A previous analysis has already indicated that the K-quantum number may remain valid to a good extent in the ridge part of the quasi-continuum spectra[9], while a new data set enables us to analyze the K-quantum number not only in the ridge but also in the valley, thus providing information from regions of higher thermal energy [10]. Motivated by these experiments, we have performed a theoretical investigation by means of the cranked shell model[11], which has been successful to describe many aspects of the warm rotating nuclei and the associated quasi-continuum gamma-ray spectra. The model describes the onset of rotational damping around $U=1$ MeV in rare-earth deformed nuclei, in good agreement with experiments. It predicts the order-to-chaos transition in the energy region of $U=1-2$ MeV[12].

We need to describe the K-quantum number and its violation associated with the chaotic mixing. For this purpose we modify the usual cranked shell model Hamiltonian by adding a new term $-J_z^2/2I_{mom}$, where J_z is the angular momentum operator along the symmetry axis (its eigenvalue corresponds to the K-quantum number) and I_{mom} is the kinetic moment of inertia [13]. This term generates rotational bands which have large values of K out of unperturbed many-particle many-hole configurations, allowing a coherent build-up of the K-values of individual particle and hole orbits involved. We then mix the unperturbed configurations, introducing a residual two-body interaction. We adopt the same surface-delta interaction (SDI) used in our previous calculations [11,12]. Note that the SDI by itself respects the K-quantum number, so that the only source violating the K-quantum number in the present model is the Coriolis force or the cranking term. On the other hand, we find that, due to the interplay of the Coriolis and of the residual interaction, with increasing excitation energy the energy levels become a mixture of many configurations of different K-quantum numbers. The mixing of K-quantum number would be instead quite weak in the whole energy region $U=0-3$ MeV investigated in the present analysis if the residual two-body interaction were neglected.

A quantitative measure of the K-mixing can be obtained by evaluating the expectation value $K_i^2=<i|J_z^2|i>$ of the operator J_z^2 and the variance $\sigma^2(K^2)_{mix,i}=<i|J_z^4|i> - K_i^4$ of K^2 distribution in an energy level i. This quantity indicates the degree of mixture of different K-values. K_i may be regarded as an average value of the K-quantum number in each level. If the K-mixing is weak, the values of K_i fluctuate from level to level since it takes significantly different values for low-K and high-K states. One can define the variance of this statistical fluctuation by $\sigma^2(K^2)_{stat} = <K_i^4>$

$-<K_i^2>^2$. This statistical variance may exceed the variance of K-mixing $\sigma^2(K^2)_{mix}$ which will be small in the weak K-mixing case. On the other hand, the opposite relation $\sigma^2(K^2)_{mix} > \sigma^2(K^2)_{stat}$ is expected when the mixing of K-quantum number is strong. Defining the K-mixing ratio $r_{K-mix} = \sigma^2(K^2)_{mix}/(\sigma^2(K^2)_{mix} + \sigma^2(K^2)_{stat})$, we may say that the K-mixing is weak for $r_{K-mix} \ll 1$, and is strong for $r_{K-mix} \sim 1$.

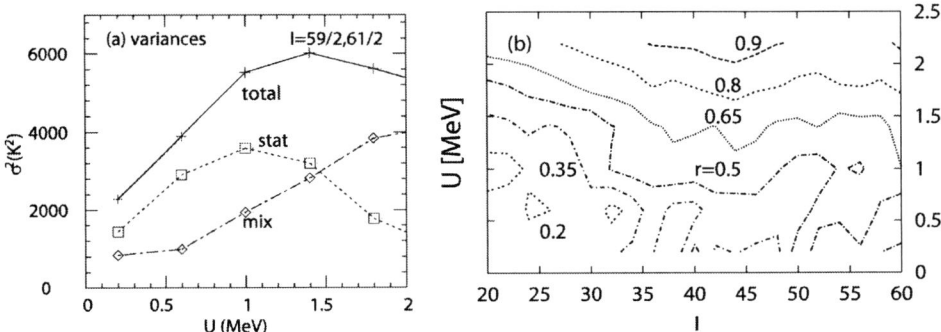

FIGURE 1. (a) The statistical and mixing variances $\sigma^2(K^2)_{stat}$ and $\sigma^2(K^2)_{mix}$ calculated at spin $I=59/2, 61/2$ as a function of the thermal excitation energy U in ^{163}Er. (b) The K-mixing ratio r_{K-mix} calculated for ^{163}Er, plotted as a function of U and spin I.

Fig.1(a) shows the variances associated with the K-mixing as a function of the thermal excitation energy U at the fixed spins $I=59/2, 61/2$ in ^{163}Er. It is seen that the K-mixing evolves with increasing U in the interval of $U=1-2$MeV. Fig.1(b) exhibits the K-mixing ratio r_{K-mix} as a function of U and the spin I. It is seen that the K-mixing ratio varies from ~0.2 to ~0.9 as U increases from zero to $U\sim 2$ MeV. The K-mixing ratio is around 0.4-0.7 at $U\sim 1.5$ MeV. The change is rather gradual, and complete K-mixing is achieved only at $U>2$ MeV. Note that in nuclei near ^{168}Yb rotational bands exist only below around $U=1$ MeV, which is the energy where the configuration mixing becomes significant. In ^{163}Er, however, we find several high-K rotational bands at higher energy ($U=1-1.5$MeV). This is another indication that the K-quantum number is partially conserved in this energy region. It is also seen from Fig.1(b) that a weak but non-negligible spin-dependence is present especially for $I<40$, where for a given value of U the K-mixing becomes weaker as I decreases.

The gradual K-mixing and its spin dependence can be connected to the fact that the residual two-body interaction (SDI) itself does not violate the K-quantum number. Indeed the observed features are gone if we replace the SDI by a random two-body residual interaction, in which case the shell model matrix elements of the interaction are represented by Gaussian random numbers, neglecting completely the K-quantum number. In this case, the K-mixing becomes almost complete ($r_{K-mix}>0.8$) in the energy region $U>1$MeV as soon as the configuration mixing sets in. We have examined the two-body matrix elements of SDI, and found that they keep some selectivity with respect to the K-quantum number of rotational bands. Similar K-selectivity is seen also in experimental matrix elements[13,14].

The partial conservation of the K-quantum number for $U<\sim 1.5$MeV provides a qualitative explanation of the experimental finding[9] that the K-quantum number seems to be rather well preserved in the ridge part of the quasi-continuum spectra. Preliminary analysis indicates, in addition, a systematic trend towards stronger K-mixing as a function of gamma-ray energy (i.e., with increase in I and U)[10]. To make a quantitative analysis, we are currently performing a simulation analysis of the full gamma-decay cascades as was done in Ref.[15], but now on the basis of the new microscopic calculation presented here.

PRECURSOR OF ERGODIC ROTATIONAL BANDS

Recently a ridge structure associated with quasi-continuum gamma-ray spectra obtained by gating on the superdeformed(SD) rotational bands has been observed in ^{194}Hg[16]. The quasi-continuum gamma-rays originate from the thermally excited SD states which are fed in the gamma-decay cascades before reaching the discrete SD bands situated near the SD yrast line. The SD-gated quasi-continuum ridge exhibits unusual features; the width of the ridge is significantly small, FWHM ~10 keV, compared to the typical value, FWHM~20-30 keV, found in the normal deformed (ND) rare-earth nucleus ^{168}Yb. In addition the effective number of paths obtained by the fluctuation analysis method is anomalously large, N_{path}~50-200, compared to the corresponding number N_{path} ~20-30 in ^{168}Yb[17,18].

The cranked shell model calculation performed for SD ^{192}Hg shows[19] that the shell structure associated with the angular momentum alignment of the cranked Nilsson single-particle orbits near the Fermi level is quite different in SD ^{192}Hg from that in ND ^{168}Yb. One can theoretically estimate the number of rotational bands N_{band}, which can be compared with N_{path} in the ridge assuming that the ridge is formed only by rotational bands with similar moment of inertia. It is predicted that N_{band} ~100-150 in the spin interval I=30-50 if one considers only rotational transition among SD states, and N_{band}=20-150 with some ambiguity if one takes into account decay-out transitions to normal deformed states[20]. These numbers are significantly larger than the corresponding number N_{band}~20-30 in ND ^{168}Yb. It is also possible to evaluate the width of the gamma-gamma correlation, which is about 5-20 keV depending on the excitation energies. These numbers are in agreement, at least qualitatively, with the observed features.

It is tempting to relate these features to the "ergodic rotational bands"[21,22]. The collective motion that exhibits the ergodic rotational bands may be regarded as a peculiar form of nuclear rotation. Usually one assumes that each discrete rotational band has a fixed and simple intrinsic configuration, which is rather independent of rotational spin. This is not true for the ergodic rotational bands to which chaotically mixed, and spin-dependent intrinsic configurations are assigned. Nevertheless the rotational band structure is formed. This is also in contrast with the phenomenon of rotational damping, where configuration mixing leads to fragmentation of collective E2 transitions. The ergodic bands have been proposed as a theoretical hypothesis[20], and later their possible existence has been discussed in detail in connection with hyperdeformed nuclei[21], which are however difficult for experiments.

One may consider three conditions for the occurrence of the ergodic rotational bands. The first is the strong configuration mixing. Otherwise ergodicity or chaos loses sense, and the usual regular rotational bands are restored if the mixing is weak. The second is that the rotational damping width Γ_{rot} is smaller than the average level spacing D. If this is realized, the rotational E2 transitions are not fragmented in spite of the strong configuration mixing, and rotational bands are formed. This second condition is hardly met unless the special mechanism called the "motional narrowing" [23] takes place. The third condition is therefore the motional narrowing.

We have looked into the microscopic cranked shell model calculations to check these conditions.

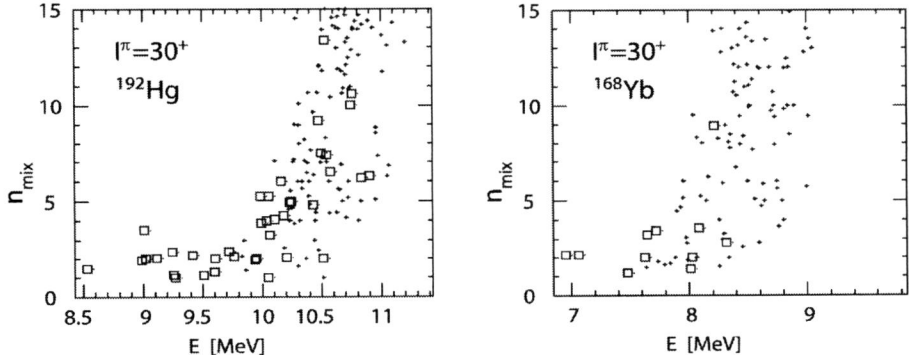

FIGURE 2. The mixing number n_{mix} calculated for individual energy levels of the cranked shell model, plotted as a function of the excitation energy. The square symbol represents levels which exhibit rotational band structure, whereas the small crosses are levels with the rotational damping. Left and right panel is for SD ^{192}Hg and ND ^{168}Yb, respectively.

A measure of the configuration mixing may be given by the mixing number $n_{mix,i}=(\Sigma_\mu x_{i\mu}^4)^{-1}$ where $x_{i\mu}$ denote the wave function coefficients of the energy level $|i\rangle=\Sigma_\mu x_{i\mu}|\mu\rangle$ expanded in terms of the many-particle many-hole cranked shell model basis states $|\mu\rangle$. Assuming that the wave function coefficients x_{ik} are distributed uniformly, $n_{mix,i}$ becomes an effective number of basis states involved in a given level i. Fig.2 shows examples of the calculated mixing number of individual levels plotted as a function of excitation energy. The states that form rotational bands are marked with a square. Fig.2 indicates that in the case of ^{192}Hg there exist many levels forming rotational bands, but at the same time nearly half of these levels have large mixing number n_{mix}~4-10. This is in contrast to the case in ^{168}Yb, where almost all the levels with n_{mix}>4 exhibit the rotational damping and the rotational bands are formed by the levels with small n_{mix}~1-3. Thus a sizable fraction of the rotational bands in ^{192}Hg are accompanied by significant configuration mixing although $n_{mix} = 4$-10 may not be large enough to qualify for ergodicity.

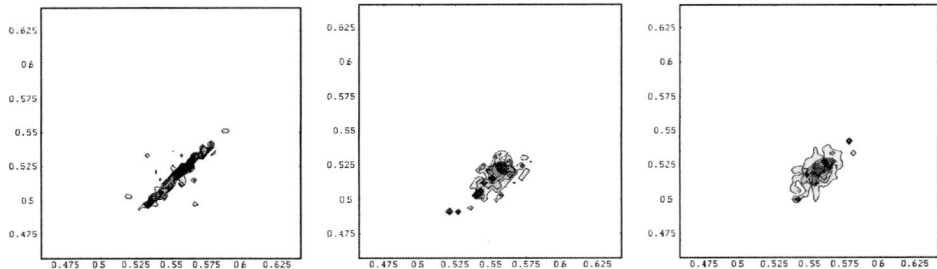

FIGURE 3. Gamma-gamma correlation strength distribution for consecutive two-decay step I=32-30-28 calculated for ^{192}Hg, plotted as a contour in the $E_{\gamma 1}$-$E_{\gamma 2}$ plane. The energy axis is in unit of MeV. The left panel shows the strength distribution for the lowest 50 levels for each parity and signature. The center and right panels are for the next and the third energy bins, respectively.

The occurrence of the motional narrowing mechanism is confirmed by plotting the gamma-gamma correlation strength distribution and by looking into how it evolves with the excitation energy (Fig.3). The distribution along the diagonal axis in the $E_{\gamma 1}$-$E_{\gamma 2}$ plane is reduced with increasing the excitation energy, indicating the narrowing.

Concerning the second condition $\Gamma_{rot} < D$, the average level spacing D can be evaluated on the basis of the cranked shell model calculation. Taking for instance U=1.5 MeV, which is relevant to the rotational bands with large n_{mix}, a typical number for D is ~several keV. Note however that this quantity decreases exponentially with U. The rotational damping width Γ_{rot} depends more weakly on the energy U and the spin I. At spin I=20, for example, we evaluate Γ_{rot} ~10keV. Thus the condition $\Gamma_{rot} < D$ is not met. It should be noted that there is a different kind of level spacing, which is also relevant to the configuration mixing. Since the mixing is caused by residual two-body interaction, direct coupling takes place among selected configurations differing by the occupation of only one or two orbits. The density ρ_2 of levels reflecting this selection is smaller than the total level density. The level spacing $D_2 = 1/\rho_2$ is relevant for the two-body interaction. The configuration mixing sets in for $v > D_2$ where v is the typical matrix element of the two-body interaction. The spreading width due to the mixing is given by $\Gamma_{comp} = 2\pi v^2/D_2$. On the basis of the cranked shell model calculation, we evaluate D_2~20 keV at U~1.5 MeV. If we compare with the above values of Γ_{rot} and D_2, we see that the rotational damping width is smaller than D_2, i.e. $\Gamma_{rot} < D_2$.

We then see that the condition $\Gamma_{rot} < D$, not satisfied in the present case, is replaced by $\Gamma_{rot} < D_2$. One can examine this situation by using the analytic expressions [23] for the rotational damping width $\Gamma_{rot} = 8\Delta\omega^2/\Gamma_{comp}$ and the motional narrowing condition $\Gamma_{comp} > 2\Delta\omega$ (where $\Delta\omega$ is the dispersion of the rotational frequency of different rotational bands). It is then possible to prove the relation $\Gamma_{rot} < 2D_2 \sim D_2$ for the onset region of mixing. One can also show that the same argument leads to the relation $v > \Delta\omega$ between the residual matrix elements v and the dispersion of the rotational frequency $\Delta\omega$. If we consider two rotational bands interacting through the two-body interaction with the condition $v > \Delta\omega$, the two bands repel strongly and E2 transition does not show strong fragmentation even at the band crossing. In other words, rotational bands survive even if the interaction is present. This mechanism of forming rotational bands may be regarded as a phenomenon precursory of the ergodic rotational bands. The condition $\Gamma_{rot} < D_2$ is much weaker than the condition $\Gamma_{rot} < D$ for the proper ergodic bands, and can be more easily realized, as in the case of ^{192}Hg.

REFERENCES

1. Bohr, A., and Mottelson, B.R., *Nuclear Structure*, Vol.2, Benjamin 1975.
2. Bohigas, O., and Weidenmüller, H.A., Ann. Rev. Nucl. Part. Sci. **38**, 421 (1988).
3. Weidenmüller, H. A., Nucl. Phys. **A520**, 509c (1990).
4. Zelevinsky, V., Brown, B.A., Frazier, N., Horoi, M., Phys. Rep. **276**, 85 (1996).
5. Brody, T.A. et al., Rev. Mod. Phys. **53**, 385 (1981). Mehta, M.L. *Random matrices*, 2nd ed. Academic Press, 1991.
6. Barrett, B.R. et al., Phys. Rev. **C45**, R1417 (1992).
7. Brockstedt, A. et al., Nucl. Phys. **A571**, 337 (1994).
8. Hagemann, G.B., et al., Nucl. Phys. **A618**, 199 (1997).
9. Bosetti, P. et al., Phys. Rev. Lett. **76** 1204 (1996).
10. Leoni, S. et al., in preparation.
11. Matsuo, M. et al., Nucl. Phys. **A617**, 1 (1997).
12. Matsuo, M. et al., Nucl. Phys. **A620**, 296 (1997).
13. Matsuo, M. et al., in preparation.
14. Døssing, T. et al., Nucl. Phys. **A649**, 370c (1999).
15. Bracco, A. et al., Phys. Rev. Lett. **76**, 4484 (1996)
16. Lopez-Martens, A. et al., in preparation.
17. Herskind, B. et al., Phys. Rev. Lett. **68**, 3008 (1992).
18. Døssing, T. et al., Phys. Rep. **268**, 1 (1996).
19. Yoshida, K., and Matsuo, M., Nucl. Phys. **A636**, 169 (1998).
20. Yoshida, K., Matsuo, M., and Shimizu, Y.R., Nucl. Phys. **A696**, 85 (2001).
21. Mottelson, B. R., Nucl. Phys. **A557**, 717c (1993).
22. Åberg, S., Z. Phys. **A358**, 269 (1997).
23. Lauritzen, B., Døssing, T. and Broglia, R.A., Nucl. Phys. **A457**, 61 (1986).

Order to Chaos Properties of the Decay-out Gamma Rays from Superdeformed Bands

T. Døssing*, A.P. Lopez-Martens[†], T. L. Khoo**, T. Lauritsen** and S. Åberg[‡]

*The Niels Bohr Institute, Blegdamsvej 17, DK2100 Copenhagen Ø
[†]CSNSM-IN2P3-CRNS, Orsay
**Argonne National Laboratory, Physics Division
[‡]Mathematical Physics, Lund Institute of Technology

Abstract. Based on GOE sparse matrices, a model for decay-out of superdeformed bands is formulated, with focus on the degree of chaoticity of the spectrum of normally deformed states, to which the superdeformed band couples at decay-out. By means of the *effective dimensionality* parameter, the spectrum may be varied between the two limiting situations of complete order and complete chaos. The model is applied to the measured distribution of transition strength of decay-out γ-ray lines in ^{194}Hg, and it is found that the normally deformed states should be closer to the chaotic than the ordered situation, with GOE-type spectral correlations extending over at least 10 levels.

INTRODUCTION

For heavy nuclei, (mass number above 100), only few probes give access to ensembles of discrete quantum states, within limited regions of energy and angular momentum. These are depicted on figure 1.

Especially, the precise regulation of neutron velocities makes it possible to obtain very detailed information on neutron resonance states with low angular momentum and with excitation energy just above the neutron separation energy. Neutron resonance states are the prototype of chaotic quantum states, the description of which initiated random matrix theories [1].

Addressing a quite different region, present γ-ray detection techniques can resolve elaborate level schemes forming near-yrast rotational bands in deformed nuclei. The investigations of nearest neighbor spacings by J.Garrett *et.al.* [2] reveal an ordered situation, much closer to independent selection of energies than to energies resulting from the diagonalization of Gaussian Orthogonal (GOE) random matrices.

In between the two regions of near-yrast bands and neutron resonances, one finds the decay-out of superdeformed (SD) rotational bands, and the related decay-back of fission isomers. Here, the probe is a specific state, which is shielded from the bulk of the spectrum by a barrier in the deformation coordinate. But the superdeformed state energy is fixed, and it will only accidentally be on resonance with one state of the normally deformed (ND) spectrum. Yet, it probes the spectrum of normally deformed states, since the decay-out of the SD band proceeds via small admixtures of ND states acquired by the coupling through the barrier [3, 4, 5].

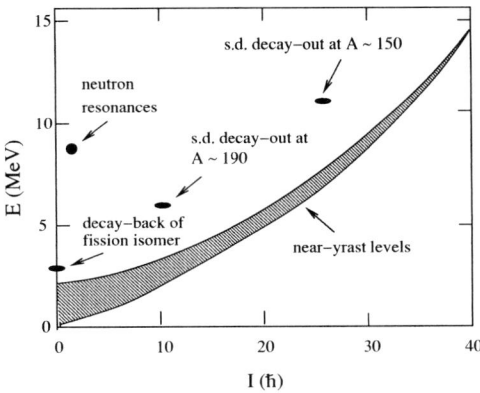

FIGURE 1. The regions of excitation energy and angular momentum where sequences of discrete states are observed are shown together with the decay-out points superdeformed states.

MODEL BASED ON SPARSE MATRICES

A realistic interpolation between the two extremes of spectra of completely ordered or chaotic states can be obtained by diagonalizing *sparse matrices*. Recently, the spectral properties of sparse matrices were thoroughly investigated by Jackson *et. al.* [6].

To describe the decay-out of SD bands, we apply sparse matrices for the ND states, illustrated in figure 2. One of these states acts as a *doorway state* for the coupling of the SD band through the barrier. The model depicted on figure 2 is equivalent to the model [7] used to describe ND states with varying degree of chaoticity, coupled to a SD band, applying scaled GOE matrices. Moving from discrete non-interacting states to the full GOE with increasing scaling factor, the decay-out of the band was found to increase drastically, described as *chaos assisted tunneling*.

Sparse matrices have full GOE values on the diagonal, while a fraction of off-diagonal matrix elements are given by GOE, and the rest are set to zero. The sparcity α, that is the fraction of non-zero matrix elements, is not the most descriptive parameter, which is instead the *effective dimensionality* $d_{eff} = N\alpha$, with N being the basis size. The effective dimensionality gives a measure of how many states are coupled to a typical basis state.

In the present calculations, the SD state is added to the sparse matrix of ND states, close to the middle of the spectrum, with a weak coupling matrix element V_t connecting the SD state to a doorway ND state. V_t is chosen to be very small relative to the average spacing of ND states. This ensures that the states resulting from the full diagonalization will have one state of predominately SD nature, with small components of ND states.

Presently, we are interested in the strength distribution of these components of ND nature in the SD state, as it may be revealed by the strength distribution of decay-out lines down into low-lying ND states. The ND basis states of the sparse matrix are viewed as many-particle-many hole states. Assuming that the low lying states have a certain structure, they will have selection rules to the ND basis states surrounding the SD band and the doorway state. As in reference [7], we shall apply an extreme ansatz

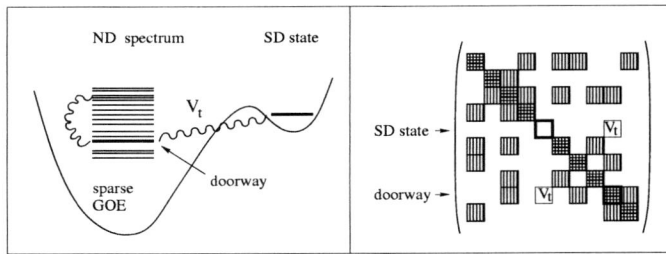

FIGURE 2. Left hand side: Schematic illustration of the deformation energy curve showing two minima, one with the superdeformed state, and another with the doorway state surrounded by other states of normal deformation. Right hand side: illustration of the sparse matrices applied, with the SD state as number 4 from the top and the doorway as number 2 from the bottom. In the example shown, the sparsity is $\alpha = 1/3$ and the effective dimensionality $d_{eff} = 3$.

for these selection rules, assigning one and only one state of the basis to each of the low-lying states. With this ansatz, the relative distribution of transition matrix elements will simply be the relative distribution of admixtures of ND states in the SD state.

The SD state is surrounded by N basis ND basis states, out of which one is the doorway state. Of these, N_{trans} will be transition matrix elements down to states in the energy interval where the SD-decay out transitions are observed, and the strongest N_{visib} of these can reliably be identified in the experiment. Consequently, we will have:

$$N > N_{trans} > N_{visib}$$

CALCULATIONS

The input to the calculations consist of the intensities of the strongest decay-out lines [8] in the nucleus ^{194}Hg. They are emitted in the first step of the decay-out cascade, directly from the SD band state. But the daughter states can only be resolved and linked to the yrast ND states for very few of them.

The distribution of the 19 visible strengths above the 3σ limit can be conveniently displayed by the *cumulative distribution*, as was also done for [9] γ-strengths investigated in neutron resonance reactions. The experimental cumulative distribution is shown as the staircase functions on figure 3, with the strengths normalized to the low detection limit, and compared to two different calculations arising from diagonalizations of many sparse matrices. The basis size is $N = 800$, and 500 simulations are carried out to obtain the smooth curves on figure 3. In each simulation, N_{trans} transitions are randomly selected, and the strongest N_{visib} lines are identified, and normalized to the detection limit, midway between the weakest of the visible and the strongest of the non-visible lines

To make a quantitative comparison between data and calculations, one may calculate the likelihood function. If there is no particular correlation between the visible strengths, each strength will have a certain probability $P(W)$, which can be inferred from the cumulative distribution. The combined probability for the observed strengths $W_1 > W_2 >$

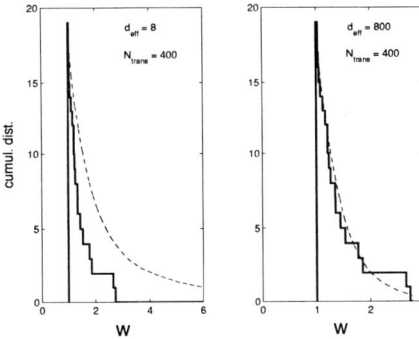

FIGURE 3. Cumulative distribution of the 19 strengths, staircase, are shown together with two examples of cumulative distributions evaluated on the basis of sparse random matrices, dashed lines.

$\cdots > W_{N_{visib}}$ is then given by the *likelihood function*:

$$L_{(d_{eff}, N_{trans})}(W_1, W_2, \cdots W_{N_{visib}}) = P(W_1)P(W_2) \cdots P(W_{N_{visib}})$$

Alternatively, one may evaluate the square deviation from the smooth cumulative distribution to the staircase function for the data, and equivalently to all the individual simulations. On this basis, we define the *cumulative comparison* as the fraction of the simulations for which the data have less deviation from the smooth averaged curve. For the examples shown on figure 3, we find $C_{cumul} = 0.008$ for $d_{eff} = 8$, $N_{trans} = 400$ (left hand part of figure 3) and $C_{cumul} = 0.75$ for $d_{eff} = 800$, $N_{trans} = 400$ (right hand part of figure 3). These numbers quantify the impression one gets from figure 3: the smooth curve for $d_{eff} = 800$ is close to the data. $d_{eff} = 800$ corresponds to the full GOE.

Figure 4 shows perspective plots of the likelihood and the cumulative comparison for the 19 lines in ^{194}Hg. Allowing for factors of about 2 down from the maximum, (corresponding to a FHWM of a Gaussian distribution), one infers from figure 4 that the 19 lines are selected from more than $N_{trans} > 200$ transitions and that the effective dimensionality is larger that $d_{eff} > 12$.

The results obtained in the present analysis of the transition strength distribution can be put in perspective by comparing to the results of the correlation length $L_c^{(0.1)}$, which measures the number of levels in the spectrum for which a GOE-like rigidity will persist [6]. Another perspective on the result is provided by comparing to the fluctuation analysis of the decay-out spectra [10].

Table 1 displays a fine accordance between the number of transitions given by the fluctuation analysis and the number of transitions extracted from the strongest transitions out of about 500. The 19 transitions represent the top of the iceberg of all transitions, and staying with that metaphor, one can say that the analysis shows an agreement between the tip and the submerged part of the iceberg.

For $d_{eff} \geq 12$, the distribution of strengths is close to a Porter-Thomas distribution, and the quite large correlation length $L_c^{(0.1)} \geq 10$ tells that one would have to make a

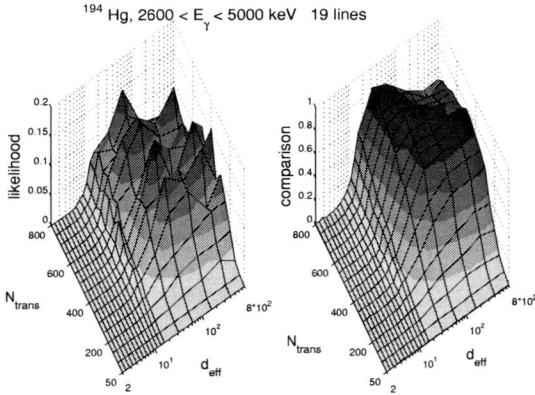

FIGURE 4. Perspective plots of the likelihood and the cumulative comparison for the 19 lines in ^{194}Hg are shown as functions of the effective dimensionality d_{eff} and the number of transitions N_{trans}.

TABLE 1. The number of transitions and effective dimensionality correlated with the spectral rigidity and the number of transitions determined from the fluctuation analysis

N_{trans} (from decay strengths)	d_{eff}	$L_c^{(0.1)}$	N_{trans} (from fl. analysis)
≥ 200	≥ 12	≥ 10	530^{+250}_{-200}

complete sample of more than 10 ND states around the SD state to be able to observe deviations from the GOE with respect to spectral correlations.

On this basis we conclude that the states around angular momentum $I \approx 10\hbar$ and energy above the yrast line $U \approx 4$ MeV in ^{194}Hg are found to be closer to the chaotic than the ordered situation.

Acknowledgement: This research is partially supported by the U. S. Dept. of Energy under contract No. W-31-109-ENG-38.

REFERENCES

1. for a review, see T. Guhr *et. al.* Phys. Rep. **299** (1998) 190.
2. J.D. Garrett *et. al.*, Phys. Lett. **B392** (1997) 24.
3. E. Vigezzi *et. al.*, Phys. Lett. **B249** (1990) 163.
4. J.-Z. Gu and H. Weidenmüller, Nucl. Phys. **A660** (1999) 197.
5. D. M. Cardamone *et. al.*, Phys. Rev. Lett **91** (2003) 102502.
6. A. D. Jackson *et. al.*, Nucl. Phys. **A 687** (2001) 405.
7. S. Åberg, Phys. Rev. Lett **82** (1999) 299.
8. A. P. Lopez-Martens *et. al.*, Nucl. Phys **A647** (1999) 217.
9. H. E. Jackson *et. al.*, Phys. Rev. Lett. **17** (1966) 656.
10. A. P. Lopez-Martens *et. al*, Phys. Rev. Lett. **77** (1996) 1707.

Nuclear Alignment in Projectile Fragmentation as a Tool for Moment Measurements

G. Georgiev*, I. Matea*, J.M. Daugas[†], M. Hass**, R. Astabatyan[‡],
L.T. Baby**, D.L. Balabanski[§], G. Bélier[†], D. Borremans[¶],
F. de Oliveira Santos*, G. Goldring**, H. Goutte[†], P. Himpe[¶],
M. Lewitowicz*, S. Lukyanov[‡], V. Méot[†], G. Neyens[¶],
Yu.E. Penionzhkevich[‡], O. Roig[†] and M. Sawicka[∥]

GANIL, BP 55027, 14076 Caen Cedex 5, France
[†]*CEA/DIF/DPTA/PN, BP 12, 91680 Bruyères le Châtel, France*
**The Weizmann Institute, Rehovot, Israel*
[‡]*FLNR-JINR, Dubna, Russia*
[§]*Faculty of Physics, St. Kliment Ohridski University of Sofia, 1164 Sofia, Bulgaria*
[¶]*University of Leuven, IKS, Celestijnenlaan 200 D, 3001 Leuven, Belgium*
[∥]*IFD, Warsaw University, Hoża 69, 00681 Warsaw, Poland*

Abstract. The application of the <u>T</u>ime <u>D</u>ependent <u>P</u>erturbed <u>A</u>ngular <u>D</u>istribution (TDPAD) method to study isomeric states produced and oriented in projectile-fragmentation reactions provides the opportunity to perform nuclear-moment measurements in a wide range of neutron-rich nuclei, unaccessible by other means. An absolute necessity for the application of the TDPAD technique is a spin-aligned ensemble of nuclei. The preliminary results from a recent application of this method on ^{61m}Fe and ^{54m}Fe at GANIL, Caen, France showed that a significant increase of the amount of the observed alignment, compared to our previous measurement on ^{67m}Ni and ^{69m}Cu, can be obtained. Some experimental details, concerning the conservation of the reaction obtained alignment, are discussed.

INTRODUCTION

The nuclear magnetic and quadrupole moment observables are highly sensitive probes for the single-particle and collective structure of nuclear states. They are often used as stringent tests for nuclear models. However, their measurements in the neutron-rich side of the nuclear chart is hampered by the production mechanisms of the nuclear states and the necessity of obtaining them in a spin-oriented manner.

One of the very powerful methods for measuring of nuclear moments of isomeric states is the TDPAD, which has been widely used in fusion-evaporation reactions [1]. Its first application on isomeric states produced in high-energy ($\sim$ 500 MeV/u) projectile fragmentation has been performed on the case of ^{43m}Sc [2], where a significant amount of spin-alignment (up to 35%) has been observed. The first experiment in which the TDPAD technique after projectile fragmentation was applied on an unknown case [3], was performed in 1999 at GANIL, Caen, France. A significantly lower amount of alignment was observed compared to the case of Ref. [2]. In a subsequent experiment, which took place in December 2002, the specificities of intermediate-energy projectile fragmentation were taken into account, leading to an increase of a factor of ∼ 10 of the experimentally observed alignment. Some experimental details and the changes of the setup, that led to this increase, will be discussed hereafter.

EXPERIMENTAL DETAILS

An axially-symmetric spin-aligned nuclear ensemble, emersed in a constant magnetic field, rotates around $\vec{B}$ with Larmor frequency ω_L, that is directly proportional to the gyromagnetic factor of the nuclear state $\left(\omega_L = -\frac{g\mu_N B}{\hbar}\right)$. This rotation causes a modulation of the decay curve of the isomeric state, which can be monitored by γ-ray detectors positioned in a plane perpendicular to the magnetic field direction axis. For two detectors, placed at angles of θ and $\theta + \pi/2$ one can obtain the standard $R(t)$ function

$$R(t,\theta) = \frac{I(t,\theta) - I(t,\theta + \pi/2)}{I(t,\theta) + I(t,\theta + \pi/2)} = \frac{3A_2 B_2}{4 + A_2 B_2} \cos 2(\theta - \omega_L t - \alpha), \qquad (1)$$

where A_2 is the angular distribution coefficient, B_2 is the orientation parameter and α is the initial phase of the oscillations, that depends on the deviation angle of the secondary beam passing through the separator and on the gyromagnetic factor of the nuclear state [4, 3]. The experimental determination of the $R(t)$ function provides information both on the g factor of the nuclear state and on the amount of spin-alignment.

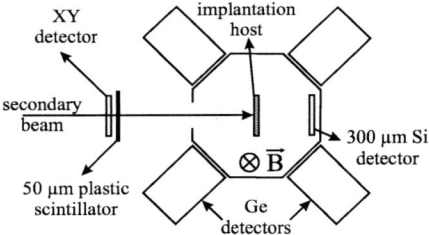

FIGURE 1. Schematic drawing of the experimental setup.

The TDPAD measurement on ^{61m}Fe and ^{54m}Fe was performed at the GANIL, Caen, France. The 54.7 MeV/u ^{64}Ni beam was sent to a 98 mg/cm² ^{9}Be target. The nuclei of interest were selected using the LISE fragment separator. A 102 mg/cm² Be wedge-shaped degrader was used in order to purify the secondary beam. A position-sensitive Si XY detector in front of the vacuum chamber (see Fig. 1), together with 300 μm Si ΔE detector behind the implantation point, were used for the initial beam tuning and identification, during which the implantation crystal was removed. After the selection of the nuclei of interest the position sensitive detector was taken out of the beam line and a 50 μm plastic scintillator was introduced in its place. It was kept constantly there and served for giving the $t = 0$ signal for the time-decay curves. A 0.5 mm high-purity annealed Cu foil was used as an implantation host, fixed in the center of a vacuum chamber. It was positioned between the poles of an electromagnet, that provided a constant magnetic field in vertical direction. The $R(t)$ function was observed using four single-crystal Ge detectors positioned in the horizontal plane.

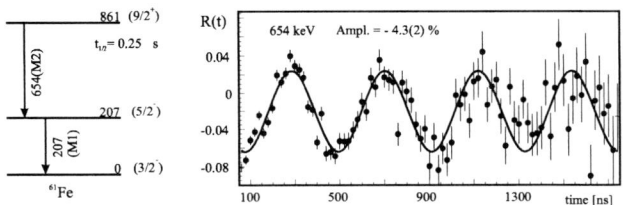

FIGURE 2. Level scheme of ^{61}Fe below the isomer (left). An example of the $R(t)$ function for the 654 keV transition is given on the right. The line is a result of a χ^2 fit to the data using Eqn. (1)

A clear TDPAD pattern was observed both for the 654 keV and for the 207 keV transitions lying below the isomeric state in ^{61}Fe (see Fig. 2). The oscillations in the two $R(t)$ functions had the same frequencies and opposite amplitudes, in agreement with the initially assumed multipolarities of the transitions. From their analysis the preliminary value of $g(^{61m}\text{Fe}) = -0.229(4)$ was derived. It is in a good agreement with the expectation for a neutron $g_{9/2}$ state and will be discussed elsewhere [5]. In the following section we will address the issues related to the spin-alignment obtained in the projectile-fragmentation reaction, its conservation up to the implantation point and the setup changes between the previous experiment [3] and the present one that led to a factor of about 10 increase of the experimental signal.

ALIGNMENT IN PROJECTILE-FRAGMENTATION REACTIONS

According to the "participant - spectator" model the spin-alignment in projectile fragmentation depends on the momentum distribution of the nuclei transmitted through the fragment separator [6]. At the wing of the momentum distribution one expects to obtain the highest (negative) alignment, which should go through zero and change its sign going towards the center (see Fig. 3).

Following these model predictions we performed the following settings:

- ^{61}Fe in the upper wing of the momentum distribution. Two scans were made, with and without "micro-bunch suppression" (MBS), see further for more details;
- ^{61}Fe in the center of the momentum distribution, using the MBS;
- ^{54}Fe in the lower wing of the momentum distribution with MBS.

The primary beam at GANIL arrives at the target position in packages of about 2 ns length with a period of $\sim$ 100 ns, determined by the cyclotron frequency ($\sim$ 10 MHz). The trigger of the data-acquisition system (DAQ) was set in a way that a passage of a heavy ion through the plastic scintillator provided a start signal and opened a coincidence window of $3\mu s$. Any γ-ray, registered by the Ge detectors within this time window, was accepted as a valid event. Otherwise the event was rejected and the acquisition system was cleared. The combination of beam time-structure and DAQ trigger leads to two possible cases of "wrongly" correlated events:

1. During the dead time of the DAQ a nucleus in an isomeric state can be implanted in

the host without giving a start signal. An arrival of a subsequent ion can give a start and open a coincidence window, within which the γ-rays from the previous ion are detected.

2. An arriving heavy ion gives a trigger for the DAQ without any γ-ray being detected. Another nucleus, arriving within the coincidence window, gives a γ-ray, which is detected.

Both of these cases give wrong timing signal for the detected γ-ray. The probability of such false coincidences depends on the intensity of the secondary beam and on the lifetime of the isomeric state. For an isomeric state with half-life of ∼ 250 ns the useful time window is about 1 μs. Therefore, in our case, we suppressed 9 out of every 10 bunches, actually eliminating the probability of having false coincidences. This procedure we call the "micro-bunch suppression". As a result, an increase of the experimentally-observed alignment of a factor ∼3 was observed. The results are presented in left part of Fig. 3 with the two points at the upper wing of the momentum distribution. Having the "wrong" events as described above with just random time signal one would expect to decrease the observed alignment with approximately the probability of registering these type of events. In our case, with a count rate of about 80 000 i/s (without MBS), we estimated the probability of having events of type 1 or 2 to be about 11 %. This clearly cannot account for the observed difference in the alignment assuming random false coincidences. A possible explanation can be that the period of the Larmor precession for ^{61m}Fe (∼ 420 ns) is a multiple of the cyclotron period (∼ 100 ns), leading to a possible superposition of events with exactly opposite phases.

In order to derive the absolute value of the experimentally observed alignment we performed Monte-Carlo (GEANT) simulations using the realistic setup, from which we obtained the alignment-decrease due to the geometrical factor of our setup ∼ 0.81. No corrections were made for possible alignment loss due to relaxation effects or non-substitutional implantation. The experimental values are in good qualitative agreement with the expectation of the "participant-spectator" model both for the ^{61}Fe and ^{54}Fe cases. However, in order to obtain a quantitative correspondence one had to divide the calculated values by a factor of 1.8.

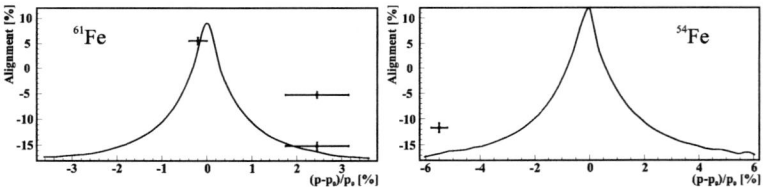

FIGURE 3. Comparison of the experimentally observed alignment with model calculations [7, 8] for ^{61m}Fe (left) and ^{54m}Fe (right). The horizontal error-bars are due to the momentum acceptance of the LISE separator for each of the setting.

In projectile fragmentation one inevitably introduces a flight-path between the target and the implantation host along the fragment separator. If the detected ions are not fully stripped, the interaction of the randomly oriented electron spin, coupled to the nuclear spin, with any external magnetic field can cause a significant decrease up to a complete loss of the orientation [9]. Therefore, only fully stripped nuclei should be selected.

In our previous experiment [3], a 300 μm Si detector was used for the ion identification and for giving the $t = 0$ signal. As a result about 60% of the nuclei were arriving at the implantation point not fully stripped and respectively their spin-alignment was destroyed. In the here presented experiment the Si detector was replaced by a 50 μm plastic scintillator. This led to a significant decrease of the electron pick-up of the ions passing through it. We estimated that at least 98% of the nuclei were implanted in fully-stripped condition. Another difference between the two experiments which can lead to electron pick-up is the implantation host. In the previous one we used a stack of Cu foils fixed together, without glue or other type of contact material. This might have result in a small gap between them of the order of a μm. In the later experiment this ensemble was replaced by a single foil of 0.5 mm. As a result of the above mentioned changes a significant increase of the alignment was observed.

CONCLUSIONS

The control of the reaction-produced alignment and its conservation is an important point for the successful application of the TDPAD technique in projectile-fragmentation reactions. With our present experience most of the parameters are well understood, however there is still a significant factor missing in order to obtain a quantitative agreement with the model. More investigation are necessary in order to clarify its origin. The present results pave the way for further determination of electromagnetic moments of neutron-rich isomeric states.

ACKNOWLEDGMENTS

We are grateful for the technical support received from the staff of the GANIL facility. This work has been partially supported by the Access to Large Scale Facility program under the TMR program of the EU, under contract nr. HPRI-CT-1999-00019 and the INTAS project nr. 00-0463. We are grateful to the IN2P3/EPSRC French/UK loan pool for providing the Ge detectors. G.N., D.B. and P.H. acknowledge the financial support of the FWO-Vlaanderen. The WI work was supported by the Israel Science Foundation.

REFERENCES

1. G. Goldring and M. Hass, *in: "Treaties on Heavy-Ion Science"* (Ed: D. Allan Bromley, Plenum Press, New York, 1985), Vol. 3, p. 539.
2. W.-D. Schmidt-Ott *et al.*, Z. Phys. **A350**, 215 (1994).
3. G. Georgiev *et al.*, Journ. of Physics **G28**, 2993 (2002).
4. G. Neyens, R. Nouwen, and R. Coussement, Nucl. Instr. and Meth. in Phys. Res. **A340**, 555 (1994).
5. I. Matea *et al.*, (2004), to be published.
6. K. Asahi *et al.*, Phys. Rev. **C43**, 456 (1991).
7. H. Okuno *et al.*, Phys. Lett. **B335**, 29 (1994).
8. J. Daugas *et al.*, Phys. Rev. **C63**, 64609 (2001).
9. K. Vyvey *et al.*, Phys. Rev. **C62**, 034317 (2000).

Coulomb Breakup of Neutron-Rich Oxygen Isotopes

C. Nociforo[a], R. Palit[b], P. Adrich[d], T. Aumann[b], K. Boretzky[a],
D. Cortina-Gil[b], U. Datta Pramanik[b], Th. W. Elze[c], H. Emling[b],
H. Geissel[b], M. Hellström[b], N. Iwasa[b], K. L. Jones[b*], J. V. Kratz[a],
R. Kulessa[d], Le Hong Khiem[a], A. Leistenschneider[b], G. Münzenberg[b],
P. Reiter[e], C. Scheidenberger[b], H. Scheit[f], H. Simon[g], K. Sümmerer[b],
S. Typel[b], E. Wajda[d], W. Walus[d], H. Weick[b]

[a]*Institut für Kernchemie, Johannes Gutenberg Universität, Mainz, Germany*
[b]*Gesellschaft für Schwerionenforschung, Darmstadt, Germany*
[c]*Institut für Kernphysik, Johann Wolfgang Goethe Universität, Frankfurt, Germany*
[d]*Instytut Fizyki,Uniwersytet Jagelloński, Kraków, Poland*
[f]*Max-Planck Institut für Kernphysik, Heidelberg, Germany*
[g]*Institut für Kernphysik, Technische Universität Darmstadt, Darmstadt, Germany*

Abstract. The ground state structure of the near drip-line ^{23}O nucleus and neutron-rich ^{17,21}O isotopes has been investigated at GSI via exclusive measurements of one-neutron Coulomb breakup reactions with a Pb target using secondary beams of the corresponding projectiles at 400-600 MeV/u. Differential cross sections for electromagnetic dissociation have been measured and compared with the calculated ones in a direct breakup model. The cross section $d\sigma/dE^*$ (excitation energy E^*), further differentiated according to the observed (excited) states of the remaining nucleus, provides spectroscopic information on the ground state of these nuclei.

INTRODUCTION

Radioactive beam facilities allow to access the neutron drip line in case of very light nuclei and to perform reaction studies. The oxygen chain is an example. Neutron-rich oxygen isotopes, which have been recently produced via fragmentation at intermediate and relativistic energies, have attracted a lot of interest since the neutron drip line was found at neutron number N=16 rather than at N=20 [1,2]. Single-particle properties of weakly bound nuclei breaking up into neutron+core have been successfully studied via electromagnetic (e.m.) excitations [3], especially when the core excited states could be identified [4]. They have provided information on the quantum numbers and spectroscopic factors of the ground state configuration for nuclei with low neutron emission threshold ($S_n \approx 1$ MeV) [4,5]. For such systems, the non-resonant e.m. strength observed has been explained by using the direct breakup model [3]. Here, we investigate the possibility to extract spectroscopic information also for systems with larger neutron separation energy ($S_n \approx 4$ MeV) such as the ^{17,21}O

---

* Present address: Rutgers University, New Brunswick, USA

nuclei. In particular, results for the ground state of ^{23}O nucleus (S_n=2.74 MeV), whose spin assignment is controversial at present [6,7,8,9], will be presented and discussed.

EXPERIMENTAL METHOD FOR COULOMB BREAKUP

The reaction mechanism of electromagnetic dissociation is well understood [10]. Indicating with $N_{E1}(E^*)$ the number of equivalent dipole photons absorbed at excitation energy E*, the differential Coulomb cross section for the one-neutron breakup channel of a nucleus with mass number A can be written as

$$\frac{d\sigma}{dE^*}(I_c^\pi) = \frac{16\pi^3}{9\hbar c} N_{E1}(E^*) \sum_{nlj} C^2 S(I_c^\pi, nlj) \sum_m \left| \langle q | \frac{Ze}{A} r Y_m^1 | \varphi_{nlj}(r) \rangle \right|^2 \quad (1)$$

for a residual nucleus (A-1) with spin and parity I_c^π. In our case, the superposition of different single particle components $|\varphi(r)\rangle$ in the ground state wave function of the A nucleus, i.e. $|v_{nlj} \otimes ^{A-1}O(I_c^\pi)\rangle$, with the outgoing neutron wave function $\langle q|$ into the continuum is considered. The weight of each component is given by the factor $C^2 S(I_c^\pi, nlj)$, which can be considered as the corresponding spectroscopic factor. $N_{E1}(E^*)$ can be calculated in semiclassical approximation [10]. The measurement of the cross section $d\sigma/dE^*(I_c^\pi)$ was facilitated as follows.

The neutron-rich oxygen beams were produced via fragmentation using primary ^{40}Ar beams at 500-720 MeV/u provided by the SIS synchrotron at GSI. A Be production target was used at the entrance of the Fragment Separator (FRS), which was set to transmit secondary beams of different unstable isotopes to the LAND experimental set up. Details of the experimental setup are given in Ref. [5]. The combined information of energy loss and time of flight measurements allows the ion identification event by event in front and behind the secondary Pb (or C) target. Due to the high energy of the incoming beam, the fragments were emitted at very forward angles with velocities near to that of the corresponding projectile, and detected (resolution $\Delta Z \approx 0.2$, $\Delta A \approx 0.3$) with efficiency almost equal to 1. The high energy neutrons emitted in forward direction were detected with high efficiency (~0.9) in the neutron detector LAND. A highly segmented γ-detector centered at the target position allowed coincidence measurements between the decay products of the projectile and the γ-rays emitted by the fragments, which, due to the high energy, were boosted in forward direction. In the ^{23}O experiment, the γ-detector covering the forward hemisphere consisted of 144 CsI crystals in order to minimize the Doppler broadening effect. A 4π Crystal Ball, consisting of 160 NaI detectors surrounding the target, was used instead in the experiment with the other oxygen isotopes. The excitation energy of the nucleus of interest can be reconstructed using the invariant mass method [5] by measuring the four-momenta of all breakup products, i.e. fragment $^{A-1}$O and neutron. After the subtraction of the corresponding nuclear contribution from the total differential cross section, the e.m. cross section, further differentiated according to the observed (excited) states of the remaining nucleus, can be extracted.

DISCUSSION OF EXPERIMENTAL RESULTS

The main results arising from a comparison with the direct breakup model are discussed for the studied oxygen isotopes. The case of the nucleus $^{17}O(J^{\pi}_{g.s.}=5/2^+)$ has been considered for testing the model. In our data, a negligible contribution from excited states of the ^{16}O core has been observed after breakup of ^{17}O. The distribution obtained for the differential e.m. cross section is well reproduced assuming a valence neutron in a $d_{5/2}$ state, as known. In a preliminary analysis, approximating $\langle q|$ as a plane wave, a value $C^2S(0^+)=0.8$ has been extracted. In our $^{21}O(J^{\pi}_{g.s.}=5/2^+)$ analysis, contributions to the ground state built by a $d_{5/2}$ neutron coupled to $0^+,2^+$ and 4^+ states of ^{20}O have been found. From γ-γ coincidence measurements, the transitions $4^+\rightarrow 2^+$ (E_γ=1.9 MeV) and $2^+\rightarrow 0^+$ (E_γ=1.7 MeV) in ^{20}O were clearly observed, as shown in Fig. 1, and a spectroscopic factor in agreement with the result of a shell model calculation [11] was found.

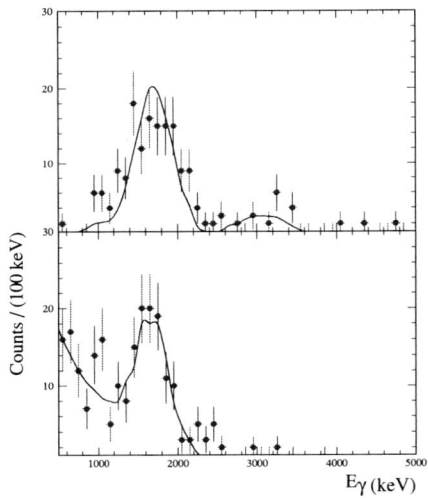

FIGURE 1. Left: γ-γ cascade ($4^+\rightarrow 2^+$ at E_γ=1.9MeV and $2^+\rightarrow 0^+$ at E_γ=1.7MeV) measured in coincidence with neutron+^{20}O for the Pb target. The curves are the response functions simulated with the GEANT code [12] taking into account Doppler corrections.

In a single particle model, the last (bound) neutron in the nucleus ^{23}O is expected to occupy the s orbital according to the conventional level sequence, but according to the results of an interaction cross section measurement in ^{23}O, a $(s_{1/2})^2(d_{5/2})^{-1}$ configuration has been proposed [8]. An exclusive experiment such as ours can elucidate the situation. According to known experimental levels from γ-spectroscopy of ^{22}O [13,14], our preliminary γ-analysis yields an upper limit of core excited states of ^{22}O of about 20-25% in the ^{23}O ground state. Its predominant configuration is found to be $\left|v_{nlj}\otimes ^{22}O(0^+)\right\rangle$. In consequence, the quantum numbers of the valence neutron uniquely determine the ^{23}O ground state spin and parity. For the model calculations, an $s_{1/2}$ state has been assumed for the last neutron in the ^{23}O ground state configuration, but also the case with the neutron in a $d_{5/2}$ state has been considered for comparison. The calculated results are shown in Fig. 2, and compared to the data (full points). The solid (thin) and dashed lines represent the result of a direct breakup model calculations in plane wave approximation assuming $J^{\pi}_{g.s.} = 1/2^+$ and $J^{\pi}_{g.s.} = 5/2^+$, respectively. The ground state wave function of the ^{23}O has been calculated in a Woods-Saxon (WS) potential. The depth parameter V_0 has been chosen in order to reproduce the experimental value of the neutron separation energy in ^{23}O (V_0=43.2 MeV for $J^{\pi}_{g.s.} = 1/2^+$ and V_0=40.8 MeV for $J^{\pi}_{g.s.} = 5/2^+$) fixing the diffuseness at a=0.7 fm and the radius parameter r_0=1.25 fm. The s-wave calculation shows the presence of a peaked dipole strength at low excitation energy but the distribution obtained is too broad.

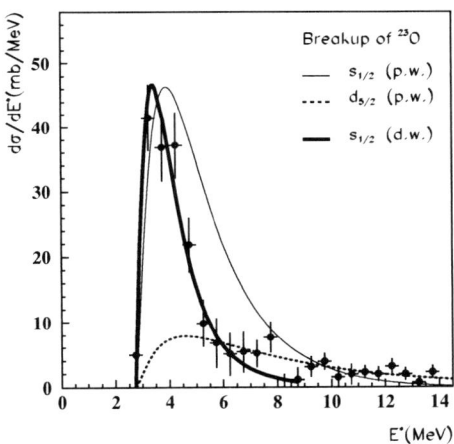

FIGURE 2. Differential e.m. cross section calculated as a function of ^{23}O excitation energy for dipole excitations at 422 MeV/u in plane wave approximation assuming a neutron in s- (thin line) and d- (dashed line) wave, respectively. The bold line is the result of a preliminary calculation in distorted wave approximation. The experimental data are represented by points.

This pattern appears different from that of weakly bound nuclei [4,5] and also from the ^{17,21}O cases previously discussed. The discrepancy with the theoretical calculations seems to be due to the fact that the final state interaction between the neutron and the core in this system plays an essential role. In order to describe the data, the optical potential model of Ref. [15] has been considered for distorting the outgoing wave function $\langle q|$. With the parameter sets of the optical potential we used, even though they do not take into account any data of neutron-rich nuclei, good agreement is obtained assuming $J^{\pi}_{g.s.} = 1/2^+$. The result is plotted in Fig.2 as bold curve. Our preliminary analysis results in a spectroscopic factor near unity, in agreement with the shell model prediction and thus supporting the shell closure at N=16 for the oxygen isotopes [16]. The value of the Coulomb cross section in ^{23}O case, extracted for integrating up to 15 MeV excitation energy, is $\sigma_{e.m}= 101(7)$ mb. Such a value is higher than the one measured for lighter oxygen isotopes whose valence neutrons occupy the $d_{5/2}$ orbital, but, it is not as large as those measured in case of well known halo nuclei, like ^{11}Be [5], which have a very diffuse tail of the ground state wave function. Here, a strong dependence on the depth parameter in the real part of the optical potential has been observed, most likely due to the high neutron separation energy value. Further investigations of the parameter dependence of the optical potential are necessary in order to extract a more precise value of the spectroscopic factor.

CONCLUSIONS

Exclusive measurements of one-neutron Coulomb breakup reactions of ^{17,21,23}O nuclei have been performed at 400-600 MeV/u. Spectroscopic information deduced for ^{17,21}O confirm shell model predictions and are in agreement with those provided by different spectroscopic probes. For ^{23}O, we observe that ^{22}O is emitted mainly in the ground state. The measured e.m. cross section has been compared with the results of a direct breakup model. The comparison shows the importance of the use of an appropriate optical potential that takes into account the final state interaction. Good agreement with the data using a phenomenological optical model is obtained. The presence of a peaked dipole transition probability into the continuum near the neutron emission threshold suggests a predominant s-wave component in the ^{23}O ground state with a spectroscopic factor near unity.

REFERENCES

1. Ozawa, A., et al., *Nucl. Phys.* A **65**, 411 (2000).
2. Sakurai, H., et al., *Phys. Lett.* B **448**, 180 (1999).
3. Nakamura, T., et al., *Phys. Rev. Lett.* **83**, 1112 (1999).
4. Datta Pramanik, U., et al., *Phys. Lett.* B **551**, 63 (2003).
5. Palit, R., et al., *Phys. Rev.* C **68**, 034318 (2003).
6. Sauvan, O., et al., *Phys. Lett.* B **491** 1 (2000).
7. Cortina-Gil, D., this proceeding.
8. Kanungo, R., Tanihata, I., Ozawa, A., *Phys. Lett.* B **512**, 261 (2001).
9. Brown, B.A., Hansen, P.G., Tostevin, J.A., *Phys. Rev. Lett.* **90**, 159201 (2003).
10. Winther, A., and Alder, K., *Nucl. Phys.* A **319**, 518 (1979).
11. Brown, B.A., *Prog. Part. Nucl. Phys.* **47**, 517 (2001).
12. GEANT, Cern Library Long Writup W5013 (1994).
13. Thirolf, P., et al., *Phys. Lett.* B **485**, (2000).
14. Sorlin, O., *Nucl. Phys.* A **685** 186c (2001) and this proceeding.
15. Chadwick M.B., and Young, P.G., *Nucl. Sci. Eng.* **123**, 1 (1996).
16. Otsuka, T., this proceeding.

Continuum Response and Reaction in Neutron-Rich Be Nuclei

Takashi Nakatsukasa*, Manabu Ueda[†] and Kazuhiro Yabana**

*Department of Physics, Tohoku University, Sendai 980-8578, Japan
[†]Akita National College of Technology, Akita 011-8511, Japan
**Institute of Physics, University of Tsukuba, Tsukuba 305-8571, Japan

Abstract. We study $E1$ resonances, breakup and fusion reactions for weakly bound Be nuclei. The absorbing-boundary condition (ABC) is used to describe both the outgoing and incoming boundary conditions. The neutron continuum plays important roles in response and reaction of neutron drip-line nuclei.

INTRODUCTION

At the drip line, since the separation energy becomes close to zero, theoretical studies of nuclear structure and reaction require continuum wave functions. We have recently investigated an efficient and comprehensive method of treating the continuum [1, 2, 3, 4, 5]. This is practically identical to the one called "Absorbing Boundary Condition (ABC) method" in the chemical reaction studies [6]. The method allows us to calculate the continuum wave functions with the outgoing asymptotic behavior of many-body systems.

The essential trick for the treatment of the continuum in the ABC method is to allow the infinitesimal imaginary part in the Green's function, $i\varepsilon$, to be a function of coordinate and finite, $i\varepsilon(\mathbf{r})$. The $\varepsilon(\mathbf{r})$ should be zero in the interacting region and be positive outside the physically relevant region of space. Wave functions obtained using the ABC method are meaningful only in the interacting region. However, this provides all the necessary information to obtain the scattering matrix.

In order to understand how the ABC method is functioning in later applications, let us consider the potential scattering of a particle. The scattering wave function is given by

$$|\psi_{\mathbf{k}}^{(+)}\rangle = |\mathbf{k}\rangle + \frac{1}{E - H + i\varepsilon} V |\mathbf{k}\rangle \equiv |\mathbf{k}\rangle + |\psi_{\text{scat}}^{(+)}\rangle. \tag{1}$$

The scattering amplitude, $f(\Omega)$, is usually defined by its asymptotic behavior

$$\psi_{\mathbf{k}}^{(+)}(\mathbf{r}) \to \exp(i\mathbf{k}\cdot\mathbf{r}) + f(\Omega)\frac{\exp(ikr)}{r}, \qquad (r \to \infty), \tag{2}$$

but can be also written in a form

$$f(\Omega) = -\frac{m}{2\pi\hbar^2}\langle \mathbf{k}'|V|\psi_{\mathbf{k}}^{(+)}\rangle = -\frac{m}{2\pi\hbar^2}\int d\mathbf{r}\exp(-i\mathbf{k}'\cdot\mathbf{r})V(\mathbf{r})\psi_{\mathbf{k}}^{(+)}(\mathbf{r}), \tag{3}$$

where Ω is the direction of $\mathbf{k}'$ and $|\mathbf{k}'| = |\mathbf{k}|$. Equation (3) implies that the $f(\Omega)$ can be determined by the scattering wave function, $\psi_\mathbf{k}^{(+)}(\mathbf{r})$, in the interacting region where $V \neq 0$. In other words, the $\psi_\mathbf{k}^{(+)}(\mathbf{r})$ outside the interacting region is not needed to determine the scattering properties. This is why we are allowed to add the absorbing potential, $-i\varepsilon(\mathbf{r})$, to the Hamiltonian as long as the $\psi_\mathbf{k}^{(+)}(\mathbf{r})$ in the interacting region is correctly described.

In this paper, we present examples of the ABC method. The method well works in the time-dependent and time-independent approaches of mean-field and few-body models.

APPLICATION OF ABSORBING BOUNDARY CONDITION (ABC)

Time-dependent-Hartree-Fock (TDHF) calculation in the 3D coordinate space

In studies of giant resonances, effects of the continuum has been treated in the random-phase approximation (RPA) with Green's function in the coordinate space [7]. However, it is very difficult to directly apply the method to deformed nuclei because construction of the Green's function becomes a difficult task for the multi-dimensional space. We have shown that the ABC method is very useful to treat the electronic continuum in deformed systems, such as molecules and clusters [1]. We have also investigated the applicability of the ABC in studies of nuclear response calculations [2, 3]. In this section, we discuss properties of the giant dipole resonance (GDR) in nuclei at the $N = Z$ line, ^{8}Be, and at the neutron drip line, ^{14}Be.

We use the ABC in the time-dependent Hartree-Fock (TDHF) calculations on a three-dimensional (3D) coordinate grid. The time evolution of the TDHF state, $\det\{\phi_i(\mathbf{r},t)\}$, is computed by applying the electric dipole ($E1$) field to the Hartree-Fock (HF) ground state:

$$i\frac{\partial}{\partial t}\phi_i(\mathbf{r},t) = \left(-\frac{1}{2m}\nabla^2 + V_{\text{HF}}(\mathbf{r},t) + V_{\text{ext}}(\mathbf{r},t) - i\varepsilon(\mathbf{r})\right)\phi_i(\mathbf{r},t), \quad i = 1 \sim A, \quad (4)$$

$$V_{\text{ext}}(\mathbf{r},t) = kr_\alpha \left\{\frac{1}{2}(1-\tau_z)e - \frac{Ze}{A}\right\}\delta(t), \quad r_\alpha = x,y,z, \quad (5)$$

where k should be small enough to validate the linear response approximation. We calculate the expectation values of the $E1$ operator as a function of time, then Fourier transforming to get the energy response [1, 2, 3]. Since all frequencies are contained in the initial perturbation, the entire energy response can be calculated with a single time evolution.

We use the Skyrme energy functional of EV8 [8] with the SGII parameter set. For the time evolution of the TDHF state, we follow the standard prescription [9]. The model space is a sphere whose radius is 22 fm. The $i\varepsilon(r)$ is zero in a region of $r < 10$ fm, while it is non-zero at $r > 10$ fm. The TDHF single-particle wave functions are discretized on a rectangular mesh in a 3D real space. Now, we can solve Eq. (4) together with the vanishing boundary condition at $r = 22$ fm. Time evolution is carried out up to $T = 30$ $\hbar$/MeV.

The density distribution of the ground states of ^{8}Be and ^{14}Be both possess a prolate superdeformed shape. The $E1$ oscillator strengths are shown in Fig. 1. Here, we use a smoothing parameter of $\Gamma = 0.5$ MeV. The deformation splitting of the GDR peak is as large as about 10 MeV. Since the ground-state shape is almost identical between the two isotopes, average peak positions are similar. However, the peak width is very different between the isotopes. In ^{14}Be at the neutron drip line, the double-peak structure is almost smeared out in the total strength (thick line). On the other hand, we observe prominent double-peak structure in ^{8}Be. The significant peak broadening in ^{14}Be may attribute to the small neutron separation energy and the Landau damping.

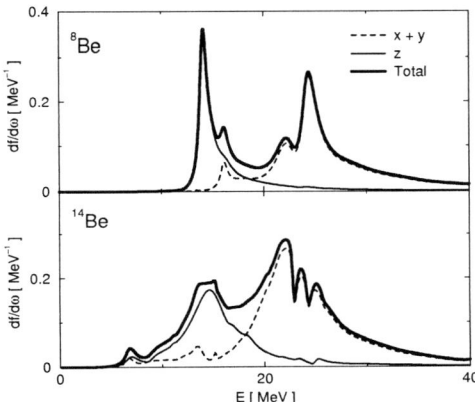

FIGURE 1. Calculated $E1$ strength in ^{8,14}Be. Thin solid and dashed lines indicate the response to dipole fields parallel and perpendicular to the symmetry axis, respectively. Thick line shows the total strength.

Calculation of breakup reaction in the coordinate space

Let us consider a reaction of a projectile, composed of core (C) plus neutron (n), on a target nucleus (T). Denoting the projectile-target relative coordinates by $\mathbf{R}$ and the neutron-core relative coordinates by $\mathbf{r}$, the Hamiltonian of this three-body system with the ABC is expressed as

$$H = -\frac{\hbar^2}{2\mu}\nabla_\mathbf{R}^2 - \frac{\hbar^2}{2m}\nabla_\mathbf{r}^2 + V_{nC}(\mathbf{r}) + V_{nT}(\mathbf{r}_{nT}) + V_{CT}(\mathbf{R}_{CT}) - i\varepsilon(\mathbf{R},\mathbf{r}) \quad (6)$$

where μ and m are the reduced masses of projectile-target relative motion and neutron-core relative motion, respectively. V_{nC}, V_{nT}, V_{CT} are the interaction potentials of constituent particles. The $i\varepsilon$ vanishes in the region where the nuclear interactions are active.

The wave function may be expressed as a sum of the Coulomb wave, $\psi^{(+)}(\mathbf{R})$, in the incident channel and the scattered wave.

$$\Psi^{(+)}(\mathbf{R},\mathbf{r}) = \psi^{(+)}(\mathbf{R})\phi_0(\mathbf{r}) + \Psi_{\text{scat}}(\mathbf{R},\mathbf{r}) \quad (7)$$

where $\phi_0(\mathbf{r})$ is the ground state of the projectile, described as a n-C bound state. The Ψ_{scat} satisfies the following inhomogeneous equation,

$$\{E + e_0 - H\}\Psi_{\text{scat}}(\mathbf{R},\mathbf{r}) = \{V_{nT}(\mathbf{r}_{nT}) + V_{CT}(\mathbf{R}_{CT}) - V_C(\mathbf{R})\}\psi^{(+)}(\mathbf{R})\phi_0(\mathbf{r}), \qquad (8)$$

where V_C is the Coulomb distorting potential, E is the bombarding energy, and e_0 is the ground-state energy of the projectile. One should note that the right hand side of Eq. (8) is a localized function if we can neglect the difference between V_{CT} and V_C at large R. In [5], we have studied a deuteron breakup reaction and compared our results with those of the continuum discretized coupled channel (CDCC) calculation. Readers may refer to [5] for numerical details. In this paper, we report the application to a nuclear breakup reaction of ^{11}Be on ^{12}C.

The ^{10}Be-n potential is taken as a Woods-Saxon shape whose depth is set so as to produce the 2s orbital binding energy. We adopt the optical potentials for ^{10}Be-^{12}C and for n-^{12}C. The radial region up to 30 fm and 50 fm are used for R and r, respectively. The $i\varepsilon(R,r)$ is non-zero in the region of 20 fm $< R <$ 30 fm and in that of 25 fm $< r <$ 50 fm. The n-^{10}Be relative angular momenta are included up to $l = 3$.

In the left panel of Fig. 2, we show the elastic breakup cross sections of ^{11}Be-^{12}C reaction. The filled circles are the result of the ABC calculation and the open circles for the eikonal calculation. The elastic breakup cross section is substantially larger than that in the eikonal approximation at lower incident energy. The failure of the eikonal approximation is apparent at the incident energy below 50 MeV/A. There, the quantum-mechanical treatment is required for the three-body continuum.

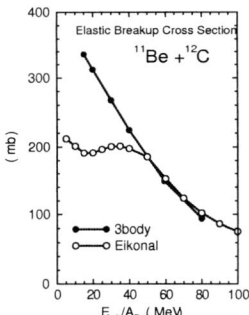
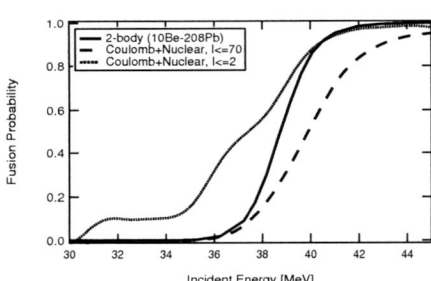

FIGURE 2. (Left) The elastic breakup cross section in ^{11}Be-^{12}C reaction. Quantum calculation with ABC (closed circles) is compared with the eikonal calculation (open).
(Right) Calculated fusion probability for ^{11}Be+^{208}Pb as a function of the projectile-target incident energy for head-on case ($J=0$). Different truncation for n-C partial waves is taken for a dotted line ($l \leq 2$) and for a dashed line ($l \leq 70$). The solid line is the fusion probability for ^{10}Be+^{208}Pb without a halo neutron.

Time-dependent-wave-packet calculation for fusion probability

So far, we have used the ABC to simulate the outgoing boundary condition, $i\varepsilon$ being non-zero outside the interacting region. In this section, we set the $i\varepsilon$ non-zero inside the

Coulomb barrier, to simulate the incoming boundary condition.

We consider a fusion reaction of ^{11}Be on ^{208}Pb near the Coulomb barrier energy. Again, ^{11}Be is described as a weakly-bound system of a neutron and ^{10}Be core. The target is now ^{208}Pb. All the nuclear potentials are taken to be real and of the Woods-Saxon type. The $i\varepsilon(R_{CT})$ is also of the Woods-Saxon shape and corresponds to the imaginary part of the C-T optical potential which describes the fusion between core and target. See [10] for details. The time-dependent Schrödinger equation,

$$i\frac{\partial}{\partial t}\Psi(\mathbf{R},\mathbf{r},t) = H\Psi(\mathbf{R},\mathbf{r},t), \qquad (9)$$

where H is given by Eq. (6), is solved with an initial wave function

$$\Psi(\mathbf{R},\mathbf{r},t=0) = \exp(-iKR)\Phi_0(\mathbf{R})\phi_0(\mathbf{r}), \qquad (10)$$

where $\Phi_0(\mathbf{R})$ is a Gaussian wave packet. The initial wave packet, $\Phi_0(\mathbf{R})\phi_0(\mathbf{r})$, exists in a region beyond the range of nuclear interaction of V_{nT} and V_{CT}. The wave packet is going to approach to the target, then if the ^{10}Be core penetrates the Coulomb barrier, the wave function disappears because of absorption by the imaginary potential $i\varepsilon(R_{CT})$. The ratio of the flux loss against the initial flux gives the fusion probability. In this calculation, since the final destination of neutron is irrelevant, the calculated fusion probability contains both complete and incomplete fusions.

The energy projection [10] leads to the fusion probability shown in Fig. 2 (right). When we include the n-C partial angular momenta only up to $l = 2$, the result indicates a strong fusion enhancement at sub-barrier energies, compared to the two-body fusion calculation of ^{10}Be and ^{208}Pb. However, when we enlarge our model space to include $l \leq 70$, this enhancement disappears and we even have suppression. This is consistent with similar study for fusion suppression obtained for ^{11}Be and ^{40}Ca [11]. The results suggest necessity of high partial waves for the n-C relative motion to describe correct dynamics of neutron breakup and transfer under the strong Coulomb field.

REFERENCES

1. Nakatsukasa, T., and Yabana, K., *J. Chem. Phys.*, **114**, 2550–2561 (2001).
2. Nakatsukasa, T., and Yabana, K., "Resonance and continuum states in weakly-bound systems," in *Proceedings of the 7th International Spring Seminar on Nuclear Physics: Challenges of Nuclear Structure*, edited by A. Covello, World Scientific, Singapore, 2002, pp. 91–98.
3. Nakatsukasa, T., and Yabana, K., *Prog. Theor. Phys. Suppl.*, **146**, 447–451 (2002).
4. Yabana, K., Ueda, M., and Nakatsukasa, T., *Prog. Theor. Phys. Suppl.*, **146**, 329–337 (2002).
5. Ueda, M., Yabana, K., and Nakatsukasa, T., *Phys. Rev. C*, **67**, 014606 (9 pages) (2002).
6. Seideman, T., and Miller, W., *J. Chem. Phys.*, **97**, 2499–25124 (1992).
7. Shlomo, S., and Bertsch, G., *Nucl. Phys. A*, **243**, 507–518 (1975).
8. P. Bonche, P. H., H. Flocard, *Nucl. Phys. A*, **467**, 115–135 (1987).
9. H. Flocard, M. W., S.E. Koonin, *Phys. Rev. C*, **17**, 1682–1699 (1978).
10. Yabana, K., Ueda, M., and Nakatsukasa, T., Time-dependent wave-packet approach for fusion reactions of halo nuclei (2003), to be published in Nucl. Phys. A; Preprint nucl-th/0301053.
11. Yabana, K., *Prog. Theor. Phys.*, **97**, 437–450 (1997).

Spectroscopy of very heavy elements

Rauno Julin

Department of Physics, PB 35 (YFL), FIN-40014 University of Jyväskylä, Finland

Jyväskylä-Liverpool-GSI-Saclay-Argonne-Münich-Helsinki collaboration

Abstract. Data from in-beam spectroscopic experiments using recoil gating and recoil-decay-tagging (RDT) techniques carried out at the Jyväskylä Accelerator Laboratory (JYFL) for ^{254}No, ^{252}No, ^{250}Fm and ^{255}Lr are presented and discussed.

Introduction

Coulomb energy of the heavy nuclei with $Z > 100$ is so large that in the liquid drop picture these nuclei should be unstable against spontaneous fission. However, it has been shown theoretically that the nuclear shell correction energy is large enough for creating an island of spherical superheavy elements around $Z = 114, N = 184$. The stability of known α-decaying nuclei with $Z > 100$ is supposed to originate from the shell effects in a deformed nucleus. It is important to verify the predicted deformation and moment of inertia as well as alignments in these nuclei experimentally.

Properties of superheavy elements can be calculated using the Nilsson–Strutinsky shell-correction approach. The success of such approaches is closely related to the knowledge of the single-particle levels near the Fermi surface. Current theoretical models give different predictions of the proton and neutron magic numbers beyond $Z = 82$ and $N = 126$. Spectroscopic studies of the heaviest odd-mass nuclei are therefore of importance in order to determine the ordering and energies of the single-particle levels in this region.

Most of the scarce experimental information on the structure of heaviest nuclei has been obtained in α-decay studies [1]. Important information from single-particle properties has been obtained by using transfer reactions [2]. Excited states up to $I^{\pi} = 8^+$ in ^{256}Fm have been seen in the β decay of the isomeric 8^+ state of ^{256}Es [3]. Coulomb excitation has been used to populate excited states of ^{248}Cm up to a 5.1 MeV 30^+ state [3].

In the present contribution, results from in-beam γ-ray spectroscopic studies of heavy actinide nuclei, ^{254}No [4], ^{252}No [5], ^{250}Fm and ^{255}Lr produced in cold fusion-evaporation reactions are presented. An in-beam electron spectroscopic study of ^{254}No [6] will also be discussed. The experiments were carried out in the Accelerator Laboratory of the Department of Physics of the University of Jyväskylä (JYFL).

Production cross-sections

The small production cross-sections make any kind of detailed spectroscopic studies of heavy elements extremely difficult. They are produced with available stable beams and targets in heavy-ion induced fusion-evaporation reactions. Due to fission, the production rates decrease rapidly with the proton number of the compound system, being down to 10 nb for example for the ^{40}Ar + ^{208}Pb reactions.

However, by using the doubly magic projectile ^{48}Ca and Pb, Bi or Hg targets, exceptionally high cross-sections of cold fusion-evaporation reactions are obtained. Especially the fusion of two doubly magic nuclei in the ^{208}Pb(^{48}Ca,2n)^{254}No reaction leads to an anomalously high production cross-section of about 2 μb providing a unique opportunity for an in-beam experiment on a transfermium nuclide. In the similar ^{48}Ca induced reactions on the ^{204}Hg, ^{206}Pb and ^{209}Bi targets, ^{250}Fm, ^{252}No and ^{255}Lr are produced with cross-sections of 300 – 1000 nb. For the conventional in-beam spectroscopic measurements these yields are still far too low but are above the limit of sensitivity obtained in recent in-beam experiments, where the novel Recoil-Decay-Tagging (RDT) method has been employed.

RDT measurements at RITU

The gas-filled recoil separator RITU (Recoil Ion Transport Unit)(Fig. 1) has been designed to separate residues of fusion-evaporation reactions from beam particles and other reaction products, especially fission [7]. The position sensitivity of the focal plane Si detector of RITU enabled the recoils to be correlated with their subsequent particle decay (so far α decay). Very recently in 2003 a new sophisticated focal plane detector system GREAT consisting of segmented Si and Ge detectors has been commissioned at RITU.

In the in-beam γ-ray experiments at RITU the Jurosphere and SARI arrays and recently in 2003 a JuroGam array have been used to detect prompt γ rays at the target area. The Jurosphere array consisted of 25 Compton-suppressed Ge-detectors (15 Eurogam Phase 1, 10 Nordball and/or TESSA detectors) and had a photo-peak efficiency of 1.5 – 1.7% for 1.3 MeV γ rays. The SARI array was a combination of four Ge-clover detectors (without BGO shields) in a close geometry. The photo-peak efficiency of SARI was 1.7 % when operated in add-back mode. The JuroGam array commissioned in April 2003 consists of 43 Eurogam Phase 1 type of Compton suppressed detectors and has an efficiency of 4 % at 1.3 MeV.

The SACRED spectrometer was used to obtain in-beam electron spectra from heavy nuclei in RDT measurements at RITU (Fig. 2) [8]. In SACRED, electrons emitted from the target into backward angles are guided by the solenoid field and distributed over a Si detector (diameter 2 cm) which is divided into 25 independent pixels enabling to detect $e^- - e^-$ coincidences from a cascade of converted transitions.

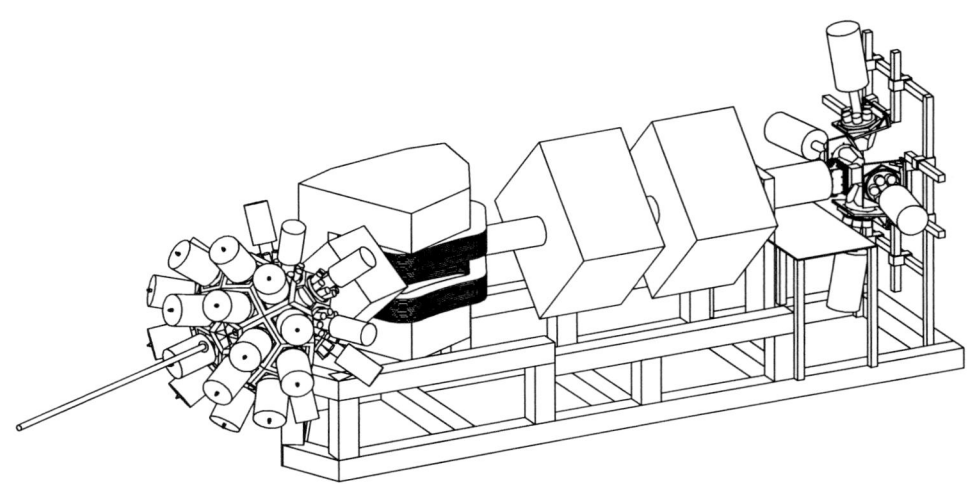

FIGURE 1. The RITU separator combined with a Ge-detector array around the target

The SACRED Electron Spectrometer

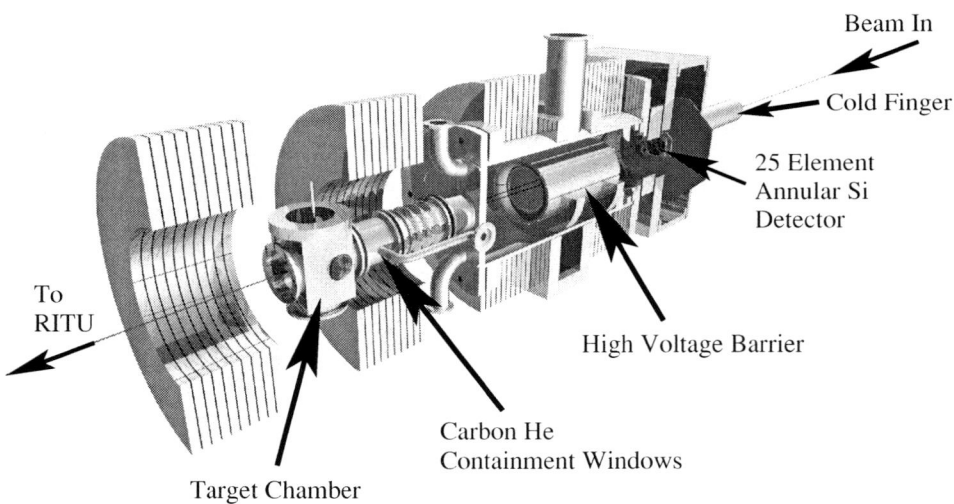

FIGURE 2. The SACRED electron spectrometer designed for the use in conjunction with the RITU gas-filled recoil separator

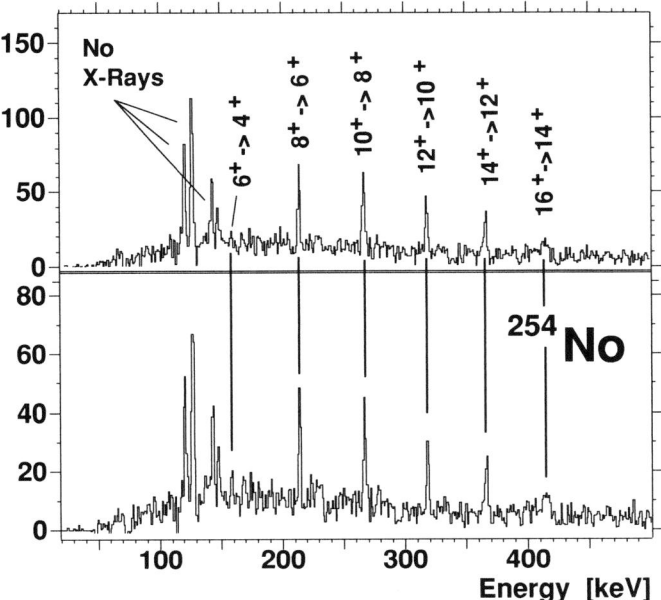

FIGURE 3. A singles γ-ray spectrum (upper) from the ^{48}Ca + ^{208}Pb reactions gated with fusion-evaporation residues. The same γ-ray spectrum but in addition, tagged with the ^{254}No α decays (lower). The SARI 4 clover-detector array was used.

Prompt gamma rays from ^{254}No, ^{252}No and ^{250}Fm

The first RDT measurement for ^{254}No at RITU was carried out by using the ^{208}Pb(^{48}Ca,2n)^{254}No reaction and an array of 4 Clover detectors (SARI array) at the target area [4]. A resulting γ-ray spectrum gated with fusion products detected at the RITU focal plane is shown in the upper panel of Fig. 3. The lower panel of Fig. 3 shows a γ-ray spectrum in coincidence with fusion products identified as ^{254}No nuclei on the basis of recoil-α correlations revealing that basically the only open evaporation channel is the 2n-channel.

In addition to the No X-rays, transitions having energies of 158.9, 214.1, 267.2, 318.2, 266.5 and 414.0 keV were observed and assigned to originate from ^{254}No. The first five of these transitions were observed, for the first time, in a similar tagging experiment at ANL [9]. The pattern of the γ-ray peaks in the spectra reveals that the corresponding transitions form a cascade, obviously of E2 transtions in ^{254}No. The spin assignments are based on a fit of the kinematic moment of inertia. The γ-ray transitions from the states with $I < 6$ are not seen due to their internal conversion.

This ^{254}No experiment was very recently repeated by employing the 43 Ge-detector JuroGam array. A preliminary recoil gated spectrum is shown in Fig. 4. The yrast line is seen up to $I = 20$ and there are peaks in the spectrum which must represent depopulation of sidebands in ^{254}No. The analysis of these data is in progress.

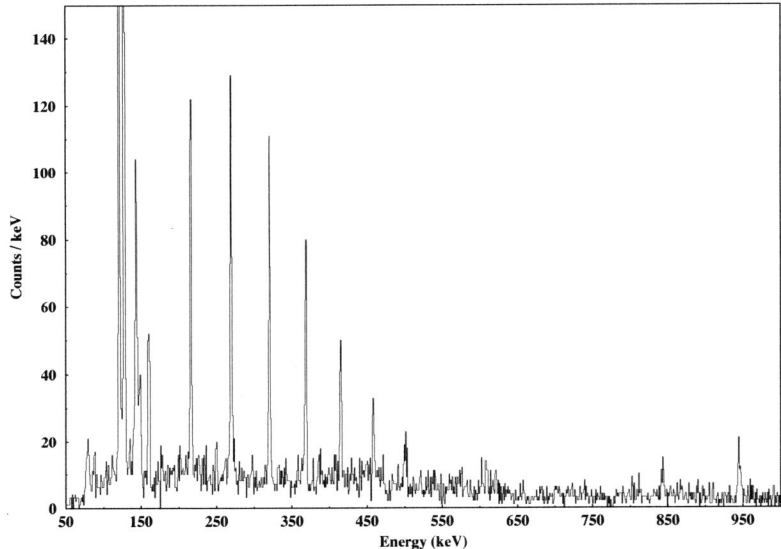

FIGURE 4. A preliminary recoil gated γ-ray spectrum for ^{254}No detected by the JuroGam array at RITU.

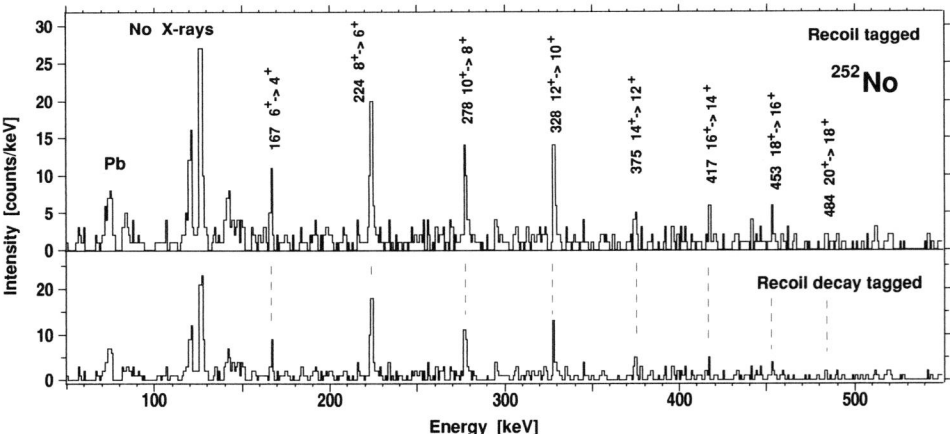

FIGURE 5. A recoil gated singles γ-ray spectrum (upper) from the ^{48}Ca + ^{206}Pb reactions. The same spectrum but, in addition, tagged with the ^{252}No α decays (lower).

An experiment similar to that for ^{254}No was carried out for ^{252}No by using the Jurosphere array and the ^{206}Pb(^{48}Ca,2n)^{252}No reaction [5]. The resulting recoil gated and RDT γ-ray spectra are shown in Fig. 5. The quality of these spectra is much better than of those for ^{254}No in Fig. 3, albeit the reaction cross-section is only 300 nb. This is because of the lack of Compton suppression of the SARI array.

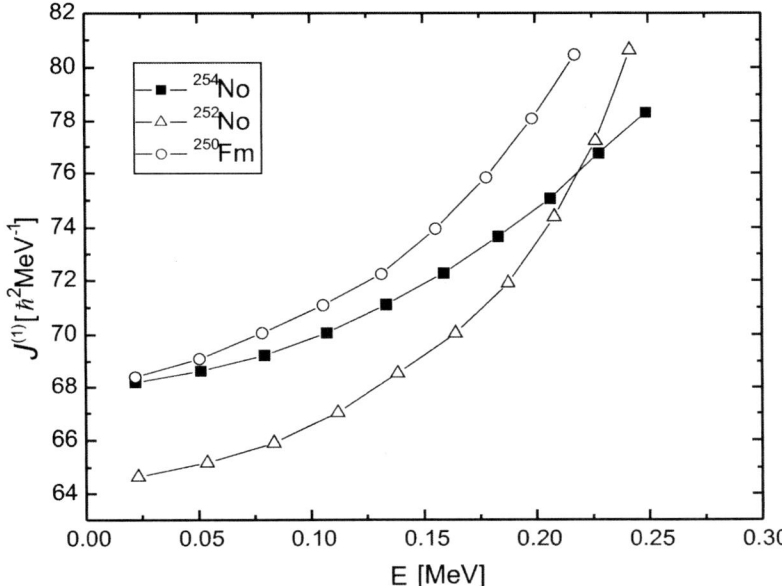

FIGURE 6. Kinematic moments of inertia for ^{250}Fm, ^{252}No and ^{254}No extracted from the measured γ-ray energies

The Jurosphere array was further employed in an RDT experiment to collect γ rays from the ^{204}Hg(^{48}Ca,2n)^{250}Fm reaction. Spectra similar to those for ^{252}No were obtained for ^{250}Fm.

The observation of discrete γ-ray lines of a rotational cascade of transitions up to $I = 20$ in ^{254}No, ^{252}No and ^{250}Fm reveals that these trans-fermium nuclei are deformed and can in rotation compete against fission up to at least that spin. The kinematic moment of inertia values for these nuclei derived from the observed transition energies are about half of the rigid rotor value and are slightly increasing with spin (Fig. 6), obviously due to gradual alignment of quasi-particles. For ^{252}No the extracted values increase more rapidly at high spin indicating a more dramatic alignment of quasi-particles. The kinematic moment of inertia values for ^{250}Fm are almost identical to the ^{254}No ones at low spin but then follow the alignment pattern of ^{252}No at higher spin.

It is possible to extract the ground state deformation parameter β_2 from the extrapolated energy of the 2_1^+ state using global systematics [10, 11]. The value we derived for ^{254}No is $\beta_2 \approx 0.27$, which is in good agreement with the values calculated using the macroscopic-microscopic method [12, 13]. A β_2 value similar to ^{254}No is obtained for ^{250}Fm. The extracted value of $\beta_2 = 0.26$ for ^{252}No indicated that ^{252}No is less deformed than ^{254}No and ^{250}Fm.

In recent relativistic Hartree–Bogoliubov calculations the strength of the Gogny force in the pairing channel had to be decreased compared to the lighter nuclei, in order to reproduce the moment of inertia of ^{254}No [14].

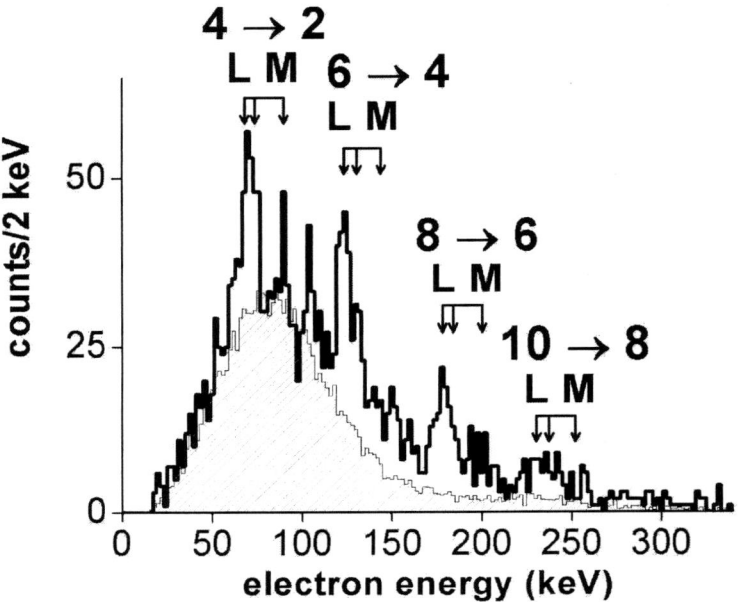

FIGURE 7. A conversion-electron spectrum tagged by ^{254}No recoils. The hashed area shows a simulated spectrum of electrons from M1 transitions of high K bands in ^{254}No [6].

Conversion electrons from ^{254}No

The SACRED conversion-electron spectrometer was used to measure prompt conversion electrons from the ^{208}Pb(^{48}Ca,2n)^{254}No reactions. In a resulting recoil gated spectrum shown in Fig. 7, electron peaks originating from transitions between the low-spin yrast states in ^{254}No are seen. In a careful analysis of the prompt recoil-gated electron-electron coincidence spectra it was found out that the broad distribution under these electron peaks is not due to random events but consists of high-multiplicity events, obviously originating from cascades of highly converted M1 transitions within rotational bands built on high-K states in ^{254}No [6].

^{255}Lr experiment

An RDT experiment similar to that for the even-mass No nuclei was carried out to study excited states in ^{255}Lr. The ^{209}Bi(^{48}Ca,2n)^{255}Lr reaction at a bombarding energy of 221 MeV was used. About 2200 correlated recoil-α pairs were observed corresponding to a reaction cross-section of about 300 nb. In the prompt recoil-gated and α-tagged γ-ray spectra, no clear candidate for γ-ray transitions in ^{255}Lr was observed. However, the observation of strong Lr X-ray peaks in these spectra implies that the yrast transitions in ^{255}Lr are strongly converted. Consequently, when comparing the ^{255}Lr spectra with the ^{252,254}No spectra, it is obvious that the decay along the yrast line of ^{255}Lr does not proceed via a cascade of E2 transitions but rather via a cascade of M1 transitions. This decay might well be more complicated depending on the coupling of the odd proton i.e. on the orbital the odd proton occupies and on the related g-factor.

Recent developments

In the in-beam experiments the maximum current of the ^{48}Ca beam on the target is limited by the counting rate of the Ge detectors. A large contribution of this rate has been due to scattering of the beam in the He gas and the gas window situated upstream of the target. For suppressing this background a differential pumping system near the target is now available for the JuroGam experiments going on.

As in an α decay excited states of the final nuclei are also fed and as there are long-living isomeric states expected in heavy nuclei, a sophisticated focal plane spectrometer is needed at RITU. For these purposes a novel GREAT spectrometer system has been designed by the UK physicists and is now available for the 2003 campaign.

For E2 transitions below 230 keV and M1 transitions below 440 keV, internal electron conversion dominates over the γ-ray emission in nuclei with $Z \approx 102$. Therefore, especially in the study of odd-mass heavy nuclei detection systems for combining in-beam conversion-electron spectroscopie and γ-ray spectroscopie methods should be developed.

ACKNOWLEDGMENTS

This work is supported by EPSRC (UK), the EU 5th framework IHP – Access to Research Infrastructure (HPRI-CT-1999-00044) and IHP – RTD (HPRI-CT-1999-50017) programmes and the Academy of Finland under the Finnish Centre of Excellence Programme 2000-2005.

REFERENCES

1. S. Hofmann, *Rep. Prog. Phys.* **61** (1998) 639.
2. I. Ahmad et al., *Phys. Rev. Lett.* **39** (1977) 12.
3. M. R. Schmorak, *Nucl. Data Sheets* **57** (1989) 515.
4. M. Leino et al., *Eur. Phys. J.* **A6** (1999) 63.
5. R.-D. Herzberg et al., *Phys. Rev.* **C65** (2001) 014303.
6. P. A. Butler et al., *Phys. Rev. Lett.* **89** (2002) 202501.
7. M. Leino et al., *Nucl. Instr. Meth.* **B99** (1995) 653.
8. P. A. Butler et al., *Nucl. Instr. Meth.* **A381** (1996) 433.
9. P. Reiter et al., *Phys. Rev. Lett.* **82** (1999) 509.
10. L. Grodzins, *Phys. Lett.* **2** (1962) 88.
11. S. Raman et al., *At. Data Nucl. Data Tables* **42** (1989) 1.
12. Z. Patyk et al., *Nucl. Phys.* **A533** (1991) 132.
13. S. Cwiok et al., *Nucl. Phys.* **A573** (1994) 356.
14. A. V. Afanasjev et al., *Phys, Rev.* **C67** (2003) 024309.

Proton Emitter Studies using the Argonne Fragment Mass Analyzer

PJ Woods

School of Physics, The University of Edinburgh, Edinburgh EH9 3JZ, UK

Abstract. The paper reports on experimental developments for studies of decays of proton radioactive nuclei on the Argonne Fragment Mass Analyzer. The results illustrate that there is a clear proton decay rate sensitivity to the unpaired neutron orbital in highly deformed odd-odd nuclei consistent with the Influential Spectator Model of Ferreira and Maglione. No such sensitivity is observed for the proton decay rates of spherical odd-odd nuclei studied here. The first ever study of proton decay from a nucleus, ^{135}Tb, produced via 1p6n fusion evaporation is also reported.

INTRODUCTION

Many new proton emitters have been discovered in the last decade [1]. In particular the discovery of proton decay from the highly deformed rare earth nuclei ^{131}Eu and ^{141}Ho [2] has provoked great theoretical and experimental efforts. The highly deformed nature of ^{131}Eu was confirmed with the first observation of proton decay fine structure [3], and the Recoil Decay Tagging (RDT) technique [4] was used to demonstrate the more modest deformation of ^{141}Ho [5, 6]. These results, combined with a successful modelling of proton decay spectroscopic factors in near spherical nuclei [7, 8], represented a significant broadening and deepening of our knowledge of proton emitters and proton emission. The present paper will report on some of the more recent experimental developments, and proton radioactivity measurements performed at Argonne using the Fragment Mass Analyzer (FMA). The results of RDT studies performed at Argonne on proton emitters are reported in [9].

EXPERIMENTAL DEVELOPMENTS ON THE FMA

The development of a new intense ECR ion source for the ATLAS accelerator facility encouraged us to explore the sensitivity limits of our experimental set-up. In particular, we aimed to increase our sensitivity to small cross-sections (eg 1p5n and 1p6n fusion evaporation channels) in order to access the most remote regions of the proton drip-line. Such an approach has the added benefits of increasing precision where measurements are required to discriminate between different theoretical approaches, and enables more detailed features to be revealed, such as isomers, or fine structure.

A new design of double-sided silicon strip detector (DSSD) was developed to encompass most of the region behind the focal plane of the FMA while maintaining

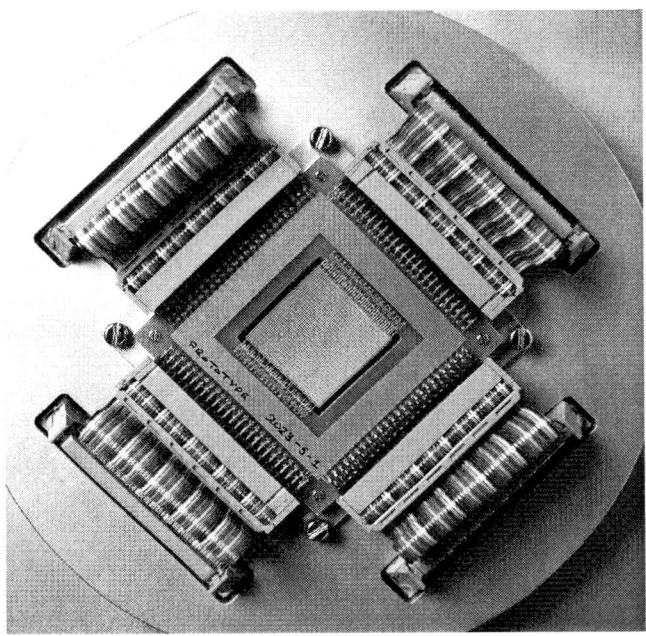

FIGURE 1. The new 80*80 DSSD used behind the focal plane of the Argonne FMA for proton decay studies.

a high granularity for recoil-decay correlation performance. The detector consists of 80*80 strips with a pitch of 400μm (see figure 1) and is consequently larger and more segmented than the original prototype DSSD developed for the Daresbury Recoil Separator [10].
The much higher recoil detection efficiency is particularly valuable for RDT studies where experiments are rate limited by forward angle Ge detector rates. The higher granularity means higher implantation rates can be tolerated for proton decay searches.

The great variation in energy and beam species available from the ATLAS accelerator are a major benefit for our experimental programme. In particular it enables us to choose inverse kinematics to access the nucleus of interest. This can significantly increase the recoil transmission efficiency through the FMA. Inverse kinematics are particularly useful for RDT experiments since the target position has to be displaced upstream of the FMA in order to accommodate Gammasphere - this compensates for what otherwise would be a significant loss in sensitivity. The downside of inverse kinematics is increased levels of scattered beam at the focal plane of the FMA. Following the approach developed at JAERI, a split anode was constructed for the first electric dipole of the FMA. This was successfully commissioned in 2002 resulting in significant reductions in scattered beam rates ensuring that even in extreme kinematical scenarios the recoil rate is dominated by evaporation residues.

The standard electronic response of the Edinburgh/RAL pre-amplifier shaping amplifier system [11] used to instrument the DSSD is limited to decays occurring $\geq 6\mu s$ after heavy ion implantation onto the same strip. This compares with a recoil flight time $\leq 1\mu s$. While in most cases of proton decay this is not a key issue, there are instances where improved detection efficiency for a few short-lived ($t_{1/2} \sim \mu s$) proton emitters can significantly enhance a measurement [12]. A delay line shaping system [13] was successfully introduced to instrument all 160 strips of the DSSD in parallel with the existing instrumentation system. Proton decays from ^{113}Cs could be seen down to $\sim 1.0 \mu s$ following implantation. This system was then used to make a high precision half-life measurement of the short-lived isomeric proton transition from ^{150m}Lu [14, 15]. The energy resolution of the delay line system was found to be comparable, but slightly inferior, to that obtained with the Edinburgh/RAL electronics system. The delay line system gives at least equivalent performance to the digital signal processing approach [16] in this application. The delay line system has a significantly lower per channel cost, requires no specialist expertise for setting-up, and there is no limit imposed on the search time range. Our experiments now utilise the delay line and Edinburgh/RAL systems in parallel for search experiments where the energy and half-life of the proton decay is unknown.

PROTON DECAY MEASUREMENTS ON THE FMA

^{117}La

Evidence was reported by Soramel et al. for the ground state proton decay of ^{117}La [17]. Subsequently an additional high spin isomeric transition was reported from the same experiment [18]. We made a higher precision measurement of the proton decay of ^{117}La and confirmed the existence of the ground state transition indicating a high degree of prolate deformation, $\beta \sim 0.3$, with the decay occurring from either a $3/2^+$ or $3/2^-$ configuration [19]. However, despite the better statistics and experimental running conditions, no evidence was found for the reported proton decay of a $9/2^+$ isomer.

^{130}Eu

The nucleus ^{131}Eu is currently thought to be the most deformed known proton emitter. Recently, a theoretical framework has been developed for proton decays from highly deformed odd-odd nuclei by Ferreira and Maglione [20]. This demonstrates that the proton decay rate can be affected by the nature and orientation of the neutron orbit even though it remains inert during the transition. With the permission of the originators one may refer to this as being the Influential Spectator Model (ISM). ^{130}Eu is predicted to have the same proton configuration and high deformation as its neighbour ^{131}Eu [21]

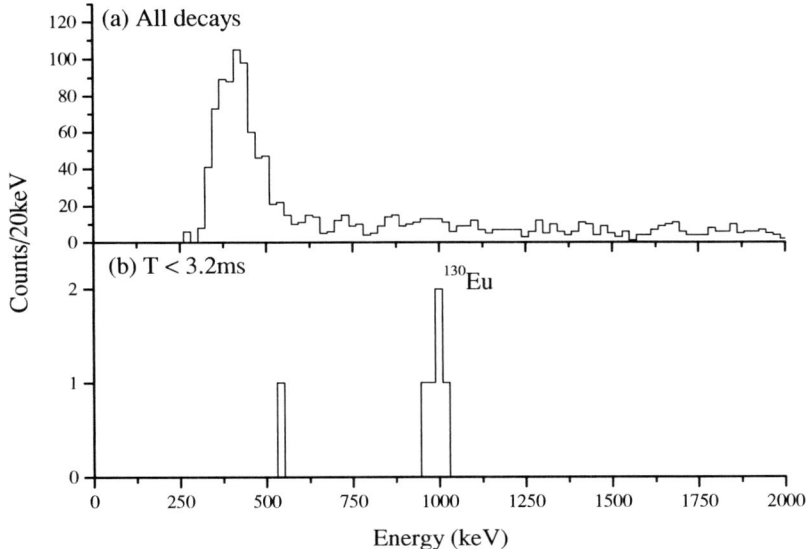

FIGURE 2. (a) All decay events from the reaction ^{78}Kr + ^{58}Ni, restricted by slits to A = 130 recoils. (b) Decay events occurring within 3.2 ms of an implant in the same DSSD pixel. The proton decay peak of ^{130}Eu is clearly visible.

and is an ideal candidate to explore this theoretical model. Figure 2 shows a proton decay energy spectrum indicating the existence of ^{130}Eu produced via a 1p5n fusion evaporation channel [22].

Figure 3 shows a comparison of the measured half-life with calculations performed by Davids based on the ISM. The $K_p=3/2^+$ proton orbital can couple with either the $K_n=1/2^+$ or $7/2^-$ neutron orbitals. It can be seen that the $7/2^-$ neutron orbital is not in agreement with theory but both K = 1^+ and 2^+ states produced by coupling to the $K_n=1/2^+$ orbital are in good agreement with theoretical predictions. Ironically, the proton decay rate gives information on the neutron Fermi surface beyond the proton drip-line even though the neutron itself remains undisturbed in the transition. This is a further illustration of the unique physics insights that can be obtained through proton decay studies.

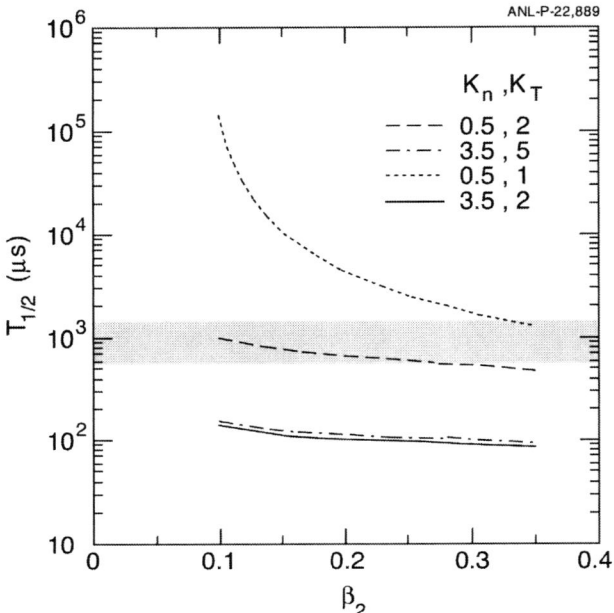

FIGURE 3. The measured proton decay half-life of ^{130}Eu compared with calculations for proton emission from deformed odd-odd nuclei

^{135}Tb

^{135}Tb was produced by bombarding a ^{92}Mo target with ^{50}Cr beams. Figure 4 shows clear evidence for the proton decay of ^{135}Tb representing the first example of a proton decaying isotope produced via the 1p6n evaporation channel [23]. Figure 5 shows a comparison of the measured half-life with different proton orbital configurations assuming a high prolate deformation. It can be seen that good agreement can be obtained for a $7/2^-$ proton configuration. The success of this experiment suggests it will be possible to access all of the proton emitting elements between Z=51-83 with the present experimental set-up.

^{164m}Ir and ^{170m}Au

Proton decays were observed from ^{164m}Ir and ^{170m}Au by bombarding ^{92}Mo and ^{96}Ru targets with ^{78}Kr beams. In each case the proton emitter was produced via a 1p3n evaporation channel. The proton half-lives were in both cases consistent with emission from $h_{11/2}$ orbitals, and the spectroscopic factors were in excellent agreement with low seniority shell model calculations [20, 7]. In the case of ^{164m}Ir similar results were also obtained by Kettunen et al. [24]. The results are therefore consistent with a spherical

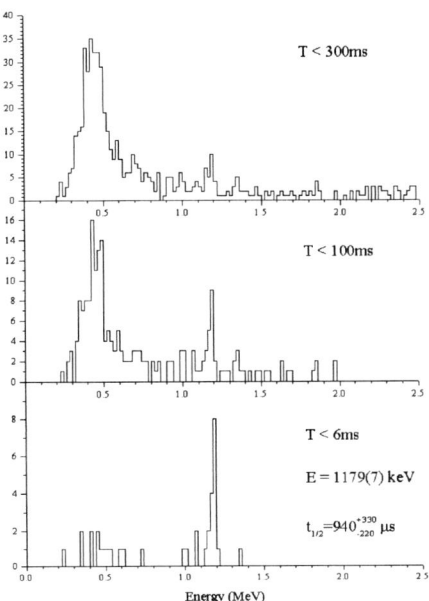

FIGURE 4. These spectra show the clear emergence of the proton decay of ^{135}Tb at short time intervals. This is the first observation of proton decay from a nucleus produced via a 1p6n evaporation channel.

shape and show no clear sensitivity to the neutron configuration.

SUMMARY

A number of changes to the FMA proton decay set-up have been successfully implemented. The increased sensitivity of the system has been used to explore small cross-section and short-lived proton decays. The emphasis in the coming period will be to explore the structure of proton emitters using the RDT technique on the Gammasphere array. As I write this article, exciting new results have been obtained for ^{145,146}Tm including the first observation of γ-decays observed in coincidence with proton radioactivity presented at this Conference [25]. Such experiments entail a intrinsically holistic approach to the study of the structural and decay properties of proton emitters.

ACKNOWLEDGMENTS

The author represents here the proton decay collaboration at Argonne. The measurements referred to have been performed over a period of time and have involved different

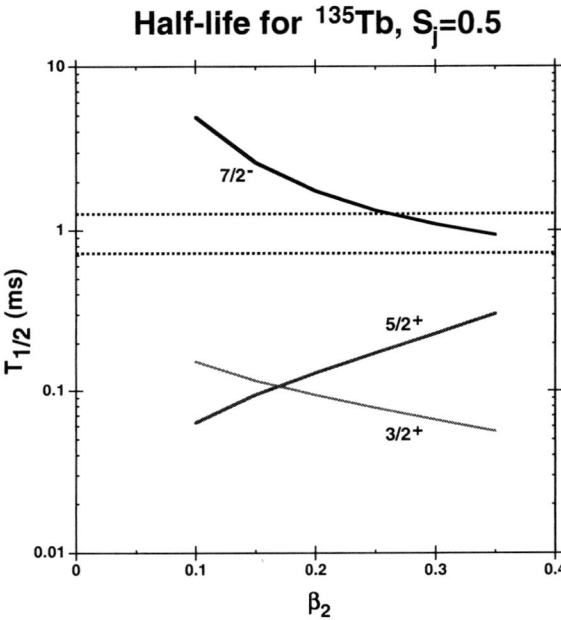

FIGURE 5. A comparison of the measured proton decay half-life of ^{135}Tb with calculations assuming a highly prolate deformed shape.

people, all of whom I wish to thank. In particular I would thank Cary Davids, Darek Seweryniak, Andreas Heinz, Bill Walters, Tom Davinson, Hassan Mahmud, Andrew Robinson and Peter Munroe.

REFERENCES

1. PJ Woods, Procon '99 Proceedings (1999) 34
2. CN Davids et al., PRL 80, (1998) 1849
3. AA Sonzogni et al., PRL 83 (1999) 1116
4. ES Paul et al., PRC 51 (1995) 78
5. D Seweryniak et al., Procon '99 Proceedings (1999) 112
6. D Seweryniak et al., PRL 86 (2001) 1458
7. CN Davids et al., PRC 55, (1997) 2255
8. PJ Woods et al., Annu. Rev. Nucl. Part. Sci. 47 (1997) 541
9. D Seweryniak, Proceedings of Procon 2003, 161, Padova (Ed. E Maglione and F Soramel) AIP Press.
10. PJ Sellin et al., Nucl. Inst. Meth. A311 (1992) 217
11. SL Thomas et al., Nucl. Inst. Meth. A288 (1988) 212
12. M Karny et al., PRL 90 (2003) 012502
13. P Wilt, private communication
14. R Grzywacz et al., Proceedings of Procon 2003, 259, Padova (Ed. E Maglione and F Soramel) AIP Press.
15. T Ginter et al., PRC 61 (2000) 014308

16. A Robinson et al., accepted paper Phys. Rev. C.
17. F Soramel et al., Procon '99 Proceedings (1999) 68
18. F Soramel et al., PRC 63 (2001) 031304(R)
19. H Mahmud et al., PRC 64 (2001) 0131303(R)
20. E Maglione and LS Ferreira, PRC 61 (2000) 047307
21. P Moller et al., At. Data Nucl. Data Tables 66 (1997) 13
22. H Mahmud et al., Eur. Phys. J. A15 (2002) 85
23. CN Davids et al., paper in preparation.
24. H Kettunen et al., Acta. Phys. Pol. B32 (2001) 989
25. D Seweryniak, private communication.

Tracking dissipation in capture reactions

T. Materna[*], V. Bouchat[*], V. Kinnard[*], F. Hanappe[*], O. Dorvaux[†],
C. Schmitt[†¶], L. Stuttgé[†], K. Siwek-Wilczynska[**], Y. Aritomo[‡#],
A. Bogatchev[‡], E. Prokhorova[‡] and M. Ohta[§]

[*]*PNTPM, Université Libre de Bruxelles, Brussels, Belgium*
[†]*Institut de Recherches Subatomiques, Strasbourg, France*
[**]*Warsaw University, Warsaw, Poland*
[‡]*Flerov Laboratory of Nuclear Reactions, Joint Institute for Nuclear Research,
Dubna, Moscow region, Russian Federation*
[§]*Kobe University, Kobe, Japan*
[¶]*present address: GSI, Darmstadt, Germany*
[#]*and Tokyo University, Tokyo, Japan*

Abstract. Nuclear dissipation in capture reactions is investigated using backtracing. Combining the analysis procedure with dynamical models, the difficult and long-standing problem of competition and mixing of quasi-fission and fusion-fission is solved for the first time. At low excitation energy a new protocol able to handle low statistics data gives access to the prescission neutron multiplicity in two different systems ^{48}Ca + ^{208}Pb, Pu. The results are in agreement with a domination of fusion-fission in the case of ^{256}No and an equal mixing of quasi-fission and fusion-fission in the case of Z=114. The nature of the relevant dissipation is determined as one-body dissipation.

INTRODUCTION

Introduced by Kramers [1] very early after the discovery of nuclear fission, the role of the dissipation was only recognized in the 1970s with the emergence of deep inelastic collisions. Its nature, one-body dissipation (OBD) or two-body dissipation (TBD) and its magnitude and evolution with different parameters as the shape and the temperature, is still a matter of controversy.

A lot of experimental data obtained in most of the cases by the observation of the pre- and postscission emission of particles or γ-rays have been devoted to the determination of the dissipation. But until recently, depending on the experiments but also deeply on the models used to extract the dissipation coefficient (Kramer's γ coefficient), a dispersion of the results covering at least two orders of magnitude is observed. Different behaviours are extracted for the evolution with the temperature and no definitive conclusions can be drawn.

Even if coherent experimental data [2] are selected and analysed by the same dynamical model [3], a large spreading of the values for the dissipation coefficient is observed. For example, in figure 1 adapted from reference [4], the evolution with the temperature of the dissipation coefficient is presented for different reactions leading to very different fissioning systems. As pointed out by the authors, on one hand, for true fusion-fission systems ($Z_1Z_2 \ll 1600$), the deduced γ coefficient ranges from 2 to 10 and is clearly compatible with OBD. No particular evolution with the temperature is observed. On the other hand, for systems with $Z_1Z_2 > 1600$, the friction coefficient values are considerably larger and clearly not compatible with OBD. This puzzling behavior for these systems is assumed to be due to TBD or to the expected mixing of mass-symmetric fragments coming from the different reaction mechanisms of capture reactions: quasi-fission and fusion-fission. Indeed, no separation between the two mechanisms was available and only the mean value of the multiplicity for neutron pre- and postscission emission could be obtained using the classical χ^2 minimization.

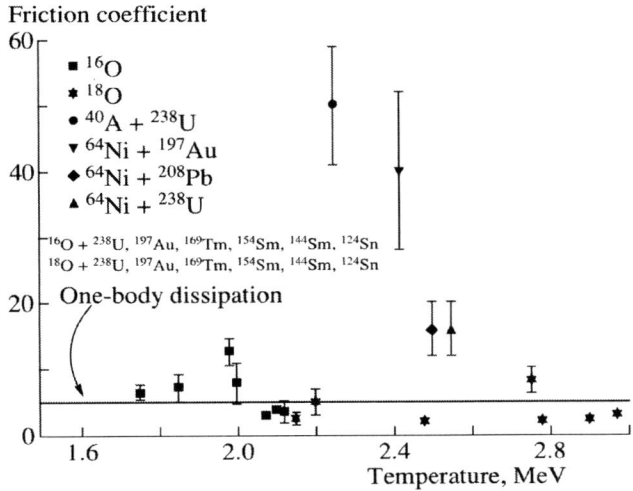

FIGURE 1. Evolution of the dissipation coefficient as a function of the temperature for different reactions. Data from Hinde *et al* [2].

In this report we will show how a powerful analysis protocol, the backtracing [5], which is able to produce not only mean values but also correlations and distributions, can help us to solve this long-standing problem.

Recently applied to the Ni + Pb [6] and Ca + Th [7] reactions leading to isotopes of Z=110, the backtracing procedure allowed us for the first time to clearly disentangle, at least intuitively, the contributions or quasi-fission and fusion-fission in the neutron prescission multiplicity distribution. A complete description of the backtracing procedure and its application to our case is described in references [6,7,8].

A SIMPLE CASE: FUSION-FISSION LEADING TO ^{126}Ba

In order to validate the backtracing application to the determination of the distribution of pre- and postscission neutron multiplicity, we will first present results obtained for the ^{28}Si + ^{98}Mo reaction at 204 MeV.

This system has been investigated [8] at the VIVITRON, IReS, Strasbourg, using, as for all the data presented here, the DEMON neutron detector associated to parallel plates or CORSET set-ups for the detection of the reaction fragments.

In such a low-mass system, obtained in a reaction with a low Z_1Z_2 product, only the fusion-fission mechanism is expected to contribute to the mass-symmetric fragment distribution.

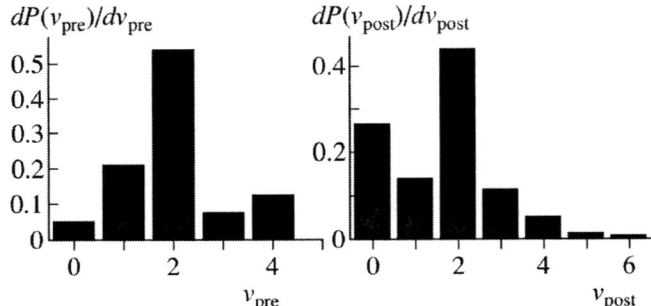

FIGURE 2. Pre- and postscission neutron multiplicity distribution for ^{126}Ba obtained by backtracing.

Figure 2 shows the distributions of the pre- and postscission neutron multiplicities as obtained by the backtracing protocol. It must be noted that the mean values obtained here are in complete agreement with those deduced from the conventional χ^2 minimization. For instance the mean values of the neutron prescission multiplicity are 2.52 and 2.54 for the χ^2 and the backtracing, respectively (see figure 3). Figure 3 presents, in addition to the backtraced distribution, the theoretical results obtained using the dynamical model of Pomorski et al [9].

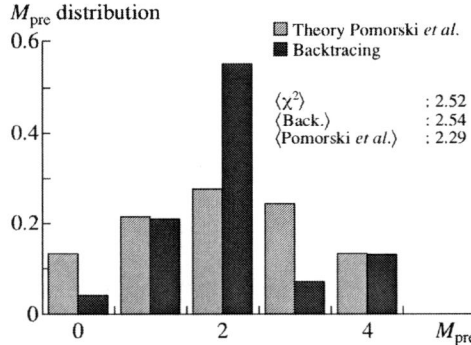

FIGURE 3. Comparison between the experimental prescission neutron multiplicity obtained by backtracing for the ^{28}Si + ^{98}Mo reaction at 204 MeV and the prediction of the model of Pomorski [8].

This model, based on the resolution of the one-dimensional Langevin equation, is able to describe only the fusion-fission process and is thus well adapted to the ^{126}Ba system. The agreement observed between experimental data and model calculations is obvious. In particular, the zero-neutron multiplicity channel is well reproduced. The model uses OBD (wall and window formula) and the agreement confirms that, at least when only the fusion-fission process is concerned and even if the model is not perfect – e.g. no dependence on temperature is included - there is no need to introduce TBD to reproduce the experimental data. To our knowledge, this is the first time that the neutron prescission multiplicity distribution has been experimentally observed and compared with such a good agreement with dynamical calculations.

A MORE COMPLEX CASE: Z=110

In this case, the fused Z = 110 nucleus is obtained by two different entrance channels (^{40}Ca or ^{58}Ni projectiles on ^{232}Th or ^{208}Pb targets) leading to the same high excitation energies ranging from around 60 MeV to more than 160 MeV. For such (super)heavy systems, the competition between quasi-fission and fusion-fission is expected to populate the symmetric part of the fragment mass distribution and until now there was no real way to disentangle these two contributions at these excitation energies. The experiments were carried out at SARA, Grenoble [6,7].

Neutron pre- and postscission multiplicities (mean values) were first obtained by χ^2 minimization and, as usually, led to the same difficulties as before: large values for γ are deduced using dynamical models. It must be noted that, when it was possible, our experimental results were compared to and found to be in complete agreement with those of Hinde *et al* [2]. In a second step, backtracing was applied and provided us with neutron pre- and postscission multiplicity distributions at least for the highest excitation energy for both systems. Indeed backtracing required very high statistics and has been applied only to these two cases.

Figure 4 presents the backtracing results for Ni + Pb (left) and Ca + Th (right). In both cases, the neutron pre- and postscission correlations exhibit two well-defined regions corresponding essentially to two different distributions for the prescission neutrons. Intuitively, we can think that each separated distribution can be attributed to each of the two capture processes. As expected, a low mean value (of the order of 4) of the neutron prescission multiplicity distribution can be associated to quasi-fission (faster mechanism), a larger mean value (~ 7) to fusion-fission (slower mechanism).

Then, using the same HICOL + DYNSEQ code as the one used by K. Siwek-Wilczynska to deduce the dissipation coefficient for the Hinde *et al* data, we can reproduce our experimental backtraced correlations by two different scenarii: quasi-fission and fusion-fission (see rectangles and squares on fig. 4). They correspond to different angular momentum ranges obtained in HICOL by the comparison with the experimental mass distribution. In both systems, the same dissipation coefficients are needed to reproduce the experimental distributions (γ=5 for quasi-fission and ranges from 5 to 11 for fusion-fission). This spectacular and first-time agreement with OBD (or with a value between OBD and two times OBD) reconciles completely these data with those corresponding to fusion-fission only and clearly supports the conclusion that, as soon as one is able to distinguish between quasi-fission and fusion-fission, no discrepancy remains and OBD is large enough to reproduce the experimental data.

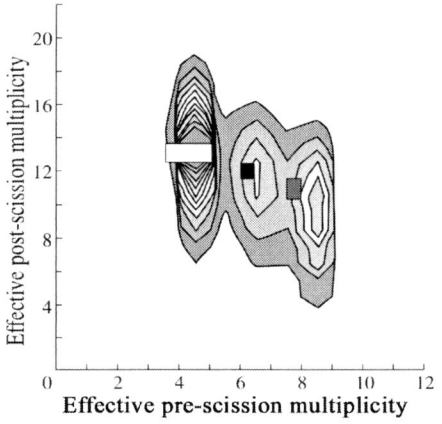 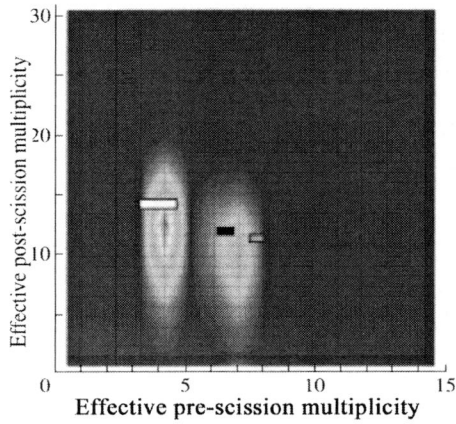

FIGURE 4. Pre- and postscission neutron multiplicity correlations for the systems ^{58}Ni + ^{208}Pb (left) and ^{40}Ca + ^{232}Th (right) at 186 MeV and 166 MeV excitation energy, respectively. Results of calculations using HICOL + DYNSEQ from Siwek-Wilczynska et al [3] are also shown. The rectangle to the left represents quasi-fission (30<l<120, l is angular momentum) with only one-body dissipation. The two squares stand for fusion-fission (0<l<30) with OBD (left) and two times OBD (right).

COMPARISON WITH LANGEVIN EQUATION MODELS

Figure 5 shows the distribution of the prescission multiplicity for the Ni + Pb system.

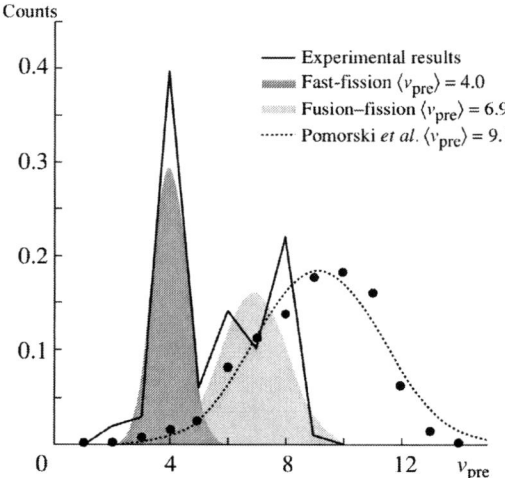

FIGURE 5. Comparison between the experimental prescission neutron multiplicity distribution obtained by backtracing for the ^{58}Ni + ^{208}Pb system at 186 MeV excitation energy (black curve) and the prediction of the Pomorski model [8] (points). The full Gaussian curves are a fit to the experimental distribution, whereas the dashed Gaussian one is a fit to the model.

Two free Gaussian curves have been fitted on the experimental backtraced distribution (full line) and are now considered to represent the distribution to be attributed to the quasi-fission and fusion-fission mechanisms. Calculations for the fusion-fission part, performed by Schmitt [10] using the Pomorski model, are also given in this figure (dotted curve). One can note that the mean value and the width obtained in this model are slightly too high, but the agreement can be considered as satisfactory with this OBD (only) symmetric fission model if one takes into account that it does not contain the dissipation in the entrance channel and that no dependence of friction on shape nor temperature is considered. These points are under consideration.

Recently, a three-dimensional Langevin equation model has been developed and applied to the dynamics of capture reactions, in particular, in the superheavy region [11]. A first comparison of these calculations with our experimental result for the Ni + Pb reaction is presented in figure 6 [12].

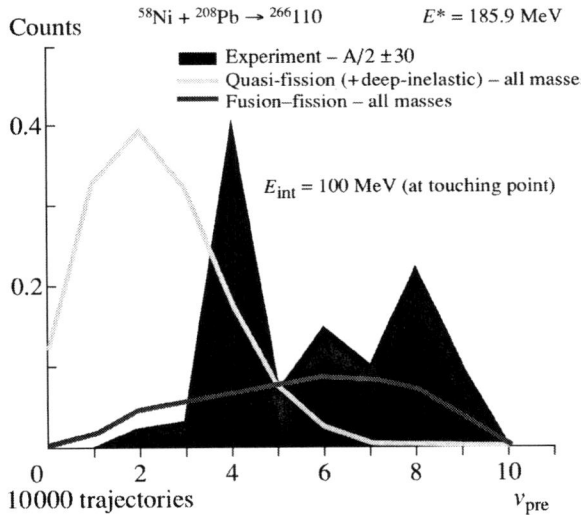

FIGURE 6. Comparison between the experimental prescission neutron multiplicity distribution obtained by backtracing for the ^{58}Ni + ^{208}Pb system at 186 MeV excitation energy for fragment masses in the range of A/2± 30 (full black curve) and the prediction of the Aritomo model [11] for quasi-fission and fusion-fission for the whole mass range of deep-inelastic, quasi-fission and fusion-fission processes.

If one takes into account that these preliminary calculations include all the events associated not only with quasi-fission and fusion-fission but also with deep-inelastic processes, the overall agreement is satisfactory. It seems obvious that, as soon as the experimental mass and total kinetic energy cuts are included in the calculations, the agreement will be better. Indeed, deep-inelastic processes are known to be low prescission multiplicity events and the symmetric fusion-fission process corresponds to the largest multiplicity.

AT LOW EXCITATION ENERGY

Low excitation energies, which means the ones required for the superheavy synthesis, represent a very serious problem in the determination of the dynamical aspects of capture reactions. Indeed, regardless of the decreasing of the cross-section, which is already a severe limitation, the prescission multiplicity will also drop and can represent a real difficulty for the backtracing since it requires very high statistics for the neutron experimental observables.

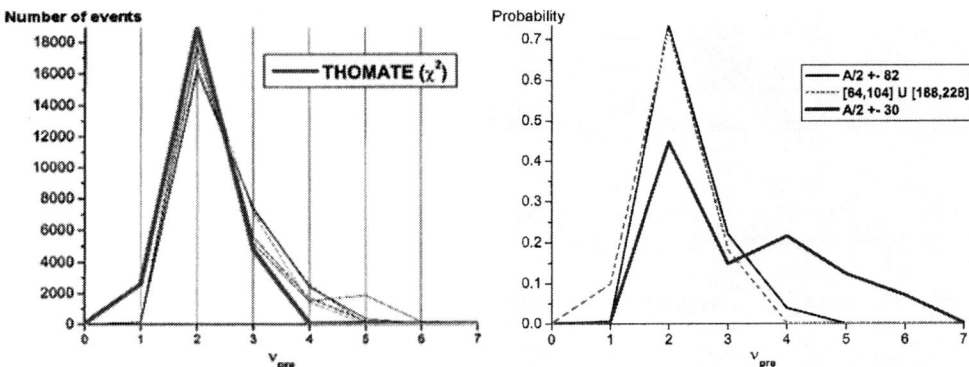

FIGURE 7. Prescission neutron multiplicity distributions for Ca + Pb (left) and Ca + Pu (right). In the Pu case, two distributions appear, a one-bump distribution corresponding to quasi-fission events with mass around 208 and a two-bump distribution corresponding to symmetric-mass fragments.

A new protocol (THOMATE), able to handle low statistics data [13], has been developed and applied to ^{48}Ca + ^{208}Pb, Pu at 40 MeV of excitation energy. The neutron prescission multiplicity distributions obtained in both cases are presented on figure 7 for the symmetric part of the mass distribution. In the ^{256}No case, as expected, the prescission multiplicity distribution exhibits only one bump, confirming the domination of the fusion-fission process. In the Z=114 case, the distribution shows two bumps in agreement with 50-50% of quasi-fission and fusion-fission. A first preliminary comparison with Langevin calculations using OBD is in agreement with the experimental results [12].

ACKNOWLEDGEMENTS

This work has been partly supported by INTAS 97-11929 and 00-655. It's also a pleasure to thank all our colleagues from the DEMON-CORSET collaboration for all the experimental work performed at the U400 of FLNR, Dubna and at the VIVITRON, Strasbourg.

REFERENCES

1. H. Kramers, Physica **7**, 284 (1940).
2. D. Hinde *et al.*, Phys. Rev. **C45**, 1229 (1992).
3. K. Siwek-Wilczinska *et al.*, Phys. Rev. **C51**, 2054 (1995).
4. J. Wilczinski *et al.*, Phys. Rev. **C54**, 325 (1996).
5. P. Désesquelles *et al.*, Nucl. Phys. **A604**, 183 (1996).
6. L. Donadille *et al.*, Nucl. Phys. **A656**, 259 (1999).
7. B. Benoit, Ph.D. Thesis, Univ. of Brussels, Brussels, 2000.
8. E. de Góes Brennand, Ph.D. Thesis, Univ. of Brussels, Brussels, 2000.
9. K. Pomorski *et al.*, Nucl. Phys. **A679**, 25 (2000).
10. C. Schmitt, Ph.D. Thesis, Univ. of Strasbourg, Strasbourg, 2002.
11. Y. Aritomo, Proc. Int. Symp. *"Exotic Nuclei, EXON-2001"*, Baïkal Lake, Irkutsk, Russia, July 24-28, 2001, ed. Yu.E. Penionshkevich and E.A. Cherepanov, World Scientific 2002, p. 106.
12. Aritomo and T. Materna, work in progress.
13. T. Materna, Ph.D. Thesis, Univ. of Brussels, Brussels, 2003 and submitted to NIM.

Deformation of the very neutron-deficient rare-earth nuclei produced with the SPIRAL ^{76}Kr radioactive beam and studied with EXOGAM + DIAMANT

N. Redon[*], A. Prévost[*,†], D. Guinet[*], Ph. Lautesse[*], M. Meyer[*], B. Rossé[*], O. Stézowski[*], P.J. Nolan[**], C. Andreoiu[**], A.J. Boston[**], M. Descovich[**], A.O. Evans[**], S. Gros[**], J. Norman[**], R.D. Page[**], E.S. Paul[**], G. Rainovski[**], J. Sampson[**], G. de France[‡], J.M. Casandjian[‡], Ch. Theisen[§], J.N. Scheurer[¶], B.M. Nyakó[∥], J. Gál[∥], G. Kalinka[∥], J. Molnár[∥], Zs. Dombrádi[∥], J. Timár[∥], L. Zolnai[∥], K. Juhász[††], A. Astier[†], I. Deloncle[†], M.G. Porquet[†], R. Wadsworth[‡‡], P. Raddon[‡‡], Y. Lee[‡‡], A. Wilkinson[‡‡], P. Joshi[‡‡], J. Simpson[§§], D. Appelbe[§§], D. Joss[§§], R. Lemmon[§§], J. Smith[¶¶], D. Cullen[¶¶], A. Brondi[***], G. La Rana[***], R. Moro[***], E. Vardacci[***] and M. Girod[†††]

[*]*IPN Lyon, IN2P3/CNRS, Université Claude Bernard Lyon-1, F-69622 Villeurbanne Cedex, France*
[†]*CSNSM Orsay, IN2P3/CNRS, Bat 104, F-91405 Orsay Campus, France*
[**]*Oliver Lodge Laboratory, University of Liverpool, P.O. Box 147, Liverpool L69 7ZE, United Kingdom*
[‡]*GANIL, B.P. 55027, F-14076 Caen Cedex, France*
[§]*DAPNIA/SPhN CEA-Saclay, F-91191 Gif sur Yvette Cedex, France*
[¶]*CEN Bordeaux-Gradignan, Le Haut Vigneau, F-33170 Gradignan, France*
[∥]*Institute of Nuclear Research, H-4001 Debrecen, Hungary*
[††]*Institute of Mathematics and Informatics, University of Debrecen, H-4001 Debrecen, Hungary*
[‡‡]*Department of Physics, University of York, Heslington, York Y01 5DD, United Kingdom*
[§§]*CRLC, Daresbury Laboratory, Daresbury, Warrington WA4 4AD, United Kingdom*
[¶¶]*Department of Physics, University of Manchester, Manchester M13 9PL, United Kingdom*
[***]*Dipartimento di Scienze Fisiche dell' Universita and INFN, Complesso Universario di Monte S. Angelo, Via Cintia, 80126 Napoli, Italy*
[†††]*CEA/DAM Ile de France, 91680 Bruyères le Châtel, France*

Abstract. The structure of the very neutron-deficient rare-earth nuclei has been investigated in the first experiment with the EXOGAM gamma array coupled to the DIAMANT light charged particle detector using radioactive beam of ^{76}Kr delivered by the SPIRAL facility. Very neutron-deficient Pr, Nd and Pm isotopes have been populated at rather high spin by the reaction ^{76}Kr + ^{58}Ni at a beam energy of 328 MeV. We report here the first results of this experiment.

INTRODUCTION

Mapping the understanding major regions of deformation away from closed shells is an important aspect of nuclear structure physics. Adding protons above Z = 50 while removing neutrons from N = 82 produces a major region of deformation, and experiments have succeded in studying the nuclei approaching, but not yet reaching, the peak of this deformation. The use of radioactive beams will open up this exotic region of very neutron-deficient rare-earth nuclei and allow the probing of nuclei around the peak of this systematic feature.

The theoretical proton drip-line and corresponding axial deformations of the last proton-stable nuclei are shown in Figure 1, obtained in Hartree-Fock-Bogolioubov (HFB) self-consistent calculations using the Gogny effective nucleon-nucleon interaction. In these calculations, ^{130}Sm and ^{126}Nd are the lightest proton-bound samarium and neodynium isotopes, i.e. lying at the proton drip-line. Maximal ground-state deformation in this mass region is predicted with $\beta \sim 0.40$.

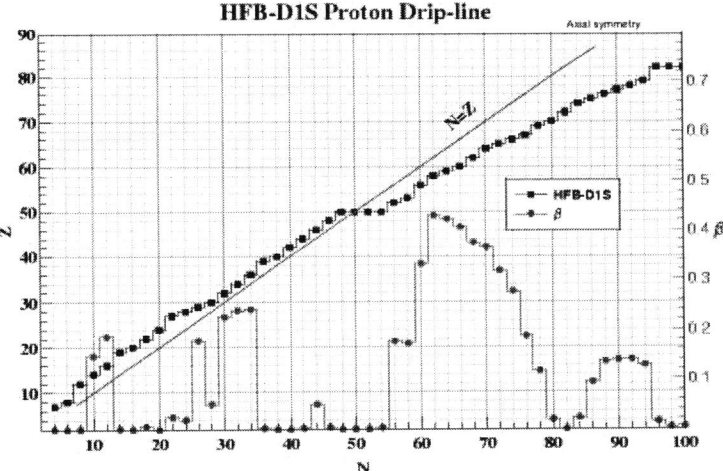

FIGURE 1. The calculated proton drip-line (squares) and corresponding axial deformations (circles) for these nuclei, obtained in HFB calculations. Note that maximal ground-state deformation is calculated for $Z \approx 60$, $N \approx 65$ drip line nuclei. For instance, ^{130}Sm$_{68}$ and ^{126}Nd$_{66}$ are predicted on the drip-line with a deformation of $\beta \sim 0.4$.

This value of ground state deformation is the same as that deduced for superdeformed cerium and neodynium isotopes, the lightest of which being ^{129}Ce and ^{130}Nd. Thus, studies of the very neutron-deficient nuclei in this mass region will allow us to answer some interesting questions concerning the behaviour of the deformation when approaching the proton drip-line. The deformation trend may be followed by examining the energies of the lowest transitions of rotational bands in these nuclei.

EXPERIMENT

To reach very neutron-deficient exotic nuclei located near the proton drip-line, ^{76}Kr + ^{58}Ni fusion-evaporation reaction has been used. The target consisting of a ^{58}Ni foil (~ 1 mg/cm^2) was bombarded by a 328 MeV neutron-deficient radioactive beam of ^{76}Kr delivered by the CIME cyclotron from the SPIRAL facility at GANIL. During the experiment, the beam intensity was about 5×10^5 particle/s.

The γ-rays were detected with an early implementation of the EXOGAM spectrometer [1, 2] consisting of 6 full-sized EXOGAM germanium segmented clover detectors and 2 small germanium segmented clover prototypes. Due to the half-life of the ^{76}Kr radioactive beam ($T_{1/2} = 14.8$h) and its very low intensity ($\sim 10^5$ reduction compared to stable-ion beams), it is a challenge to reduce the background radiation. The DIAMANT detector system [3] has been coupled to EXOGAM in order to measure the charged-particle evaporation residues. It was composed of 56 CsI scintillators, 24 assembled in a chessboard detector at forward angles and two rings of 16 detectors located near 90° relative to the beam axis.

The data were recorded in an event-by-event mode with the requirement that one Ge detector firing in coincidence with at least one DIAMANT cell. A total of 4×10^5 events were collected.

RESULTS

During the experiment, more than 10 channels were opened with a cross-section greater than $8 - 10$ mb. Thanks to the coincident spectra with evaporated charged particles detected with DIAMANT (see Figure 2), four channels have been clearly identified : 4p (^{130}Nd), 3p (^{131}Pm), α2p (^{128}Nd), α3p (^{127}Pr) as shown in Figure 3.

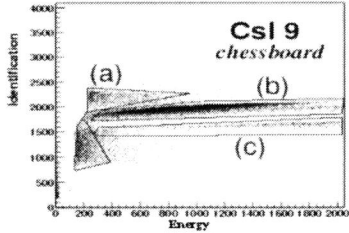

FIGURE 2. Identification matrix of one chessboard cell of DIAMANT with outlines corresponding to : (a) γ-rays; (b) protons; (c) α-particles. The Identification variable is extracted from the rising time of the signal obtained in the DIAMANT detector.

From the rising time of the signal obtained in each DIAMANT detector, the Identification variable is extracted and can be presented as a function of the energy as shown in Figure 2. From this matrix, we can separate the different particles detected in DIAMANT. In Figure 2, the outlines correspond to : (a) γ-rays induced by the radioactive beam implanted; (b) at least one evaporated proton; (c) at least one evaporated α-particle.

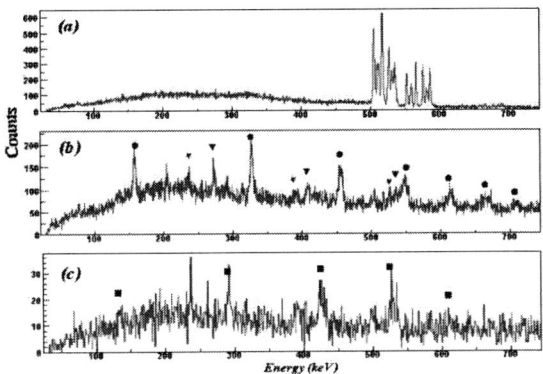

FIGURE 3. γ-ray Doppler corrected spectra gated by particle identified with DIAMANT as shown in Figure 2 : (a) spectrum gated by the γ-rays detected in the DIAMANT detector; (b) spectrum gated on at least one proton, (c) spectrum gated on at least one α-particle. The γ-rays belonging to the different nuclei are marked : ^{130}Nd (full circle), ^{131}Pm (triangle down), ^{127}Pr (full star) and ^{128}Nd (full square).

By gating on the outlines in the Identification matrices of DIAMANT, we obtain three γ-ray spectra presented in Figure 3. The γ-ray spectrum in Figure 3a have been obtained by gating on the outline (a) and show the radioactivity due to the ^{76}Kr beam. These γ-rays are emitted while the beam is stopped and if we apply a Doppler correction, they are splitted in six peaks corresponding to the six angles of the clover crystals. The two groups of six peaks in the spectrum (a) correspond to the decay ^{76}Kr ($T_{1/2}$ = 14.8h) → ^{76}Br ($T_{1/2}$ = 16.2h) → ^{76}Se (stable), in particular the 511 keV line coming from the β-decay and the 559 keV line ($2^+ \rightarrow 0^+$ transition in ^{76}Se).

The selection obtained with DIAMANT allows us to extract the different fusion-evaporation channels as shown in Figure 3b and 3c. Indeed, we have observed yrast band of four known nuclei : ^{130}Nd [4] (up to state 18^+), ^{131}Pm [5] (up to state $23/2^-$), ^{127}Pr [5] (up to state $23/2^-$) by gating on at least one evaporated proton detected in DIAMANT and ^{128}Nd [6] (up to state 10^+) by gating on at least one evaporated α-particle detected in DIAMANT.

In Figure 3b, one can see the γ-transitions of the yrast band in ^{130}Nd [4] only up to 14^+ (159, 326, 454, 548, 613, 664 and 704 keV) but if we extend the spectrum, we can observed two more transitions (up to 18^+). In the same spectrum, three transitions of the yrast band in ^{131}Pm [5] (273, 407 and 538 keV) and three transitions of the yrast band in ^{127}Pr [5] (237, 390 and 527 keV) have also been observed. In Figure 3c five transitions of the yrast band in ^{128}Nd [6] (134, 292, 424, 530 and 611 keV) have been observed in coincidence with at least one α-particle. In these two spectra, all the radioactivity lines have disappeared.

These nuclei are predicted by the ALICE evaporation code [7] with a cross section from $100-200$ mb (for the first ones) down to ~ 30 mb (for the last ones). We have not enough statistics to obtain clean $\gamma-\gamma$ data except for the most produced nucleus ^{130}Nd and to observe new states in these nuclei. However, the γ-multiplicity spectra obtained for the different outlines in the DIAMANT matrices show that the γ-multiplicity is high enough for the fusion-evaporation residues to use $\gamma-\gamma$ coincidences.

CONCLUSION

During this experiment, we have demonstrated that the coupling of EXOGAM with DIAMANT constitutes an excellent device to remove the background due to the radioactivity of the beam. We were able to observe γ-transitions in ^{128,130}Nd, ^{131}Pm and ^{127}Pr up to relatively high-spin states. These nuclei are the most neutron-deficient isotopes known in this region. This is the first time that, using fusion-evaporation reaction induced by radioactive beam, we observe so neutron-deficient nuclei at high spin.

ACKNOWLEDGMENTS

The authors would like to thank all those involved in the setting up and commissioning of EXOGAM and DIAMANT, in particular the persons in charge the EXOGAM detectors, G. Voltolini and J. Ropert, and of the experimental area, A. Vigot. We thank also the accelerator staff for the good quality of the radioactive beam. This work has been supported by the European Community-Access to Research Infrastructure action of the Improving Human Potential Programme, contract No HPRI-CT 1999-00066.

REFERENCES

1. Azaiez, F, *Nucl. Phys.*, **A654**, 1003c–1008c (1999).
2. Simpson, J., Azaiez, F, De France, G., Fouan, J., Gerl, J., Julin, R., Korten, W., Nolan, P.J., Nyakó, B., Sletten, G., Walker, P.M., and the EXOGAM collaboration, *Heavy Ion Physics*, **11**, 159 (2000).
3. Scheurer, J.N., Aiche, M., Aleonard, M.M., Barreau, G., Bourgine, F., Boivin, D., Cabaussel, D., Chemin, J.F., Doan, T.P., Goudour, J.P., Harston, M., Brondi, A., La Rana, G., Moro, R., Vardaci, E., Curien, D., *Nucl. Inst. and Meth. In Phys. Res.*, **A385**, 501–510 (1997).
4. Hartley, D.J., Reviol, W., Riedinger, L.L., Jin, H.Q., Smith, B.H., Yoder, N., Zeidan, O., Galindo-Uribarri, A., Sarantites, D.G., LaFosse, D.R., Wilson, J.N., Mullins, S.M., *Phys. Rev.*, **C63**, 024316 (2001).
5. Parry, C.M., Boston, A.J., Chandler, C., Galindo-Uribarri, A., Hibbert, I.M., Janzen, V.P., Joss, D.T., Mullins, S.M., Nolan, P.J., Paul, E.S., Regan, P.H., Vincent, S.M., Wadsworth, R., Ward, D., Wyss, R., *Phys. Rev.*, **C57**, 2215–2221 (1998).
6. Petrache, C., Lo Bianco, G., Bazzacco, D., Kröll, Th., Lunardi, S., Menegazzo, R., Nespolo, M., Pavan, P., Rossi Alvarez, C., Axiotis, M., de Angleis, G., Farnea, E., Marginean, N., Napoli, D.R., Blasi, N., *Eur. Phys. J.*, **A12**, 139–141 (2001).
7. Blann, M., Report Overlaid ALICE COO 3494-29, University of Rochester (unpublished).

Chaos in the Nucleus: SD Decay-out and Masses

Sven Åberg[1], Thomas Døssing[2], Henrik Olofsson[1], Ingemar Ragnarsson[1], Corina Andreoiu[3], Claes Fahlander[3] and Dirk Rudolph[3]

[1] *Mathematical Physics, Lund Institute of Technology, P.O. Box 118, SE 221 00 Lund, Sweden*
[2] *The Niels Bohr Institute, Blegdamsvej 17, DK2100 Copenhagen Ø, Denmark*
[3] *Department of Physics, Lund University, PO Box 118, SE 221 00 Lund, Sweden*

Abstract. Some consequences of chaos in the nucleus are discussed. The decay-out from super-deformed band occurs in a region where normal-deformed states may be chaotic, and it is shown how the distribution of decay-out matrix elements may reveal the degree of chaos. The decay-out is mediated by doorway states that are not resolved in the A=190 region, but experimentally seen in ^{59}Cu. A model for the decay-out in the A=60 region is set up and applied to ^{59}Cu. It is discussed how errors in nuclear mass formula may be related to chaotic motion, and a theoretical model is set up and applied to chaotic masses.

INTRODUCTION

While regular energy bands and distinct selection rules characterize the yrast region of heavy, deformed nuclei [1], chaotic behavior can be connected to the excitation region of 7-8 MeV (the neutron resonance region). Obviously, chaos sets in with increasing excitation energy somewhere in this excitation energy interval, and it is a fundamental issue to find out where and how, and what are the experimental consequences.

The onset of chaos in an excited interacting many-body system like the nucleus is suggested to be caused by the residual two-body interaction rather than by the shape and structure of the one-body potential [2,3]. Since the level density increases exponentially with excitation energy, the relative importance of the residual interaction increases, and the degree of chaoticity increases with increasing excitation energy. Generally, chaos is expected to set in when the residual two-body interaction is of the size of the typical energy distance between many-body states that are directly coupled by the two-body interaction [2,3], i.e. when $V_{residual} \approx d_{2p-2h}$. From theoretical calculations, based on a schematic [2] or a more realistic two-body interaction [4], it has been suggested that chaos sets in already 2-3 MeV above the yrast line in heavy deformed nuclei. In the next section we discuss how the onset of chaos strongly enhances the tunneling probability associated with the decay-out from superdeformed (SD) rotational bands, and how actual measurements of matrix elements connecting the SD states with normal-deformed (ND) states may reveal the degree of chaos.

We then show how the observed SD decay-out in ^{59}Cu is theoretically well described by a model that incorporates detailed nuclear structure information with a two-body random matrix model.

In the last section nuclear masses, or rather the error in nuclear mass formulas, are studied. In spite of efforts over decades the overall deviation of about 0.7 MeV between the best-calculated nuclear masses and experimental data seems to be difficult to overcome. We discuss the suggestion that this deviation is due to chaos in the nucleus [6,7].

DECAY-OUT OF SUPERDEFORMED BANDS

In the A=150 and A=190 regions the decay-out from SD states is very fragmented and only a few cases of discrete γ-transitions have been observed. In lighter nuclei the level density is generally smaller and the decay-out may be observed in one or a few discrete γ-transitions. We shall first discuss the decay-out in heavy nuclei and then in lighter nuclei, specifically in ^{59}Cu where the decay-out is known in great detail.

The A=190 Region – Chaos-Assisted Tunneling [9,10]

The decay-out from superdeformed states in the A=150 and A=190 regions occurs at excitation energies between 3 and 5 MeV. This provides a unique possibility to use the decay-out process as a measure of chaos in the nucleus. By considering large amplitude collective motion, utilizing e.g. the generator coordinate method (GCM), the coupling between SD and ND states may be calculated. Bonche et al [8] include states calculated in HF + BCS and constrained in β_2 deformation, and could obtain a strong vibrational coupling between SD and ND states. The ND states involved in these couplings may be considered as *doorways* to the decay-out [9]. The doorway states may be subsequently mixed with all other ND states by the residual two-body interaction. Consequently, the degree of mixing, i.e. the degree of chaos, becomes important for the decay-out.

In the extreme case of no mixing, i.e. completely *regular states*, the decay-out must proceed via the doorway states. The density of such states is quite low [8], $\rho_{doorway} \sim 1$ MeV^{-1}, i.e. the energy distance from the SD state to the nearest doorway state is quite large, of the order of $0.25/\rho_{doorway} \sim 0.25$ MeV. If we denote the level density of ND states at the excitation energy E_{exc} (the energy distance from the lowest ND state to the SD state) as $\rho_{ND}(E_{exc})$ there are on the average in total $N \approx 0.25 \rho_{ND}/\rho_{doorway}$ ND states in the energy interval between the SD state and the closest doorway state. If we denote the coupling (tunneling) matrix element between the SD and doorway state as V_t, the SD wave-function coupled with the doorway state can be estimated as

$$|SD'\rangle = |SD\rangle + \frac{V_t}{N \cdot d}|ND\rangle, \qquad (1)$$

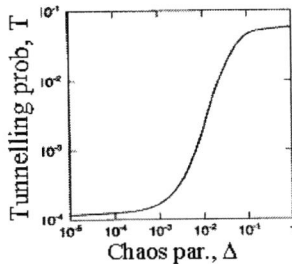

FIGURE 1. Chaos-assisted tunneling. The tunneling probability increases by a factor N (the number of ND states per doorway state) when the character of the ND states changes from order to chaos. In this model calculation $N=400$. From Ref. [9].

where d is the average level distance between ND states, $d = 1/\rho_{ND}$, and $|ND>$ in this case is the (normal-deformed) doorway state. The tunneling probability in the regular case thus becomes

$$T_{reg} = |\langle SD'|ND\rangle|^2 = \frac{V_t^2}{N^2 \cdot d^2}. \quad (2)$$

In the other extreme of full mixing, i.e. *full chaos*, all ND states are completely mixed (in some energy interval), and the doorway coupling strength is smeared out on all ND states. Each ND state thus obtains a coupling matrix element with the SD state, $V_t' \approx \pm V_t/\sqrt{N}$. The SD state can now mix with a very close-lying ND state at an energy distance $d/2$:

$$|SD'\rangle = |SD\rangle + \frac{2V_t/\sqrt{N}}{d}|ND\rangle \quad (3)$$

implying almost resonance tunneling. The tunneling probability thus becomes

$$T_{chaos} = |\langle SD'|ND\rangle|^2 = \frac{4V_t^2}{N \cdot d^2}. \quad (4)$$

That is, chaos implies an enhancement of the tunneling probability by a factor

$$\frac{T_{chaos}}{T_{reg}} = 4N = \frac{\rho_{total}}{\rho_{ND}}. \quad (5)$$

At excitation energies where the decay-out occurs the level density of ND states is very large, giving an enhancement factor of the order of $10^4 - 10^5$! This huge statistical enhancement of tunneling probability, caused by the onset of chaos, has been denoted as *chaos-assisted tunneling* [9] (see fig.1), and might trigger the decay-out from SD states.

The distribution of sizes of matrix elements connecting the SD state and the ND states is sensitive to the degree of chaos in the ND states. In the limit of full chaos this distribution becomes the Porter-Thomas distribution, while in the other extreme of fully regular states, most matrix elements will be zero, and the decay will occur via the doorway states in a few strong transitions, schematically shown in the left-hand part of fig.2. The intermediate situation of a mixed phase space of chaos and order has been simulated in model calculations [9] and is shown in the right-hand part of fig.2. By comparing such model calculations to actually measured decay-out strengths, one may determine the degree of chaos in the excited ND states around the SD state [10].

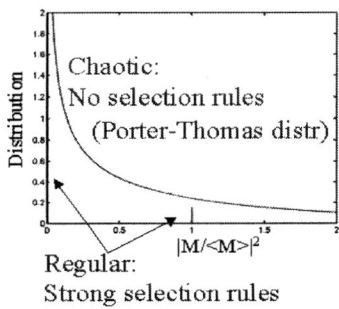
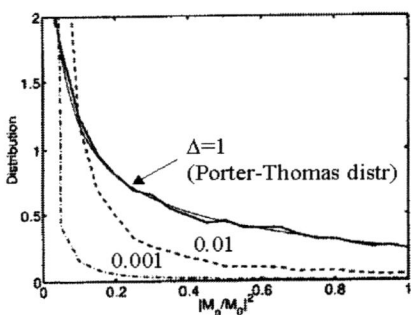

FIGURE 2. Distribution of normalized decay-out matrix elements. The left-hand part schematically shows the two limiting situations of regular ND states, where strong selection rules imply that most matrix elements are zero and a few are strong, and chaotic ND states (Porter-Thomas distr.). The right-hand part shows a calculation [9] where the degree of chaos in the ND states has been smoothly changed from full chaos, Δ=1, to smaller degrees of chaos, Δ=0.01 and 0.001.

The A=60 Region – Direct Coupling to Doorway States [11,13]

Contrary to the previously discussed A=190 (and A=150) regions, the decay-out from SD states is rather little fragmented in lighter nuclei. A good example is ^{59}Cu as can be seen in fig.3 where the experimentally known [11] decay-out properties are shown (see also Ref.[12]). The SD band crosses the yrast ND states at angular momentum $I_C \approx 34/2$ and the decay-out occurs from the 29/2 state. A large part of the decay-out intensity is observed (90 %). In the inset of fig.3, the intensity in the SD band is seen to drop from 100% to 0% in two transitions. In the first step the intensity is fragmented into four I=25/2 states. One of these states can be considered as a continuation of the SD band and the other three as doorway states, mediating the decay-out. The subsequent decay from these four 25/2 states occurs in two energy groups, one with γ–ray energies around 2 MeV and one around 3-4 MeV.

In order to understand the decay-out in ^{59}Cu with its detailed experimental information [11] we set up a theoretical model (see Ref.[5]). The different rotational states, denoted as $|\mu, \varepsilon; I^\pi\rangle$, are calculated from an accurate mean field calculation (Cranked Nilssson-Strutinsky) where parameters have been adjusted to obtain a good description of ^{59}Cu [5]. In this calculation each rotational state, specified by its configuration μ, angular momentum and parity (I^π), is minimized with respect to deformation, ε and γ. In this way the 32 lowest positive-parity states are calculated at each angular momentum. These states constitute the Hilbert space in the next step of the calculation. A residual interaction, $\hat{V}_{res}$, is introduced with matrix elements between many-body states:

$$\langle \mu', \varepsilon'; I^\pi | \hat{V}_{res} | \mu, \varepsilon; I^\pi \rangle = V_{2p2h} \cdot T_{coll}. \tag{6}$$

V_{2p2h} is a 2-body random matrix interaction with an rms-value of 100 keV that accounts for the configuration change. Since the deformation of the two eigen-states

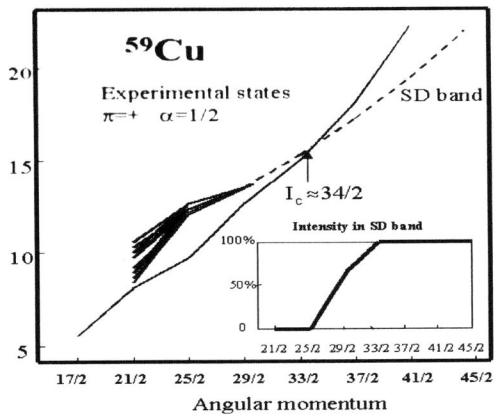

FIGURE 3. Observed [11] states and transitions in ^{59}Cu relevant for the decay-out of the SD band. The dashed lines connect SD states. The inset shows the SD intensity vs. angular momentum of the initial state in the transition.

may be different we introduce T_{coll} that accounts for possible collective deformation changes between the states μ' and μ. T_{coll} is calculated from a WKB estimate of the tunneling through the barrier separating the two states μ and μ', where standard values are employed for the mass parameter and the zero-point energy. This gives a variation of T_{coll} between 0.015 and 1. The suppression of matrix elements due to tunneling is most important for the explanation of the delayed decay-out from I_C=33/2 to 29/2 [5].

The effective many-body hamiltonian is then diagonalized for each angular momentum value yielding mixed wave-functions:

$$|\alpha;I^\pi\rangle = \sum_\mu a_{\alpha\mu}|\mu;I^\pi\rangle. \qquad (7)$$

Finally, transition matrix elements for stretched E2 transitions are calculated for the spin sequence I=37/2, 33/2, 29/2, 25/2, 21/2 and 17/2. For in-band transitions standard B(E2) values are obtained from calculated quadrupole deformations, and range between 14 and 78 W.u. for the ND states and 200 W.u. for the SD states. For transitions involving configuration changes by 1-particle-1-hole the matrix elements are generally assumed to be 5 W.u., and otherwise zero. The E2-cascade could then be simulated starting with an initial distribution of 100% in the yrast SD state at I=37/2. Sampling the result from 100 independent runs yields the γ-ray energy intensity distributions shown in fig.4. The general fragmentation of radiation strength, and the division into two main groups of γ-ray energies, one around 2 MeV and one around 3.5 MeV, are well reproduced by the theory. The general drop of intensity of the SD band at I=29/2 and 25/2 is also nicely reproduced by the calculations.

In order to test the degree of chaos in ^{59}Cu at the SD decay-out we obtain the nearest neighbours distribution (NND) at I=25/2 and I=21/2 from the calculated many-body spectrum as shown in fig. 5a. The 25/2 states exhibit a quite regular behaviour with the NND being close to a Poisson distribution. The NND of the 21/2 states is remarkably different compared to the 25/2 states, and seems to be much closer to a chaotic (GOE) behaviour. In fig.5b the distributions of matrix elements from transitions I=29/2 to 25/2 and I=25/2 to 21/2 are shown, and the higher spin is again showing a more regular behaviour than the lower spin. This is a quit unexpected behaviour since states at neighbouring spins usually have the same dynamics. One expla nation may be that some of the considered rotational bands terminate at I=21/2, i.e. there are fewer states to be involved in the mixing at I=25/2 than at I=21/2 [13].

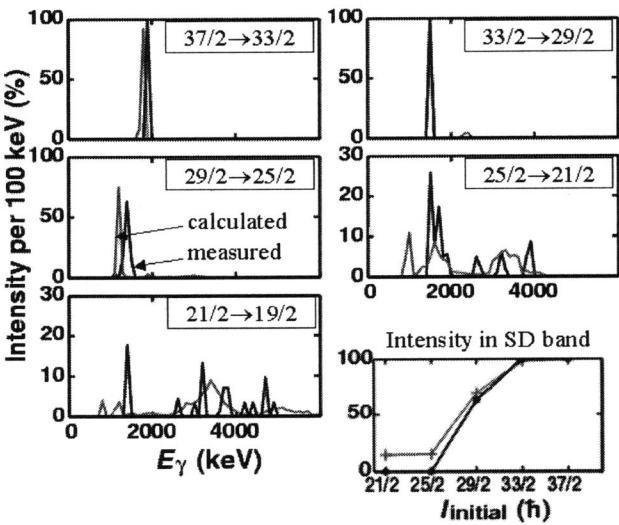

FIGURE 4. Calculated and measured [11] intensity of γ–rays in the SD decay of ^{59}Cu separated for different decay steps. The lower right figure shows calculated and measured intensity in the SD band. From Ref.[5].

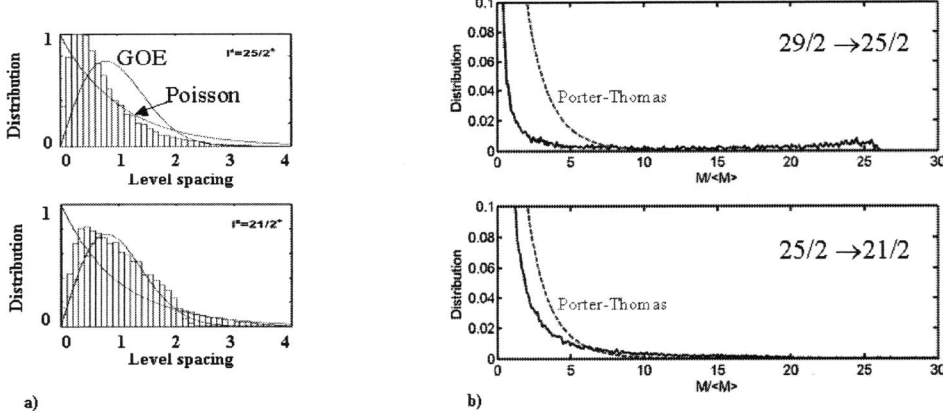

FIGURE 5. Statistical analysis of calculated data in ^{59}Cu. In a) nearest neighbor energy distributions are shown for the calculated energies at I=25/2 and 21/2. The results are compared to the limiting cases of full chaos (GOE) and regularity (Poisson), respectively. In b) calculated distributions of B(E2) matrix elements are shown for the 29/2 to 25/2 γ-ray transitions (upper part) and the 25/2 to 21/2 transitions (lower part). The chaotic situation implies a Porter-Thomas distribution. From Ref.[13].

IS ERROR IN MASS FORMULA DUE TO CHAOS? [14]

In a recent paper by Bohigas and Lebeouf [6] it was suggested that the error in nuclear mass formula might be due to chaotic contributions in the nuclear ground state. We shall discuss the underlying theory [6,15] and a possible model for the description of chaos in the ground states [14].

The value of the measured nuclear mass excess can be divided into the three different terms,

$$m(N,Z)/c^2 = E_{L.D.} + E_{shell} + E_{rest}, \qquad (8)$$

where the first term corresponds to a smoothly varying term, the liquid-drop energy, the second term appears due to shell effects, and the third term is the deviation (*error*) in the theoretical description of the nuclear mass in a mean field model. Extensive mass calculations in the whole periodic system by Möller et al [16], utilizing the Folded-Yukawa potential and the finite range droplet model, gives typical values for these three terms as

$$E_{L.D.} \approx 8A \text{ MeV}, \qquad E_{shell}^{rms} \approx 3 \text{ MeV}, \qquad E_{rest}^{rms} \approx 0.7 \text{ MeV}, \qquad (9)$$

i.e. the error of the binding energy in a medium heavy nucleus is less than one part per thousand.

In periodic orbit theory the fluctuating part of the level density can be expressed as a sum over classical periodic orbits. This can be calculated also for a non-interacting fermi gas of nucleons, giving [15]

$$E_{osc}(A) = \int_0^{e_F} \rho_{osc}(e,\beta) \cdot e \cdot de = 2\hbar^2 \sum_p \sum_{r=1}^{\infty} \frac{A_{p,r}}{r^2 \tau_p^2} \cos(rS_p/\hbar + v_{p,r}), \qquad (10)$$

where ρ_{osc} is the fluctuating part of the single-particle level density, e_F the Fermi energy, $A_{p,r}$ are amplitudes that are determined by the stability of the classical periodic orbit, S_p the classical action of a primitive periodic orbit p, r is the repetition of the periodic orbit, the phase $v_{p,r}$ is the Maslov index, and the period time for the orbit is $\tau_p = \partial S_p / \partial E$. This expression is badly convergent, while the second moment can be obtained analytically giving [15]

$$\left\langle E_{osc.}^2 \right\rangle = \frac{\hbar^2}{2\pi^2} \int_0^{\infty} \frac{d\tau}{\tau^4} K_D(\tau) \qquad (11)$$

where the form factor, K_D, can be approximated to zero for times shorter than the shortest periodic orbit, $\tau < \tau_{min}$, and else it is constant and equal to the Heisenberg time, $K_D \approx \tau_H$, for a regular system, and $K_D \approx \tau$ for a chaotic system. This gives the rms value of the fluctuating part of a completely regular system as:

$$E_{regular}^{rms} = \sqrt{<E_{regular}^2>} = \sqrt{\frac{\hbar^2}{6\pi^2} \frac{\tau_H}{\tau_{min}^3}} \approx 2.8 \text{ MeV}, \qquad (12)$$

and correspondingly for a completely chaotic system,

$$E^{rms}_{chaos} = \sqrt{<E^2_{chaos}>} = \sqrt{\frac{\hbar^2}{6\pi^2} \frac{1}{\tau_H^2}} \approx \frac{2.78}{A^{1/3}} \text{MeV}. \quad (13)$$

In eqs. (12) and (13) relevant values for the nucleus have been estimated [6]. The *regular* contribution compares well to the calculated shell energy, see eq. (9). The variation with mass number of the *chaotic* contribution is however found to be well reproducing the error in the mass formula, see Fig.6.

We set up a simple model to describe the chaotic contribution to the nuclear mass [14]. The model Hamiltonian of the mean field has one regular and one chaotic term,

$$\hat{H}_{meanfield} = \hat{H}_{regular} + \delta \cdot \hat{H}_{chaotic}, \quad (14)$$

where the regular part corresponds to a regular mean field, e.g. the deformed Nilsson model, and the chaotic mean field is modeled by a random matrix model. The strength parameter δ is fitted to data. If we fit to the expression eq.(13) we find that $\delta = 2/A^{5/6}$ MeV, while a fit to the errors in the mass formula gives $\delta = 4.2/A$ MeV that is shown in fig.6. Mainly the ground-state mass obtains a sizeable fluctuation due to the chaotic part of the Hamiltonian; the excitation spectrum is found to fluctuate very little [14].

It is an open and most important question to what extent chaos is responsible for error in nuclear mass formulas. The above treatment of regular and chaotic contributions separately requires that they are uncorrelated. However, errors and shell contributions in the Möller mass calculation gives a small correlation of −16% [7]. Also notice there is a worse agreement of the nuclear mass calculation for light nuclei than for heavy nuclei. Contributions from correlations beyond the mean field presumably explain a part of the error in mass formula.

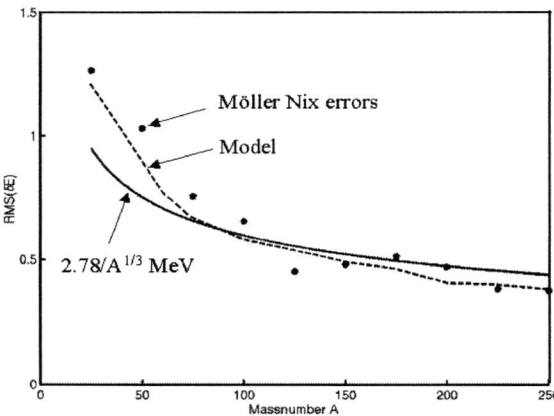

FIGURE 6. RMS value of the chaotic contribution to the fluctuating part of the mass gives 2.78/A1/3 MeV [6] (solod line) is compared to errors in the mass formula by Möller and Nix [15]. The dashed line shows the behavior of the Model described in the text with the strength $\delta = 4.2/A$ MeV. From [14]

SUMMARY

We have shown that decay-out of the SD band provides a unique possibility to *measure* chaos in the excited nucleus. The chaotic situation, when normal-deformed states around the SD state are completely mixed, implies a huge enhancement of the tunneling probability (chaos-assisted tunneling).

We pointed out that the decay-out of the SD band in ^{59}Cu is mediated by doorway states with a fairly small mixing to surrounding ND states. Details of the decay-out could be well understood from a theoretical model of the excited many-body states. The chaotic properties of the ND states at the decay-out were suggested to be small, while lower spin and decay indicated more chaos.

Finally, we discussed the possibility that chaos contributes to the ground state mass, and that (a part) of the error in the mass formula is due to chaos. This type of error will indeed be very difficult to overcome.

By these examples we have stressed that chaos is most relevant for understanding the nucleus. We also emphasize that studies of chaos in the nucleus enhance our understanding of chaos in general in interacting many-body systems.

ACKNOWLEDGEMENTS

This research was supported by the Swedish Research Council.

REFERENCES

1. J.D. Garrett et al, Phys. Lett. **B392** (1997) 24.
2. S. Åberg., Phys. Rev. Lett. **64** (1990) 3119.
3. P. Jacquod and D. Shepelyansky, Phys. Rev. Lett. **79** (1997) 1837.
4. M. Matsuo et al, Nucl Phys **A620** (1997) 296.
5. C. Andreoiu et al, Phys. Rev. Lett., to appear 2003.
6. O. Bohigas and M. Leboeuf, Phys. Rev. Lett. **88** (2002) 092502-1.
7. S. Åberg, Nature **417** (2002) 499.
8. P. Bonche et al, Nucl. Phys. **A519** (1990) 509.
9. S. Åberg, Phys. Rev. Lett. **82** (1999) 299; Nucl. Phys. **A249** (1999) 227.
10. Th. Døssing et al, these proceedings, and to be publ.
11. C. Andreoiu et al, Eur. Phys. J. **A 14** (2002) 317.
12. C. Andreoiu et al, these proceedings.
13. S. Åberg et al, to be publ.
14. H. Olofsson, Master Theses, Math. Phys., Lund 2003; H. Olofsson and S. Åberg, to be publ.
15. P. Leboeuf and A.G. Monastra, Ann. Phys. 297 (2002) 127.
16. P. Möller et al, Atom. Data Nucl. Tables **59** (1995) 185.

Decay-out of ^{151}Tb Yrast Superdeformed Band and Shape Coexistence

G. Duchêne[*], J. Robin[*], A. Odahara[*,†], Th.Byrski[*], F.A. Beck[*], P.J. Twin[**], P. Bednarczyk[*,‡], D. Curien[*], S. Courtin[*], O. Dorvaux[*], B. Gall[*], P. Joshi[*,§], A. Nourreddine[*], E. Pachoud[*,¶], I. Piqueras[*], J.P. Vivien[*,‖], K. Zuber[††], N. Adimi[‡‡], D.E. Appelbe[§§], A. Bracco[¶¶], B. Cederwall[***], D.M. Cullen[**], S. Ertürck[**], G. de France[†††], S.L. King[**], A. Korichi[‡‡‡], K. Lagergren[***], S. Leoni[¶¶], G. Lo Bianco[§§§], A. Lopez-Martens[‡‡‡], S. Lunardi[¶¶¶], B. Million[¶¶], E.S. Paul[**], C. Petrache[§§§], N. Redon[@], A. Saltarelli[§§§], J. Simpson[§§], O. Stézowski[@] and R. Venturelli[¶¶¶]

[*]*Institut de Recherches Subatomiques, UMR 7500, CNRS-IN2P3 et Université Louis Pasteur, F-67037 Strasbourg Cedex 2, France*
[†]*Present address : Nishiniteck University, Japan*
[**]*Oliver Lodge Laboratory, University of Liverpool, P.O. Box 147, Liverpool L69 7ZE, UK*
[‡]*Present address : GSI, Darmstadt 11, Germany*
[§]*Present address : Department of Physics, University of York, York Y010 5DD,UK*
[¶]*Present address : Canberra Eurisys Company, Strasbourg, France*
[‖]*Deceased*
[††]*H. Niewedniczanski Institute of Nuclear Physics, Krakow, Poland*
[‡‡]*USHB, Alger, Algărie*
[§§]*CLRC Daresbury Laboratory, Daresbury, Warrington WA4 4AD, UK*
[¶¶]*Dipartimento di Fisica, Universita' di Milano and INFN, sez di Milano, Italy*
[***]*KTH, Royal Institute of Technology, Physics Department Frescati, Frescativägen 24, S-104 05 Stockholm, Sweden*
[†††]*GANIL, BP 55027, 14076 CAEN Cedex 5, France*
[‡‡‡]*CNSM Orsay, IN2P3-CNRS, bât. 104, 91405 Orsay Campus, France*
[§§§]*Dipartimento di Fisica, Universita' di Camerino and INFN, sez. di Perugia, Italy*
[¶¶¶]*Dipartimento di Fisica, Universita' di Padova and INFN, sez. de Padova, Italy*
[@]*IPNL Lyon, IN2P3-CNRS, Université Claude Bernard Lyon-1, 69622 Villeurbanne Cedex, France*

Abstract. Linking transitions between the superdeformed (SD) and the normal deformed (ND) wells have been searched in ^{151}Tb nucleus. Two experiments of 5 and 17 days have been performed with EUROBALL IV. Transitions of 2818 keV and 3748 keV with intensities of about 1 % relative to the yrast SD band have been observed. Their decay-out properties are discussed in the text. In addition the eight known SD bands have been extended towards higher rotational frequencies where orbital crossings are observed. For the first time, weakly populated collective ND structures, likely triaxial, similar to the ones recently identified in ^{152}Dy, ^{153}Ho and ^{155}Er nuclei have been observed in ^{151}Tb. The SD and ND structures are interpreted in the frame of Woods-Saxon theoretical calculations.

INTRODUCTION

Since the discovery in 1986 of the first superdeformed (SD) band in ^{152}Dy[1] more than 100 SD bands have been identified in the A $\sim$ 150 and 190 mass regions. Despite strong experimental efforts the decay-out path to known normal deformed (ND) states could be established for only few of them. In particular in the A $\sim$ 150 mass region only two cases were published, a double step decay in ^{149}Gd[2] and very recently the decay-out of ^{152}Dy yrast SD band [3]. The interest to determine the decay-out path of SD bands down to known ND states is to establish the excitation energy and spin of each SD state which imposes constraints on the theoretical models particularly in case of identical bands.

Due to the peculiar properties of his SD bands, the ^{151}Tb nucleus has been chosen for that study. The yrast (SD1) and the two first excited (SD2 - SD3) SD bands are strongly populated: SD1 was found to carry $\sim$ 2 % of the total ^{151}Tb decay flux and the SD2 and SD3 bands 50 % and 35 % of SD1 intensity, respectively. In addition, SD2 band is identical within 2.5 keV to ^{152}Dy yrast SD band [4] and decays about 6-8 $\hbar$ lower than SD1 [5]. The feeding pattern into ^{151}Tb ND states of the SD1-SD3 bands decay flux has been studied in EUROGAM I and II data [5 - 9] and the quadrupole moments for SD1 and SD2 have been measured [10].

Two experiments have been performed at EUROBALL IV, one of 5 days and the second one of 17 days. The preliminary data analysis results are presented below. In the first section the SD to ND linking transitions and their properties are given, in the second section the behavior of the known SD bands at extreme rotational frequencies is reported and in the last section shape coexistence in ^{151}Tb is evidenced.

SEARCH FOR SD TO ND LINKING TRANSITIONS

With the aim to identify SD to ND linking transitions in ^{151}Tb nucleus, an experiment has been run at EUROBALL IV in may 1999. The ^{27}Al beam delivered by the IReS, Strasbourg VIVITRON accelerator at 155 MeV was impinging a stack of two thin ^{130}Te targets. These are the best experimental conditions to populate via the 6n evaporation channel a cold ^{151}Tb nucleus at high spin. The targets of $\sim$ 500$\mu g/cm^2$ thickness each was evaporated on Au backings ($\sim$ 250$\mu g/cm^2$ thickness) which were facing the beam. To avoid the target evaporation under beam heat the Te material was covered on the other side by a thin Au layer ($\sim$ 40$\mu g/cm^2$ thickness). Quadruple-and higher-fold data were collected on tape when at least three raw Ge detectors and 16 inner ball BGO counters were fired simultaneously. Up to 850 Mega events corresponding to long γ-ray cascades were registered. As it was the first experiment of EUROBALL after its installation at Strasbourg, several technical problems were encountered. For each Ge channels, two energy ranges are provided, 0 - 4 MeV and 0 - 20 MeV corresponding to calibrations of 0.5 keV/channel and 2.5 keV/channel, respectively. The 0 - 20 MeV range could not be used as the data above 4 MeV were not Compton suppressed. Due to the large non linearity above 3 MeV of the energy calibration for most of the detectors, we have limited our investigations below 3.2 MeV in the 0 - 4 MeV range spectra. The data analysis performed by A. Odahara led to the discovery of a single 2815 keV linking

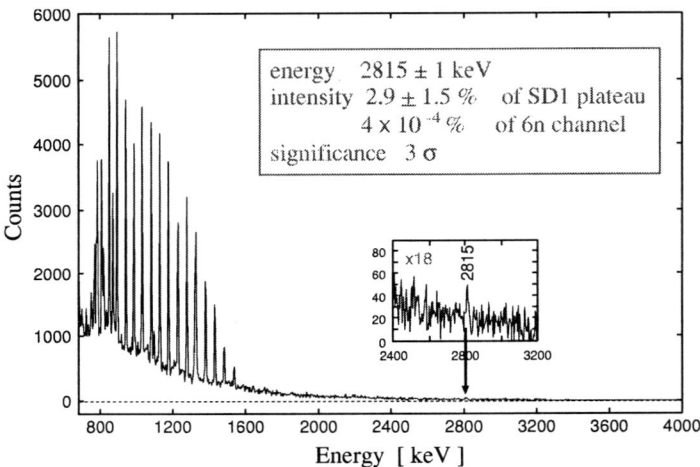

FIGURE 1. Spectrum triple gated on the yrast SD band of ^{151}Tb nucleus obtained with the data from 1999. The insert shows the linking transition to ND states of 2815 keV.

transition (see Fig.1). Its intensity relative to the plateau intensity of band SD1 was estimated to be 2.9 ± 1.5 %

A triple-gated spectrum double-conditioned on the SD transitions and with a third gate on the 2815 keV link candidate was produced. The corresponding background spectrum was obtained by double gates on SD lines and wide gates on left and right sides of the 2815 keV peak. After proper subtraction, the resulting spectrum exhibited SD peaks with a small statistical significance. The 769 keV and 811 keV SD transitions were missing suggesting that 2815 keV decays from the SD state fed by the 854 keV intra SD band transition. No further information could be extracted from the data due to lack of statistics.

A second experiment of 17 days was performed late 2001 with the same reaction at the same energy. The beam intensity was increased by a factor ~ 2. The triggering conditions were modified to enable also the observation of shorter γ cascades like ND rotational bands (see corresponding section). Data were recorded when at least 6 raw Ge detectors and 10(13) inner ball BGO counters were fixed simultaneously. A total of 7 Giga events were stored containing about 5.3 times more SD events.

The data analysis confirmed the existence of a 2818 keV γ-ray and led to the discovery of a new 3748 keV linking transition (see Fig.2). These peaks were observed in both double- and triple-gated spectra independently of the type of background subtracted (total projection, fold-1 and fold-2) and of the energy calibration (1, 2.5 or 4 keV/channel). Their intensities relative to the plateau of SD1 band were determined from a triple-gated spectrum and were found to be both equal to 0.9 (2) %. These values amount to about 2.10^{-4} of the 6n channel cross-section.

We have tentatively deduced the asymetry A of both transitions in order to extract indications on their multipolarity natures. A is defined as the ratio of the peak intensity measured at two different angles. We have projected the triple-gated data on the clover

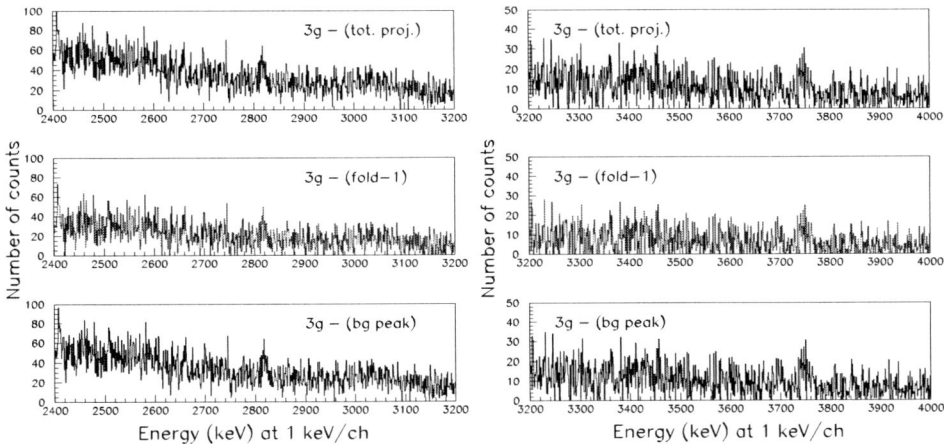

FIGURE 2. Spectra triple gated on SD1 with different background subtraction (see text) showing the 2818 keV (left panel) and 3748 keV (right panel) SD to ND linking transitions.

detectors ($\theta \sim 75°$) or on the tapered and cluster detectors ($\theta \sim 40°$). A is normalised by the photopeak efficiciency $\varepsilon_p\omega$ of the corresponding parts of EUROBALL. For known dipole and quadrupole transitions of the ND level scheme observed in the triple-gated spectrum conditionned by SD1 band energies, A values equal 0.58(2) and 1.11(2), respectively. The A values obtained for the 2818 keV and 3748 keV linking transitions are 0.7 ± 0.2 stat. ± 0.1 syst. and 0.5 ± 0.2 stat. ± 0.1 syst. These results suggest that both the transitions could be dipoles.

In the same spirit of the data analysis performed in 1999, we have produced triple-gated spectra double-conditionned by SD1 band energies with a third gate on one of the linking transitions. A background subtraction similar to the one performed with the 1999 data was realised. In contrary to the tentative assignment of the decaying SD state for the 2818 keV linking transition using the data from 1999, the 811 keV SD1 peak is observed in the present data. The 769 keV transition is still missing which suggests that the 2818 keV transition decays from the second known SD state of ^{151}Tb yrast SD band. Both the 769 keV and 811 keV SD1 peaks are missing in the spectrum gated by the 3748 keV linking transition which indicates that this γ ray decays the third known SD state of band SD1.

Due to lack of statistics, the precise entry point in the ND level scheme could not be firmly established. Further investigations are still in progress.

THE KNOWN ^{151}TB SD BANDS AT EXTREME ROTATIONAL FREQUENCIES

The properties of the second well in ^{151}Tb nucleus are pretty well known. Eight SD bands have been assigned to this nucleus [11]. The yrast band is built on a $\pi 6^3 \otimes \nu 7^2$

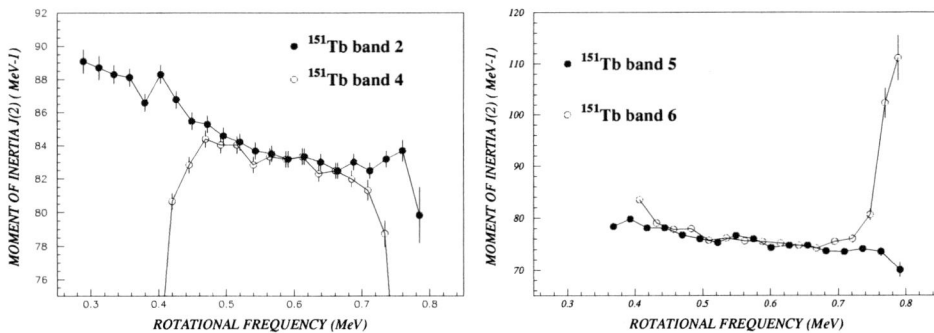

FIGURE 3. Dynamical moment of inertia $\mathscr{J}^{(2)}$ of the signature partner SD bands SD4 and SD2 (left panel) and SD5 and SD6 (right panel).

intruder orbital configuration. Three excited SD bands, SD2 to SD4, correspond to a proton excitation and show dynamical moments of inertia $\mathscr{J}^{(2)}$ similar to the one of the yrast SD band of ^{152}Dy involving the $\pi 6^4 \otimes \nu 7^2$ intruder configuration. Four excited SD bands, SD5 to SD8, are based on neutron excitations. Their dynamical moment of inertia $\mathscr{J}^{(2)}$ are similar to the one of ^{150}Tb yrast SD band with the $\pi 6^3 \otimes \nu 7^1$ intruder configurations. The bands SD5 and SD6 as well as SD7 and SD8, are signature partners.

With the present high-statistics data set most of these bands have been extended at both high and low rotational frequencies. As an example, three, two and three new transitions have been observed at the top of bands SD4, SD5 and SD6, respectively. Two new transitions have been added at the bottom of band SD4. The deduced dynamical moments of inertia $\mathscr{J}^{(2)}$ shown in Fig.3 present strong deviations above $\hbar\omega = 0.7$ MeV versus the $J^{(2)}$ tendancy observed at lower rotational frequency. Such observations are of great importance as they enable to test the theoretical models in extreme conditions of spin, at the limit of fission.

Band SD4 corresponds to the promotion of a proton from the natural parity orbital [301]1/2, $\alpha = + 1/2$ to the second [770]1/2 intruder orbital. In Fig.3 left panel, the dynamical moment of inertia of band SD4 is compared to the one of band SD2. They are identical in the rotational-frequency range 0.45 MeV $< \hbar\omega <$ 0.7 MeV. However the new $\mathscr{J}^{(2)}$ points from this work lead to two sharp downbends at both ends of the frequency range probably due to the occurence of orbital crossings. The proton routhian (Fig. 4 left panel) obtained from cranked deformed Wood-Saxon calculations [12] does not show such crossings for the [301]1/2, $\alpha = + 1/2$ orbital up to $\hbar\omega = 0.8$ MeV. However, at very large rotational frequency the nucleus deformation is expected to change which could influence all deformation parameters, even the γ parameter, with a possible onset of triaxiality.

The dynamical moments of inertia of bands SD5 and SD6 are shown in Fig.3 right panel. The $\mathscr{J}^{(2)}$'s exhibit a downbend and a sharp upbend at $\hbar\omega \sim 0.75$ MeV for SD5 and SD6, respectively. These signature partner SD bands have been assigned the $\pi 6^3 \otimes \nu 7^2([402]5/2)^{-1}$ configuration. The [402]5/2 orbital however does only show

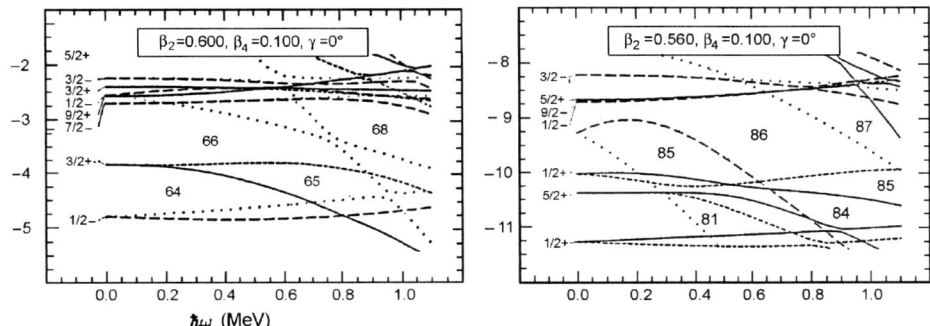

FIGURE 4. Single-particle energy calculated for the proton configurations of band SD4 (left panel) and for the neutron configurations of bands SD5 and SD6 (right panel).

a sharp crossing at extremely high rotational frequency for $\hbar\omega \sim 1$ MeV (Fig.4 right panel). Deformation change at very high spin could reduce the rotational frequency of the crossing for the [402]5/2 orbital. Therefore theoretical calculations are presently performed for the eight SD bands of ^{151}Tb to check the previous configuration assignments by exploring $\lambda = 2, 3, 4$ deformation degrees of freedom for a better understanding of the occurence of strong orbital interactions around $\hbar\omega \sim 0.7$ MeV.

SHAPE COEXISTENCE IN ^{151}TB NUCLEUS: DISCOVERY OF ND ROTATIONAL BANDS

The nucleus ^{152}Dy is the archetype of shape coexistence. The four first states of the nucleus are of vibrational nature. On top of them single-particle configurations develop at higher spins in parallel with a well deformed rotational band labelled A [13]. More recently two new ND rotational bands labelled B and C were discovered and studied [14,15]. Bands A, B and C were interpreted as triaxial rotational bands with deformation parameters $\beta_2 = 0.30, \beta_4 = -0.004$ and $\gamma = 18°$.

Revisiting recently EUROGAM II data on ^{151}Tb nucleus, two rotational bands labelled A and B have been discovered. In the EUROBALL IV data of 1999 they could be observed but with less statistics and a much worse peak-to-background ratio. Therefore, we decided to impose less severe triggering conditions in the 2001 experiment in order to enable the selection of shorter γ-ray cascades. Bands A and B were clearly observed (Fig.5 top and middle panels) and a new band labelled C was discovered (Fig.5 bottom panel) [16]. These bands are weakly populated, about 4 times less than the ^{151}Tb yrast SD band ($I_A = 0.45\%, I_B = 0.30\%$ and $I_C = 0.40\%$) with average γ-ray spacings of 65 keV, 60 keV and 67 keV, respectively. The dynamical moments of inertia of ^{151}Tb bands A and B are compared to the one of ^{152}Dy band A (Fig. 6 left panel) and a similar plot is given in Fig.6 right panel for bands C of both nuclei. The $\mathscr{J}^{(2)}$ values are very similar to their corresponding ^{152}Dy bands in the frequency range 0.35 MeV $\leq \hbar\omega \leq 0.5$ MeV for bands A and B and 0.5 MeV $\leq \hbar\omega \leq 0.7$ MeV for band C which suggests that the new

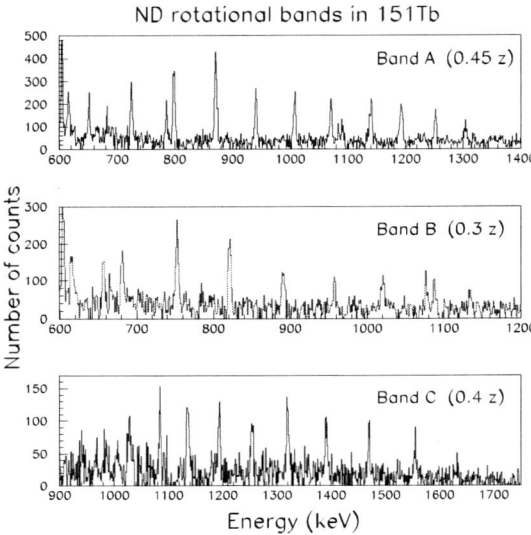

FIGURE 5. ND collective rotational band A, B and C observed for the first time in ^{151}Tb. their intensity relative to the 6n channel cross-section is given in percentage in brakets.

collective rotational bands of ^{151}Tb are built on similar configurations than bands A and C of ^{152}Dy which are triaxial. Theoretical calculations using a deformed Wood-Saxon potential are currently undertaken for orbital assignments and interpretation.

No decaying transitions to the single-particle states have been evidenced yet. However, the triple-gated spectra on the collective bands A and B exhibit only γ-ray transitions from the known level scheme below 35/2 $\hbar$. Band A feeds most probly the states 33/2$^+$ and 35/2$^+$ and band B the states 31/2$^+$, 31/2$^-$ and 27/2$^+$.

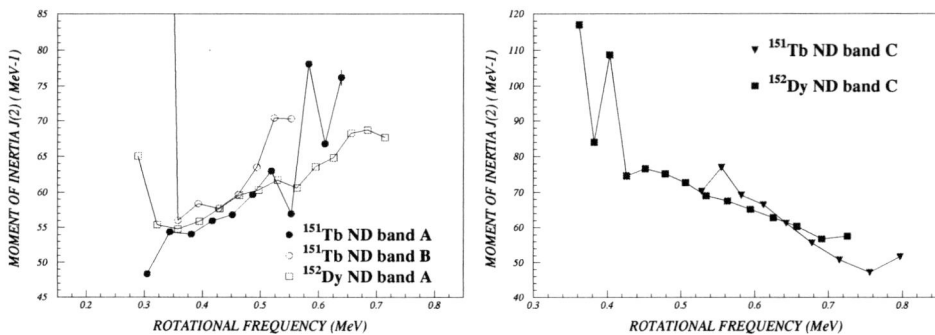

FIGURE 6. Dynamical moment of inertia of ^{151}Tb bands A and B are compared to the one of ^{152}Dy band A (left panel). A similar plot is given for bands C of both nuclei in the right panel.

CONCLUSION

Two EUROBALL IV experiments of 5 and 17 days have been run in 1999 and 2001, respectively. Two SD to ND linking transitions of 2818 keV and 3748 keV have been observed for the yrast SD band of ^{151}Tb. The 2818 keV and 3748 keV transitions which have been tentatively assigned a dipole nature, decay out from the SD states fed by the 811 keV and the 854 keV intra-band transitions, respectively. The precise entry point in the ND level scheme of ^{151}Tb nucleus could not be firmly established and further investigations are still in progress.

The known SD bands have been extended at lower and higher rotational frequencies. Above $\hbar\omega = 0.7$ MeV the dynamical moment of inertia of bands 4 and 5 show down-bends and the one of band 6 presents an upbend. These structures test the theoretical models at the limit of fission. They could be the signature at very high spin of a change in deformation including an onset of triaxiality.

Three ND rotational bands A, B and C have been discovered in ^{151}Tb leading to prolate-oblate shape coexistence in this nucleus. Bands A and B (C) of ^{151}Tb are built on a configuration similar to that of band A (C) in ^{152}Dy which suggests that they present a triaxial deformation as the ^{152}Dy ND collective bands. Bands A and B feed ND states in the vicinity of spin $31/2\hbar$.

ACKNOWLEDGMENTS

This work was supported by the European EUROVIV contract number HPRI-CT-1999-00078

REFERENCES

1. Twin, P.J., et al., Phys. Rev. Lett. **57**, 811 (1986)
2. Finck, C., et al., Phys. Lett. **B 467**, 15 (1999)
3. Lauritsen, T., et al., Phys. Rev. Lett. **88**, 042501 (2002)
4. Byrski, T., et al., Phys. Rev. Lett. **64**, 1650 (1990)
5. Curien, D., et al., Phys. Rev. Lett. **71**, 2559 (1993)
6. Beck, F.A., et al., Nucl. Phys. **A 557**, 67c (1993)
7. Petrache, C.M., et al., Nucl. Phys. **A 579**, 285 (1994)
8. Petrache, C.M., et al., Phys. Scripta **T56**, 299 (1995)
9. Finck, C., Ph. D. Thesis, Université Louis Pasteur, Strasbourg, France, CRN 97-18 n° 2632
10. Finck, C., et al., Eur. Phys. **J A 2**, 123 (1998)
11. Kharraja, B., Ph. D. Thesis, Université Louis Pasteur, Strasbourg, France, CRN 94-26 n° 1806
12. El Aouad, N., et al., Nucl. Phys. **A 676**, 155 (2000)
13. Nyako, B.M., et al.,Phys. Rev. Lett. **56**, 2680 (1986)
14. Smith, M.B., et al., Phys. Rev. **C61**, 034314 (1999)
15. Appelbe, D.E., et al., Phys. Rev. **C66**, 044305 (2002)
16. Robin, J., et al., *Linking transition and exotic shapes in* ^{151}*Tb*, Symposium on "Nuclear Structure Physics with EUROBALL: achievements 1997-2002", Orsay, march 19-20, 2002

Determining the Excitation Energy, Spin, and Parity of Levels in the Superdeformed Bands of ^{152}Dy

T. Lauritsen*, M.P. Carpenter*, R.V.F. Janssens*, T.L. Khoo*, P. Fallon[†], B. Herskind**, D.G. Jenkins*, F.G. Kondev*, A. Lopez-Martens[‡], A.O. Macchiavelli[†], D. Ward[†], K.S. Abu Saleem*, I. Ahmad*, R. Clark[†], M. Cromaz[†], T. Døssing**, J.P. Greene*, F. Hannachi[‡], A.M. Heinz*, A. Korichi[‡], G. Lane[†], C.J. Lister*, P. Reiter*[§], D. Seweryniak*, S. Siem*[¶], R.C. Vondrasek* and I. Wiedenhöver*[∥]

*Argonne National Laboratory, Argonne, Illinois 60439, USA.
[†]Lawrence Berkeley National Laboratory, Berkeley, California 94720, USA
**Niels Bohr Institute, DK–2100, Copenhagen, Denmark
[‡]C.S.N.S.M, IN2P3-CNRS, bat 104-108, F-91405 Orsay Campus, France
[§]Ludwig-Maximilians-Universität, Munich, Germany
[¶]University of Oslo, Oslo, Norway.
[∥]Florida State University, Tallahassee, Florida 32306.

Abstract. The excitation energy, spin and parity of the yrast superdeformed (SD) band in ^{152}Dy have been established. The excitation energy of the lowest observed level in this band is found to be 10,644 keV with an assigned spin and parity of 24^+. In addition, nine transitions of dipole character have been identified, which connect the excited superdeformed band 6 in ^{152}Dy to the yrast SD band. The excitation energy of the lowest level in the excited SD band 6 is 14,239 keV. The spin and parity of this state has been determined to be either 29^- or 31^-. The measured properties of SD band 6 are consistent with an interpretation in terms of a rotational band built on an octupole vibration. A comparison with an RPA calculation by Nakatsukasa et al. suggests that the spin of the lowest SD band 6 level is 31^-.

LINKING SD BAND 1 TO THE NORMAL YRAST LINE

The first superdeformed (SD) band was observed in 1986 in the nucleus ^{152}Dy [1]. A sequence of nineteen almost equally spaced γ rays revealed the presence of a SD nucleus with an axis ratio of nearly 2:1. Since then $\sim$175 SD bands have been found in the A=150 and A=190 SD mass regions. However, the energy, spin and parity of only a handful of SD bands in the A=190 region have been established, the first being in ^{194}Hg from a Gammasphere experiment performed in 1995 [2].

Following an experiment peformed in 1999 while Gammasphere was cited at Argonne National Laboratory, a 4011 keV transition was observed in conicidence with the yrast SD band of ^{152}Dy, however, the statistics were not sufficient to determine whether this transition directly linked the SD band to known states. Consequently, a much longer experiment (12 days) was performed with Gammasphere at Lawrence Berkeley National

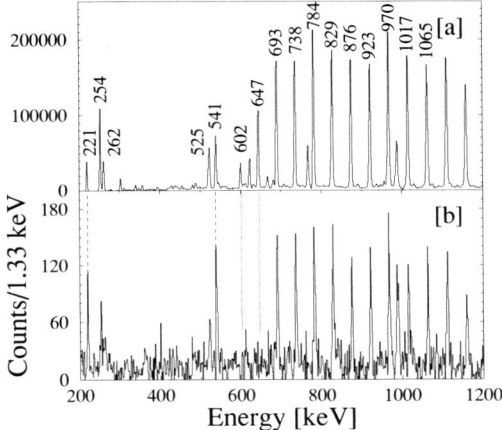

FIGURE 1. [a] Spectrum from pairwise coincidence gates in the yrast SD band of ^{152}Dy. [b] Spectrum obtained from setting pairwise gates on a SD line and the 4011 keV transition. All SD transitions listed above (except that of 647 keV) were used as gates.

Laboratory utilizing the reaction ^{108}Pd(^{48}Ca,4n)^{152}Dy at 191 MeV (mid–target).

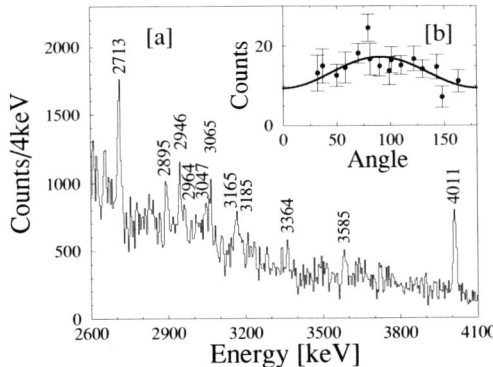

FIGURE 2. [a] High-energy portion of the spectrum in coincidence with ^{152}Dy SD transitions. [b] Angular distribution of the 4011 keV transition.

Fig. 1a shows the spectrum obtained by placing pairwise coincidence gates on transitions in the yrast SD band in ^{152}Dy. At higher γ–ray energies, shown in Fig. 2a, several candidates for decay–out transitions are clearly seen. The coincidence spectrum obtained by placing pairwise gates on SD lines and on the 4011 keV transition is presented in Fig. 1b. It is clearly seen that the 647 and 602 SD lines (see Fig. 3) are not in coincidence with the 4011 keV γ ray, whereas the 693 keV and higher SD lines are. This unambiguously establishes that the 4011 keV transition originates from the SD level fed by the 693 keV line. Of the normal yrast transitions in the spectrum [3], the 221 and

541 keV γ rays have the full intensity of the SD band (0.8(3) and 0.9(2), respectively), whereas there is no indication of the normal 967 keV transition (the intensity is 0.3(3)), which should be detectable despite the proximity of the 970 keV SD γ ray. A comparison with the spectrum in Fig. 1a shows no new peaks with an area larger than 3 standard deviations. This strongly suggests that the 4011 keV γ ray feeds directly into the 27⁻ yrast state in a single step. This establishes the excitation energy of the SD level fed by the 693 SD line as 11,893 keV – as shown in the partial level scheme of Fig. 3.

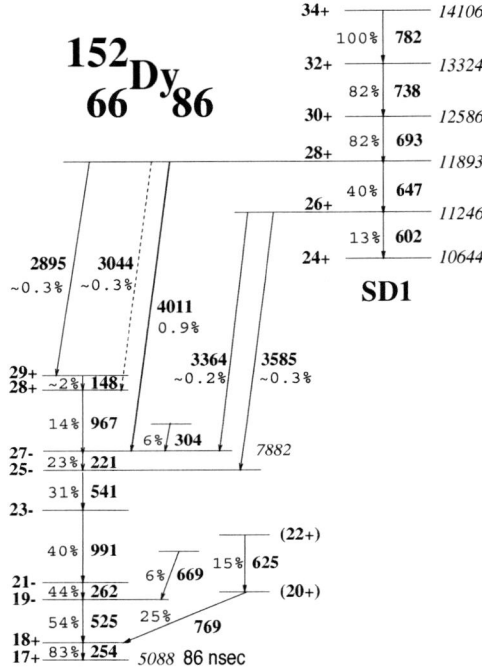

FIGURE 3. Partial level scheme of ^{152}Dy showing the lowest part of the yrast SD band and normal states to which the SD band mainly decays.

To determine the spin of the 11,893 keV SD level an angular distribution analysis of the 4011 keV transition was performed. The intensity of this γ ray, as a function of polar angle, is presented in Fig. 2b. Using the functional form $W(\theta) = A_0(1 + A_2 P_2(cos\theta) + A_4 P_4(cos\theta))$ [4], the angular distribution coefficients were determined to be: A_2=−0.35(12) and A_4=−0.02(16) – consistent only with a stretched or anti–stretched dipole character [4]. Thus, based on the 4011 keV transition, the feeding SD level must have a spin of either 26 or 28 $\hbar$.

Based on Weisskopf estimates [5], transitions of E1 character are expected to be nearly two orders of magnitude faster than those with M1 multipolarity. In neutron capture experiments, indeed, the E1 transitions have been shown to dominate [6]. Therefore, it is most likely that the 4011 keV transition is of E1 character and, thus, a positive parity is assigned to the SD band. This assignment is also supported by theoretical expectations [7].

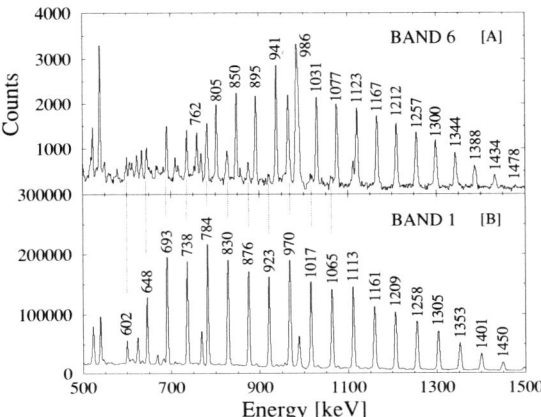

FIGURE 4. [A] Spectrum from triple coincidence gates on lines in SD band 6 of ^{152}Dy. [B] Spectrum obtained from setting pairwise gates on clean SD lines in band 1 of ^{152}Dy [8].

With the energy of the SD band determined by the 4011 keV line, four additional γ rays in the 3 MeV region of figure 2 can be placed in the level scheme as direct links between the SD band and the normal states. They are included in figure 3. All four of these additional links are very weak and, thus, it is difficult to place coincidence gates on them. However, in a spectrum of pairwise gates placed on the 693 keV line and on clean SD transitions above it, the 2895, 3044, 3364 and 3585 keV γ rays are clearly present, whereas a similar spectrum with a gate on the 647 keV line only shows the two transitions with the highest energy, *i.e.*, the 2895 and 3044 keV γ rays are absent. Thus, the latter two transitions emanate from the 11893 keV SD level, and the two others are associated with the deexcitation of the SD state directly below it at 11,246 keV. These weaker one–step linking transitions also resolve any remaining ambiguity concerning the spins of the SD band members. Only when 28^+ is assigned to the 11893 SD level are the multipolarities of the 2895 and 3364 keV lines reasonable: M1 and E1, respectively. A 26^+ assignment would result in respective M3 and E3 multipolarities which are improbable, be it only because of the competition with the in-band, highly collective 602 keV γ ray.

The 4011 keV one-step decay line carries only 0.9(2)% of the intensity of the SD band. The quadrupole moment of the SD band in ^{152}Dy, 17.5(2) eb [9], gives a partial lifetime of the 647 keV in-band transition of 66 fs and a partial lifetime of the 4011 keV transition of 2.9 ps, equivalent to a strength in Weisskopf units (W.U.) of $\approx 2 \times 10^{-6}$. Just as in the A=190 mass region [2], the decay–out transition is very retarded.

OCTUPOLE EXCITATIONS IN THE SUPERDEFORMED WELL

Evidence for octupole vibrations in SD nuclei was first reported in the case of ^{190}Hg [10, 11, 12, 13, 14] where SD band 2 was found to deexcite into the yrast SD band via

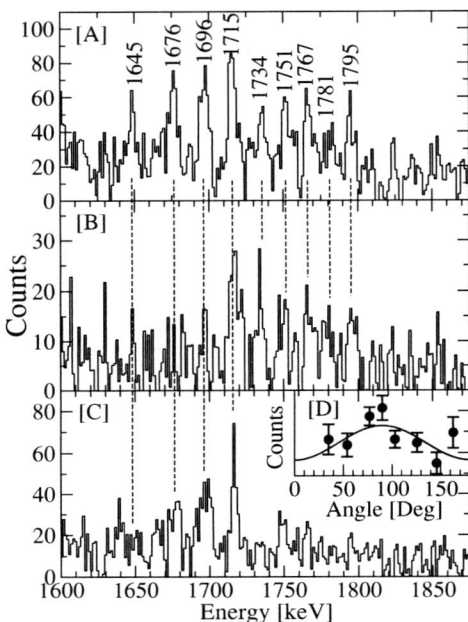

FIGURE 5. [A] Summed coincidence spectrum obtained by placing gates on clean SD band 6 high energy transitions and SD band 1 low energy transitions. The nine transitions linking SD band 6 to band 1 are marked with their energies. [B] As A, but requiring the 830 keV transition in band 1. [C] As A, but requiring the 895 keV transition in band 6 and any SD band 1 transitions below 876 keV. [D] Angular distribution of the sum of the intensities in the 1676, 1696 and 1715 keV linking transitions vs. the polar angle of the Gammasphere detectors.

E1 transitions with rather low energy ($\sim$ 800 keV) and high transition rates. Similar evidence has also been reported for ^{194}Hg and $^{196-198}$Pb (see references in [15]).

The SD minima in both the A $\sim$ 150 and A $\sim$ 190 regions are calculated (see references in [15]) to be soft with respect to octupole deformation because of the presence of intruder orbitals ($j_{15/2}$ neutrons and $i_{13/2}$ protons) near the Fermi surface, where they are close to levels of opposite parity differing by three units ($\Delta l = 3$) in angular momentum ($g_{9/2}$ neutrons and $f_{7/2}$ protons). In fact, based on RPA calculations, Nakatsukasa et al. [16] proposed that most low excitations in A$\sim$190 even-even SD nuclei are associated with octupole vibrations. Remarkably, while compelling evidence exists in the A $\sim$ 190 region, the situation is quite different near A $\sim$ 150. There is some evidence for interband transitions only for a single band in both ^{150}Gd and ^{152}Dy [17, 18]. In the latter nucleus, five excited bands are known [18], and it is one of the weakest of those (band 6) that has been proposed to decay into the yrast SD band, *i.e.* the transitions of the yrast SD band were observed to be in coincidence with the γ rays of SD band 6, but the transitions linking the two bands were not observed. Nevertheless, based on this fragmentary evidence an interpretation in terms of an octupole vibration

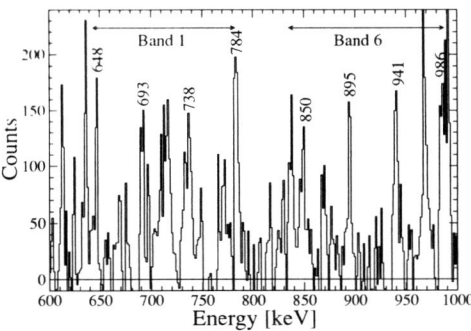

FIGURE 6. Summed coincidence spectrum obtained by placing gates on the 1696 keV linking transition and clean lines in SD band 6.

was proposed in Ref. [19]. The other ^{152}Dy excited SD bands are understood in terms of proton and neutron excitations across the Z=66 and N=86 SD shell gaps [19, 18].

Using the current dataset, transitions linking SD band 6 to band 1 were discovered [15] and confirming the previous ascertion that band 6 decays to band 1 (see Fig 4). In fact, 53(8)% of the decay of band 6 proceeds through band 1. Fig. 5A presents the high energy part of the coincidence spectrum of Fig. 4A. Nine weak transitions can be seen from 1645 to 1795 keV. Coincidence gates were placed on the 1696 keV transition together with relevant lines in SD band 6: the resulting spectrum is given in Fig. 6. It clearly shows only transitions in SD band 6 with energies $E_\gamma \geq 850$ keV, *i.e.* the 805 keV and 762 keV γ rays of the sequence are clearly missing. Under the same coincidence conditions, only transitions of the yrast SD band with $E_\gamma \leq 784$ keV are observed. This unambiguously establishes the ordering proposed in the level scheme of Fig. 7. Additional supporting evidence is given in Fig. 5. In panel B, the γ cascades are required to pass through SD band 6 *and* include the 830 keV band 1 transition: the decay–out transitions below 1715 keV are clearly absent. Conversely, in panel C, the γ cascades are required to pass through the lower part of band 1 *and* include the 895 keV band 6 transition. Now the lower decay-out transitions are present, but the upper ones, from 1734 keV on, are missing. An angular distribution analysis of the three linking transitions at 1676, 1696 and 1715 keV finds negative A_2 coefficients in every case (-0.9(4), -0.3(3) and -0.3(3), respectively). If the yields of these three lines are added up and analyzed together, the combined A_2 coefficient is determined to be -0.5(2) which is consistent with those expected for stretched or anti-stretched $\Delta I=1$ transitions (-0.24).

RPA calculations by Nakatsukasa *et al.* [19] interpret SD band 6 as an octupole excitation with signature $\alpha = 1$. At zero frequency, the band is characterized by $K = 0$, but K-mixing is significant at the frequencies of interest here because of the Coriolis force. Experiment and calculations are compared in Fig. 8, where the Routhian of band 6 with respect to the yrast SD band is given as a function of the rotational frequency. The figure presents the lowest octupole excitation (dashed line), and the first 1p–1h configuration (solid line). The calculations reproduce the magnitude and evolution with frequency of the $\mathfrak{I}^{(2)}$ moment of inertia satisfactorily (see Fig. 3 in [19]). From Fig. 8, it is clear that the excitation energy and the evolution of the Routhian with frequency are best

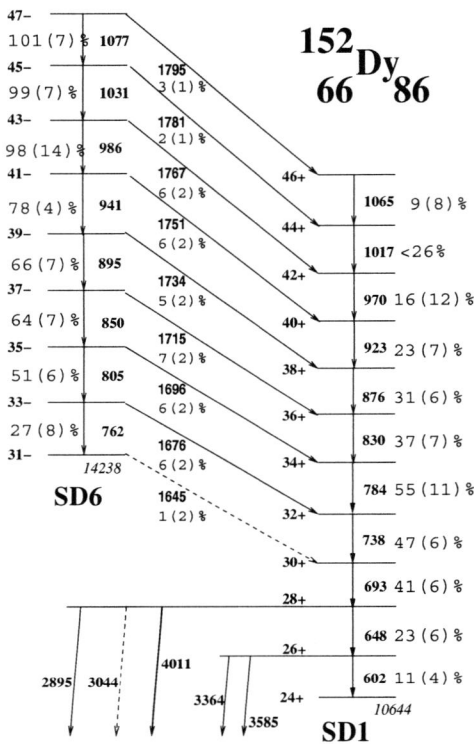

FIGURE 7. Partial level scheme of ^{152}Dy showing the lowest part of SD band 6, the lowest part of the yrast SD band 1 and the transitions that link the yrast SD band 1 to the normal states [8]. The transition intensities are with respect to the strongest lines in SD band 6 (the 1031 and 1077 keV lines). The intensities shown for band 1 are from the feeding by band 6 only.

reproduced when the interband transitions are considered to be of the J+1→J type. This argues for the spin assignment given in Fig. 7.

SUMMARY

In summary, a high statistics experiment with Gammasphere has allowed for the linking of the yrast superedeformed band with previously identified states in ^{152}Dy. As a result, the excitation energies, spins and parities of the levels in this SD band have been determined for the first time. The lowest identified level in the band is at 10,644 keV with a spin/parity assignment of 24^+. In addition, nine transitions have been observed linking SD band 6 with the SD yrast band. These transitions are found to be of dipole character. As a result the excitation energy of the lowest level in SD band 6 is at 14,239 keV with a spin/parity assignment of either 29^- or 31^-. The deduced B(E1) rates are consistent with the theoretical interpretation that band 6 is built on an octupole vibrational state.

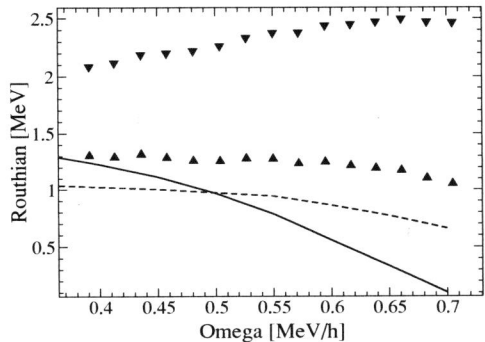

FIGURE 8. Routhians of band 6 with respect to band 1 as a function of rotational frequency. The up (down)–triangles are the data with the high (low) spin assignments to band 6. The lines are the result of the RPA calculations [19] for negative–parity states with signature $\alpha = 1$. The dashed line characterizes the lowest SD excitation associated with a octupole vibration. The solid line likewise shows the lowest 1p–1h excitation which, according to [19], corresponds to SD band 2.

ACKNOWLEDGMENTS

Valuable suggestions by and discussions with K. Hauschild, A. V. Afanasjev and T. Nakatsukasa are appreciated. This work was supported in part by the U.S. Dept. of Energy, under Contract No. W–31–109–ENG–38 and DE–AC03–76SF00098, and the Danish Natural Science Foundation.

REFERENCES

1. P. J. Twin *et al.*, Phys. Rev. Lett. **57**, 811 (1986).
2. T. L. Khoo *et al.*, Phys. Rev. Lett. **76**, 1583 (1996).
3. B. Hass *et al.*, Nucl. Phy. **A362**, 254 (1981).
4. E. Mateosian and A. W Sunyar, At. Data Nucl. Data Tables **13**, 391 (1974).
5. J. M. Blatt and V. F. Weisskopf, Theoretical Nuclear Physics, Wiley, New York, 1952.
6. L. M. Bollinger and G. E. Thomas, Phys. Rev. **C2**, 1951 (1970).
7. I. Ragnarsson and S. Åberg, Phys. Lett. **B180**, 191 (1986).
8. T. Lauritsen *et al.*, Phys. Rev. Lett. **88**, 042501 (2002).
9. D. Nisius *et al.*, Phys. Lett. **B392**, 18 (1997).
10. B. Crowell *et al.*, Phys. Lett. **B333**, 320 (1994).
11. B. Crowell *et al.*, Phys. Rev. **C51**, R1599 (1995).
12. A. N. Wilson *et al.*, Phys. Rev. **C54**, 559 (1996).
13. H. Amro *et al.*, Phys. Lett. **B413**, 15 (1997).
14. A. Korichi *et al.*, Phys. Rev. Lett. **86**, 2746 (2001).
15. T. Lauritsen *et al.*, Phys. Rev. Lett. **89**, 282501 (2002).
16. T. Nakatsukasa *et al.*, Phys. Rev. **C53**, 2213 (1996).
17. P. Fallon *et al.*, Phys. Rev. Lett. **73**, 782 (1994).
18. P. J. Dagnall *et al.*, Phys. Lett. **B335**, 313 (1994).
19. T. Nakatsukasa, K. Matsuyanagi, S. Mizutori, and W. Nazarewicz, Phys. Lett. **B343**, 19 (1995).

Investigation of Dipole Bands in the ^{142}Gd region with EUROBALL

R.M. Lieder*, A.A. Pasternak*† and E.O. Podsvirova*†

*Institut für Kernphysik, Forschungszentrum Jülich, D-52425 Jülich, Germany
†A.F. Ioffe Physical Technical Institute RAS, RU-194021 St. Petersbourg, Russia

Abstract. Lifetimes have been measured for dipole bands in ^{142}Gd and ^{141}Eu. The resulting $B(M1)$ and $B(E2)$ values are compared with calculations in the framework of the TAC (Tilted Axis Cranking) and SPAC (Shears mechanism with Principal Axis Cranking) models. The dipole bands can be interpreted as magnetic rotational bands with a significant contribution of collectivity.

INTRODUCTION

Magnetic rotations (MR) [1, 2, 3] are expected in regions close to magic numbers if high-j nucleons are available and the deformation is small. Also for the $N \approx 82$ region MR bands are predicted and a search for such bands has been carried out for the nuclei ^{142,143}Gd and ^{141}Eu. High-spin states in these nuclei have been studied with EUROBALL III using the ^{99}Ru(^{48}Ti,ypxn) reaction at a beam energy of 240 MeV. Several dipole bands have been observed in each of these nuclei. The four dipole bands found in ^{142}Gd have been interpreted as MR bands [4] from a comparison of their features with predictions of the Tilted Axis Cranking (TAC) model [3]. Such features are the angular momentum I as function of the rotational frequency $\hbar\omega$ and the ratio of reduced transition probabilities $B(M1)/B(E2)$ resulting from the branching ratio λ of the crossover and cascade transitions in the dipole bands. Two of the dipole bands in ^{142}Gd are considered to have $\pi h_{11/2}^2 \otimes \nu h_{11/2}^{-2}$ and $\pi h_{11/2}^1 \otimes \pi g_{7/2}^{-1} \nu h_{11/2}^{-2}$ configurations The other two originate from those by the breakup of a second $h_{11/2}$ neutron-hole pair.

The dipole bands in ^{143}Gd and ^{141}Eu have similar features and some of them may result from the $\pi h_{11/2}^2 \otimes \nu h_{11/2}^{-2}$ band in ^{142}Gd by subtraction of a proton and a neutron hole, respectively. A proof for the interpretation of these dipole bands as MR bands may result from lifetime measurements and the determination of $B(M1)$ and $B(E2)$ values.

EXPERIMENTAL RESULTS

With EUROBALL IV at IReS Strasbourg a lifetime experiment using the Doppler-shift attenuation method (DSAM) has been carried out for the nuclei ^{142,143}Gd and ^{141}Eu to determine the $B(M1)$ and $B(E2)$ values of their dipole bands. To minimize the sidefeeding lifetimes, a ^{114}Sn(^{32}S,2p2n) reaction at a beam energy of 160 MeV has been used. As target a selfsupporting metallic ^{114}Sn foil of 8 mg/cm^2 thickness

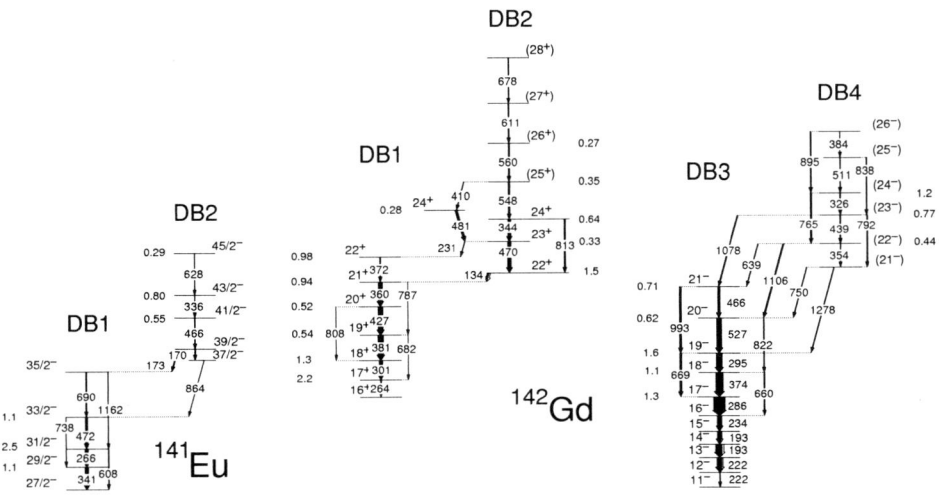

FIGURE 1. Partial level schemes of dipole bands in ^{142}Gd and ^{141}Eu. The lifetimes are given in ps.

with an enrichment of 71.1% has been used. The initial recoil velocity is $v/c = 2.2\%$ for this reaction. EUROBALL IV (consisting of 14 CLUSTER, 26 CLOVER and 30 individual Compton-suppressed Ge detectors) was equipped with an inner BGO ball and the particle detector array EUCLIDES to increase the selectivity for the reaction channel of interest. Events were recorded when ≥ 5 unsuppressed Ge detectors fired in coincidence. Approximately $4.3 \cdot 10^9$ high-fold γ-events have been collected. The energy and efficiency calibrations of the EUROBALL IV array have been carried out with a ^{152}Eu source.

To extract lifetimes the data have been sorted into several γ-γ-γ-cubes for which the γ-ray energies from any detector were stored into the first two axes and the energies from detectors placed under a certain angle with respect to the beam direction in the third axis of the respective cube. The DSAM line shape analysis was hence carried out in doubly-gated coincidence spectra. The detector rings at average angles of 133° (10 CLUSTER detectors), 103° (13 CLOVER detectors) and 77° (13 CLOVER detectors) have been used, because the corresponding spectra have the highest statistics. These cubes contain $54 \cdot 10^9$, $54 \cdot 10^9$ and $59 \cdot 10^9$ γ-γ-γ-events, respectively. The energy and time calibrations, the presort and the sort of the γ-γ-γ-cubes has been carried out with the software package *Ana* [5]. Since the instrumental lineshapes and efficiencies, respectively, of the various detector rings differ from each other they have been determined separately for each ring. No limits on the BGO fold were set and no particle selection was made.

The analysis of experimental DSAM lineshapes was carried out using updated versions of the Monte-Carlo codes COMPA, GAMMA and SHAPE [6]. The "wide gate below" technique has been applied to obtain sufficient statistics in the analysed spectra. Therefore, sidefeeding has to be taken into account. The time distributions of the side

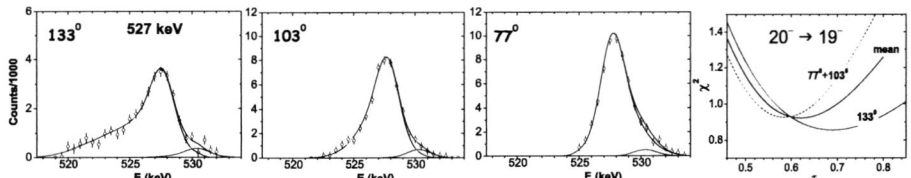

FIGURE 2. Lineshape analysis for the 527 keV $20^- \to 19^-$ transition of DB3 in ^{142}Gd. The lifetimes have been obtained in a χ^2 analysis. The avarage value with its statistical uncertainty is $\tau = 0.62 \pm 0.13$ ps.

feeding have been calculated with Monte-Carlo methods and the relevant parameters were determined by the investigation of γ-ray multiplicity spectra for $^{142-146}$Gd [7]. In some cases the spectra for the symmetric angles of 77° and 103° have been added to improve the statistics. The resulting DSAM lineshapes are then symmetric. The uncertainties result from statistical contributions and from uncertainties related to the cascade- and side-feeding pattern as well as from the uncertainty of the stopping power for the recoils. The statistical uncertainties are obtained by a χ^2 analysis taking into account the variable parameters of all peaks in the fitted multiplet. Intensities, important for taking into account cascade feeding, have been evaluated from the fit of the DSAM lineshapes in the spectra.

Sofar lifetimes have been deduced for two dipole bands in ^{141}Eu and four dipole bands in ^{142}Gd. In figure 1 partial level schemes of ^{141}Eu and ^{142}Gd for these bands are shown and the lifetimes are given. As an example, the DSAM lineshape analysis for the 527 keV $20^- \to 19^-$ transition of the dipole band DB3 in ^{142}Gd is displayed in figure 2.

DISCUSSION

Calculations have been carried out in the framework of the TAC model [3] and a semi-classical model based on the Clark and Macchiavelli approach [2]. The MR bands result when the angular momentum vectors of the involved particles and holes $\vec{j}_1$ and $\vec{j}_2$ are initially oriented approximately perpendicular to each other. A few semiclassical scenarios are considered to calculate the features of dipole bands. In the limiting case, when $\vec{j}_1$ is aligned along the symmetry (z) axis and $\vec{j}_2$ along the rotation (x) axis, considered in the framework of the principal axis cranking (PAC) coupling scheme [8], the angular momentum vector $\vec{R}$ of the collective rotation is parallel to the x-axis. In the case when the total spin is produced solely by the shears effect (the angular momentum of the band is increased by the gradual alignment of the individual nucleon spins along the total angular momentum vector) and the collective rotation is absent, only the angle between the angular momentum vectors $\vec{j}_1$ and $\vec{j}_2$ is considered and a semiclassical description is possible [2]. Subsequently, collective rotation has been added to this model [9] in the framework of a uniform core rotation coupling scheme, when the vectors $\vec{R}$ and $\vec{I} = \vec{j}_1 + \vec{j}_2$ are parallel to each other, corresponding to the TAC model [3].

We have chosen another approach in which the shears mechanism is combined with

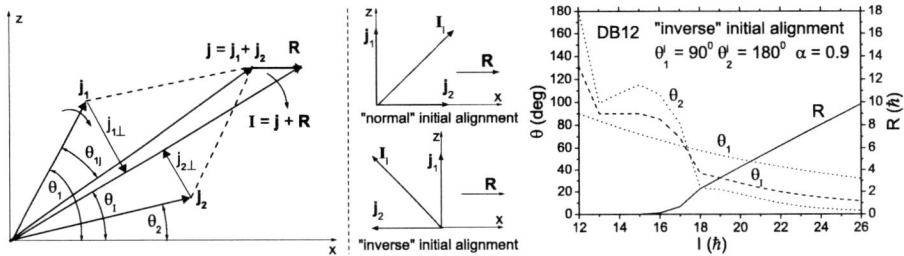

FIGURE 3. Left portion: Angular momentum coupling scheme used in the SPAC model. Middle portion: Two different initial orientations of the angular momentum $\vec{j}_2$ are depicted. Right portion: Evolution of the angles θ_1, θ_2, θ_I and R with spin for the "inverse" initial alignment.

PAC and this semiclassical model has been called SPAC model. The vector coupling scheme is depicted in the left portion of figure 3. The total energy contains collective and quasiparticle contributions: $E(I,\theta_1,\theta_2) = C(I) R^2(I,\theta_1,\theta_2) + V_2 P_2(\theta_1 - \theta_2) + const$; where $C(I) = (2J)^{-1}$, $P_2(\theta_1 - \theta_2) = [3\cos^2(\theta_1 - \theta_2) - 1]/2$ and $R(I,\theta_1,\theta_2) = \sqrt{I^2 - (j_1\sin\theta_1 + j_2\sin\theta_2)^2} - j_2\cos\theta_2 - j_1\cos\theta_1$. Generally, for each value of I the angles $\theta_1(I)$ and $\theta_2(I)$ can be found from the minimization condition $(\frac{\partial^2 E}{\partial\theta_1\partial\theta_2})_I = 0$. Subsequently, the total energy $E(I)$ and transition probabilities can be caculated as functions of I. Taking into account the classical approximation of Clebsch-Gordon coefficients the reduced M1 transition probability is $B(M1; I \to I - 1) = \frac{3}{8\pi}\mu_\perp^2$, with $\mu_\perp = g_1^* j_{1\perp} - g_2^* j_{2\perp}$, $j_{1\perp} = j_1\sin(\theta_1 - \theta_I)$, $j_{2\perp} = j_1\sin(\theta_I - \theta_2)$, $g_1^* = g_1 - g_R$, $g_2^* = g_2 - g_R$ and $g_R = Z/A$. The reduced E2 transition probability is $B(E2; I \to I - 2) = \frac{15}{128\pi}[Q_{eff}\sin^2\theta_{1j} + Q_{coll}\cos^2\theta_I]^2$, where Q_{eff} and Q_{coll} are quasiparticle and collective transitional quadrupole moments. The collectivity of the core is assumed to be I dependent and can in the band crossing region be described by the Boltzmann functions: $C(I) = (2J)^{-1} = C_f + (C_i - C_f)/\{1 + exp[(I - I_{tr})/\Delta]\}$ and $Q_{coll}(I) = Q_f + (Q_i - Q_f)/\{1 + exp[(I - I_{tr})/\Delta]\}$.

Generally, any initial orientation of the angular momentum vectors $\vec{j}_1$ and $\vec{j}_2$ can be considered. Particularly, the two symmetric initial orientations of $\vec{j}_2$, resulting from a rotation by $\mathfrak{R}_z(\pi)$, generate two different dipole bands. The corresponding alignment schemes are shown in the middle portion of figure 3 and are labelled "normal" and "inverse" initial alignment. In case of the normal initial alignment $\theta_2 = 0$ and $\theta_1(I)$ can be found applying the one-dimensional minimization condition $(\frac{dE}{d\theta_1})_I = 0$. For the inverse initial alignment the required two-dimensional search of the energy minimum can be replaced by the one-dimensional minimization condition $(\frac{dE}{d\theta_2})_I = 0$ if the model function $\theta_1(I) = \theta_1^i exp[(1 - I/I_i)\alpha]$ is used, where α is a parameter describing how fast full alignment is achieved. It turned out that the dipole band based on the inverse initial alignment is energetically favoured as compared to the normal one. For the inverse solution, the evolution of the angles θ_1, θ_2, θ_I as well as R as funcion of spin is shown

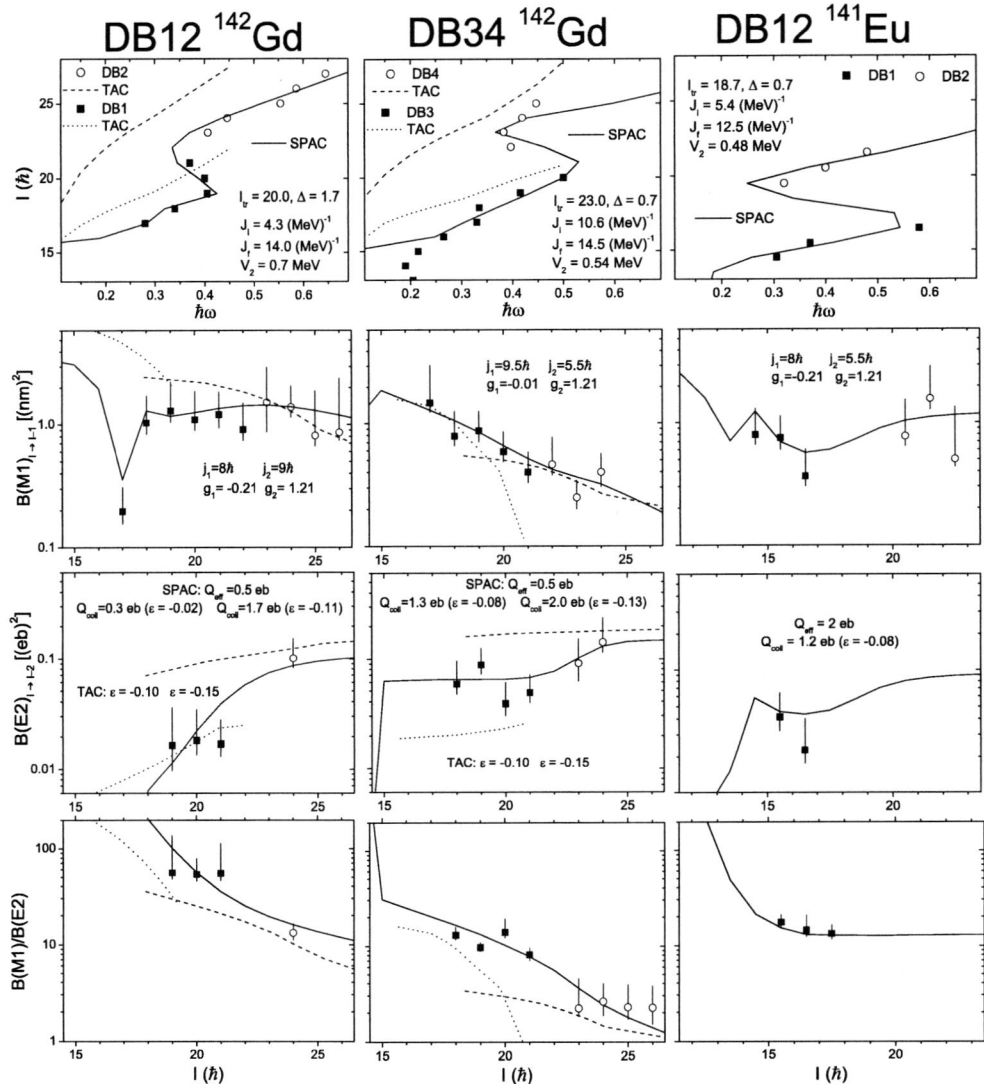

FIGURE 4. Comparison of experimental results for dipole bands in ^{142}Gd and ^{141}Eu with those of calculations in the framework of the TAC and SPAC models. First row: angular momentum I vs. rotational frequency $\hbar\omega$. Second row: $B(M1)$ values. Third row: $B(E2)$ values. Fourth row: $B(M1)/B(E2)$ ratios. The TAC calculations are shown as dotted lines for DB1 and DB3 ($\varepsilon_2 = -0.10$) and as dashed lines for DB2 and DB4 ($\varepsilon_2 = -0.15$). SPAC results for the "inverse" initial alignment are shown as full lines.

in the right portion of figure 3.

In figure 4 the experimental results for the dipole bands in ^{142}Gd and ^{141}Eu are compared with calculations in the framework of the TAC and SPAC models. In the first row the angular momentum I is plotted vs. the rotational frequency $\hbar\omega$. In the following rows $B(M1)$ and $B(E2)$ values as well as $B(M1)/B(E2)$ ratios are plotted as function of spin. The TAC model calculations for ^{142}Gd [4] are based on the above mentioned four and six particle-hole configurations for DB1/DB3 and DB2/DB4, respectively. The TAC calculations can quite well reproduce the experimental results except for DB1. Results of the SPAC model calculations for ^{142}Gd and ^{141}Eu for the inverse initial alignment are shown. Here, the $\pi h^2_{11/2} \otimes \nu h^{-2}_{11/2}$ and $\pi h^1_{11/2} \otimes \pi g^{-1}_{7/2} \nu h^{-2}_{11/2}$ configurations have been considered for DB1/DB2 and DB3/DB4 in ^{142}Gd, respectively. For ^{141}Eu a $\pi h^1_{11/2} \otimes \nu h^{-2}_{11/2}$ configuration has been assumed. At the spin I_{tr} (cf. figure 4 first row) an increase of the collectivity in the respective dipole bands without a configuration change is assumed. This allows to explain the backbendings which can be seen in the first row of figure 4. An argument for the absence of a configuration change is that the experimental $B(M1)$ values do not increase between the respective bands. Also calculations assuming a band crossing have been made with the SPAC model taking into account band mixing [10] but they can particularly only poorly reproduce the angular momentum as function of the rotational frequency in the backbending region.

The comparison of the experimental results with those of the TAC and SPAC models indicate that the dipole bands in ^{142}Gd and ^{141}Eu can indeed be considered as magnetic rotational bands with a significant contribution of collectivity.

ACKNOWLEDGMENTS

The work was partly supported by IB of BMBF at DLR, Germany under the WTZ contract RUS 99/191 and by the EU under the TMR contract ERBFMRXCT970123 and the LSF contract HPRI-CT-1999-00078. The support of the colleagues from the EUROBALL collaboration during the experiment and enlightening discussions with S. Chmel and H. Hübel on the SPAC model are gratefully acknowledged.

REFERENCES

1. Hübel, H., et al., Z. Phys. A, **358**, 237 (1997)
2. Clark, R.M., and Macchiavelli, A.O., Ann. Rev. Nucl. Part. Sci., **50**, 36 (2000)
3. Frauendorf, S., Rev. Mod. Phys., **73**, 463 (2001)
4. Lieder, R.M., et al., Eur. Phys. J. A, **13**, 297 (2002)
5. Urban, W., Manchester University, Nuclear Physics Report 1991-1992, p. 95
6. Lieder, R.M., et al., Eur. Phys. J. A, accepted for publication (2003)
7. Pasternak, A.A., et al., Laboratori Nazionali di Legnaro Annual Report 2001, p. 44, web ed.: www.lnl.infn.it and to be published in Eur. Phys. J. A
8. Dönau, F. and Frauendorf, S., Proc. Conf. on High Angular Momentum Properties of Nuclei, Oak Ridge 1982, ed. N.R. Johnson, Nucl. Sci. Res. Conf. Series V4, Harwood, New York, 1983, p. 143
9. Macchiavelli, A.O., et al., Phys. Lett. B, **450**, 1 (1999)
10. Cooper, J.R., et al., Phys. Rev. Lett, **87**, 132503 (2001)

Precise ft-value Measurement for the Superallowed $0^+ \to 0^+$ β Decay of ^{22}Mg

V.E. Iacob, J.C. Hardy, M. Sanchez-Vega, R.G. Neilson, A. Azhari,
C.A. Gagliardi, V.E. Mayes, L.Trache and R.E. Tribble

Cyclotron Institute, Texas A&M University, College Station, Texas

Abstract. Very accurate measurements of the half-life, *3.8755(12) s*, and the branching ratio, *0.5315(12)*, are reported for the superallowed $0^+ \to 0^+$ β-decay of ^{22}Mg. The precision in the branching ratio was achieved by the efficiency calibration of an HPGe detector to *0.15%* precision for γ-ray energies between 50 and 1400 keV. The *ft*-value extracted from the measurement, corrected for small calculated radiative effects is *ℱt=3071(9) s*. With this measurement, we open up a new series of superallowed emitters with $T_z = -1$ to precision measurements. This series will permit direct tests of the calculated nuclear-structure-dependent correction terms essential to the determination of the vector coupling constant, G_V, and of V_{ud}, the up-down quark-mixing element of the Cabibbo-Kobayashi-Moskawa (CKM) matrix. As the top row of the CKM matrix currently fails the unitarity test by more than two standard deviations, measurements to improve the precision with which V_{ud} is known are of paramount importance.

INTRODUCTION

Measurements of *ft*-values for superallowed $0^+ \to 0^+$ nuclear β-decays offer the most precise value for G_V, the vector coupling constant, provided that some small corrections (of order 1%) are well accounted for. This value, along with the muon coupling constant, permits the determination of the up-down quark-mixing element, V_{ud}, of the Cabibbo-Kobayashi-Mskawa (CKM) matrix element, an essential ingredient in the Standard Model. A recent review of the best-known values used in the CKM matrix points out that it fails the unitarity test in its first row by more than two standard deviations [1,2]. Before such a result can be seen as signaling a failure of the Standard Model, it must be made statistically more definitive, and measurements that would increase the reliability of the CKM matrix elements are welcome. The main contributor to the top-row test is V_{ud}, carrying a weight of about 95% (V_{ud} = 0.9740±0.0005), and refined superallowed beta-decay measurements can increase the precision of G_V and implicitly that of V_{ud}. Since the uncertainty in V_{ud} is dominated by the small correction terms used in the derivation of G_V, the theoretically calculated corrections need to be validated. One possible way of doing this is to experimentally test the calculations for previously unstudied nuclei that cover a wide range of calculated values for the corrections.

In this paper we present the measurements performed to extract a highly accurate *ft*-value for the superallowed β-decay of ^{22}Mg. The extraction of an *ft*-value for any beta transition requires knowledge of the transition energy, Q_{EC} (needed to calculate the statistical rate function *f*), and the partial half-life for the particular transition of interest; the latter is obtained from the half-life of the β-emitter and the branching ratio for that transition. The measurements involved are simple in principle but, because of the high precision required, all possible sources of systematic error have to be addressed. To achieve results that are meaningful for the unitarity test, the precision of the measurements should be ~0.1% or better.

THE EXPERIMENTS

A radioactive ^{22}Mg beam was produced from the ^{1}H(^{23}Na, 2n)^{22}Mg reaction: a 28 *A* MeV ^{23}Na primary beam produced by the K500 cyclotron bombarded a LN$_2$-cooled hydrogen gas target at about 1.6 atm. The ejectiles were separated by the MARS recoil spectrometer at Texas A&M University, yielding a ^{22}Mg beam of ~10^4 atoms/s at 23 *A* MeV with purity greater than 99.6%. This beam was extracted into air and then, after passing through a 0.3 mm thick BC-104 scintillator and a stack of aluminum degraders, it was implanted in the double-coated aluminized mylar tape (75 μm-thick, 25.4mm wide) of a fast tape-transport system [3]. Because of differences in stopping powers for the weak remaining contaminants, the purity of the implanted ^{22}Mg ions was found to be significantly better than 99.9%.

Both the branching-ratio and the half-life experiments were carried out in cyclic measurements. A typical cycle involved the collection of radioactivity, its transport to a well-screened low background region (about 90 cm away from the beam-line collection point) and the detection of the β- and/or γ-rays. The beam was on only during the collect time, and the heavy ions passing through the scintillator were scaled for each cycle. The typical *collect-move-detect* time intervals used were 5s-180ms-5s and 5s-180ms-80s for the branching-ratio and half-life measurements respectively. Data were acquired until the desired statistical accuracy was obtained.

The Branching-Ratio Experiment

In the branching ratio experiment we measured the β$^+$ particles with a 1.0-mm-thick BC-104 scintillator and the γ-rays with a 70% coaxial HPGe detector. The tape-transport system delivered the collected sample to a spot located between the β and the γ detectors, which were placed facing one another at 4 mm and 15 cm, respectively, from the tape. For each detection cycle we detected and scaled the implanted heavy-ions, and measured the subsequent β-γ coincidences (or γ singles); all events were time-tagged relative to the beginning of the detect interval.

The decay of ^{22}Mg is characterized by several important features (see Fig. 1). Since the β-decay of ^{22}Mg (ground state: $J^\pi=0^+$) populates ^{22}Na (ground state $J^\pi=3^+$), the beta transition to the ground state of ^{22}Na is second forbidden. For this reason, it is possible to determine branching ratios for the β-decay solely from the relative intensities of the β-delayed γ-rays. However, because we are aiming at 0.1% precision in the superallowed branching ratio, we require that same precision in the measured

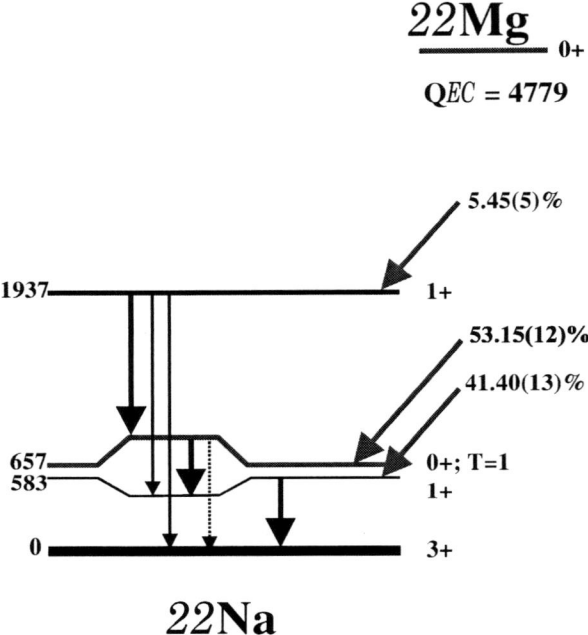

FIGURE 1. Decay scheme for the β^+-decay of ^{22}Mg

relative γ intensities. This demands an efficiency calibration of unprecedented accuracy for the γ-ray detector over a relatively wide energy range (from 74 to 1937 keV) to cover all the γ-rays involved in the decay. Moreover, it demands special care in the acquisition of the low energy 74-keV γ-rays as those signals from our detector exhibit a wide range of rise-times. Another critical aspect in the analysis of the coincident events arises from the 583-keV state in ^{22}Na whose half-life is 245 ns. This required a wide β-γ coincidence window (2 μs) and an appropriate correction for the events decaying outside this window.

The desired precision, along with the relatively short distance between the HPGe detector and the source (15 cm) require precise accounting for coincident-sum peaks: the γ-ray spectrum observed in coincidence with β^+ particles (see Fig.2) provides a visual impression of the importance of coincident-sum peaks. This becomes particularly critical in the case of the 511+74-keV sum-peak, which cannot be resolved from the 583 keV γ-ray.

The efficiency calibration of the HPGe was performed with "standard" sources and with several key sources we prepared ourselves. The latter and some of the former exhibited simple γ cascades with no side feeding and with transitions having small internal conversion components. We measured thirteen individual sources of ten

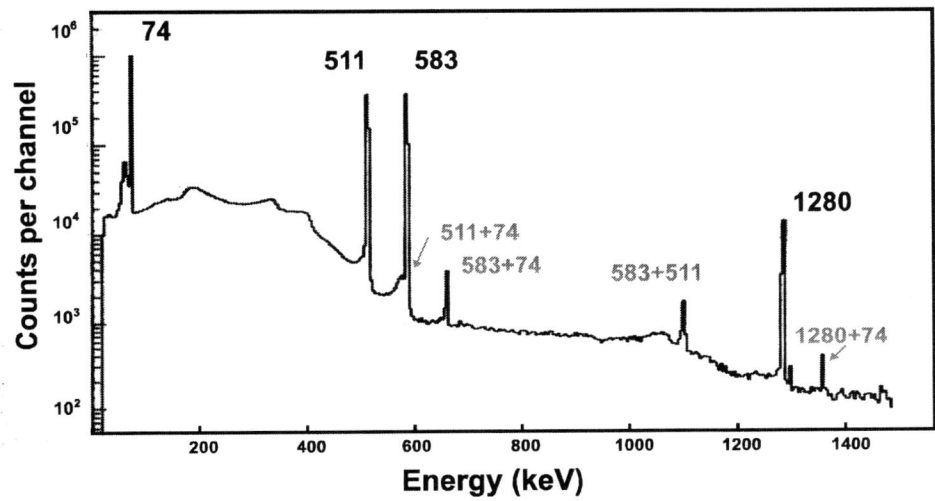

FIGURE 2. Spectrum of g-rays observed in the β-decay of ^{22}Mg, as observed in coincidence with the b particles; the HPGe detector was situated at 15 cm relative to the source; each peak is marked with its energy in keV; sum peaks are also indicated

radioisotopes: ^{48}Cr, ^{60}Co, ^{88}Y, ^{108m}Ag, ^{109}Cd, ^{120m}Sb, ^{133}Ba, ^{134}Cs, ^{137}Cs and ^{180m}Hf. Two ^{60}Co sources were specially prepared by the Physikalisch-Technische Bundesanstalt [4] with activities certified to 0.06%. We prepared three ourselves: ^{48}Cr and ^{120m}Sb were produced in the ^{1}H(^{50}Cr,p2n)^{48}Cr and ^{120}Sn(p,n)^{120m}Sb reactions with beams from our K500 cyclotron, and ^{180m}Hf was produced by neutron activation of enriched ^{179}Hf, with thermal neutrons from the Texas A&M reactor. All other sources were acquired from commercial suppliers. The calibration measurements were performed for a source-detector distance of 15 cm, as used in the on-line measurement of the ^{22}Mg decay. Interpolation between the measured points was accomplished with the aid of Monte Carlo calculations performed with the electron and photon transport code CYLTRAN (from the Integrated Tiger Series of codes [5]). The calibration procedure is detailed elsewhere [6,7] and yielded a relative efficiency of 0.15% precision between the ^{22}Mg rays at 74 and 583 keV.

To control possible systematic errors, we collected data at four different counting rates, allowing us to test for the possible presence of dead-time effects. No statistically significant differences were observed. Even so, to be safe, we eliminated from the final data set all cycles recorded with the highest counting rate. Although the relative efficiencies depend only weakly on the source-detector distance, we monitored the sample positioning of our fast tape-transport system for all cycles: we evaluated the ratio of β-γ counts to the number of implanted ^{22}Mg ions and retained only those cycles having this ratio within a small acceptable range. The results of our analysis are summarized in Table 1. They are consistent with the previously known values but have a much higher accuracy.

TABLE 1. Relative intensities of the β-delayed γ-rays and deduced branching ratios for the β-decay of ^{22}Mg

E_γ [keV]	I_γ	E_x (^{22}Na) [keV]	I_β [%]
74	58.36(6)	583	41.40(13)
583	100.00(19)	657	53.15(12)
1280	5.40(7)	1937	5.45(5)
1354	0.015(3)		
1937	0.032(3)		

The Half-life Experiment

In the half-life experiment, the collected activity was placed by our tape-transport system in the center of a 4π proportional gas detector. At our shielded counting location, the background in that detector was ~0.5 counts/s. We recorded the β$^+$-events in a multiscaler whose time-width and channel-advance was controlled by a clock with time-base accurate to 5 ppm. As with the branching ratio measurement, we monitored the precision of the positioning of the collected activity by obtaining the ratio of the total number of detected betas versus the total number of implanted heavy-ions. Even though a less precise positioning has almost no impact on the extracted half-life, we retained only those cycles having that ratio within a small acceptable range.

The signals from the β-detector were amplified, then sent to a fast discriminator and finally to a non-retriggerable gate generator whose signals have a well defined and stable width. That width was chosen to be much longer than any dead time in the upstream modules, thus enforcing a well-defined non-extendable dead time. This dominant dead time was continuously monitored during the experiment and was later used in the data analysis to account exactly for the associated losses.

The experiment was divided into sub-runs, allowing for frequent changes in the acquisition-parameters that might affect the results of the measurement: detector bias, discriminator threshold, dominant dead time. The relatively long (80s) detect time, which exceeds 20 half-lives of ^{22}Mg, allowed us to get very good control over the background and thus to sensitively monitor the possible intrusion of long-lived impurities. However, to test even more sensitively for such impurities, we performed an additional sub-run with collect/detect times set to 60/160 s; these long time-intervals particularly favor the collection of long-lived impurities relative to ^{22}Mg, whose activity reaches saturation after several half-lives (~15s). The analysis of this sub-run proved that the only long-lived impurity implanted in the tape was ^{21}Na ($t_{1/2}$=22.5 s), which had an initial activity 1.7 10^{-4} that of ^{22}Mg; the decay of this impurity was taken into account in all subsequent analyses.

We recorded a total of more than 54 million decay events in about 3000 collect/detect cycles with 80s detect time. A decay spectrum can be seen in Fig. 3, for which all the cycle spectra in the sub-runs with 10-µs dead-time were added together.

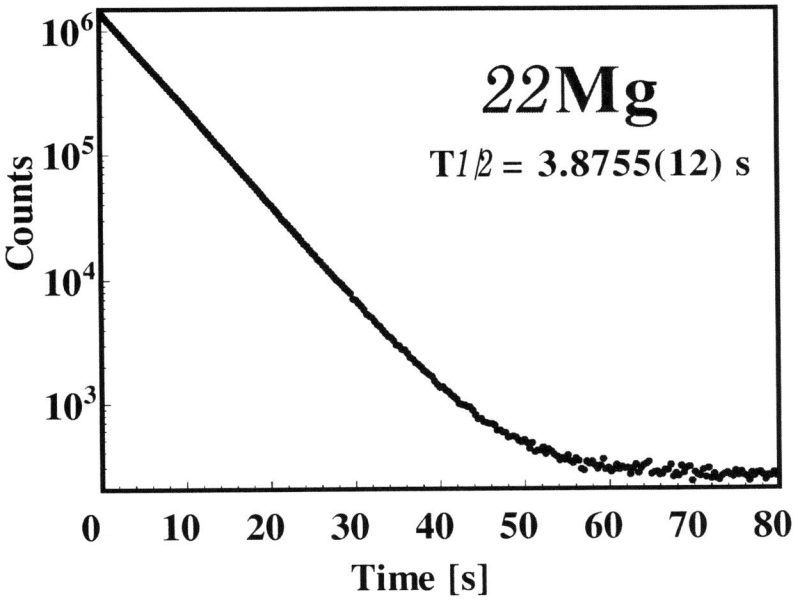

FIGURE 3. Total decay spectrum of ^{22}Mg β^+-particles

One should notice that the decay is followed over nearly 4 decades: the ratio of the first channel to the background is ~5000. This feature allowed us to extract a very accurate half-life for the β^+-decay of ^{22}Mg.

To extract the half-life of ^{22}Mg, each sub-run was analyzed with two different maximum-likelihood procedures: (i) the sum of all dead-time corrected cycle-spectra was processed in a single fit; and (ii) all individual cycle spectra were fitted simultaneously with a common half-life but with individual amplitudes. The later procedure is more computer-intensive but has a fit function that incorporates the exact dead-time effects; whereas the former one corrects (to a good approximation) the cycle-data itself for dead-time losses before they are combined into a single sum-spectrum for fitting. The two procedures gave concordant results over all sub-runs. Fig. 4 presents the scatter of the fitted half-lives over various sub-runs: the half-life shows no systematic bias related to the detector bias, discriminator threshold or dominant dead time.

To test for the presence of possible short-lived impurities we removed from the analysis the contribution from the first three seconds of the counting period and refitted the remainder; then we repeated the procedure, removing the first six seconds, and so on. The fitted half-life was stable versus this procedure too (see Fig. 5).

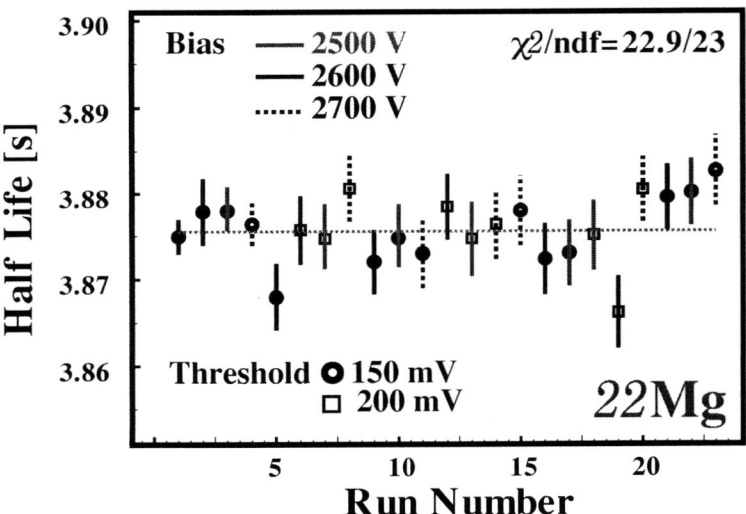

FIGURE 4. Intercomparison of the results obtained with various settings in the electronics. There is no evidence of systematic errors introduced by the acquisition chain.

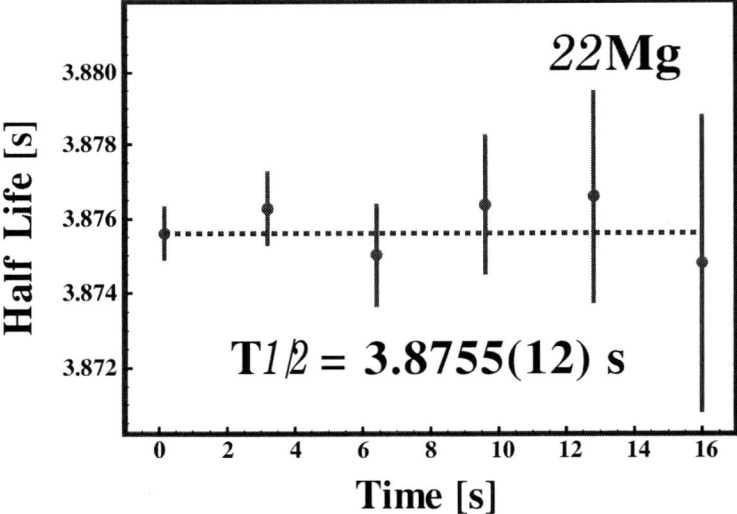

FIGURE 5. Test for systematic errors possibly caused by unidentified short-lived impurities of by rate-dependent counting losses. The abscissa represents the time period at the beginning of the counting cycle for which the data is omitted from the fit.

Finally, to further consolidate the results, particularly the fitting procedures, all the consistency tests were repeated on a parallel artificially generated data set, mimicking the parameters of all real sub-runs but with known half-life, dead time and background.

The retrieval of the decay parameters used in the generation of the artificial data further confirmed the accuracy of the fitting methods. The weighted average over all sub-runs gave 3.8755(12) s for the half-life of ^{22}Mg.

The $\mathcal{F}t$-value

The branching ratio and the half-life reported in this paper allow the extraction of a very precise partial half-life for the superallowed transition between ^{22}Mg and ^{22}Na. This value, along with a recently updated Q-value [9] and structure-dependent correction term [1], yields a final result of *$\mathcal{F}t=3071(9)$ s*. This is in perfect agreement with the weighted average $\mathcal{F}t$ extracted from the nine well-known cases studied to date [1,2]. An improvement in the mass of ^{22}Mg to sub-keV precision would drop the error in the $\mathcal{F}t$-value to ±7 s. Such a measurement is expected to be performed in the near future.

CONCLUSION

The excellent agreement between the $\mathcal{F}t$-value extracted from the measurements reported in this paper and the previously known average allows us to confirm the values of the theoretically calculated corrections for ^{22}Mg. With these measurements, we open up a new series of superallowed emitters with $T_z = -1$ to precision measurements. This series will permit direct tests of the calculated nuclear-structure-dependent correction terms and is expected to sharpen the value of the vector coupling constant, G_V, and that of V_{ud}, the up-down quark-mixing element of the Cabibbo-Kobayashi-Moskawa (CKM) matrix.

ACKNOWLEDGEMENTS

This work was supported by the U.S. Department of Energy under Grant No. DE-FG03-93ER40773 and by the Robert A. Welch Foundation.

REFERENCES

1. I.S. Towner, J.C. Hardy, Phys. Rev. C **66**, 035501 (2002)
2. I.S. Towner, J.C. Hardy, J. Phys. G: Nucl. Phys. **29**, 197 (2003)
3. J.C. Hardy *et al.*, Progress in Research, Cyclotron Institute, Texas A&M University (1998-1999), p. V-20
4. E. Schonfeld *et al.*, J. Appl. Rad. Isot. **56**, 215 (2002)
5. J.A. Halbleib, T.A. Mehlhorn, Nucl. Sci. Eng. **92**, 338 (1986).
6. J.C. Hardy *et al.*, J. Appl. Rad. Isot. **56**, 65 (2002)
7. R.G. Helmer *et al.*, Nucl. Instr. and Meth. in Phys. Res. A **511**, 360 (2003)
8. V.T. Koslovsky *et al.*, Nucl. Instr. Meth. A **401**, 289 (1997) and Nucl. Phys. **A405**, 29 (1983)
9. J.C. Hardy *et al.*, Phys. Rev. Lett. 91, 082501 (2003).

Beta decay of ^{76}Sr using the Total Absorption Spectrometer "Lucrecia" at ISOLDE-CERN

E. Nácher[*], A. Algora[*,†], B. Rubio[*], J.L. Taín[*], M.J.G. Borge[**],
D. Cano-Ott[‡], S. Courtin[§], Ph. Dessagne[§], D. Escrig[**], L.M. Fraile[¶,||],
W. Gelletly[††], A. Jungclaus[**,‡‡], G. Le Scornet[||], F. Maréchal[§], Ch. Miehé[§],
E. Poirier[§] and O. Tengblad[**]

[*]*IFIC, Valencia, Spain*
[†]*MTA ATOMKI, Debrecen, Hungary*
[**]*IEM, CSIC, Madrid, Spain*
[‡]*CIEMAT, Madrid, Spain*
[§]*IReS, Strasbourg, France*
[¶]*Univ. Complutense, Madrid, Spain*
[||]*ISOLDE-CERN, Geneva, Switzerland*
[††]*Univ. of Surrey, Guildford, U.K.*
[‡‡]*UAM, Madrid, Spain*

Abstract. A new Total Absorption Spectrometer (TAS) called "Lucrecia" has been installed at ISOLDE (CERN) to investigate the β-decay of some nuclei with A$\approx$70-80 in the vicinity of the N=Z line. In this work we report on the decay of the N=Z nucleus ^{76}Sr, measured with a TAS for the first time. The Gamow-Teller strength distribution B(GT) for this decay is presented and compared with theoretical calculations.

INTRODUCTION

The N$\approx$Z, A$\approx$70-80 region of the nuclear chart is characterized by different shape effects such as strong deformation in the ground state, shape transitions and shape coexistence. These effects are exhibited by the light Sr isotopes, which evolve from sphericity at N=50 to large deformation (presumably prolate) at N=40. ^{76}Sr, with N=Z=38, is the most deformed nucleus in the region [1].

According to Hamamoto *et al.* [2] and Sarriguren *et al.* [3], one can study the deformation of the ground state of a particular nucleus by measuring the B(GT) distribution of its β-decay. In these references the authors calculate the B(GT) distribution for various nuclei in the region assuming different deformations for the ground state. In some cases, the results differ markedly with the shape of the ground state of the parent, especially for the light Kr and Sr isotopes. We used the calculations of Ref. [3] to compare with our experimental B(GT) for the case of ^{76}Sr decay. From this comparison we were able to confirm the strong deformation of the ground state of ^{76}Sr already seen in [1], but using β-decay for the first time. An indication of prolate deformation was already given in [4], but this was based on a less complete set of data. In this work we will be able to give the first conclusive evidence confirming the prolate shape of the ^{76}Sr ground state.

A precise determination of the B(GT) distribution is required for such studies, and this is far from trivial. Traditional high resolution techniques, based on the use of high purity Germanium (HPGe) detectors to measure the γ-rays emitted after the β-decay, often fail to detect significant but very fragmented strength at high excitation energy in the daughter nucleus. This is mainly due to three factors: the low photo-peak efficiency of HPGe detectors for high energy γ-rays, the high fragmentation of the B(GT) at high excitation energy, and the fragmentation of the gamma de-excitation of the levels in the daughter through many different gamma cascades. Together they cause the so-called *Pandemonium effect* [5]: many weak cascades de-exciting levels at high energy can remain undetected leading to large systematic errors in the determination of the B(GT). This is the reason why, even although Ref. [4] gives the first indication of the sign of the deformation of the ^{76}Sr ground state, one must determine the B(GT) distribution more accurately over the whole Q_{EC} window to provide a conclusive proof.

The alternative method to measure the B(GT) distribution avoiding these systematic errors is the Total Absorption Spectroscopy technique. The basis of this method is the detection of the entire gamma cascades rather than individual gamma rays. For this purpose one needs a high efficiency detector with acceptable resolution for gammas such as the inorganic scintillators NaI(Tl) or BaF_2. Furthermore, this detector must have a geometry as close as possible to 4π to absorb the complete cascade energy.

THE EXPERIMENT

With the aim of measuring the β-decay of nuclei far away from the stability line a Total Absorption Spectrometer called "Lucrecia" has been installed at the ISOLDE mass separator at CERN. It consists of a large NaI(Tl) crystal of cylindrical shape (L=∅=38 cm) with a cylindrical hole (∅=7.5 cm) at right angles to the symmetry axis. The purpose of the hole is twofold: on the one hand it allows the beam pipe (coming from the separator) to enter up to the centre of the crystal, thus allowing on-line activity of very short half-life (>5 ms) to be measured. On the other hand it allows us to place ancillary detectors inside for the detection of the positrons (β^+-decay), electrons (β^--decay) or X-rays (EC process) produced in the decay. In our case we used a plastic scintillator to detect the positrons and a Germanium telescope (planar+coaxial) to detect X-rays and γ-rays. Covering the whole setup there is a shielding box (2.1 m×1.2 m×1.2 m) made of four different layers: borated polyethylene (10 cm), lead (5 cm), copper (2 cm) and aluminium (2 cm). This shielding stops a large fraction of the room background (mainly neutrons and γ-rays) and reduces the background counting rate in the NaI crystal to 1.4 kHz with proton beam on the ISOLDE target. In Figure 1 a schematic view of the detector setup placed inside the shielding box is shown.

In order to produce the isotopes of interest ($^{76-80}$Sr), a 52 g/cm^2 Nb target was bombarded with a 1.4 GeV proton beam. The intensity was chosen to produce a counting rate of $\approx$3.5 kHz in our NaI crystal. In order to ionise Sr selectively, a fluoridation technique was used [6]. A certain amount of CF_4 gas was introduced into the cavity of a hot surface ion source, and SrF$^+$ molecular ions were produced and extracted with a 60 kV potential. The High Resolution Separator (HRS) magnet was set to select mass 95, which corresponds to ^{76}Sr^{19}F$^+$, in our beam line. In this way we could get rid of the

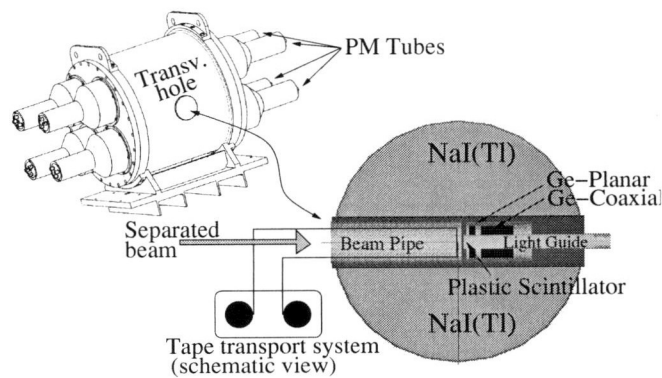

Figure 1. Detector setup for the ^{76}Sr measurement. In the upper-left part we have a 3D plot of the NaI(Tl) cylinder and its photo-tubes, and in the lower part we see a transverse cut of the main crystal and the ancillary detectors, as well as a schematic view of the tape where the separated ^{76}Sr beam is implanted.

isobaric contaminant ^{76}Rb as a primary product.

The radioactive beam was steered to our detector setup and implanted in an aluminised mylar tape which was moved every 15 seconds to remove the daughter activity ($T_{1/2}(^{76}Sr) = 8.9$ s, $T_{1/2}(^{76}Rb) = 36.8$ s). During this 15 s cycle we measured the decay of the implanted radioactive source. The X-rays coming from the EC processes were measured in the Ge planar detector. The positrons produced in the β^+-decay were detected in the plastic scintillator. Finally, the γ-rays following the decay (either by β^+ or by EC) were viewed by the NaI(Tl) crystal.

DATA ANALYSIS AND RESULTS

In ideal conditions, if the TAS had 100% efficiency over the whole energy range, the experimental spectrum measured in the NaI(Tl) cylinder would be the β intensity distribution $I_\beta(E_x)$ convoluted with the energy resolution of the crystal. In reality the detector does not have 100% efficiency and the spectrum is modified by the response function of the detector. The procedure to unfold the experimental data in order to obtain the $I_\beta(E_x)$ distribution is not trivial. The impossibility of inverting the response matrix of the detector implies the need to solve the inverse problem. In Ref. [7] there is a systematic study of three methods applied to our specific problem of the TAS data. We have used the Expectation Maximization algorithm [8] to obtain the $I_\beta(E_x)$ by unfolding our experimental data.

Once the β intensity distribution $I_\beta(E_x)$ is obtained, we only need to correct the numbers by the Fermi function $f(Q_{EC} - E_x)$, which carries the information on the phase space available in the final state and the Coulomb interaction that participates in the decay (Q_{EC} value from Ref. [9]). In this way we arrive at the quantity of interest: the Gamow-Teller strength distribution B(GT). Figure 2 shows the experimental B(GT) obtained in this experiment for the decay of ^{76}Sr. The graph has been split into two

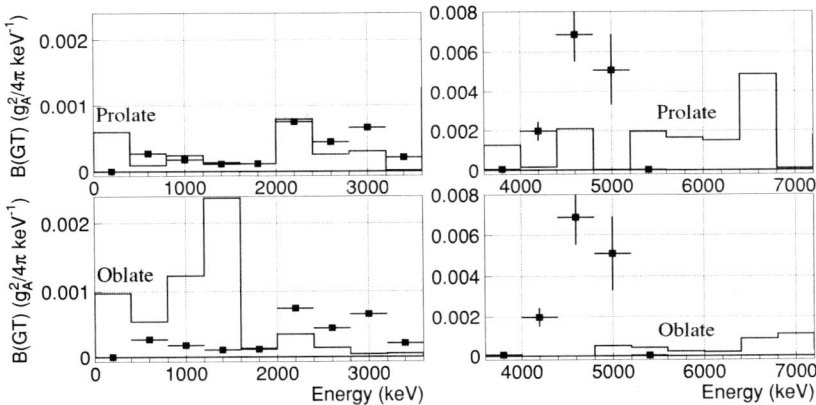

Figure 2. Average B(GT) of the ^{76}Sr→^{76}Rb decay over 400 keV energy bins, as a function of excitation energy in the daughter nucleus. It has been divided into two energy ranges with two different scales. In the upper panels the experimental results (squares with error bars) are compared with the theoretical calculations [3] (solid lines) for prolate shape. In the lower panels the comparison is with the oblate shape.

different energy ranges in order to adjust the vertical scale in the most convenient way. At the same time we have plotted the theoretical calculation of Ref. [3] for both prolate (β_2=0.41) and oblate (β_2=-0.13) shapes for the ground state of ^{76}Sr. These deformation parameters β_2 correspond to the two minima in the energy surface vs deformation plot. The agreement of the experimental results with the prolate shape calculation is very good over the energy range 0-3.6 MeV. However this does not happen with the oblate shape calculation. On the other hand, in the high energy range (3.6-7.2 MeV) we observe a resonance in the experimental B(GT) between 4 and 5 MeV. The theoretical B(GT) shows a resonance beyond 4 MeV only in the prolate case, but rather higher in energy. We think that we are dealing with the same resonance in both cases as the integrated B(GT) between 4 and 5 MeV in experiment agrees with the integrated B(GT) between 5.5 and 6.5 MeV in the prolate shape calculation.

An alternative way to look at the results is to plot the accumulated B(GT) versus the excitation energy in ^{76}Rb as shown in Figure 3. Looking at this plot it is even clearer that the calculation for the prolate shape agrees much better than the oblate case. This confirms the strong deformation ($\beta_2 \approx 0.4$) of the ground state of ^{76}Sr already seen in an earlier in-beam experiment [1], but using β-decay for the first time, and it also gives the first definite experimental evidence of the prolate sign of the deformation, confirming what was indicated in [4].

SUMMARY AND CONCLUSION

A new Total Absorption Spectrometer, "Lucrecia", is working successfully at ISOLDE. It is particularly useful to extract the B(GT) distribution in the β-decay for most of the nuclear species that can be produced at this separator facility.

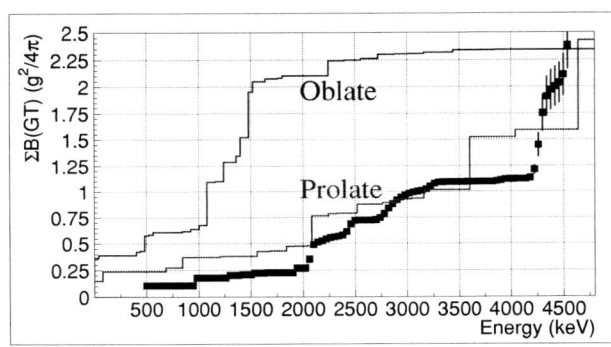

Figure 3. Accumulated B(GT) as a function of the excitation energy in the daughter nucleus. The experimental results from this work (squares) are compared with the theoretical calculations [3] (solid lines) assuming prolate and oblate shapes for the ^{76}Sr ground state.

In this work we present the results obtained for an experiment devoted to measuring the B(GT) distribution in the decay of the N=Z isotope ^{76}Sr. When we compare our experimental results with the theoretical calculations of Ref. [3] (see Figures 2 and 3) we conclude that the ground state of ^{76}Sr is strongly prolate ($\beta_2 \approx 0.4$), in agreement with theoretical predictions [10, 11] and with previous experimental work [1]. Furthermore, this experiment validates the method to deduce the deformation, including the sign, from the comparison of the experimental and the calculated B(GT) since the ^{76}Sr ground state is a very clean case (free of shape admixtures).

REFERENCES

1. C.J. Lister *et al*, *Phys. Rev.*, **C42**, R1191 (1995).
2. I. Hamamoto *et al*, *Z. Phys.*, **A353**, 145 (1995).
3. P. Sarriguren *et al*, *Nucl. Phys.*, **A658**, 13 (1999).
4. C. Miehé *et al*, in *New Facet of Spin Giant Resonances in Nuclei*, ed. by H. Sakai *et al*, 140 (1997).
5. J.C. Hardy *et al*, *Phys. Lett.*, **71B**, 307 (1977).
6. H.L. Ravn *et al*, *Nucl. Instr. and Meth.*, **123**, 131 (1975).
7. D. Cano Ott, Ph.D. thesis, Universidad de Valencia (2000).
8. A.P. Dempster *et al*, *J. R. Statist. Soc.*, **B39**, 1 (1977).
9. G. Sikler Ph.D. thesis, Univ. Heidelberg (2003), *to be published in Nucl. Phys A* (2004).
10. P. Bonche *et al*, *Nucl. Phys.*, **A443**, 39 (1985).
11. A. Petrovici *et al*, *Nucl. Phys.*, **A605**, 290 (1996).

Recent results at the $N = Z$ line with GASP and EUROBALL

E. Farnea

INFN Sezione di Padova, Padova, Italy

Abstract. Valuable information on the validity of the isospin symmetry was obtained by studying nuclei close to the $N = Z$ line with the GASP and EUROBALL γ-ray spectrometres coupled to ancillary devices. Here a few selected results on the study of mirror nuclei are presented, together with an estimate of the isospin mixing probability through the measurement of a forbidden $E1$ transition in ^{64}Ge.

INTRODUCTION

In the past few years, a wealth of relevant information on $N \approx Z$ nuclei has been collected, thanks to the efforts of many research groups all around the world. Given the occupation symmetry between proton and neutron orbitals, implying a rather large superposition of the wave functions, phenomena arising from the valence particle structure are enhanced in $N \approx Z$ nuclei.

An attractive topic which can be investigated in these nuclei is the study of the isospin symmetry and of its possible breaking. It should be pointed out that the impact of these studies is not limited to nuclear physics. For instance, as discussed in [1, 2], a good understanding of the mechanism of isospin mixing in nuclei close to the $N = Z$ line is necessary in order to perform reliable corrections in deriving the Fermi constant G_V from the Log ft-values of superallowed Fermi β decays. This topic is closely related to the question of the unitarity of the Cabibbo-Kobayashi-Maskawa (CKM) matrix and to physics beyond the standard model. Spectroscopic studies of $N = Z$ nuclei are relevant in this matter since, in principle, the isospin mixing probability can be determined using isospin-forbidden γ-transitions. In fact, given the structure of the charge operator, in the long wavelength limit $E1$ transitions are forbidden in $N = Z$ nuclei between states of equal isospin and have equal strength in mirror nuclei [3]. Attributing deviations from this rule to the mixing between states of different isospin, induced by the Coulomb interaction, it is possible to estimate the isospin mixing probability.

An alternative way of investigating the isospin symmetry is the study of the so-called Coulomb Energy Differences (CED) in mirror nuclei, that is nuclei with interchanged numbers of neutrons and protons, and in isospin multiplets. Following the charge independence of the strong interaction, the expected experimental signature is the appearance of nearly identical decay schemes in pairs of mirror nuclei. While the effect of the Coulomb interaction on the nuclear mass and binding energy is large, the bulk of it cancels when considering differences between excitation energies in mirror pairs, that is the

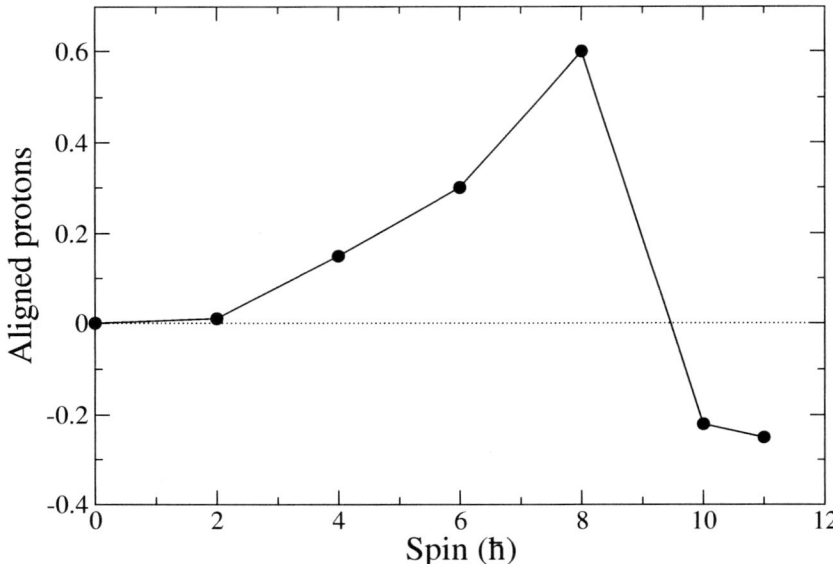

FIGURE 1. Number of maximally aligned proton pairs in the ^{50}Fe-^{50}Cr mirror pair. See text for details.

CEDs, which can be studied with first order perturbation theory [4, 5].

Experimentally, a great difficulty in studying medium-mass and heavy $N \approx Z$ nuclei is due to the fact that they lie out of the stability line. Using fusion-evaporation reactions between stable beams and targets, the nuclei of interest are typically populated with low cross sections (below 1 mb), with a huge background of other undesired reaction products. Thus, in order to study these nuclei high detection efficiency is needed, combined with large selectivity. This can be obtained using a 4π array of germanium detectors, coupled with ancillary devices. Two notable examples of this kind of set-up are GASP coupled with ISIS and the n-Ring and EUROBALL coupled with EUCLIDES and the n-Wall. In the following, we will present a few selected results obtained with these two arrays.

THE ^{50}FE-^{50}CR MIRROR PAIR

The ^{50}Fe-^{50}Cr mirror pair has been studied [4] in an experiment performed at the Laboratori Nazionali di Legnaro, using the ^{28}Si+^{28}Si fusion-evaporation reaction at 110 MeV beam energy and a 0.8 mg/cm^2 target with a 15 mg/cm^2 Au backing to measure transition probabilities though the DSAM technique. The experimental set-up comprised the EUROBALL III array [6] to detect γ rays, coupled to the ISIS light charged particle detector [7] and to the n-Wall [8].

Previous studies of rotating mirror nuclei in the $1f_{7/2}$ shell suggest that the alignment mechanism invoked to explain backbending lead to a qualitative understanding of the

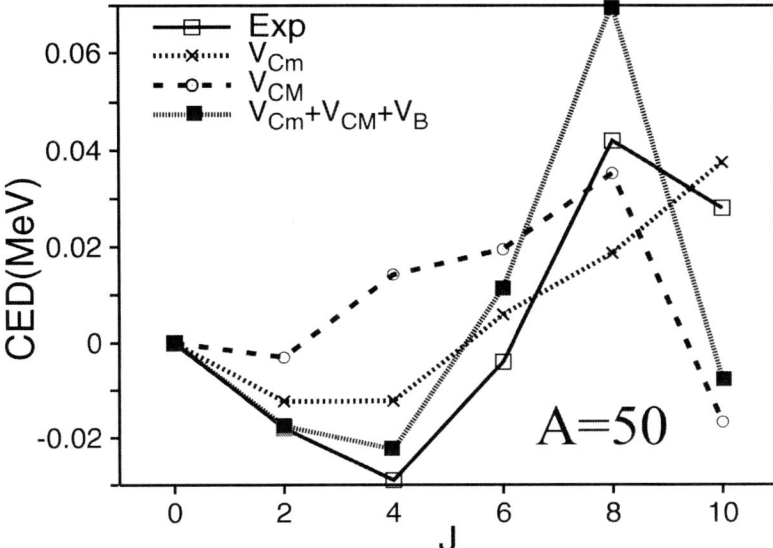

FIGURE 2. Experimental CEDs for the $A = 50$ mirror pair, together with the individual terms needed to reproduce the experimental data: the monopole term V_{Cm}, the multipole term V_{CM} and the sum of V_{Cm} and V_{CM} with an isospin non-conserving term V_B [5].

observed energy patterns [9, 10]. When a pair of protons aligns from a $J = 0$ state, its contribution to the Coulomb energy decreases, following the decrease in the superposition of the individual wavefunctions. When a pair of neutrons aligns instead, no effects on the Coulomb energy are expected. Thus, the difference in the expectation value of the operator:

$$A = \left[(a_\pi^+ a_\pi^+)^{J=6} (a_\pi a_\pi)^{J=6} \right]^0 \qquad (1)$$

counting the number of maximally aligned proton pairs in a shell model context should reflect the trend in CED for the ^{50}Fe-^{50}Cr mirror pair. Such difference is shown in Fig. 1 and correctly reproduces the trend of the experimental CEDs. However, as discussed in [4], shell model calculations fail to reproduce the details of the experimental CEDs. Actually, as shown in [5], two effects should be considered in order to obtain a better agreement between the experiment and the calculations. On one hand, it is necessary to consider not only the multipole part of the Coulomb field, corresponding to the curve labelled V_{CM} in Fig. 2, but also the monopole part (V_{Cm}) or, in other words, the effects generated by differences between the radii of the charge distribution. On the other hand, the nuclear interaction should include also an isospin non-conserving term (V_B). Considering these effects, the agreement between experimental results and calculations is satisfactory.

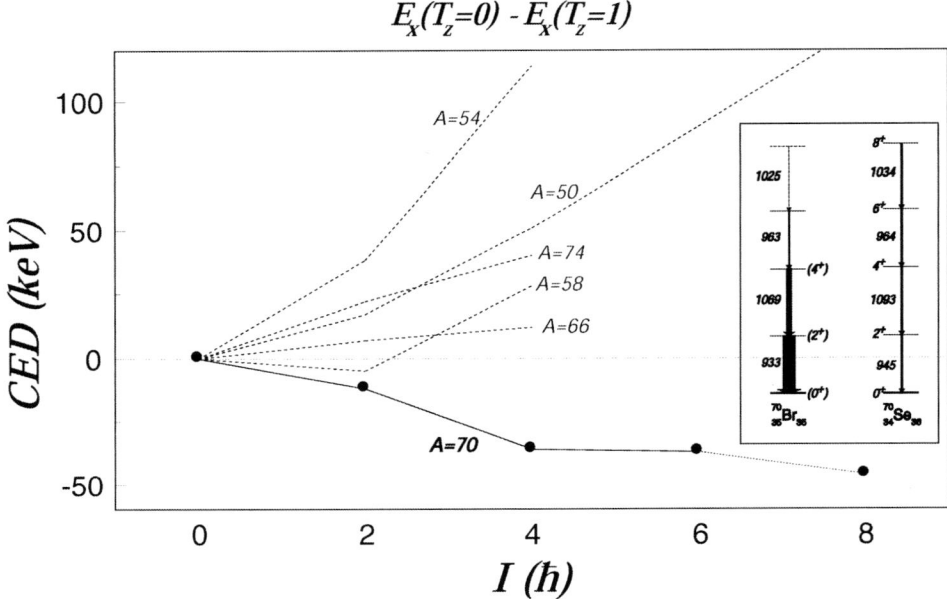

FIGURE 3. Comparison between the CEDs for the $A = 70$ and other known cases. The partial decay schemes of ^{70}Br and ^{70}Se are shown in the inset.

ISOBARIC ANALOGUE STATES IN ^{70}BR AND ^{70}SE

The $N = Z$ nucleus ^{70}Br is the lightest stable bromine isotope, being its $N = Z - 1$ neighbour ^{69}Br proton-unbound [11]. Given the close proximity to the proton drip-line, one expects a rather diffused proton distribution in ^{70}Br, which should reflect itself in the behaviour of the CEDs for the $A = 70$ triplet.

Since no excited states have been identified in ^{70}Kr to date, the excitation energies in ^{70}Br can only be compared with those in ^{70}Se, which is the isobar with $N = Z + 2$ ($T_z = +1$). Excited states in these nuclei have been identified in two experiments [12], the first one performed with GASP+ISIS+n-Ring at the Laboratori Nazionali di Legnaro and the second one with EUROBALL coupled to the EUCLIDES Si-ball [13] and with the n-Wall at the Institut de Recherches Subatomiques in Strasbourg. In both measurements, the reaction ^{40}Ca(^{32}S, pn)^{70}Br at 90 MeV beam energy was used. The 1 mg/cm^2 target was backed in both cases with 12 mg/cm^2 of gold.

The partial decay schemes of ^{70}Br and ^{70}Se are shown in Fig. 3, together with the resulting CEDs. One can notice that the behaviour of the CEDs in the $A = 70$ case is actually quite different from the behaviour of the CEDs in lighter nuclei, since only in the $A = 70$ case the CEDs decrease with the increasing spin. As discussed thoroughly in [12], this can be associated with two effects, both of which are related to the increased proton radius for dripline nuclei. The first effect, known as Thomas-Ehrman shift, is a reduction of the Coulomb repulsion due to the spatial extension of the proton wave

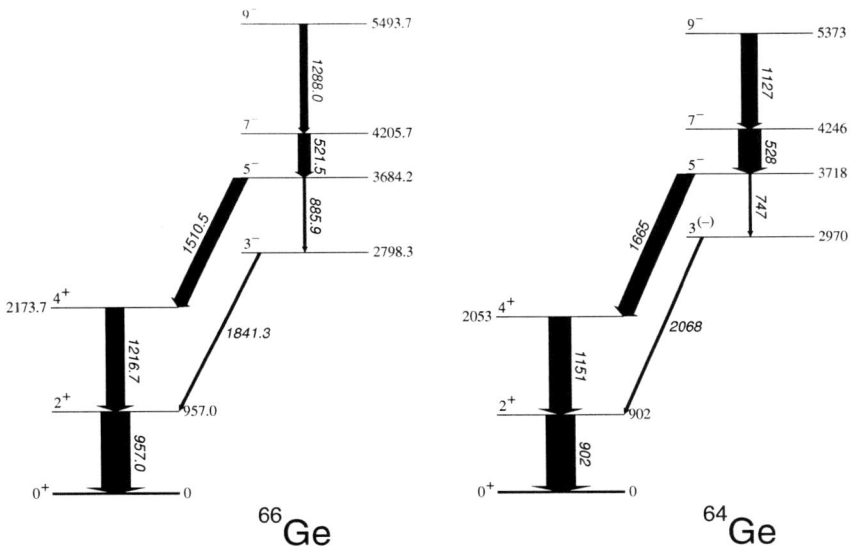

FIGURE 4. Partial decay schemes for the ^{64}Ge and ^{66}Ge nuclei.

function [14, 15]. The second effect is a decrease of the nuclear two-body residual interaction due to the different radial distributions for the wave functions for neutrons and loosely bound protons. Because of the low centrifugal barriers, the radial functions of the low-l orbits will be more affected, resulting in a compression of the energy spectrum for the $N = Z$ case [16].

FORBIDDEN $E1$ TRANSITIONS IN ^{64}GE

An intense transition of 1665 keV was first reported in ^{64}Ge by Ennis and co-workers [17], to which they assigned tentatively a stretched $E1$, $5^- \rightarrow 4^+$ character (see Fig. 4). This was somewhat surprising since, assuming the isospin selection rules, in the long wavelength limit $E1$ transitions are forbidden in $N = Z$ nuclei between states of equal isospin, as mentioned already in the Introduction [3]. It was suggested by the same authors that such a strong $E1$ transition is indicative of mixing between the low-lying $T = 0$ states with the isobaric analog $T = 1$ states, and that measuring the reduced electrical dipole transition probability one could estimate the isospin mixing probability α^2, defined as [18]:

$$\alpha^2 = \frac{1}{2}\langle N = Z | T_- T_+ | N = Z \rangle \quad (2)$$

Since a firm assignment of the multipolarity of the 1665 keV transition was considered of paramount importance in order to extract the isospin mixing probability from the reduced transition probabilities, we performed a first experiment at the Laboratori

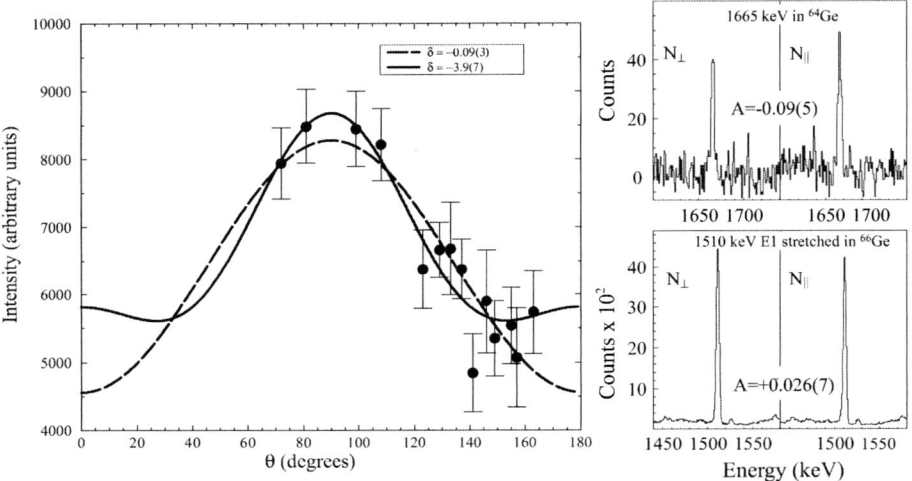

FIGURE 5. Left: angular distribution data for the 1665 keV line in ^{64}Ge. The curves corresponding to the best fit value are shown for both the solution with large $\delta = -3.9$ and with small $\delta = -0.09$. See text for details. Right: polarization correlation data for the 1665 keV line in ^{64}Ge (top) and for the corresponding transition of 1510 keV in ^{66}Ge.

Nazionali di Legnaro [19] using EUROBALL+ISIS+n-Wall in order to measure the multipolarity of the 1665 keV transition. The ^{64}Ge nuclei were populated via the reaction ^{40}Ca(^{32}S, 2α)^{64}Ge at 125 MeV beam energy. A 1 mg/cm^2 ^{40}Ca target on a 12 mg/cm^2 Au backing was used, in the attempt to extract lifetimes through the DSAM method. Spins were deduced through an angular distribution analysis, while parities were assigned through a polarization correlation analysis, using the Clover detectors of EUROBALL as Compton polarimeters [20]. The results shown in Fig. 5 suggest that the $5^- \rightarrow 4^+$ 1665 keV transition has actually a dominant $M2$ character, being the multipole mixing ratio $\delta = -3.9^{+0.7}_{-0.4}$. A solution with $\delta = -0.09(3)$ was also found in the angular distribution analysis, which would imply a stretched $M1$ character for the transition. This solution is in contrast with the systematics of the region and was discarded, being unfavoured on the basis of reduced χ^2 arguments.

Since the DSAM analysis could provide only a lower limit for the lifetime of the 5^- level, it was extracted through the RDDS method in a second experiment [19], performed at the Institut de Recherches Subatomiques, Strasbourg, using the EUROBALL array coupled to the Köln plunger device [21]. The same reaction was used at 135 MeV beam energy in the attempt to increase the production cross section for the ^{64}Ge nuclei. Given our measured lifetime for the 5^- level, $\tau = 24 \pm 3$ ps, the measured intensities and mixing ratio, the resulting reduced transition probability is $B(E1) = 2.6^{+0.9}_{-0.6} \cdot 10^{-7}$ W.u. which is an order of magnitude lower than the corresponding transition in the non-forbidden case of ^{66}Ge [22]. The isospin mixing probability has been estimated using a schematic model, describing the ^{64}Ge states as the equivalent ^{66}Ge states after the removal of a correlated pair of neutrons, with projection on pure isospin states [19]. The result, $\alpha^2 = 2.50\%^{+1.0\%}_{-0.7\%}$ is of the order of magnitude of microscopical calculations found in the literature [23, 18].

SUMMARY

The isospin symmetry has been investigated in a series of experiments close to the $N = Z$ line. The Coulomb Energy Differences were shown as a powerful method to extract information on the evolution of the microscopical nuclear structure and on the radius of the proton distribution. The case of the $A = 50$ mirror pair and of the ^{70}Br-^{70}Se analogue states were presented. Finally, the first estimate of isospin mixing probability with in-beam γ-spectroscopy techniques in ^{64}Ge was discussed.

ACKNOWLEDGMENTS

The results which have been presented were obtained thanks to the invaluable effort of a large number of persons who helped with the preparation of the experiments and with the analysis. Unfortunately here there is space left only to mention a few collaborators who provided materials for the oral presentation, namely S.M. Lenzi, S. Lunardi, G. de Angelis and T. Martínez. I am especially indebted to A. Gadea and N. Mărginean for sharing their preliminary results on ^{54}Ni.

This work was partially supported by the European Community under TMR contracts n°ERBFMBICT983126, n°ERBFMGECT980110 and n°HPRI-CT-1999-00078.

REFERENCES

1. W.E. Ormand, B.A. Brown, Phys. Rev. **C52**, 2455 (1995)
2. H. Sagawa, N. Van Giai and T. Suzuki, Phys. Rev. **C53**, 2163 (1996)
3. L. Radicati, Phys. Rev. **87**, 521 (1952)
4. S.M. Lenzi et al., Phys. Rev. Lett. **87**, 122501 (2001)
5. A.P. Zuker, S.M. Lenzi, G. Martínez-Pinedo and A. Poves, Phys. Rev. Lett. **89**, 142502 (2002)
6. J. Gerl and R.M. Lieder, *EUROBALL III, European γ-ray facility*, GSI Darmstadt (1992)
7. E. Farnea et al., Nucl. Inst. and Meth. in Phys. Res. **A400**, 87 (1997)
8. Ö. Skeppstedt et al., Nucl. Inst. and Meth. in Phys. Res. **A421**, 531 (1999)
9. J.A. Sheikh, P. Van Isacker, D.D. Warner and J.A. Cameron, Phys. Lett. **B252**, 314 (1990)
10. J.A. Sheikh, D.D. Warner and P. Van Isacker, Phys. Lett. **B443**, 16 (1998)
11. B. Blank et al., Phys. Rev. Lett. **74**, 4611 (1995)
12. G. de Angelis et al., Eur. Phys. Jour. **A12**, 51 (2001)
13. A. Gadea et al., *A new charged particle Si detector for EUROBALL*, LNL/INFN (Rep) **118/97**, 225 (1997)
14. J.B. Ehrman, Phys. Rev. **81**, 412 (1951)
15. R.G. Thomas, Phys. Rev. **88**, 1109 (1952)
16. K. Ogawa et al., Phys. Lett. **B464**, 157 (1999)
17. P.J. Ennis et al., Nucl. Phys. **A535**, 392 (1991); P.J. Ennis et al., Nucl. Phys. **A560**, 1079 (1993) (erratum)
18. J. Dobaczewski and I. Hamamoto, Phys. Lett. **B345**, 181 (1995)

19. E. Farnea *et al.*, Phys. Lett. **B551**, 56 (2003)
20. A. Gadea *et al.*, in: *Proc. Conf on Experimental Nuclear Physics in Europe*, Sevilla, Spain, June 1999, AIP Conf. Proc. 495, B. Rubio, M. Lozano, W. Gelletly, Eds., p.195 (1999)
21. A. Dewald *et al.*, in: *Selected Topics in Nuclear Structure, Proceedings of the XXV Zakopane School on Physics*, J. Styczen and Z. Stachura, Eds., Vol.2 p. 152 (World Scientific, Singapore, 1990)
22. U. Hermkens *et al.* Zeits. für Physik **A343**, 371 (1992)
23. G. Colò, M.A. Nagarajan, P. Van Isacker and A. Vitturi, Phys. Rev. **C52**, R1175 (1995)

AGATA

Dino Bazzacco

INFN Padova,Via Marzolo 8, I-35131, Padova, Italy

On behalf of the AGATA collaboration

Abstract. New accelerator facilities for radioactive-ion beams and high-intensity stable beams will start operation in a few years. They will provide interesting opportunities for exploring unknown territories of the nuclear landscape but the harsh experimental conditions require the construction of a new generation of detector arrays for gamma-ray spectroscopy built fully from germanium detectors and based on the emerging technique of gamma-ray tracking. The "Advanced GAmma Tracking Array" (AGATA), proposed in Europe, will be built out of 120/180 highly segmented high purity germanium crystals operated in position sensitive mode by means of digital data techniques and pulse shape analysis of the segment signals. AGATA will be capable of measuring gamma radiation in a large energy range (from ~10 keV to ~ 10 MeV), with the largest possible photopeak efficiency (25 % at $M_\gamma = 30$) and with a good spectral response. In particular, its very good Doppler correction and background rejection capabilities will allow to perform "standard" γ-ray spectroscopy experiments using fragmentation beams with sources moving at velocities up to $\beta \sim 0.5$. The talk reviews the status of development of the γ-ray tracking technique and the present design of AGATA.

INTRODUCTION

In the last decade, EUROBALL and GAMMASPHERE have been the highest efficiency 4π detector arrays available for in-beam γ-ray spectroscopy. These arrays are built out of ~100 Compton-suppressed high-purity germanium (HPGe) spectrometers arranged in a spherical configuration around the reaction point and provide, for 1 MeV γ-rays, a total peak efficiency (ε_{ph}) of about 10 % and a P/T-ratio of about 60 %. These figures combine to a high selectivity for identification of weak reaction branches and the two detectors have, in fact, played a major role in experimental nuclear structure, in particular in the study of high spin phenomena. The γ-ray spectroscopy community is offered now the opportunity to extend the field of nuclear structure studies exploiting the radioactive ion beams provided by the next generation of accelerator facilities of ISOL and fragmentation type. It is well realized however that, due to low beam intensities and large Doppler broadenings combined with high backgrounds, the experimental conditions at such facilities will be very challenging. To cope with them, both the total peak efficiency and the selectivity of our detector arrays must be improved. Larger efficiencies will be soon provided by the compact arrays EXOGAM and MINIBALL, but their use is, by design, restricted to low γ-ray multiplicity experiments and not too high recoil velocities. To advance further, the γ-spectroscopy community needs to develop general-purpose detection systems with much higher performance and it is a matter of fact that such an array cannot be obtained by simply increasing the number of detectors. The reason is that to obtain a good P/T ration we must surround the Ge crystals with BGO suppression shields which, however, take a consistent fraction of the solid angle and limit the total peak efficiency of the so-called 4π arrays to no more than 10 %. The solution of this problem may come from recent advances in crystal segmentation technology and digital signal processing, which make it possible to operate the germanium detectors in a position sensitive mode. The knowledge of energy and position of the interaction points within the germanium crystal allows to reconstruct the interaction history of the absorbed gamma radiation, leading to the so-called gamma-ray tracking concept [1]. This allows to build a compact array solely out of highly segmented Ge detectors omitting the BGO shields. As it is expected from simulations, an array consisting of a limited number (100-200) of such detectors will have an efficiency of up to 50 % and a P/T-ratio of 70 %. It should be remarked that an important benefit of γ-ray tracking is the possibility to determine with very good precision (~1°) the emission direction of the reconstructed transitions, allowing for an almost perfect correction of Doppler broadening effects for gammas emitted by nuclei recoiling at velocity as high as $v/c = 0.5$.

The unique capability of the γ-ray tracking arrays will allow to study rare reaction channels produced in experiments with weak beam intensities (e.g. radioactive beams), addressing many questions that are still open in nuclear structure, astrophysics and fundamental interactions. Finally, it is clear that, besides the perspective use for fundamental research, the principle of γ-ray tracking is also extremely interesting for its possible application in the field of γ-ray imaging.

GAMMA-RAY TRACKING METHODS

The feasibility of γ-ray tracking has been the subject of extensive R&D performed in the last 7 years by GRETA [2] in the USA and the TMR network "Gamma Ray Tracking Detectors" [3] in Europe. The main lines of development are outlined in the following.

Tracking algorithms. The task of γ-ray tracking is to identify the individual γ-rays and reconstruct their scattering sequence in the active volume of the detector from the knowledge of energy and spatial coordinates of the interaction. As long as single gamma events are considered, the characteristic features of the relevant interaction mechanisms can be exploited for the task of γ-ray tracking in simple and efficient ways.

Isolated low energy interaction points can be accepted as **photoelectric** absorption events upon checking the compatibility between energy and interaction depth in the detector. In the intermediate energy range the γ-rays undergo a sequence of Compton scatterings with a final photoelectric effect. For the reconstruction of such events, a figure of merit of each **Compton scattering** vertex (Fig. 1) is calculated for all permutations of the interaction points. The figure of merit is defined as the deviation between the measured energy of the interaction point and the energy calculated from the Compton formula (using the scattering angle derived from the spatial position of the involved points). The permutation with the best total merit is considered to correspond to the real scattering sequence and the event is accepted if its value is compatible with predefined limit.

For gamma ray energies above a few MeV, **pair production** events become important. As long as shower effects are negligible a strong signature of this mechanism is that the first point of interaction collects the total γ-ray energy minus the $2m_0c^2$ needed to create the pair. This is because the electron and positron are absorbed within a distance of ~1 mm, while the two annihilation photons generate other clusters of points somewhere else in the array where they are possibly detected. A simple algorithm built to exploit these features has a reconstruction efficiency of about 50 %.

The tracking algorithms mentioned above have been developed and tested using data from Monte Carlo simulations. In order to be realistic, the calculated interaction points are randomly smeared, before being used by the tracking algorithms, so as to take into account the position and energy resolution of actual detectors. It is important to remark that, due to experimental uncertainties but also because of fundamental limits, each of these algorithms will always accept some background events (corresponding e.g. to partial energy release in the detector) and reject "good" ones. The amount of accepted background can often be reduced, but this always at the cost of an increased rejection of good events: i.e. a better P/T implies a lower efficiency.

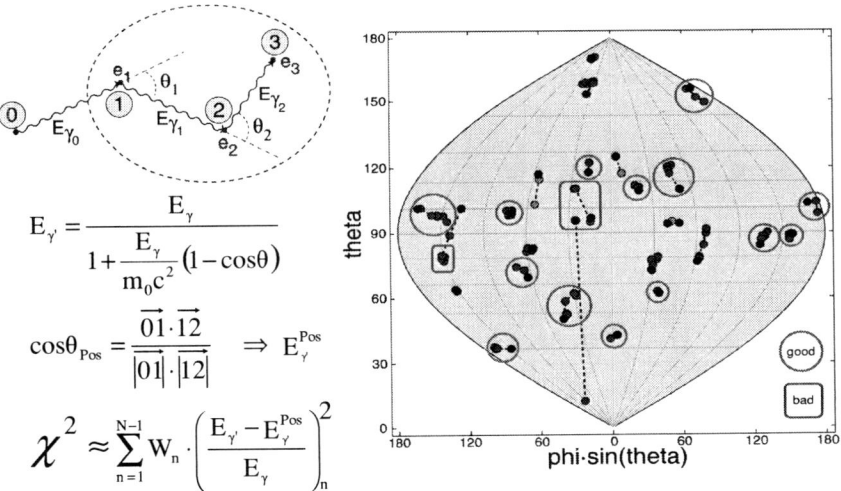

FIGURE 1. The left part shows the ingredients of γ-ray racking of a gamma ray absorbed in the detector material by means of two Compton scattering events followed by a photoelectric interaction. The right part shows the reconstruction of a high multiplicity event detected in an ideal germanium shell with an inner radius of 15 cm and a thickness of 9 cm. Of the 30 emitted gammas 27 are detected; 23 of them with full energy release. The tracking algorithms recognized correctly 14 gammas, corresponding to ~70 % reconstruction efficiency. The two badly reconstructed groups of points end up into the spectrum background.

For realistic events with several coincident γ-rays, one cannot simply apply the same considerations because there is no absolutely reliable way to determine the transition to which the individual interaction points belong. In principle, one could try all the possible ways to combine the measured points into a reasonable number of transitions but this is an extremely difficult computational task because already for medium γ-ray multiplicities the number of partitions can easily exceed 10^{20}. On the other side, it does not really make sense to try it because the limited capability of the reconstruction algorithms to distinguish good events from bad ones (e.g. points of different transitions accepted as belonging together) does not ensure that the optimum partition can be identified. The problem has therefore to be split into simpler pieces and this can be achieved in at least two ways. Using the fact that the interaction points of transitions emitted into sufficiently separated directions tend to "cluster" into spatially isolated groups, it is possible to search for such cluster and validate them as individual transitions with the methods explained above. An example of a reconstructed high multiplicity event detected by an ideal germanium shell is shown in on the right part of Fig.1. With a position resolution of 5 mm, this "ideal shell" yields ε_{ph} = 36 % and P/T = 60 %, for M_γ = 30.

Slightly smaller reconstruction efficiency are achieved with another approach, called "backtracking" [4], which starts from points with energy in the ~100 keV range (likely to be the last, i.e. photoelectric, interaction of a transition) and goes back, step by step, to the origin of the incident γ-ray looking for the correct Compton scattering vertices. Backtracking is probably more sensitive to position errors but is better suited for gamma-ray imaging purposes.

Highly segmented germanium detectors. Several laboratories are pursuing the development of highly segmented germanium detectors, both of the cylindrical and the planar configuration. The following brief summary is limited to closed-end coaxial detectors, which are believed to be more suited for the construction of 4π tracking arrays. The 36-fold segmented "hexaconical" detector for the 4π array GRETA was the first prototype to be studied in details [5] at Berkeley. In Italy, within the MARS collaboration, we are using a 25-fold segmented detector that has 6 angular sectors, 4 transversal slices and a small segment on the front face [6]. To achieve the best energy resolution, these two detectors use cold FETs but there also prototypes of segmented detectors that use simpler to handle room temperature FETs that can be placed outside the ge-crystal vacuum chamber. Two such detectors (one with 24 and one with 36 segmented) are under test at the University of Liverpool and Surrey, in the UK. At NSCL, the array SeGA (T.Glasmacher's talk, this conference) composed of eighteen 32-fold-segmented cylindrical detectors, also with warm FETs, has just entered into operation. This array is not capable of fully tracking the detected gammas; instead, the high segmentation of the germanium detectors are used to improve the energy resolution in low multiplicity experiments with fast exotic beams.

Pulse shape analysis (PSA). Necessary input data for the tracking algorithms are, for each point of interaction, three-dimensional position and deposited energy. This information is obtained from the analysis of the shapes of the signals from the segments. A characteristic feature of segmented detectors is that not only the segment that collects the net charge released by the interaction has a signal but also the neighbouring segments have transient signals with no net charge at the end. Their amplitudes increase with decreasing distance between the interaction point and the border to the neighboring segment. Using finite element analysis, signal shapes can be calculated accurately for almost arbitrary crystal geometries [7].

To test the algorithms for pulse shape analysis, simplified problems considering one or two interactions randomly distributed within only one segment are treated at present. The resulting composite signals from the segments are analyzed (decomposed) to extract the number, position, and energy of the original point(s) in various ways. Programs using a genetic algorithm (GA) determine the number of interactions correctly for 89 % of the events. The variance between reconstructed and true position is ~2 mm while that of the energies is ~4 %. The performance of the PSA algorithms has been tested with experimental data from the GRETA and MARS prototypes using tightly collimated γ-sources. A more complete test of the GA algorithms has been done with the data of the experiment described in the following.

Digital electronics and DAQ. To preserve the full signal-shape information that is needed to extract energy, timing and position of the interaction points, the electronics has to sample the preamplifier signals with fast ADCs located as close as possible to the detectors. It is not yet clear whether full pulse shape processing can be done in real time at the detector level. However, in order to reduce the amount of information transferred to the data acquisition system to a manageable value, one should be able to obtain at least a local trigger and the energy of the fired segments using the so-called Moving Window Deconvolution (MWD) [8]. This information and the subset of samples relative to the leading edge of the signals will be time stamped and transferred by high-speed fiber links to the data acquisition computers. It is foreseen that the overall data flow to the DAQ of a 4π γ-ray tracking array can be in excess of 500 Mbytes/s. Finally, after applying all necessary functions of data merging and gain matching, the pulse shape analysis will be performed and the tracking algorithms will be applied to produce "standard" events containing just energy, time and spatial direction of the reconstructed transitions.

In beam experiment. To test the performance of the algorithms developed for pulse shape decomposition and γ-ray tracking, an in-beam experiment has been performed with the 25-fold segmented MARS [6] germanium detector. The idea was to check the achieved position resolution by looking at the improvement of energy resolution of a severely Doppler broadened peak when the detected energy is corrected using the position of the first interaction point. For this purpose, the Coulomb excitation experiment ^{56}Fe, at 240 MeV, on a thin ^{208}Pb target was selected. At this bombarding energy, the spectrum contains, practically, only the ^{56}Fe $2^+ \rightarrow 0^+$ transition at 846.7 keV. The direction of the scattered ^{56}Fe ions has been detected with a precision of ~2° by 14 Si detectors. In order to get a large Doppler broadening, the MARS detector was placed as close as possible to the target (Fig. 2, right panel). A coincidence between the germanium and any of the particle detectors triggered seven digital scopes to sample (at 200 Ms/s and for 10 µs) the preamplifier signals of the 25 segments and of the inner contact. The event rate was ~1 Hz.

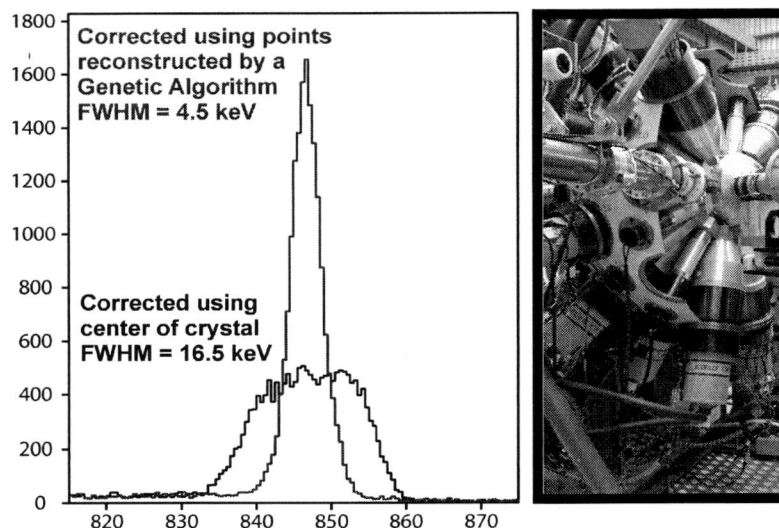

FIGURE 2. Coulex experiment with the MARS segmented detector. On the right panel the experimental setup at GASP with the detector mounted in the horizontal plane at 135° with respect to the beam axis. The left panel show the improvement of energy resolution of the 847 keV peak obtained with event-by-event Doppler shift correction using the position of the first interaction point as determined by Pulse Shape Analysis of the recorded wave forms.

The $\sim 10^5$ collected events have been processed using a Genetic Algorithm to determine the position of the interaction points inside the Ge crystal. The result of this analysis is shown in the left panel of Fig. 2. The dramatic improvement in energy resolution is in line with a detailed simulation of the experiment which gives an energy resolution of 4.2 keV, assuming a position resolution of 5 mm FWHM.

In calculating the performance of a realistic γ-tracking array, it seems therefore safe to assume that pulse shape decomposition can be done with at least a 5 mm position resolution. Further R&D will most likely improve this value so that the figures given in the following should be taken as safe values.

AGATA

The Advanced GAmma Tracking Array (AGATA) [9] project is aimed at a full scale implementation of the γ-ray tracking concept in a European context. A similar project is pursued in the USA by the GRETA [2] collaboration. The specific event that triggered the development of AGATA is the availability, in the near future, of radioactive beams from the upgraded GSI facility, from GANIL and from the SPES facility at LNL. However, the detector is being designed for a more general range of applications, e.g. also for experiments with high intensity stable beams.

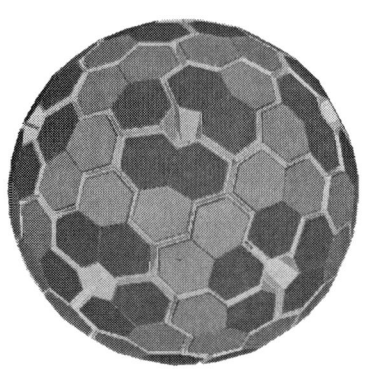
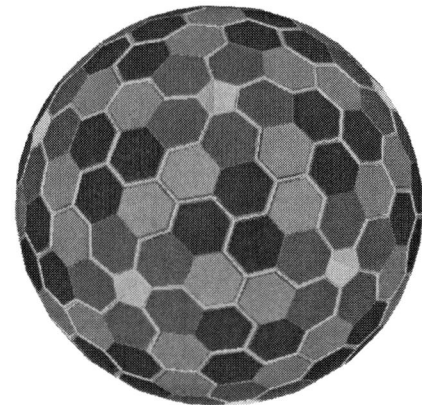

FIGURE 3. Schematic view of the two possible AGATA configurations. The smaller configuration has 120 hexagonal crystals of 2 different shapes (colour coded) that can be packed into 40 triple clusters of two different types. In the larger configuration 180 crystals of three different shapes are packed into 60 all-equal clusters.

As already mentioned, specific features of radioactive beams are limited intensities, (particularly for the most exotic nuclei); wide range of recoil velocities (from stopped to v/c≈50 %); high gamma and particle backgrounds and γ-ray multiplicities up to $M_\gamma = 30$. To cope with these conditions, a 4π γ-array with the highest efficiency, selectivity, energy resolution and capability to handle high counting rates is required. These features can only be achieved with a close packed arrangement of germanium detectors, i.e. a 4π shell built from large, highly segmented Ge crystals.

The geometric structure of AGATA is based on a geodesic tiling of the spherical surface with hexagons and 12 pentagons. Two possible configurations have been identified (see Fig. 3): a smaller one with 120 hexagons (two different shapes) and a bigger one with 180 hexagons (three different shapes). The pentagons will, most likely, not be used and, to minimize inter-detector space losses, the hexagons will be packed into clusters of three crystals. The 120 configuration will therefore have 40 clusters (of two different types) while the 180 configuration will have 60 all-equal clusters. Using the largest reasonably available germanium crystals (80 mm diameter, 90 mm length), the inner radius of the two arrays turns out to be 18 and 23 cm, while the total solid angle covered by germanium is ~72 % and ~80 % respectively.

The performance of the two candidate arrays has been simulated within the GEANT4 [10] framework including, for the sake of being realistic, the encapsulation of the crystals and their canning into triple clusters [11]. In Tab. 1 the obtained total peak efficiency and peak to total ratio are compared with the performance of the "ideal shell" and of EUROBALL. Even if a realistic detector can achieve only about 50 % of the performance of the ideal shell, the efficiency gain with respect to EUROBALL is really considerable. In particular for high γ-multiplicity experiments, this means an increase in selectivity of several orders of magnitude

TABLE 1. Performance the possible AGATA configurations at $E_\gamma = 1$ MeV.

	Number of detectors (crystals)	Amount of Germanium (kg)	ε_{ph} [P/T] % $M_\gamma = 1$	ε_{ph} [P/T] % $M_\gamma = 30$
Ideal 4π shell	1	233	65 [85]	36 [60]
AGATA 120	40(120)	212	33 [54]	19 [44]
AGATA 180	60(180)	320	38 [53]	24 [46]
EUROBALL	71(239)	210	9 [56]	6 [37]

The germanium crystals will be 36-fold segmented and the resulting 4440 or 6660 total segments will provide unprecedented position sensitivity. A key feature of AGATA is the high precision for determining the emission direction of the detected γ-rays of <1° (corresponding to an effective solid angle granularity of more than $5 \cdot 10^4$ that cannot be realistically achieved with individual germanium crystals). This granularity ensures an energy resolution better than 0.5 % for transitions emitted by nuclei recoiling at velocities as high as 50 % of the speed of light. This value is only a factor of two bigger than the intrinsic resolution of Ge detectors and is comparable with the values currently observed at 10 times smaller recoil velocity.

The demonstrator. According to the recently signed MoU, the development of AGATA will proceed in stages with a final R&D phase preceding the construction of the full array. The objective of the R&D phase is to build a subsystem of 5 triple-clusters, called the AGATA demonstrator. The demonstrator, complete of digital electronics, DAQ and full on-line processing of the digitized data, will be used for testing the γ-ray tracking concept in real experiments.

ACKNOWLEDGMENTS

The reported work is the result of a collaboration of European laboratories carried on originally within TMR Research Network "Development of γ-Ray Tracking Detectors for 4π γ-Ray Arrays" which was supported by the EC under contract ERBFMRX-CT97-0123.

REFERENCES

1. I.Y. Lee, Nucl. Instr. Meth **A422** (1999) 195
2. M.A. Deleplanque et al., Nucl. Instr. Meth. **A430** (1999) 292
3. R.M. Lieder et al, Nucl. Phys **A682** (2001) 279c.
4. J. van der Marel and B. Cederwall, Nucl. Instr. Meth. **A437** (1999) 538
5. K. Vetter et al., Nucl. Instr. Meth. **A452** (2000) 105 and 223
6. D.Bazzacco et al, LNL Ann..Rep. 2000, p 166, LNL-INFN(Rep)-178/01
7. Th. Kröll and D. Bazzacco, Nucl. Instr. Meth. **A463** (2001)227
8. A. Georgiev at al., IEEE trans. Nucl. Sci. **41** (1994) 1116
9. http://agata.pd.infn.it/documents/Agata_pub-proposal.pdf
10. S.Agostinelli et al, Nucl. Instr. Meth **A506** (2003) 250
11. E.Farnea, private communication

Study of Superdeformation in the $A \approx 60$ Mass Region; High Resolution γ-Ray Spectroscopy at EUROBALL IV with the Recoil Filter Detector and the EUCLIDES Charged Particle Detector

J. Dobaczewski[1], J.P. Vivien[2], K. Zuber[3], P. Bednarczyk[2,3,4], T. Byrski[2], D. Curien[2], G. de Angelis[5], O. Dorvaux[2], G. Duchêne[2], E. Farnea[6], A. Gadea[5], B. Gall[2], J. Grębosz[3], R. Isocrate[6], A. Maj[3], W. Męczyński[3], J.C. Merdinger[2], A. Prévost[7], N. Redon[7], J. Robin[2], O. Stézowski[7], J. Styczeń[3] and M. Ziębliński[3]

[1]*Institute of Theoretical Physics, Warsaw University, Warszawa, Poland*
[2]*Institut de Recherches Subatomiques, Strasbourg, France*
[3]*The Niewodniczanski Institute of Nuclear Physics, Polish Academy of Sciences, Kraków, Poland*
[4]*GSI, Darmstadt, Germany*
[5]*Laboratori Nazionali di Legnaro, Legnaro, Italy*
[6]*University of Padua, Padova, Italy*
[7]*University of Lyon, Lyon, France*

Abstract. High spin studies of $A \approx 60$ nuclei have been carried out with EUROBALL. The nuclei of interest were populated in the fusion-evaporation reaction with a 125 MeV ^{16}O beam bombarding a thin ^{56}Fe target. The use of the Recoil Filter Detector and the charged particle detector EUCLIDES resulted in a significant reduction of γ-line widths and facilitated a reaction channel selection. In this article, several new linking transitions of the previously known terminating band in ^{64}Zn are reported. Such bands are also observed in ^{61}Cu and ^{63}Cu. New SD bands identified in ^{61,63}Cu are discussed.

INTRODUCTION

Superdeformation in the $A \approx 60$ nuclei with $N \approx Z \approx 30$ is associated with the most rapid rotation that has been up to date observed in discrete spectroscopy. Rotational frequency deduced for the SD bands known in this region can reach as much as 2 MeV. In such extreme conditions, the standard pairing correlation plays a minor role. Instead, the influence of the $T = 0$ pairing, resistant to the Coriolis antipairing force, might be manifested. Moreover, in these relatively light nuclei the highest spins populated along the SD bands are close to the maximum aligned angular momentum. Therefore, a competition between the SD band termination and the band crossing with a high-j intruder orbital predicted at a very high rotational frequency can be examined.

The main experimental difficulty in studying the high spin phenomena in the $A \approx 60$ mass region is an excessive Doppler broadening of lines in γ-spectra measured in beam. This is due to high recoil velocity of nuclei populated via fusion-evaporation reactions and the high energy of γ-rays expected to occur. A presence of alpha reaction channels and high multiplicity of evaporated particles induces additional spread in recoil velocity and results in further worsening of the resolution.

This constrain limiting so far our knowledge on the high spin structure of $A \approx 60$ nuclei could be minimized by a) making use of the Recoil Filter Detector (RFD) in order to reduce the Doppler broadening of γ-lines and b) utilizing a charged particle detector, thus providing a reaction channel selection.

EXPERIMENTAL METHODS

The measurements were carried out at the VIVITRON accelerator in IReS, Strasbourg. Several isotopes of Ni, Cu, Zn, Ga and Ge were populated via fusion-evaporation reaction in the bombardment of a thin (0.4 mg/cm^2) ^{56}Fe target with a pulsed ^{16}O beam at 125 MeV. Gamma rays were detected in the EUROBALL IV array comprising 26 Clover detectors positioned at about 90° with respect to the beam direction and 15 Cluster detectors placed at backward angles. The forward part of EUROBALL consisting of the tapered detectors was removed in order to mount the Recoil Filter Detector. The RFD, a system of 18 heavy ion detectors distributed around the beam axis, was placed downstream at the distance of 134 cm from the target. It measured a time of flight of incoming residual nuclei produced in the reaction with respect to a beam pulse, as well as their position. In this way the RFD enabled a complete determination of the velocity vector of every recoiling nucleus. Event by event Doppler correction provided a very efficient way to reduce γ-line broadening [1] and allowed the observation of high energy γ-rays (up to 4 MeV) with a good resolution, as shown in the figure 1.

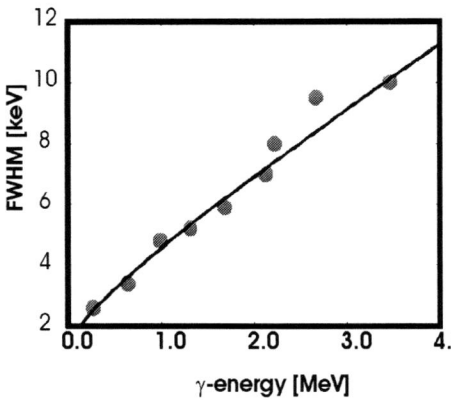

FIGURE 1. Plot of FWHM of γ-line versus γ-ray energy, obtained after event by event Doppler correction.

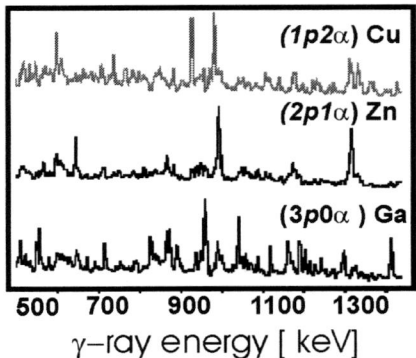

FIGURE 2. Spectra of γ-rays characteristic for different isotopes populated in the reaction. The reaction channels were selected by requiring a specific combination of charged particles detected in the EUCLIDES array.

Although the acquisition system was triggered whenever a recoil and at least two γ-rays were registered, the charged particles were simultaneously measured in the EUCLIDES array consisted of 55 silicon ΔE-E telescopes. Examples of γ-spectra selected by a specific charged particle combination are shown in the figure 2.

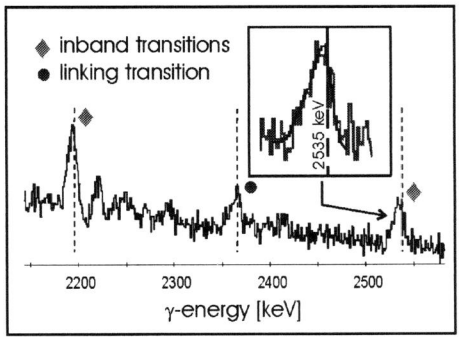

FIGURE 3. Part of a γγ-spectrum measured by Cluster detectors at 157° in coincidence with transitions from the SD band in ^{61}Cu. The dashed lines show peak positions. A narrow component of a properly Doppler corrected line is more pronounced in the case of a linking transition due to a longer (accumulative) lifetime. In the insert, the 2535 keV line fit with a trailed Gaussian curve is drawn. The line shape suggests that the 2535 keV γ-rays are emitted predominantly in the target. The estimated lifetime is shorter than 10 fs.

In the experiment, due to the employment of the RFD, short lifetimes could be estimated. The method takes advantage of the difference between the velocity vector of the recoil emitting a γ-ray while straggling inside the target and the recoil emitting that γ-ray after its escape from the target, as measured by the RFD. In the first case the insufficient Doppler correction trails the γ-line in the spectrum. A probability of the γ-emission in vacuum can be expressed by a ratio of the mean level lifetime τ and the transit time of the recoil passing through the target. Therefore τ can be estimated from a contribution of the properly Doppler corrected γ-rays to the total γ-line area. In the experiment described here, the lifetimes in the range of tens of femtoseconds could be determined. Figure 3 shows examples of line shapes corresponding to very fast γ-transitions deexciting newly identified SD band in ^{61}Cu. The lifetimes deduced in this case are shorter then 10 fs.

RESULTS

Preliminary results reveal many new high spin levels in almost all observed nuclei. The nucleus ^{64}Zn was populated in the strongest $2n2p1\alpha$ reaction channel. Prior to our work, a decay out mechanism of a terminating rotational band in ^{64}Zn reported in ref. [2] was unresolved. The band with characteristic dominant *M1* transitions between the signature partners has been interpreted as arising from an $f_{7/2}$ proton-hole excitation coupled to $g_{9/2}$ particles. Such configuration resembles a structure of states giving rise to smooth terminating bands in the Sb region or magnetic bands in Pb nuclei. However, the interpretation was incomplete due to lack of any experimental information about the excitation energy of the band. In the course of the present work we could firmly establish the connection between the band and several lower lying levels. The identified linking transitions, of predominantly *E2* character, are marked in

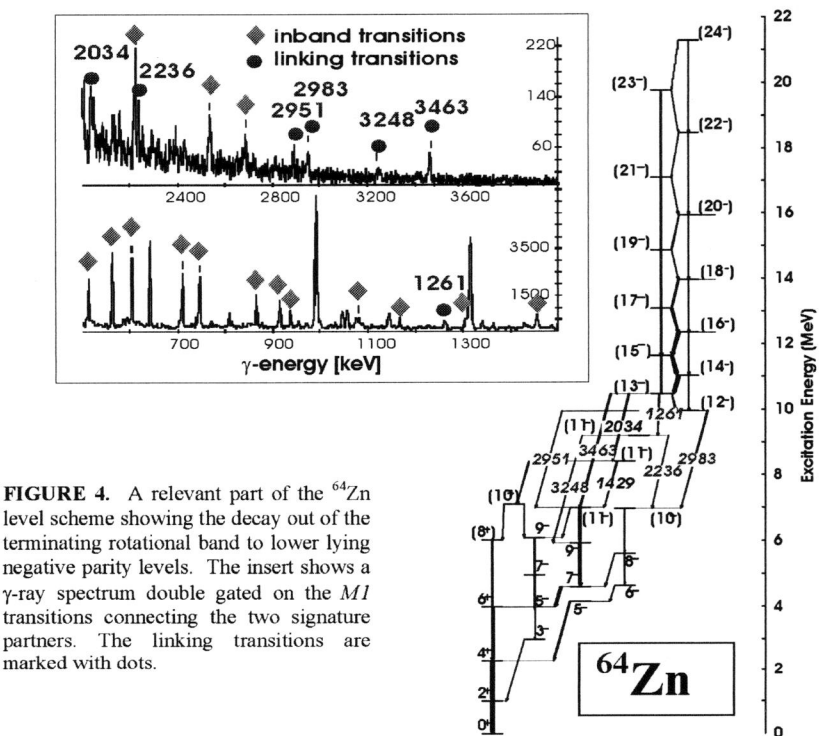

FIGURE 4. A relevant part of the ^{64}Zn level scheme showing the decay out of the terminating rotational band to lower lying negative parity levels. The insert shows a γ-ray spectrum double gated on the *M1* transitions connecting the two signature partners. The linking transitions are marked with dots.

a γ-spectrum shown in the insert of figure 4. In this figure a relevant part of the updated ^{64}Zn level scheme is displayed. The current data confirmed the previous band head spin assignment. On the other hand, the excitation energy is significantly bigger then it was assumed in ref [2]. The higher excitation energy of the band is however consistent with the position of similar deformed bands observed in ^{62}Zn [3].

Although in ^{64}Zn we observed several new high spin levels, they do not form any regular structure that could be associated with superdeformation known in several neighboring $A \approx 60$ nuclei. In ^{64}Zn due to interaction with non-collective bands the SD band based upon a 4-quasiparticle configuration is expected [4] to occur only for very high spins close to the band termination. Therefore, the experimental identification of superdeformation in ^{64}Zn will be rather difficult.

In contrast to ^{64}Zn, in ^{61}Cu and ^{63}Cu populated respectively in the $2\alpha p2n$ and $2\alpha p$ reaction channels, high spin levels constitute rotational bands similar to the known SD bands in ^{58,59}Cu and $^{60-62,65}$Zn [5]. The level scheme of ^{61}Cu and a spectrum of high energy γ-rays depopulating the new rotational band in this nucleus are presented in figure 5. Besides the rotational band in the two cupper isotopes, at lower spins, we identified a group of levels characterized by strong *M1* transitions. They resemble the band in ^{64}Zn discussed above. However, in ^{61}Cu and ^{63}Cu this excitation mode occurs at much lower energies with respect to ^{64}Zn.

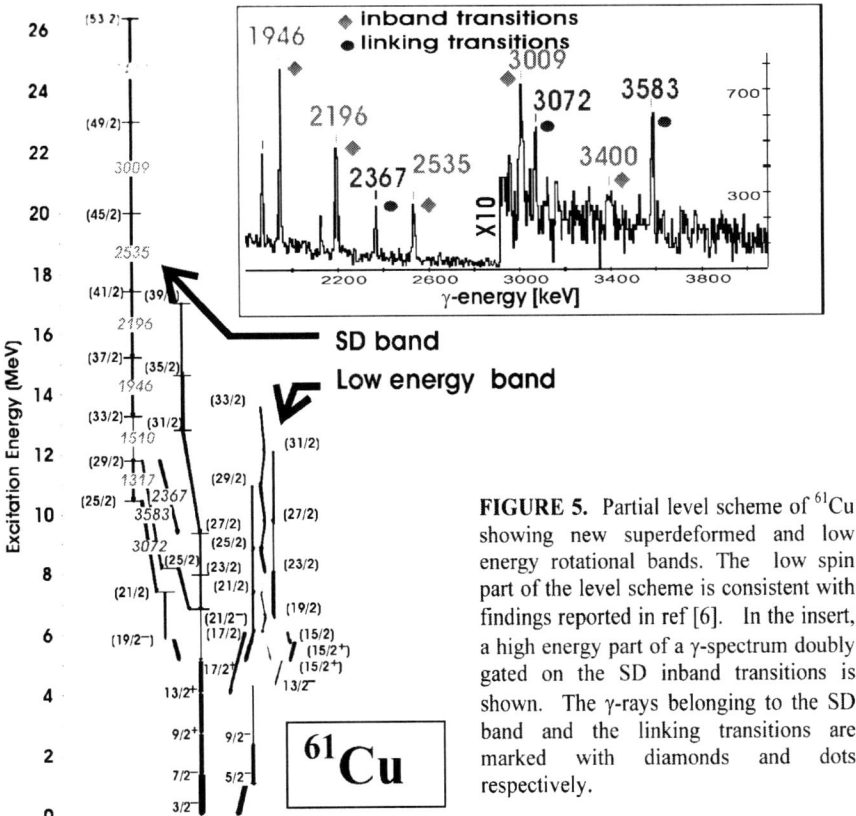

FIGURE 5. Partial level scheme of ^{61}Cu showing new superdeformed and low energy rotational bands. The low spin part of the level scheme is consistent with findings reported in ref [6]. In the insert, a high energy part of a γ-spectrum doubly gated on the SD inband transitions is shown. The γ-rays belonging to the SD band and the linking transitions are marked with diamonds and dots respectively.

A deformation associated with the high spin rotational band was estimated from a lifetime measurement. As it was already mentioned, the lifetimes in this case are below 10 fs. Such a short τ value would correspond to a superdeformation with $\beta > 0.5$. On the other hand, as displayed in figure 6, a gradual loss of collectivity with angular momentum may be the origin of changes occurring at high energies in the SD bands in ^{61,63}Cu. In both cases, a regular decrease of the dynamic moment of inertia $J^{(2)}$ with increasing rotational frequency (see figure 7) may indicate an important contribution of non-collective degrees of freedom in building states with high angular momentum. At the observed frequencies the kinematic moment of inertia $J^{(1)}$ is grater than $J^{(2)}$. This reflects a deficit of the angular momentum available for the SD configuration involving $f_{7/2}$-holes and $g_{9/2}$-particles [7].

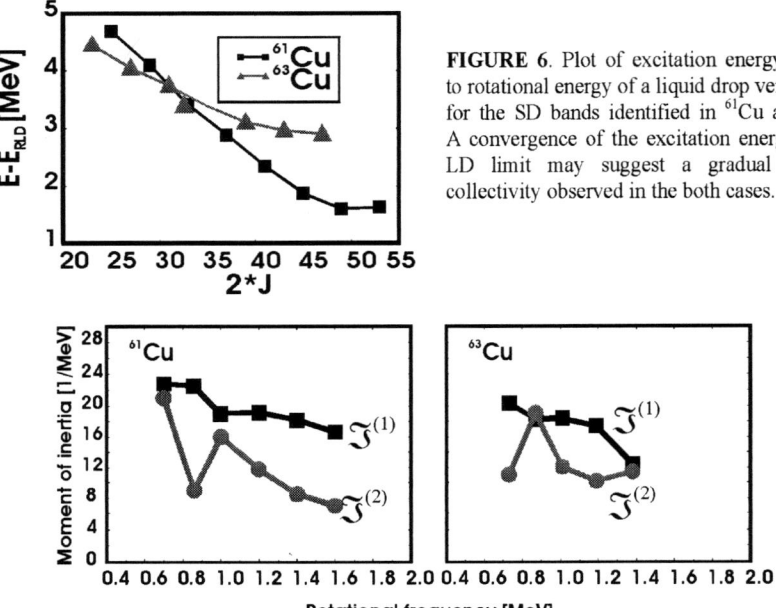

FIGURE 6. Plot of excitation energy relative to rotational energy of a liquid drop versus spin, for the SD bands identified in ^{61}Cu and ^{63}Cu. A convergence of the excitation energy to the LD limit may suggest a gradual loss of collectivity observed in the both cases.

FIGURE 7. Moments of inertia, extracted for the SD bands in ^{61}Cu and ^{63}Cu, as a function of rotational frequency.

CONCLUSION

In our study we have investigated a high spin structure of $A \approx 60$ nuclei. An ongoing analysis of this very rich data-set will provide new information on several populated nuclei. Preliminary results reveal a coexistence of a superdeformed band and a terminating band in ^{61}Cu and ^{63}Cu. In contrast, in ^{64}Zn at high spins only the terminating band dominates. The knowledge of the decay out mechanism gives the absolute excitation energy and spin of the bands, which will be important for comparison with a theory.

The variety of the observed structures show that, in this mass region, configurations involving $f_{7/2}$-holes and $g_{9/2}$-particles can give rise to distinct nuclear shapes. However, the available angular momentum is still limited by single particle degrees of freedom. Further theoretical interpretation of the results will help to better understand coupling between single-particle and collective nuclear excitations and will shed more light on the evolution of superdeformation at the limit of highest available spins and rotational frequencies.

ACKNOWLEDGMENTS

This work was supported by the Polish State Committee for Scientific Research (KBN Grant Nos. 2 P03B 118 22 and 5 P03B 014 21) and the European Commission contract EUROVIV.

REFERENCES

1. P. Bednarczyk *et al.*, Acta Phys. Polon. **B32** (2000) 747
2. A. Galindo-Uribarri *et al.*, Phys. Lett. **B422** (1998) 45
3. C.E. Svenson *et al.*, Phys. Rev. Lett. **80** (1998) 2558
4. Y. Sun *et al.*, Phys. Rev. Lett. **83** (1999) 686
5. B. Singh *et al.*, Table of Superdeformed Nuclear Bands and Fission Isomers, 3rd Edition, 2002
6. S.M. Vincent *et al.*, Phys.Rev. **C60** (1999) 064308
7. A.V. Afanasjev *et al.*, Phys.Rev. **C59** (1999) 3166

Angular Momentum Population in Projectile Fragmentation

Zs. Podolyák*, K.A. Gladnishki*†, J. Gerl**, M. Hellström**‡, Y. Kopatch**,
S. Mandal**, M. Górska**, P.H. Regan*, H.J. Wollersheim**,
K.-H. Schmidt** and O. Yordanov for the GSI-ISOMER collaboration**

*Department of Physics, University of Surrey, Guildford, GU2 7XH, UK
†Faculty of Physics, University of Sofia, BG-1164 Sofia, Bulgaria
**GSI, Planckstrasse 1, Darmstadt D-64291, Germany
‡Department of Physics, Lund University, Lund S-22100, Sweden

Abstract. Isomeric states in neutron-deficient nuclei around $A \approx 190$ have been identified following the projectile fragmentation of a relativistic energy ^{238}U beam. The deduced isomeric ratios are compared with a model based on the abrasion-ablation description. The experimental isomeric ratios are lower by a factor of ≈ 2 than the calculated ones assuming the 'sharp cutoff' approximation. The observation of the previously reported isomeric $I^{\pi}=43/2^-$ state in ^{215}Ra represents the current record for the highest discrete spin state observed following a projectile fragmentation reaction.

INTRODUCTION

During the last few years the application of projectile fragmentation reactions at intermediate and relativistic energies to studies of nuclei far from stability has made enormous progress. However, experimental information on the population of states as a function of angular momentum is still somewhat rare. In particular, for projectile energies above 100 MeV/nucleon, information is very scare. According to our knowledge, the first information regarding the population of isomeric states after projectile fragmentation at relativistic energies was presented by Schmidt-Ott et al. for ^{43}Sc populated in the fragmentation of 500 A·MeV ^{46}Ti beam [1]. Recently, Pfützner et al. published isomeric ratios for heavy neutron-rich and close to the stability line nuclei, populated in the fragmentation of ^{238}U [2] and ^{208}Pb [3].

Here we present results from the first systematic study on the angular momentum population in relativistic projectile fragmentation for heavy neutron-deficient nuclei.

EXPERIMENTAL DETAILS

Preliminary results from two recent experiments, performed at GSI, Darmstadt, are presented. *(a)* Neutron-deficient nuclei in the A~190 region were populated following the projectile fragmentation of a 750 MeV/nucleon ^{238}U primary beam impinging on a natural beryllium target. *(b)* Neutron-deficient nuclei around $^{216}_{89}$Ac were populated following the projectile fragmentation of a 950 MeV/nucleon ^{238}U primary beam on a natural beryllium target. In both experiments, the nuclei of interest were separated using the Fragment Separator (FRS) [4], then subsequently implanted into a catcher. The catcher was surrounded by a germanium detector array, in order to record the γ rays emitted from isomeric decays in the implanted ions. The delay of gamma-rays (with respect to the implantation time of a corresponding heavy ion) was measured up to 80 μs.

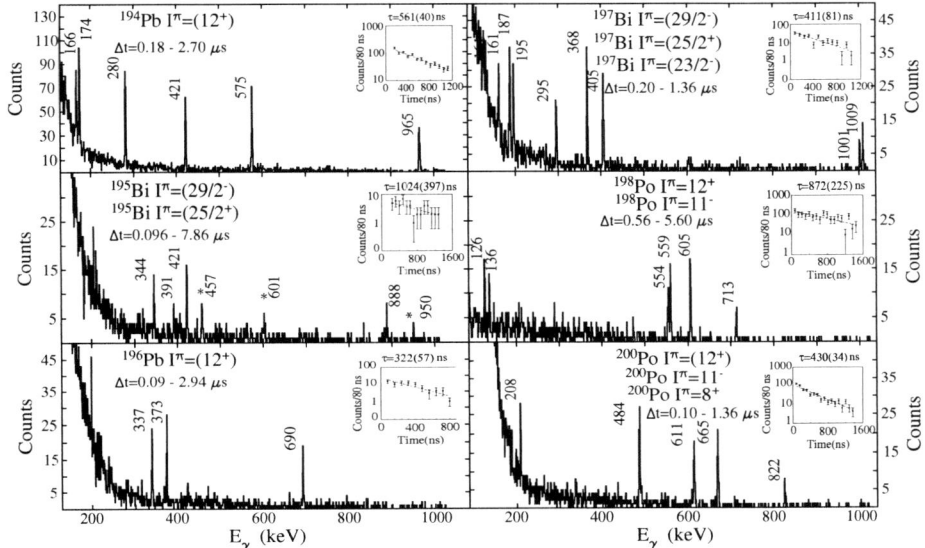

FIGURE 1. Delayed γ-ray spectra associated with ^{194}Pb, ^{195}Bi, ^{196}Pb, ^{197}Bi, ^{198}Po and ^{200}Po. The time spectra with fitted lifetimes are given in the insets.

The method is sensitive to isomers with half-lives in the range from about 100 ns up to several milliseconds. The lower limit is determined by the time of flight through the FRS (∼ 300 ns). However, as reported in previous works, if the electron conversion branch is blocked in a highly stripped ion, the effective ionic lifetime in flight is increased, which allows shorter neutral atom delay half-lifes to be measured [5, 6]. The upper limit is determined by the need to correlate the individual ions to the delayed γ rays.

RESULTS AND DISCUSSION

Isomeric ratios

Isomeric decays were identified in the ^{188}Hg, ^{192}Tl, ^{192,193,194,195,196}Pb, ^{195,197}Bi, and ^{198,200,202}Po [7]. Examples of delayed γ-ray energy spectra, as well as time spectra are presented in fig.1. The extracted isomeric ratios for the yrast $I^{\pi}=12^+$ isomers in ^{192,194,196}Pb and ^{198,200}Po are listed in Table I and shown in fig.2a.

Isomeric ratios can be determined theoretically in the framework of the abrasion–ablation model of the fragmentation reaction. In the initial abrasion phase a hot prefragment is created by removing a number of nucleons from the projectile. In the ablation

TABLE 1. Isomeric ratios of yrast metastable states produced in high energy, >100 A·MeV, projectile fragmentation. The results from the present work are preliminary.

Nucleus	J^π	Projectile	R_{exp}	R_{th}	R_{exp}/R_{th}	reference
^{205}Tl	$25/2^+$	^{238}U	0.25(5)	0.23	1.09(9)	[2]
^{205}Pb	$25/2^-$	^{238}U	0.29(5)	0.23	1.26(22)	[2]
^{206}Pb	12^+	^{238}U	0.29(5)	0.25	1.16(20)	[2]
^{208}Pb	10^+	^{238}U	0.34(9)	0.34	1.00(26)	[2]
^{175}Hf	$35/2^-$	^{208}Pb	0.025(6)	0.038	0.66(16)	[3]
^{180}Ta	15^-	^{208}Pb	0.10(3)	0.061	1.64(49)	[3]
^{200}Pt	7^-	^{208}Pb	0.30(5)	0.355	0.85(14)	[3]
^{192}Pb	12^+	^{238}U	0.14(2)	0.282	0.50(7)	present work
^{194}Pb	12^+	^{238}U	0.16(3)	0.280	0.57(11)	present work
^{196}Pb	12^+	^{238}U	0.17(4)	0.278	0.61(14)	present work
^{198}Po	12^+	^{238}U	0.089(12)	0.200	0.45(6)	present work
^{200}Po	12^+	^{238}U	0.067(12)	0.222	0.30(6)	present work

phase of the reaction, the highly excited prefragment evaporates nucleons until the final fragment is formed with an excitation energy below the particle emission threshold. Subsequently, a statistical gamma cascade proceeds down to the yrast line and then along this line to the ground state. If a long-lived state lies on this decay path, part of the cascade may be hindered or effectively stopped depending on the life-time of the isomer. The isomeric ratio is equal to the probability that gamma decay from the initial excited fragments proceeds via this isomeric states.

The angular momentum distribution can be calculated with the ABRABLA Monte Carlo code [8]. Furthermore, for a large mass difference, $\Delta A > 10$, between the projectile and the fragment this distribution can be approximated by a simple analytical formulae [9]:

$$P_I = \frac{2I+1}{2\sigma_f^2} exp\left[-\frac{I(I+1)}{2\sigma_f^2}\right], \quad (1)$$

where σ_f is called spin-cutoff parameter of the final fragments.

Given the angular momentum distribution of the final fragment, one can consider the probability that gamma decay will lead to an isomeric state of spin I_m. First, it is assumed that the initial excitation energy is well above the excitation energy of the isomer. Second, the extreme simplifying assumption is made that all states with $I \geq I_m$, and only those, decay to the isomer. A similar approach, known in the literature as the "sharp cutoff model", has been used in studies of angular momentum distributions in compound nuclei [10, 11, 12] and in fission fragments [13].

It has been shown that the approximative formulae effectively reproduces the angular momentum distribution predicted by the ABRABLA code when the mass difference between projectile and fragment is higher than 10 mass units [2], in the case of nuclei close to the stability line. However, in our case, far from stability the situation is different. The analytical formula predicts much higher angular momentum than the ABRABLA code. As shown in fig.2b, the difference increases as more 'exotic' nuclei

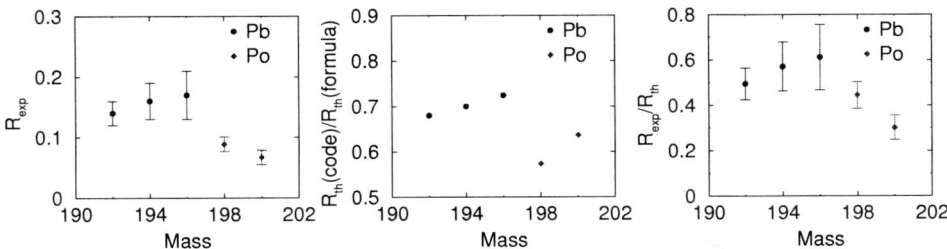

FIGURE 2. (a) Experimental isomeric ratios for the 12^+ isomers in ^{192,194,196}Pb and ^{198,200}Po. (b) The ratio of the calculated isomeric ratios using the ABRABLA code and the analytical formulae, for the same isomers. (c) The ratio of the experimental and theoretical (ABRABLA) isomeric ratios.

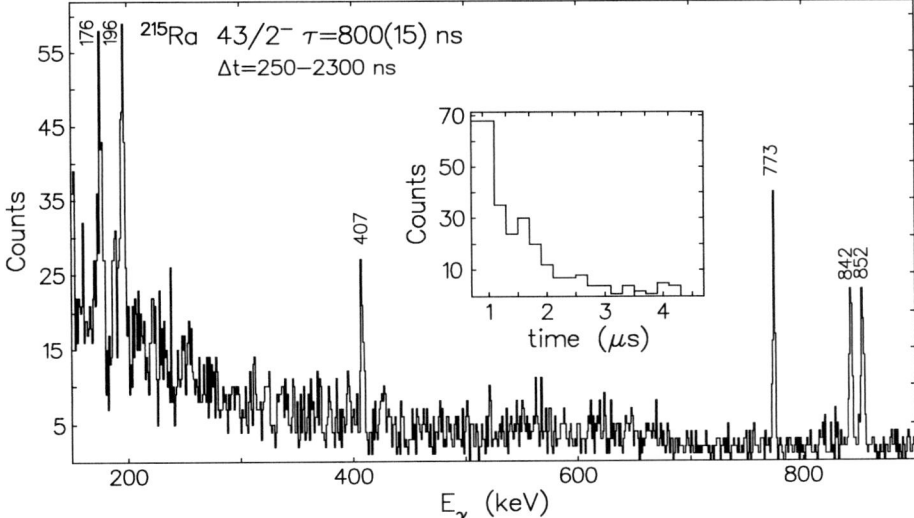

FIGURE 3. Delayed γ-ray spectra associated with ^{215}Ra. The 407 keV and 842 keV γ-ray transitions prove the population of the $43/2^-$ isomeric state [15]. The other labelled transitions are from a lower lying $29/2^-$ isomer.

are produced. Since the ABRABLA code is expected to give more reliable results, we have compared the experimental values with its predictions, as given in fig.2c and Table I. The experimental isomeric ratios are on average a factor 2 lower than the theoretical ones. In contrast, for the case of yrast isomers in nuclei close to stability the theory is in good agreement with the experiment [2, 3] (see Table I). One of the reason of the large discrepancy between theory and experiment in our case might be the extreme simplification considered, namely that all the states with spin higher than the isomer ultimately decay to the isomer. However, this is certainly not always the case, and when the level schemes are well known, the partial population can be determined By taking into account the decay properties of the states around the isomer a better agreement between the experimental and calculated angular momentum population can be obtained [14].

The $I^\pi=43/2^-$ isomer in ^{215}Ra

One of the highlights of the experiment performed with the 950 MeV/nucleon ^{238}U beam was the observation of the previously reported $I^\pi=43/2^-$ isomer in ^{215}Ra [15] (see fig.3). This represent the highest discrete spin state observed to date following a projectile fragmentation reaction, adding $4\hbar$ of angular momentum to the previous highest recorded value of $I=35/2$ [3, 5].

ACKNOWLEDGMENTS

The present paper presents results from two sets of experiments. We thank all those involved in these measurements. This work was supported by EPSRC(UK) and the EU TMR programme. Zs.P. acknowledges the receipt of an EPSRC Advanced Fellowship Award (GR/A10789/01).

REFERENCES

1. W.-D. Schmidt *et al.*, *Z. Phys.* **A350**, 215 (1994).
2. M. Pfützner *et al.*, *Phys. Lett.* **444B**, 32 (1998).
3. M. Pfützner *et al.*, *Phys. Rev.* **C65**, 064604 (2002).
4. H. Geissel *et al.*, *Nucl. Instr. Meth.* **B70**, 286 (1992).
5. Zs. Podolyák *et al.*, *Prog. Theor. Phys. Suppl.* **146**, 467 (2002).
6. M. Caamaño *et al.*, *Acta Phys.Pol.* **B32**, 763 (2001).
7. K.A. Gladnishki *et al.*, *Acta Phys.Pol.* **B33**, 2395 (2003).
8. J.-J. Gaimard and K.-H. Schmidt, *Nucl. Phys.* **A531**, 709 (1991).
9. M. de Jong, A.V. Ignatyuk and K.-H. Schmidt, *Nucl. Phys.* **A613**, 435 (1997).
10. J.R. Huizenga and R. Vandenbosch, *Phys. Rev.* **120**, 1305 (1960).
11. R. Vandebosch and J.R. Huizenga, *Phys. Rev.* **120**, 1313 (1960).
12. I.S. Grant and M. Rathle, *J. Phys.*, **G5**, 1741 (1979), and references therein.
13. H. Naik *et al.*, *Nucl. Phys.* **A648**, 45 (1999), and references therein.
14. K.A. Gladnishki *et al.*, to be published.
15. A.E. Stuchbery *et al.*, *Nucl. Phys.* **A641**, 401 (1998).

What is the Signature of $T = 0$ np Pairing in Rotating Nuclei?

Alan L. Goodman

Physics Department, Tulane University, New Orleans, Louisiana 70118, USA

Abstract. Rotation of a $T = 1$ pair band causes spin alignments and the usual large backbend in the moment of inertia at low spins. In contrast rotation of a $T = 0$ pair band (with parallel angular momenta for the neutron and proton in each pair) does not cause a backbend or upbend at any spin, even though there are spin alignments. The backbend is not delayed in the $T = 0$ pair band, rather it simply never occurs.

INTRODUCTION

There has been an intensive search for $T = 0$ neutron-proton pair correlations. The most likely candidates for np pairing are $N = Z$ nuclei, where the neutrons and protons occupy identical space-spin orbitals, and have maximum spatial overlap. But what is the distinguishing characteristic of $T = 0$ pairing? What should one look for in experiments? One possible signature for $T = 0$ pairing may be the response of the nucleus to a rotation. Rotation might affect $T = 0$ pairs in a different manner than $T = 1$ pairs. It will be demonstrated that there is a dramatic observable distinction between the rotational responses of $T = 0$ and $T = 1$ pairs.

RESPONSE OF $T = 0$ PAIRS AND $T = 1$ PAIRS TO ROTATION

Any HFB wave function has a canonical representation, in which the HFB state can be expressed as a BCS wave function, where the orbitals are the HFB canonical orbitals. Consider a non-rotating $N = Z =$ even nucleus. For the conventional $T = 1$ Cooper pairs which consist of two neutrons or two protons, the orbitals of the two paired neutrons (protons) are related by time-reversal, and the orbitals and their occupation probabilities are the same for neutron and protons. When the nucleus is rotated, the time-reversal symmetry is broken, so that the orbitals of the two paired neutrons (protons) are no longer related by time-reversal. However the orbitals and orbital occupation probabilities remain the same for neutrons and protons. This is a spin-dependent state, where the orbitals and their occupation probabilities depend upon the rotational frequency of the nucleus.

Next consider $T = 0$ Cooper pairs, in which the neutron and proton of each pair occupy *identical* orbitals, instead of time-reversed orbitals. (For each np pair in a specific orbital, there is a corresponding np pair in the time-reversed orbital, with the same occupation probability.) The orbitals and their occupation probabilities are the same for neutrons and protons. When the nucleus is rotated, the time-reversal symmetry is broken, so that the orbital of the second np pair is no longer the time-reverse of the orbital of the first np pair, and the two pairs have different occupation probabilities. Once again, it should be remembered that this is a spin-dependent state, where the orbitals and their occupation probabilities depend upon the rotational frequency of the nucleus. The essential point is that at every value of the rotational frequency, the neutron and proton in a given pair occupy exactly the same orbital. The nuclear rotation does not break these pairs.

It is also possible to have $T = 0$ Cooper pairs in which the neutron and proton of each pair occupy orbitals related by time-reversal. These pairs are not discussed here.

The HFB equation has been solved for ^{80}Zr. The model space includes the $2p_{1/2}$, $2p_{3/2}$, $1f_{5/2}$, and $1g_{9/2}$ shells. There is a closed core of ^{56}Ni. The effective interaction was calculated by T. Kuo from the Paris potential. The pair potential includes components with isospin $T = 0,1$ and angular momentum $J = 0$ to 9. Self-consistency is obtained in both the HF potential and the pair potential for each nuclear spin I.

The ground state has nn and pp pairs, where the two orbitals in each pair are related by time-reversal (i.e., anti-parallel angular momenta). Rotating this state generates a $T = 1$ pair rotational band, where the two orbitals in each pair are no longer related by time-reversal. At spin $I = 0$ there is an excited state with np ($T = 0$) pairs, where the n and p in each pair occupy the same orbital (i.e., parallel angular momenta). Rotating this state generates a $T = 0$ pair rotational band, where at each nuclear spin the neutron and proton in a given pair occupy the same spin-dependent orbital (i.e., parallel angular momenta). There is also a third rotational band, which has both $T = 0$ pairs and $T = 1$ pairs co-existing for spins $I \sim 5$, only $T = 1$ pairs at $I = 0$ and only $T = 0$ pairs at very high spins.

Next consider the pairing energy. The $T = 1$ pair band has a large (negative) pairing energy at spin $I = 0$. However as the spin increases, this band rapidly loses its pairing energy, which vanishes at $I = 26$. In contrast, the $T = 0$ pair band has a pairing energy which is approximately constant for increasing spin. (The pairing energy actually increases with I at low spins, i.e., becomes more negative.)

Why does the pairing energy behave so differently for the $T = 1$ pair band and the $T = 0$ pair band? First consider the $T = 1$ pair band. At spin $I = 0$ the two nucleons in each Cooper pair are in time-reversed orbitals, where m_z is a good quantum number, and the first nucleon has $+m$ while the second has $-m$. When this state is rotated, the Coriolis force has an opposite effect on the angular momentum vectors of the two nucleons, thereby breaking the time-reversal symmetry, and breaking the pair bond. This is the Coriolis anti-pairing (CAP) effect, which causes the rapid loss of pairing energy as the spin I increases. For $T = 1$ pairs the isospin state is symmetric, so the space-spin state must be anti-symmetric. This permits $J = 0$ pairs, which is the strongest pair mode. The rotation breaks these $J = 0$ pairs in order to realign the nucleon spins along the x rotation axis and generate the nuclear spin I.

Next consider the $T = 0$ pair band. At spin $I = 0$ the neutron and proton in each Cooper pair occupy identical orbitals, where m_z is a good quantum number. When this state is rotated, the Coriolis force has exactly the same effect on the angular momentum vectors of the two nucleons. The two spin vectors can be gradually rotated towards the x rotation axis, and at each spin I the neutron and proton in a pair will have the same wave function, thereby maintaining maximum spatial overlap. The pair is not broken by the rotation. There is no CAP effect for these $T = 0$ pairs. Therefore the pairing energy is approximately constant for increasing spin I. For $T = 0$ pairs the isospin state is anti-symmetric, so the space-spin state is symmetric. This forbids $J = 0$ pairs. Each pair has at least $J = 1$. Then each pair can contribute to the nuclear spin I by gradually realigning its spin J from the z axis to the x rotation axis, without breaking the pair.

Next we consider how rotating $T = 1$ pairs and $T = 0$ pairs generates the nuclear angular momentum. For the $T = 1$ pair band, there is a large backbend in the nuclear spin versus rotational frequency curve, with a sudden increase in the angular momentum. In contrast the $T = 0$ pair band has a gradual increase in the nuclear spin with no backbend.

Which nucleons are responsible for generating the nuclear spin? For the $T = 1$ pair band, the pf shells have a very small contribution to the spin. Almost all of the angular momentum originates from the $g_{9/2}$ shell. Now consider the contribution to the nuclear spin from two $g_{9/2}$ pairs, i.e. one nn pair and one pp pair. Since the orbitals of the nn pair are identical to those of the pp pair, the two pairs have identical responses to rotation at each spin I. At $I = 13$ the total $<J_x> = 13.49$, while these two pairs have $<J_x> = 15.35$. All of the other orbitals actually combine to have a negative contribution to the spin. The maximum $<J_x>$ for one $g_{9/2}$ pair is $9/2 + 7/2 = 8$, so that the maximum spin for two pairs is 16. This suggests that at $I = 13$ these two pairs have come close to their maximum possible spin alignment. At $I = 0$ each pair has anti-parallel nucleon spins aligned along the z deformation symmetry axis, whereas at $I = 13$ each pair has almost parallel nucleon spins aligned along the x rotation axis. This is the rotational realignment effect.

This conjecture is confirmed by considering the spin alignment $<J_x>$ for each $g_{9/2}$ neutron canonical orbital in the nn pair. (The proton spin alignments in the pp pair are identical to the neutron spin alignments.) At high spins these orbitals have spin alignments that are very close to the values for completely aligned $g_{9/2}$ orbitals. At $I = 13$, the first orbital in a pair has a 98.5% overlap with the J_x eigenstate $|g_{9/2}\ m_x = 9/2>$, and the second orbital in a pair has a 94.3% overlap with the J_x eigenstate $|g_{9/2}\ m_x = 7/2>$. This confirms that in the backbend of the $T = 1$ pair band, two $g_{9/2}$ neutrons and two $g_{9/2}$ protons are realigning their spins along the x rotation axis.

Now consider the spin alignments in the $T = 0$ pair band. The pf shells make a very small contribution to the spin. Almost all of the angular momentum comes from the $g_{9/2}$ shell. Consider the contribution to the nuclear spin from two $g_{9/2}$ np pairs, where the n and p in a specific pair have identical canonical orbitals. Therefore the n and p in a given pair respond to rotation in an identical manner. At $I = 13$ the total $<J_x> = 13.49$, while these two pairs have $<J_x> = 14.32$. All of the other orbitals combine to have a negative contribution to the spin. At $I = 13$ these two $T = 0$ pairs have a spin alignment which is not far from the maximum possible value of 16, and only 1 less than the spin of the $T = 1$ pairs at $I = 13$. At $I = 0$ each pair has parallel nucleon spins aligned along the z deformation symmetry axis, whereas at $I = 13$ each pair has parallel nucleon spins aligned along the x rotation axis. This is a rotational realignment effect.

Consider the spin alignment $<J_x>$ for the $g_{9/2}$ neutron canonical orbital in each of the two np pairs. (The proton spin alignments are identical to the neutron spin alignments.) At high spins these orbitals have spin alignments which are near the values for aligned $g_{9/2}$ orbitals. At $I = 13$, the n and p orbitals in the first pair have a 90.5% overlap with the J_x eigenstate $|g_{9/2}\ m_x = 9/2>$, and the n and p orbitals in the second pair have a 90.4% overlap with the J_x eigenstate $|g_{9/2}\ m_x = 7/2\rangle$. These orbitals are substantially, but not completely, aligned. This confirms that in the $T = 0$ pair band, two $g_{9/2}$ neutrons and two $g_{9/2}$ protons are realigning their spins along the x rotation axis. However the spin alignments in the $T = 0$ pair band do not produce an upbend or backbend in the moment of inertia. It is sometimes stated that an energy spectrum without an upbend or backbend does not contain spin alignments. However, this analysis demonstrates that the absence of an upbend or backbend in the moment of inertia does not imply the absence of spin alignments for a $T = 0$ pair band.

See Ref. [1] for additional details.

CONCLUSIONS

HFB calculations for the $N = Z$ nucleus ^{80}Zr provide a band with $T = 1$ pairs and another band with $T = 0$ pairs. For a non-rotating $T = 1$ pair state, the two nucleons in each pair have anti-parallel angular momenta, which are aligned along the z deformation symmetry axis. As expected, when this $T = 1$ pair state is rotated, the Coriolis anti-pairing effect rotates the anti-parallel spins of one nn pair and one pp pair of $g_{9/2}$ nucleons so that they become almost parallel, and aligned along the x rotation axis. This causes a collapse of the pair gap, and a large backbend in the moment of inertia between spins 8 and 14.

In sharp contrast, in the non-rotating $T = 0$ pair state, the neutron and proton in each pair have identical space-spin orbitals, so that they have parallel angular momenta, which are aligned along the deformation symmetry axis. When this $T = 0$ pair state is rotated, the Coriolis effect is therefore exactly the same for the neutron and proton in a given pair. The $g_{9/2}$ neutrons and protons in two pairs have their angular momenta rotated so that they become aligned along the rotation axis. For each nuclear spin, the angular momenta of the neutron and proton in a pair remain exactly parallel. Although rotation of the nucleus causes spin alignments, this does not break these $T = 0$ pairs. There is no Coriolis anti-pairing effect for these $T = 0$ pairs, in which the angular momenta of both nucleons are always parallel. Consequently, rotating the nucleus causes no reduction of the pair gap, and no backbending or upbending in the moment of inertia. The moment of inertia shows only a slow increase for spins up to 16, and a decrease at higher spins. Although there are spin alignments, there is no backbend or upbend in the moment of inertia.

The conclusion is that rotation of a $T = 1$ pair band causes spin alignments and the usual large backbend in the moment of inertia at low spins; whereas rotation of a $T = 0$ pair band (with parallel angular momenta for the neutron and proton in each pair) does not cause a backbend or upbend at any spin, even though there are spin alignments. The backbend is not delayed in the $T = 0$ pair band, rather it simply never occurs.

ACKNOWLEDGMENTS

This work was supported in part by the National Science Foundation.

REFERENCES

1. Goodman, A.L., *Phys. Rev. C* **63**, 044325 (2001).

Competing decay-out mechanisms of the yrast superdeformed band in ^{59}Cu

C. Andreoiu*[†], C. Fahlander*, D. Rudolph*, T. Døssing**, I. Ragnarsson[‡] and S. Åberg[‡]

*Department of Physics, Lund University, Lund S-22100, Sweden
[†]Oliver Lodge Laboratory, University of Liverpool, Liverpool L69 7ZE, UK
**Niels Bohr Institute, DK-2100 Copenhagen, Denmark
[‡]Department of Mathematical Physics, Lund Institute of Technology, S-22100 Lund, Sweden

Abstract.
This contribution reports on the decay-out of the yrast superdeformed band in ^{59}Cu. It decays by multiple γ-ray transitions into low-spin spherical states in the first minimum of the nuclear potential in ^{59}Cu, and, in competition, by emitting two prompt protons into a spherical state in ^{58}Ni. For the first time in $A \sim 60$, the role of doorway states in the course of the decay-out is discussed.

INTRODUCTION

Superdefomed bands (SD) represent a scientific evergreen in nuclear physics. They are stabilized by shell effects at high angular momentum in nuclei with a specific number of protons and neutrons. The decay-out of the SD bands involves a substantial rearrangement of nucleonic states and induces a dramatic change in deformation. Hence, this is a phenomenon of fundamental interest because it involves in a basic manner the interplay between the particle and shape degrees of freedom of the atomic nucleus. Vibrational coupling through the barrier between SD states and doorway states [1], which in turn typically are coupled to a very large number of normally deformed (ND) states, facilitates the decay-out process [2].

In order to fully characterize the superdeformed bands and their decay-out mechanisms it is necessary to determine their spin, parity, and excitation energy by observing the weak linking transitions into low-spin spherical states.

Numerous highly-deformed and SD bands have been observed in the A= 30, 40, 60, 80, and 130 mass regions [3–9], where it is relatively easy to connect the SD bands to normally deformed states because the lower level densities often result in a number of discrete decay-out transitions. In contrast, in the $A = 150, 190$ region it has not been possible to follow this decay-out in detail because the density of states at the final, smaller deformation is very high at the excitation energy of the transition. Thus, the decay is fragmented over numerous weak transitions as a statistical cascade without any discernible selection rules [10–13]. As a result, the details of the coupling of SD and ND states are masked by the statistical character of the ND states.

Contrary to the situation in heavier masses, in the mass $A \sim 60$ region, the level density is much lower leading to discrete decay-out cascades with a non-statistical character [14]. Moreover, in the $A \sim 60$ region, some of the superdeformed bands decay by prompt particle decays into spherical states in the corresponding daughter nucleus in competition with the γ decay-out mechanism. This mechanism is present in ^{58,59}Cu [15–17] and ^{56,58}Ni [18, 19]. Because the rearrangement of the small number of particles involved, more evidence to a non-statistical scenario and a selective decay-out pattern of the $N = Z$ nuclei in $A \sim 60$ nuclei was added [16, 17]. For the most recent results on prompt particle decays the reader is referred to Ref. [20].

In this contribution we report the first observation of a specific group of states which couples rather directly to the SD band, thereby acting as doorway states for the decay of the yrast SD band in the $N = Z + 1$ ^{59}Cu nucleus.

EXPERIMENT

High-spin states in ^{59}Cu were populated in two experiments performed at Argonne National Laboratory in USA. Experiment 1 used the fusion-evaporation reaction ^{28}Si+^{40}Ca at a beam energy of 122 MeV. The experimental set-up consisted of the Gammasphere array [21] comprising 103 Ge detectors, in combination with the 4π charged-particle detector array Microball [22]. High-spin states in the residual nucleus ^{59}Cu were populated via the evaporation of two α particles and one proton from the compound nucleus ^{68}Se. An experimental relative cross section of $\sigma_{rel} \approx 6\%$ was estimated. For more details see Ref. [23]. Experiment 2 consisted of the Gammasphere array involving 86 Compton-suppressed Ge detectors together with Microball. For the first time, four highly segmented $\Delta E - E$ Si-strip telescopes replacing the three most forward rings in Microball were used. The setup also involved 20 neutron detectors replacing the 20 most forward Ge detectors. Excited states in ^{59}Cu were populated using the fusion evaporation reaction ^{36}Ar+^{28}Si at a beam energy of 148 MeV, via the evaporation of one α particle and one proton from the compound nucleus ^{64}Ge. Because of the high beam energy, the relative cross section accounts for less than 2% from the total fusion cross section. For more details see Ref. [16]. Experiment 1 was the major source for detailed spectroscopy of ^{59}Cu, while experiment 2 was aiming for high-resolution in-beam particle-γ coincidence spectroscopy.

RESULTS

The yrast SD band in ^{59}Cu consists of seven stretched $E2$ transition which reach an angular momentum of $I^\pi = (57/2^+)$ at an excitation energy of $E_x = 32.0$ MeV [24]. From Doppler shift attenuation measurements of its transition quadrupole moment and assuming an axially symmetric shape a quadrupole deformation of $\beta_2 = 0.41$ was extracted. Figure 1(a) shows only the low-spin part of the yrast superdeformed band and its inclusive discrete-energy decay-out. It is connected into the low-spin states of structures 1, 2, 3 and 4 by high-energy single- or double-step transitions [23]. At $29/2^+$ the whole intensity of the SD band is divided between four stretched $E2$ transitions with energies of 1108, 1313, 1434, and 1514 keV. This initial decay is explained as

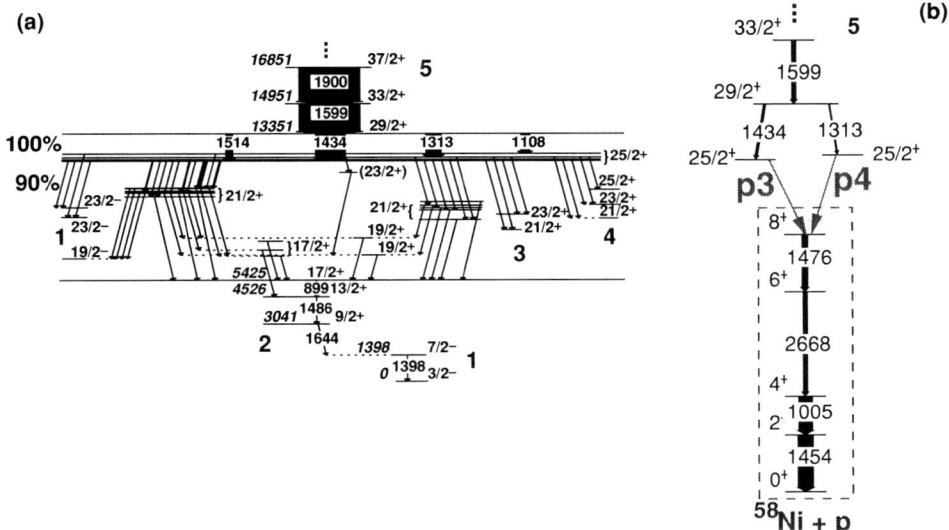

FIGURE 1. (a) Gamma decay-out of the yrast superdeformed band in ^{59}Cu. The widths of the arrows are proportional to the $B(E2)$ strengths relative to the 1599 keV transition. (b) Prompt proton decay scheme of the yrast superdeformed band in ^{59}Cu.

a fragmentation of the SD rotational strength, caused by the coupling of the SD state to three normally deformed doorway states at angular momentum $25/2^+$. The four mixed $25/2^+$ states prefer to decay predominantly to two groups of excited $21/2^+$ states, indicating a rather selective mechanism. Most of the decay paths end up at the terminating $17/2^+$ state of band 2.

Another interesting feature is that two of the $25/2^+$ states decay by two prompt protons into the 8^+ spherical state in ^{58}Ni [17]. This implies the first observation of "fine structure" for this new exotic decay mode. Figure 1(b) shows the partial level scheme of ^{59}Cu and ^{58}Ni relevant for this contribution. The proton branches for the two proton decays have been determined to $b_{p3} = 9(2)\%$ and $b_{p4} = 8(3)\%$, respectively. Including the two prompt proton decays about 90% of the SD intensity is recovered in the decay of the $25/2^+$ states. Taking into account the estimated relative cross section of ^{59}Cu in the second experiment of $\sigma_{rel} < 2\%$ and the average proton intensity of about 1%, we conclude that the experimental set-up was sensitive to proton-$\gamma\gamma$ coincidences down to a remarkable 10^{-4} level.

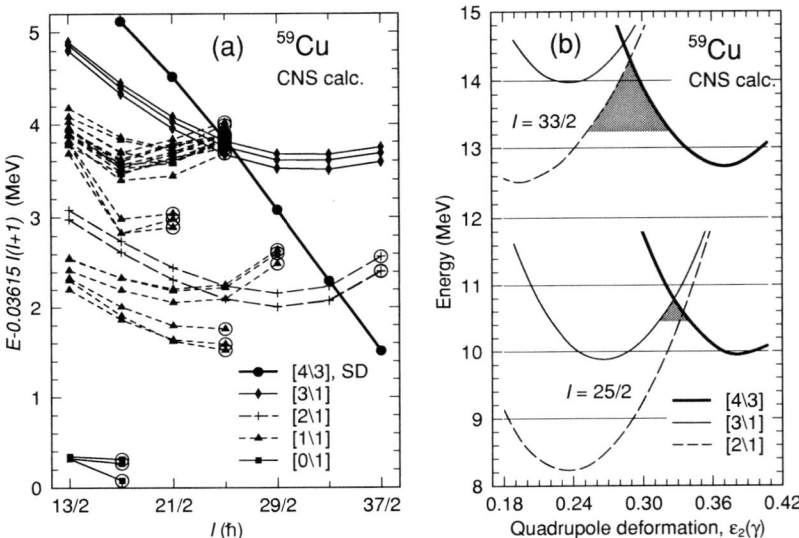

FIGURE 2. (a) Calculated energies of all rotational bands involved in the gamma decay-out mechanism of the yrast superdeformed band in ^{59}Cu. They are calculated relative to a rigid rotor and plotted as a function of angular momentum. The terminating states of some of the bands are encircled. (b) An energy cut of the lowest [4\3], [3\1] and [2\1] bands along a minimum energy path in the (ε_2, γ) plane for angular momentum of 25/2 $\hbar$ and 33/2 $\hbar$. The decay barriers are indicated by gray shadows above a zero-point energy of 0.5 MeV.

THEORETICAL INTERPRETATION

The states involved in the decay-out mechanisms in ^{59}Cu can be classified by the total number of proton and neutron holes in the $f_{7/2}$ shell, q_1, and the total number of particles in the $g_{9/2}$ shell, q_2, in a compact description [$q_1 \backslash q_2$]. Ref. [23] provides more details related to the CNS calculations in ^{59}Cu. The simulation of the positive parity states involved in the decay of the yrast SD band are plotted in more details in Fig. 2(a) which shows the calculated energies of all rotational bands involved in the gamma decay-out mechanism of the yrast superdeformed band in ^{59}Cu relative to a rigid rotor and as a function of angular momentum. They are labelled by the type of configuration they belong to, i.e., [0\1], [1\1], [2\1], [3\1], and [4\3] [25].

At $I = 33/2$ the superdeformed [4\3] band and the normally deformed [2\1] bands come close in energy [see Fig. 2(a)]. Since they only differ by a two-particle two-hole (2p-2h) excitation, one might expect a strong coupling between the superdeformed band and these [2\1] bands at this spin value leading to a decay-out already at $I = 33/2$. It seems, however, that this is prevented by a potential barrier. This is illustrated in Fig. 2(b), where the energy is shown along a straight line in the (ε_2, γ)-plane, that approximately goes through the minima of the [2\1], [3\1], and [4\3] configurations. At $I = 29/2$ no normal deformed state comes close to the superdeformed state, while at $I = 25/2$, both the three $[3\backslash 1]$ bands and several excited $[1\backslash 1]$ bands [25] lie in the same energy range as the SD band. The $[1\backslash 1]$ states differ by (at least) a 3p-3h excitation from the states of the SD band, while all three $[3\backslash 1]$ states differ by a 2p-2h excitation. There is only a small barrier [cf. Fig. 2(b)] at this spin value between the SD state and the $[3\backslash 1]$ states. This suggests a strong coupling to the $[3\backslash 1]$ states, which then act as doorway states for the decay-out of the yrast SD band in ^{59}Cu.

A theoretical model that includes a residual interaction and tunneling matrix elements between the predicted bands is described in Sven Åberg's contribution of these proceedings. A comprehensive study of the γ decay-out mechanism from the yrast SD band in ^{59}Cu will be published in Ref. [26].

CONCLUSIONS

The decay-out of the SD band in ^{59}Cu has been mapped out in great detail in experiment. About 30 linking transitions connecting the yrast SD band to the low-spin states in ^{59}Cu were depicted. Using the configuration dependent cranked Nilsson-Strutinsky formalism the active states involved in the decay-out mechanism were classified by exact and approximate quantum numbers, leading to a detailed understanding of this process. It was found that the decay-out is caused by a direct coupling of the SD states to a small number of doorway states. This is in contrast to the decay-out in heavier masses, where the coupling to the doorway states is masked by a chaotic environment of normally deformed states.

In competition with the γ decay-out mechanism, two SD states decay by two prompt protons into the same spherical state in ^{58}Ni. This was observed using a novel set-up and represents the first "fine structure" of the prompt proton decay mode.

ACKNOWLEDGMENTS

We would like to express our gratitude to all collaborators involved in the GSFMA66 and GSFMA42 experiments. This research was supported by the Swedish Research Council, the Canadian NSERC, the U.S. DOE under Contracts Nos. W-31-109-ENG-38 (ANL), DE-AC03-76SF00098 (LBNL) and DE-FG05-88ER40406 (WU), the German BMBF and the University of Liverpool.

REFERENCES

1. S. Bjørnholm and J.E. Lynn, Rev. Mod. Phys. **52**, 725 (1980).
2. S. Åberg, Phys. Rev. Lett. **82**, 299 (1999).
3. C.E Svensson *et al.*, Phys. Rev. Lett. **85**, 2693 (2000).
4. C.E Svensson *et al.*, Phys. Rev. C **63**, 061301(R) (2001).
5. E. Ideguchi *et al.*, Phys. Rev. Lett. **87**, 222501 (2001).
6. C.E. Svensson *et al.*, Phys. Rev. Lett. **79**, 1233 (1997).
7. C. Baktash *et al.*, Phys. Rev. Lett. **74**, 1946 (1995).
8. P. Nolan *et al.*, J. Phys. G **11**, L17 (1985).

9. D. Bazzacco et al., Phys. Rev. C **49**, R2281 (1994).
10. R.G. Henry et al., Phys. Rev. Lett. **73**, 777 (1994).
11. A. Lopez-Martens et al., Phys. Rev. Lett. **77**, 1707 (1996).
12. A. Lopez-Martens et al., Nucl. Phys. **A 647**, 217 (1999).
13. T. Lauritsen et al., Phys. Rev. C **62**, 044316 (2000).
14. C.E. Svensson et al., Phys. Rev. Lett. **82**, 3400 (1999).
15. D. Rudolph et al., Phys. Rev. Lett. **80**, 3018 (1998).
16. D. Rudolph et al., Eur. Phys. J. A **14**, 137 (2002).
17. D. Rudolph et al., Phys. Rev. Lett. **89**, 022501 (2002).
18. D. Rudolph et al., Phys. Rev. Lett. **82**, 3763 (1998).
19. D. Rudolph et al., Phys. Rev. Lett. **86**, 1450 (2001).
20. D. Rudolph, in Proc. *Proton-Emitting Nuclei: Second International Symposium; PROCON 2003*, Legnaro, Italy, February 2003, Eds. E. Maglione and F. Soramel, AIP Conference Proceedings **681**, p.36.
21. I.-Y. Lee, Nucl. Phys. A 520, 641c (1990).
22. D.G. Sarantites et al., Nucl. Instrum. Meth. **A381**, 418 (1996).
23. C. Andreoiu et al., Eur. Phys. J. A **14**, 317 (2002).
24. C. Andreoiu et al., Phys. Rev. C **62**, 051301(R) (2000).
25. I. Ragnarsson et al., Proc. *International Conference on Frontiers of Nuclear Structure*, Berkeley, California, July 29 - Aug. 3, 2002, AIP Conference Proceedings **656**, Eds. P. Fallon and R. Clark, p. 205; Proc. *Nuclear Structure with Large Gamma-Arrays: Status & Perspectives*, Legnaro, Italy, 2002, to be published.
26. C. Andreoiu et al., to be published in Phys. Rev. Lett.

Nuclei At Extreme Deformations

P. Fallon

Lawrence Berkeley National Laboratory, 1 Cyclotron Road, Berkeley CA 94720

Abstract. Nuclei with very large deformations provide an opportunity to study many aspects of nuclear structure: shell structure; exotic states; extreme single-particle motion (shell model); collective modes; pairing. In this talk I will discuss a few selected examples illustrating some recent developments in this area of study.

INTRODUCTION

A spectacular manifestation of extreme nuclear deformations is the superdeformed (SD) nucleus with a major-minor prolate axis ratio of ~2:1. First observed in ^{242}Am [1], through the discovery of fission isomerism and then understood a few years later as a second minimum in the fission barrier, superdeformation is now established in several nuclear mass regions ranging from A~40 through A~240. Following the observation of fission isomers there was a gap of about 25 years before the first high-spin superdeformed nucleus was discovered [2]: a wait determined by the need for multi detector high-resolution gamma-ray detector arrays. To date we have observed and studied about 250 superdeformed rotational bands throughout the nuclear chart [3].[1] With each region discovered different aspects of the physics of very deformed rotating nuclei have been emphasized; e.g. around A~150 the extreme single-particle properties were readily identified providing insight into the symmetries of superdeformed nuclei, while in the A~190 mass region the pairing correlations were seen to have a significant influence on the rotation of the superdeformed nucleus, promoting a deeper understanding of the importance of higher order pairing terms.

The existence of superdeformation is closely coupled to the concept of deformed shell gaps (secondary minima) and is a clear illustration of the role of single-particle (microscopic) effects on macroscopic properties. Superdeformed nuclei can be considered to be the deformed analog of spherical magic nuclei and are therefore a phenomenon distinct from normal deformations. They are highly polarized systems providing an opportunity to study many aspects of nuclear structure: shell structure; exotic states; extreme single-particle motion (shell model); collective modes; pairing.

---

[1] Of the approximately 250 SD bands observed most involve excitations within the second minimum and of these the vast majority are interpreted as single-particle excitations; there are a limited number of collective excitations (vibrations) and these are concentrated in heavier systems.

HYPERINTRUDER STATES

It is possible that some nuclei may attain even greater quadrupole deformations than those observed in superdeformed nuclei and the search for such hyperdeformed (HD) shapes has been a goal of nuclear structure for many years.

In a simple limit, based on a harmonic oscillator potential, superdeformed nuclei have a deformation corresponding to a prolate axis ratio c/a = 2 and hyperdeformed nuclei have c/a = 3. Both HD and SD shapes are favored due to the special symmetries associated with integer axis ratios, which generate large shell gaps that help stabilize the nuclear deformation. However, realistic nuclear potentials modify this simple picture and superdeformed nuclei are observed with axis ratios in the range c/a ~1.6 to 1.8. Similarly we may expect hyperdeformed nuclei to exhibit c/a values smaller than 3. Nevertheless, it turns out that each distinct deformed shape (whether normal-, super-, or hyperdeformed) can, in general, be correlated with the occupation of specific high-j intruder states. Normal deformed nuclei correspond to the occupation of intruders from one major oscillator shell above the spherical Fermi surface (N+1, *"normal-intruders"*), superdeformed nuclei can be correlated with the occupation of intruder states from two shells above (N+2, *"super-intruders"*), and in analogy we may expect hyperdeformed nuclei to involve intruder states from three shells above (N+3, *"hyper-intruders"*).

The proton $i_{13/2}$ intruder is a good example. It starts out at the Fermi surface for spherical nuclei around Z=100 and for normal deformed (β_2~0.2) nuclei around Z=92 where it is the *normal-intruder* state lowered by one oscillator shell due to the spin-orbit interaction. High-j states have large quadrupole moments and they are lowered even further by increasing the quadrupole deformation until at superdeformations (β_2~0.6) the $i_{13/2}$ state is at the Fermi surface for Z~66 (e.g. ^{152}Dy). At this point the $i_{13/2}$ intruder has dropped two major oscillator shells and it can be classified as a *super-intruder*. By the time we get to Cadmium, Z=48, the $i_{13/2}$ state has come down 3 major shells, and it seems natural to label this orbital a *hyper-intruder*.

Thus the key to observing hyperdeformed nuclei is to find examples where the N+3 intruders can be populated. An N+3 intruder, for example, can be "brought" to the Fermi surface by either increasing the nucleon number and/or increasing the deformation. (High angular momentum also favors the occupation of high-j, hyper-intruders states.) Cadmium and Tin nuclei lie at the top of the Z=28-50 shell, which means that the necessary N+3 hyper-intruders ($i_{13/2}$ protons and $j_{15/2}$ neutrons) approach the Fermi surface at a lower deformation than, say, a mid shell nucleus and may therefore be easier to populate. Recall that the A~190 SD nuclei have a lower deformation than those around A~150 and both correspond to the occupation of $i_{13/2}$ protons and $j_{15/2}$ neutrons, whereas these particular super-intruders are not populated in bands observed in the lower portion of the shell (i.e. with A $\leq$ 140)

Calculations by Chasman [4] indicate that several nuclei near A=100 (e.g. ^{108}Cd, ^{110}Cd, ^{112}Sn, ^{114}Sn) have significant shell correction energies for deformed shapes with c/a~2.3 and that these states are yrast around I ~ 62-65 $\hbar$. Thus there is good reason to expect that these minima can be populated and the bands observed.

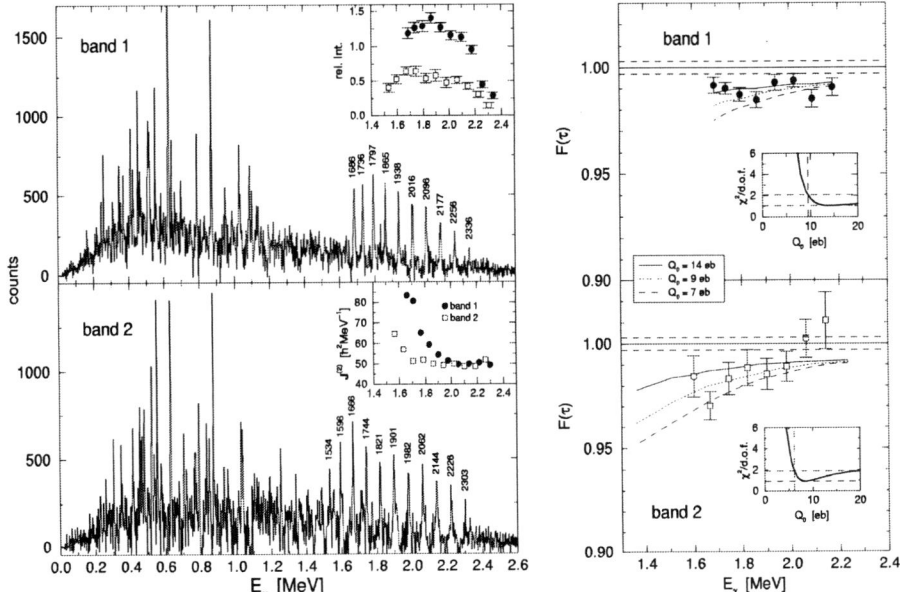

FIGURE 1. Spectra of bands 1 and 2 in ^{108}Cd. The insets show the relative intensities of the bands with respect to the total intensity of the reaction channel (upper panel), and the dynamic moments of inertia (lower panel). The values for band 1 are given by the full circles, those for band 2 by the open squares. Right: Results of the fractional Doppler shift analysis. The lines show examples for calculated $F(\tau)$ curves. The χ^2 fits for the calculated curves to the data as a function of the quadrupole moment Q_0 are shown in the insets. The horizontal dashed lines represent the minimal value χ^2 and χ^2+1. The vertical dashed lines indicate the lower limit of the quadrupole moment consistent with a 1σ error.

An experiment to search for very extended nuclear shapes in ^{108}Cd was performed on Gammasphere [7] at ATLAS using the reaction ^{48}Ca(^{64}Ni,4n) at 207 MeV. This resulted in the observation of two rotational bands [5,6] at very high angular momentum. The bands, shown in figure 1, carry about 1.4% and 0.6% of the intensity of the ^{108}Cd channel, which carries approximately 15% of the total fusion-evaporation intensity. It is estimated (assuming $J^{(1)} = J^{(2)}$ at the highest spins where $J^{(2)}$ is constant) that they extend over a spin range from about 40-60 $\hbar$. The bands have large dynamic moments of inertia, $J^{(2)}$, shown in the insert to figure 1, consistent with a very large prolate deformation of c/a ~2.

A measure of the transition state lifetimes (via the slowing down of the recoil in the target) confirmed their large deformation. Band 1 has a lower limit for the quadrupole moment of $Q_t = 9.5$eb (c/a ~1.8). Band 2 has a best-fit $Q_t = 8.5$eb (c/a ~1.7) and a lower limit of 6.2eb. In both cases a meaningful upper limit could not be established.

The bands exhibit a rapid increase in $J^{(2)}$ at the lowest frequencies. Calculations [8] suggested that the rise in $J^{(2)}$ in band 1 coincides with the occupation of the proton $i_{13/2}$ hyper-intruder state. A recent study by Dudek (private communication) also places the $i_{13/2}$ proton at the Fermi surface for Z~48 and deformations of c/a ~2.

This evidence for the occupation of the proton hyper-intruder in ^{108}Cd suggests the exciting possibility that hyperdeformed nuclei (with both $i_{13/2}$ proton and neutron $j_{15/2}$ hyper-intruders occupied) could be observed in this region. The $j_{15/2}$ hyper-intruder is predicted [9] to lie at the Fermi surface for N = 64, i.e. ^{112}Cd or ^{114}Sn, at a deformation only slightly larger (10-20%) than the ^{108}Cd bands and only 4 neutrons away. Moreover, this intruder could be occupied for N<64 (e.g. ^{110}Cd$_{62}$ or ^{112}Sn$_{62}$) if the excitation energy is increased and/or the deformation increases.

THE JACOBI TRANSITION

The Jacobi transition corresponds to a dramatic increase in deformation above some critical angular momentum. Unlike superdeformed nuclei, which owe their elongation to shell effects and can be observed at relatively low spin (spin zero in some fission-isomers), the existence of Jacobi-like configurations is solely due to the centrifugal force, and such shapes will appear only at very high spin: they should occur over a range of nuclei rather than in the small pockets characteristic of effects driven by shell-structure. The Jacobi transition depends on the liquid drop properties of nuclei.

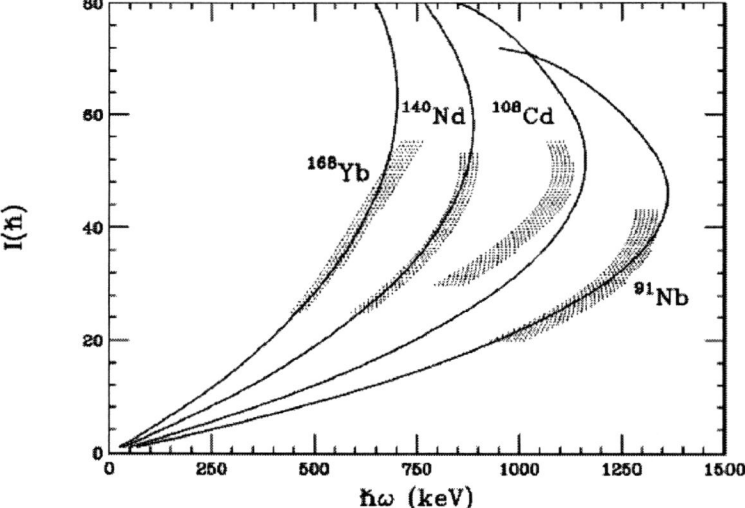

FIGURE 2. Plots of the experimentally derived spin versus rotational frequency curves, shaded areas, for various reactions given in the text. The nuclei shown are the 4n fusion-evaporation channels. Solid lines indicate results of the Thomas-Fermi calculation modified by a smooth interpolation between the low angular momentum predictions (with an empirical allowance for pairing effects) and the high angular momentum (Jacobi) regime where the unmodified calculations are likely to be adequate

The discovery of such a transition from oblate (Maclaurin-type shapes) to very deformed triaxial shapes was made in 1834 by C.G.J. Jacobi, in the context of rotating, idealized, incompressible gravitating masses. In 1961, Beringer and Knox [10]

suggested that a similar transition might be expected in the case of atomic nuclei, idealized as incompressible charged liquid drops with surface tension. Theoretical studies in 1974 [11] and 1986 confirmed this. In the A~100 mass region the available "spin window" between the onset of the Jacobi shape transition and the vanishing fission barrier attains its maximum value. It is reasonable to expect that nuclei that favor Jacobi-like shapes will also be a favorable place to search for "discrete" hyperdeformed structures because this will help the relatively small hyperdeformed shell corrections generate a stable deformed minimum.

Experiments were carried out using a ^{48}Ca beam on targets of ^{50}Ti, ^{64}Ni, ^{96}Zr and ^{124}Sn using the 8π Spectrometer at Berkeley and Gammasphere at Argonne National Laboratory. The signature of a nucleus rotating in a Jacobi-like configuration is the decrease of rotational frequency (gamma-ray energy) with increasing spin. This expectation is based on the rather universal rapid increases in an elongation, caused by the centrifugal force, once the Jacobi regime has been reached.

An analysis of the quasi-continuous gamma-spectrum was carried out [12]. It involved comparing the gamma-ray spectra for successively higher K-values in order to determine the mean gamma-ray energy at a given spin. K is the number of detectors hit and can be related to the spin of the nucleus. The spectra were processed by unfolding the Compton background and correcting for the efficiency. The results are shown in Fig. 2. The deviations in the spin versus energy curves are consistent with a rapid increase in the deformation (although other effects such as particle alignments cannot be completely ruled out) and suggest the formation of Jacobi-like shapes at the highest angular momenta observed.

STRONGLY DEFORMED TRIAXIAL NUCLEI

A region of stable triaxiality is predicted in the neutron-deficient Lu and Hf isotopes around N = 94 where cranking calculations give triaxial minima at large deformation (triaxial strongly deformed, TSD) with $(\varepsilon_2 ; \gamma) \sim (0.4; \pm 20$ degrees$)$ in the total energy surfaces. These nuclides have been studied extensively and many rotational structures that can be associated with TSD minima have been observed in several nuclides. The first proof of stable triaxiality was not the observation of these bands as such, but came rather from that fact that some bands in odd-mass Lu isotopes can be described as wobbling-phonon excitations, the best case to date is ^{163}Lu [13,14].

The wobbling phonon was predicted [15] over 30 years ago as a unique mode of excitation that can occur in a nucleus that has a stable triaxial deformation. It corresponds to the situation where the collective angular momentum R moves away from a principal axis. The rotational frequency precesses around R and the motion can be characterized as a collective (wobbling) phonon. The different rotational bands are labeled by the number of wobbling phonons, n_ω. In ^{163}Lu three TSD bands were identified as the 0-, 1-, and 2-phonon bands, with the evidence based on the characteristic properties of the inter-band transitions, for which the E2 strength of the $\Delta I = 1$ transitions was shown to exhibit the expected pattern.

Lifetime measurements in the excited TSD bands provide an important test for the wobbling interpretation. Very similar B(E2) strengths for the in-band transitions are expected, as the wobbling bands are built on the same intrinsic structure. Recently, a high-statistics experiment was carried out at the 88-Inch cyclotron using Gammasphere and the reaction ^{123}Sb(^{44}Ca,4n) at a beam energy of 190 MeV. Lifetimes of states in the proposed 0-, 1-phonon bands (TSD1 and 2) in ^{163}Lu were determined using the Doppler-shift attenuation method.

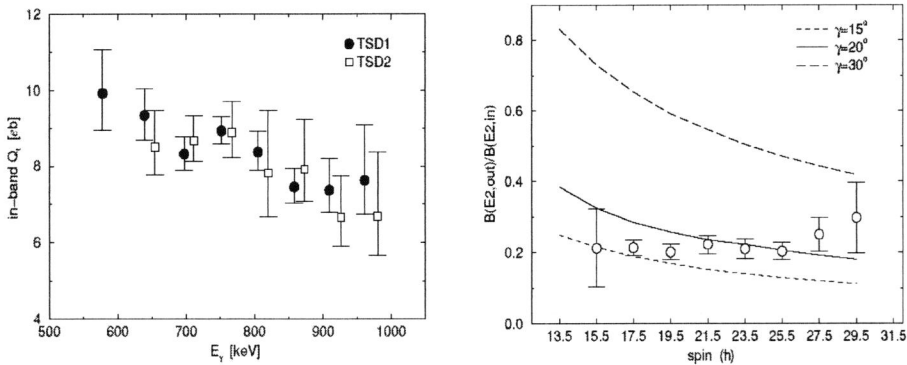

FIGURE 3. Left: Transition quadrupole moments of the in-band transitions in the ^{163}Lu triaxial strongly deformed bands. Right: Ratio of the out-band to in-band B(E2) values in comparison with particle-rotor calculations for different values of γ. A constant γ cannot reproduce the data.

The results are reported in [16] and shown in figure 3. The in-band B(E2) values and quadrupole moments for the two bands are very similar and suggest that the bands are built on the same intrinsic structure. Furthermore, the transition quadrupole moments decrease at higher spin for both bands, while the ratio of inter-band to in-band B(E2) values is constant as a function of spin. An increase in triaxiality from γ~16 to 22 degrees would explain the pronounced decrease in the in-band quadrupole moments and, at the same time, the constant B(E2) ratio for the inter-band transitions, so that a consistent overall description of the new experimental data is reached. The new results support the interpretation of the TSD bands as wobbling-phonon excitations and give an experimental handle on the triaxiality parameter.

HIGHLY DEFORMED BANDS IN A~40 NUCLEI

A major goal of these studies has been to further understand the microscopic structure of rotational collective excitations. That is, to make the connection between the deformed intrinsic states and the microscopic wave functions. However, for all but the very lightest nuclei, truncations and approximations to the theory are necessary and these must be tested by experiment. It turns out that the region around A~40 nuclei is an ideal place to carry these studies. It is a region where deformed shell gaps are characterized by intruder states ($f_{7/2}$) originating from the next higher oscillator shell, but the limited valance space is still sufficiently small to make large scale shell model calculations tractable. Figure 4 shows the single-particle energy levels for both protons and neutrons around N=Z=20. Several major deformed shell gaps are evident at, for example, N = Z = 16, 18, 20, 22, and at deformations from $\beta_2 \sim 0.4\text{-}0.6$

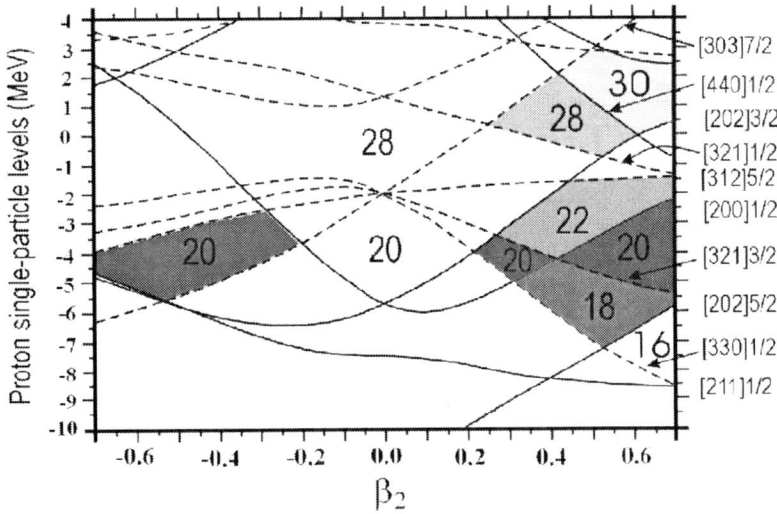

FIGURE 4. Woods-Saxon single-particle energy levels as a function of deformation. Solid (dashed) lines are used for positive (negative) parity. Their particle numbers indicates large spherical and deformed shell gaps.

Over the past few years rotational bands have been observed which correspond to deformed minima at N = Z = 18, 20, and 22 (i.e. ^{36}Ar [17], ^{38}Ar [18], ^{40}Ca [19,20] and ^{44}Ti [21]). These deformed structures are interpreted as core excitations from the sd to the pf shell. In particular the $f_{7/2}$ N=3 intruder is occupied (figure 4): ^{36}Ar has 2 protons and 2 neutrons in the Nilsson [330]1/2 state (a 4p-8h or $\pi 3^2 \nu 3^2$ configuration relative to the closed ^{40}Ca core), while the ^{40}Ca deformed band has 2 protons and 2 neutrons in the [330]1/2 state and a further 2 protons and 2 neutrons in the [321]3/2 state and is an 8p-8h, $\pi 3^4 \nu 3^4$, excitation. (The label $\pi 3^2 \nu 3^2$, for example, refers to the number of protons and neutrons in the N=3 intruder; 2-each in this case.) It is interesting to note that the lowest few states of the proposed 8p-8h band in ^{40}Ca were known previously, but it was not until one had established the full rotational band that the assignment as an 8p-8h band could be made with confidence. The band in ^{38}Ar is assigned a 4p-6h, $\pi 3^2 \nu 3^2$ configuration and in ^{44}Ti an 8p-4h, $\pi 3^4 \nu 3^4$ configuration. Despite several recent experiments no deformed states have been seen in ^{32}S that would correspond to the 4p-12h configuration, even though many calculations predict this structure to be favored.

The deformed bands in ^{36}Ar and ^{40}Ca were observed using Gammasphere and the charged-particle detector Microball [22]. In both cases it was possible to link the structures to the known level scheme (thus determining their spins, parities, and excitation energies) and to measure the state lifetimes and thus extract B(E2) values. The B(E2) values in ^{36}Ar indicate a highly collective structure with large deformations, of the order of $\beta_2\sim0.5$, consistent with a $\pi3^2\nu3^2$ configuration. The latest data on ^{40}Ca [20] also show very large B(E2) values. In this case the collectivity is seen to increase with increasing spin, suggesting a rise in deformation. At the highest spins the deformation is $\beta_2\sim0.6$ (again consistent with the proposed $\pi3^4\nu3^4$ configuration). The lower collectivity at the lowest spins is interpreted as a substantial mixing with less collective 4p-4h states.

With the essentially complete spectroscopic information now available for A~40 deformed bands these data are proving to be a critical benchmark for a wide range of theoretical models, including large scale shell models, the projected shell model, cranked Nilsson Strutinsky, relativistic mean field, Hartree-Fock, and generator coordinate methods.

ACKNOWLEDGEMENTS

The data I have discussed are the result of many peoples work and I would like to thank all my colleagues involved both in the experiments and physics discussions. This work was supported in part under DOE contract No. DE-AC03-76SF00098.

REFERENCES

1. Polikanov, S.M., et al., *Soviet. Phys. JETP* **15**, 1016 (1962); Pereligyn, V.P., et al., *Soviet. Phys. JETP* **15**, 1022 (1962).
2. Twin, P.J., Phys. *Rev. Letters* **57**, 81 (1986).
3. Singh, B., Zywina, R., Firestone, R.B, *Table of Superdeformed Nuclear Bands and Fission Isomers*, Third Edition, LBL-38004 (2002).
4. Chasman, R.R., *Phys. Rev. C* **64**, 024311 (2001).
5. Clark, R.M., *Phys. Rev. Letters*, **87**, 202502 (2001).
6. Goergen, A., et al., *Phys. Rev. C* **65**, 027302 (2002).
7. Lee, I-Y., *Nucl. Phys A* **520**, 641c (1990).
8. Lee, C-T., et al., *Phys. Rev. C* **65**, 041301 (2002).
9. Sun, Y., *private communication.*
10. Beringer, R., and Knox, W.J., *Phys. Rev.* **121**, 1195 (2001).
11. Cohen, S., Plasil, F., Swiatecki, W.J., Ann. of Phys. **82** 557 (1974).
12. Ward, D., et al., *Phys. Rev. C* **66**, 024317 (2002).
13. S.W. Ødegšard et al., *Phys. Rev. Letters*, **86**, 5866 (2001).
14. D.R. Jensen et al., *Phys. Rev. Letters* **89**, 142503 (2002).
15. Bohr, A., and Mottelson, B., *Nuclear Structure Vol. II*, Benjamin, New York, 1975, pp. 190 ff.
16. Goergen, A., et al., *to be published.*
17. Svensson, C.E., et al., *Phys. Rev. Letters* **85**, 2693, (2000); *Phys. Rev. C* **63**, 061301(R) (2001).
18. Rudolf, D., et al., *Phys. Rev. C* **65**, 034305 (2002).
19. Ideguchi, E., et al., *Phys. Rev. Letters* **87**, 222501, (2001).
20. Chiara, C.J., et al., *Phys. Rev. C* **67**, 041303(R) (2003).
21. O'Leary, C,D, et al., *Phys. Rev. C* **61**, 064314 (2002).
22. Sarantites, D.G., et al., *Nucl. Instrum. Methods Phys. Res. A* **354**, 591 (1995).

Hyperdeformed Shapes and Jacobi Transitions in ^{126}Ba

B. Herskind*, G.B. Hagemann*, G. Sletten*, Th. Døssing*,
C. Rønn Hansen*, S. Ødegård*, H. Hübel[†], P. Bringel[†], A. Bürger[†], A.
Neusser[†], A.K. Singh[†], A. Al-Khatib[†], S. B. Patel[†], A. Bracco**, S. Leoni**,
F. Camera**, G. Benzoni**, P. Mason**, A. Paleni**, B. Million**, O.
Wieland**, P. Bednarczyk[‡], F. Azaiez[‡], Th. Byrski[‡], D. Curien[‡], O. Dakov[‡],
G. Duchene[‡], F. Khalfallah[‡], B. Gall[‡], I. Piqueras[‡], J. Robin[‡], J. Dudek[‡],
N. Rowley[‡], N. Redon[§], F. Hannachi[¶], J.N. Scheurer[¶], J.N. Wilson[¶], A.
Lopez-Martens[‖], A. Korichi[‖], K. Hauschild[‖], J. Roccaz[‖], S. Siem[‖],
P. Fallon[††], I.-Y. Lee[††], A. Goergen[††], B.M. Nyakó[‡‡], A. Algora[‡‡], Zs.
Dombrádi[‡‡], J. Gál[‡‡], G. Kalinka[‡‡], D. Sohler[‡‡], J. Molnár[‡‡], J. Timár[‡‡], L.
Zolnai[‡‡], K. Juhász[§§], A. Maj[¶¶], M. Kmiecik[¶¶], M. Brekiesz[¶¶], J. Styczen[¶¶],
K. Zuber[¶¶], J.C. Lisle***, B. Cederwall[†††], K. Lagergren[†††], A. O. Evans[‡‡‡],
G. Rainovski[‡‡‡], G. de Angelis[§§§], G. La Rana[¶¶¶], R. Moro[¶¶¶], W. Gast[a],
R.M. Lieder[a], E. Podsvirova[a], H. Jäger[a], C.M. Petrache[b] and D. Petrache[b]

*The Niels Bohr Institute, University of Copenhagen, DK-2100 Copenhagen Ø, Denmark.
[†]ISKP, University of Bonn, Nussallee 14-16, D53115 Bonn, Germany.
**Dipartemento di Fisica, Universita di Milano and INFN, Italy.
[‡]IReS, IN2P3/CNRS, F-67037 Strasbourg, France.
[§]INP, Lyon, France
[¶]CENBG, Bordeaux-Gradignan, France.
[‖]CSNSM, F-91406 Orsay, France.
[††]Lawrence Berkeley Laboratory, Berkeley CA 94720, USA.
[‡‡]Institute of Nuclear Research (ATOMKI),H-4000 Debrecen, Hungary.
[§§]University of Debrecen, H-4000 Debrecen, Hungary.
[¶¶]The Niewodniczanski Institute of Nuclear Physics, Polish Academy of Sciences,
PL-31342Krakow, Poland.
***Schuster Laboratory, University of Manchester, UK.
[†††]Royal Institute of Technology, Stockholm, Sweden.
[‡‡‡]Oliver Lodge Laboratory, University of Liverpool, UK.
[§§§]Laboratori Nazionali di Legnaro INFN, Italy
[¶¶¶]Dipartimento di Fisica, Università di Napoli and INFN, Italy
[a]Institut für Kernphysik, Forschungszentrum Jülich, D-52425, Germany
[b]Università di Camerino and INFN Perugia, Italy

Abstract. Systematic searches for exotic shapes, hyperdeformation (HD) and Jacobi transitions, have been made in Hf, Nd, Ba, Xe, Sn and Cd nuclei during the last 4 years, guided by theoretical predictions. The most promising results showing patterns of rotational correlations (e.g. 2. order ridges of multiple rotational bands) are found for ^{126}Xe and ^{126}Ba when the very highest multiplicity cascades are selected by various techniques. In particular, the results on ^{126}Ba obtained in 3 different experiments, using both Gammasphere in Berkeley and Euroball IV in Strasbourg are discussed and compared to theoretical predictions and to simulations by a double-potential statistical model for population and γ-decay. A very surprising result was obtained in the last experiment at Euroball, in a full month running time, namely that the observed ridge structure depends very sensitively on the bombarding energy, which points to entrance channel effects. The analysis shows that a discrete HD yrast band intensity will be significantly less than 1 part in 10^6 of the strongest bands populated in the reaction.

INTRODUCTION

Hyperdeformation (HD) at the highest angular momentum predicted by shell model calculations [1] and the giant backbends due to Jacobi transitions predicted by the underlying liquid drop calculations [2, 3], valid at higher temperatures, have raised important questions about the limits of stability in extremely elongated nuclei prior to fission. Several early attempts were made experimentally [4, 5, 6] to search for these exotic structures, but no satisfactory verifications of discrete rotational hyperdeformed bands has been given yet. However, the rich spectroscopy information obtained following the first discovery of Superdeformation (SD) in ^{152}Dy by Peter Twin and collaborators [7], in several mass regions with an increased yield compared to states with Normal deformation (ND) has stimulated attempts to a better understanding of the population of strongly deformed rotational structures. This includes careful searches for typical rotational ridges with small spacings, coresponding to large moments of inertia, as was in fact also used and seen first in ^{152}Dy [8].

During the last years we have made several extended searches for the expected hyperdeformed rotational structures. Most promising rotational ridge structures have been observed in both ^{126}Xe and ^{126}Ba. We shall here concentrate at the Ba case, since this has so far shown the most convincing, and also the most surprising results. The data have been collected in 3 different experiments performed over a period of more than 2.5 years. A brief discussion of the early part of the work on ^{126}Ba and ^{126}Xe is given in reference [6]. A new proposal was made recently to use Gammasphere (GS) at Argonne for a possible extension of the promising results observed in ^{126}Xe by Euroball IV (EB).

Simulations of the population of SD yrast states in ^{152}Dy showed already in 1987 [10] that the enhancement of SD states over ND states at the same temperature, which could be as high as a factor of 10, had two reasons: i) the ratio of level densities in the 2 potentials $\rho_{ND}/\rho_{SD} \approx 3$ and ii) the splitting of the Giant Dipole Resonance (GDR) built on the SD states giving a factor of ≈ 3 due to the enhanced E1 transition probability in the population process from the low lying component. The enhanced E1 feeding of SD states has only recently been verified in the population of both SD-yrast and -continuum states in ^{143}Eu [11].

The same type of simulation is applied to the ^{126}Ba case (see below), with input parameters from our best estimate of potential barriers as a function of spin, GDR splitting, level densities, yrast lines, fission limits and other collective parameters. The most sensitive parameters is, not unexpectedly the fission limit in the residual nucleus. These calculations show that the intensity of ridge structures of low lying hyperdeformed bands may be in the order of 10^{-5} of the $2^+ \Rightarrow 0^+$ transition in the ground band in the same nucleus, which agrees roughly with observed ridge structures in both ^{126}Ba and ^{126}Xe [6]. With 10-20 bands of about 5 transitions along the whole ridge one would need a sensitivity significantly better than 10^{-6} to observe one single discrete hyperdeformed band. This is probably below the limit which can be obtained even by the most powerful existing γ-ray detector arrays, i.e. GS and EB. However, the most extensive experiment on ^{126}Ba at EB, discussed below, shows an unexpected bombarding energy dependence which may point to better gateways than used up to now.

THEORETICAL EXPECTATIONS

Even in the most promising cases with expected deep HD potentials it is extremely important to reach the highest angular momentum in the residual nucleus. Since evaporated particles are removing some angular momentum before the residual nucleus in question is reached after fusion, it is important to choose a cold reaction with sufficient energy to reach the highest angular momentum the nucleus can accommodate. Also for each evaporation step, there will be a fission competition. It is known that large angular momentum is best transferred to the compound nucleus with beam particles larger than A ≈ 30, and that symmetric reactions may be preferable. The fusion reaction used, ^{64}Ni + ^{64}Ni => ^{126}Ba + 2n, seems therefore ideal, although the large Doppler shifts are a disadvantage.

It is also known [12] that the highest angular momentum will exist in the mass A $\approx$ 110-140 region, where the warm nuclei are most soft towards triaxiallity with large deformations in the Jacobi transition region. This is illustrated in fig. 1, using the new Lublin-Strasbourg-Drop (LSD) model by Pomorski and Dudek [3]. A real minimum develops at $\beta_2 \geq 0.6$ in the I > 70 $\hbar$ region, and disappears again due to instability towards fission at I $\geq$ 90 $\hbar$. Similar calculations on heavier nuclei Z $\geq$ 70 show that the nucleus with increasing angular momentum jumps directly from normal deformation (ND) to fission without going through such type of Jacobi transition. We believe that the Jacobi stability for both ^{126}Ba, and ^{126}Xe is very important for the population dynamics, since these potential minima also exist at temperatures of the compound nucleus, where the population takes place, and the Jacobi transition minimum may help to make a gateway to lower lying HD discrete and ridge structures.

The potentials shown in fig. 1 can be used to make approximate estimates of the GDR strength function as a function of spin, weighted by the Boltzman factor over temperature using the expression, [9]:

$$f_{GDR} \propto \sum_{x,y} f_{GDR}(x,y) \cdot exp[(E - E_{LSD}(x,y))/T] \cdot \beta^3 |sin(3\gamma)| \qquad (1)$$

Here f$_{GDR}$ is the GDR strength function, T the temperature at (x,y), where x =

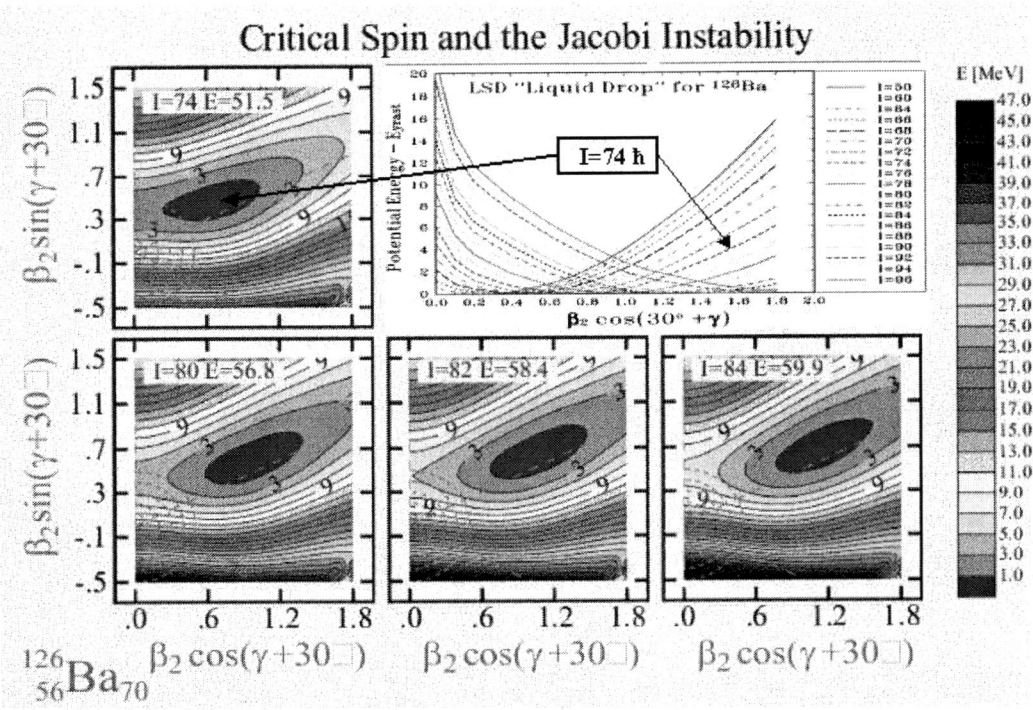

FIGURE 1. Liquid drop potentials displayed with contour lines calculated for ^{126}Ba [3] as examples in the spin region close to the fission limit. The insert (upper-right) shows the minimum of the potential extracted as function of $\beta_2 cos(\gamma + 30^o)$ for a wider region of spin values.

$\beta_2 cos(\gamma + 30^o)$ and y = $\beta_2 sin(\gamma + 30^o)$. The summation is restricted to the prolate to oblate sextant of the $\beta - \gamma$ plane only. An example for I = 72 $\hbar$, shown in fig. 2, is used in the simulation calculations discussed below. The low lying component corresponding to oscillations along the prolate axis, toward fission, is getting well isolated from the other oscillations in the Jacobi region for I $\geq$ 66 $\hbar$ and already very pronounced at I=72 as shown in fig. 2. Two of the three Lorentzian strength functions will be further split by an energy of $2 \cdot \hbar \omega$ due to Coriolis effects at the fast rotation [13, 14]. In fact, the observation of such Coriolis splitting in ^{46}Ti was recently reported in [15] and discussed at this conference [16].

At lower temperature the shell model becomes most important. The Ultimate Cranker (UC) code originally built by T. Bengtsson [17], and later modified by R. Bengtsson [18] is used to calculate the expectations for ^{126}Ba. A pronounced HD potential minimum is predicted to become yrast for I $\geq$ 70 $\hbar$. The HD potential barriers are extracted from these calculations and used also in the simulations.

An overview of the expected energies as a function of spin is shown in fig. 3 (lower panel) based on the estimates discussed above. They are shown together with differential

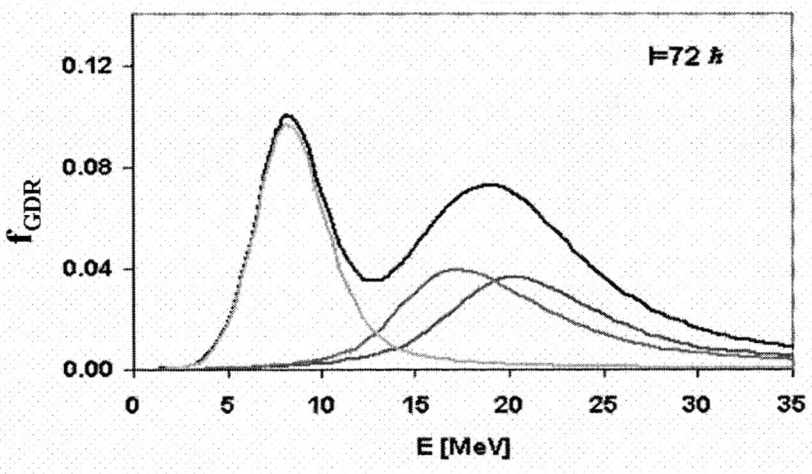

FIGURE 2. The GDR strength function at I = 72 $\hbar$ calculated on basis of the LSD potentials as shown in fig. 1, and expressed in equation 1.

cross sections for the 3 different energies used in the experiments as calculated by the model of Winther [19], which includes the collective degrees of freedom of both target and projectile. The energy loss in the target of $\approx$ 12 MeV is not included. The crossing between the fission barrier and the binding energy of the last decaying neutron occurs at $\approx$ 73 $\hbar$, leaving a window of $\approx$ 70-73 $\hbar$ to be the most important for the population of HD structures in ^{126}Ba.

Simulations

A statistical Monte Carlo simulation of the population and decay properties of ND and HD states of ^{126}Ba is applied, using the earlier developed model which was successful in describing the population and decay of superdeformed states in ^{152}Dy and ^{143}Eu [10, 20, 21, 11]. Many systematic features, e.g. excitation functions, intensity patterns, and the effect of high energy gating could be qualitatively described in the two nuclei.

In ^{126}Ba we have used the model especially to study the dependence on: i) the energy of the low-energy component of the GDR function (see fig. 2), ii) the excitation energy and spin of the entry states and iii) the high spin fission limit for the residual nucleus. Some of these results are displayed in the 3 upper panels of fig. 4, using the results of "best guess" parameter choices in the 3 lower panels. It is seen in the upper panels that the ridge structure observed becomes stronger when: i) the energy of the low component of the GDR is smaller, ii) the energy of the entry states are lower and iii) especially, when the fission limit and corresponding spin distribution of the entry states increases. The latter effect is very strong, indeed, with intensities of the ridge structures from close to zero at a spin limit of 68 $\hbar$, increasing to $\approx$ 8000 counts when the spin limit is increased

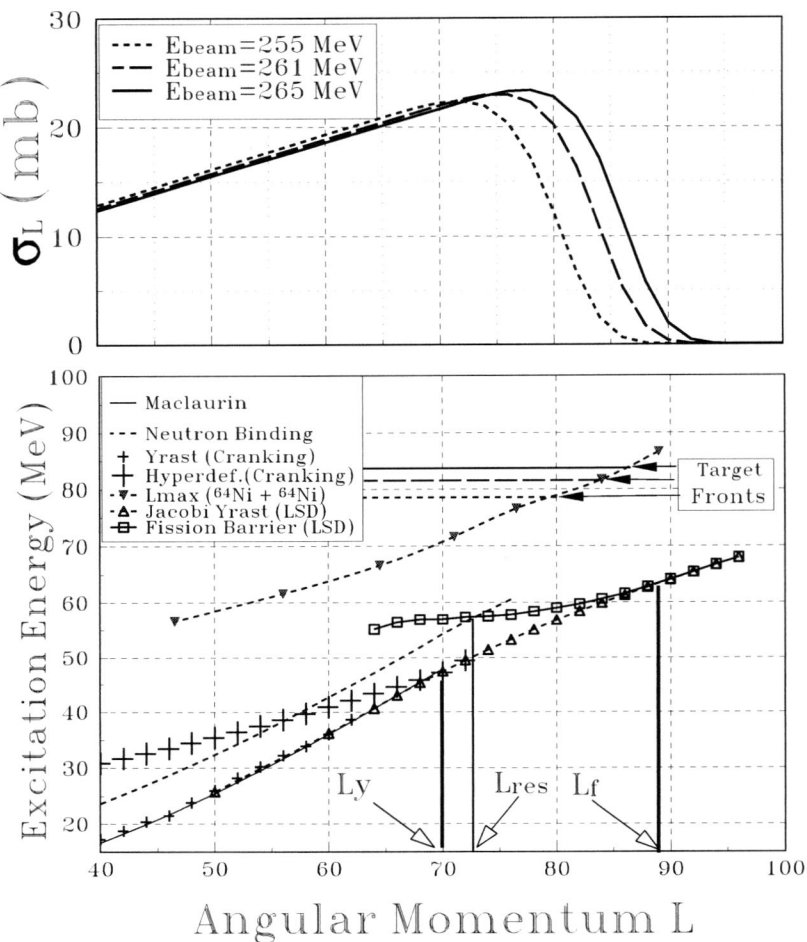

FIGURE 3. Population and decay scheme for the compound nucleus ^{128}Ba populated in the ^{64}Ni + ^{64}Ni reaction. The differential cross sections are shown in the upper panel for the 3 different energies used in different experiments, as calculated by the grazing model of ref. [19]. In the lower panel the corresponding excitation energies are shown for the front of the 500 mg/cm^3 ^{64}Ni target together with other relevant parameters for the population, as function of the angular momentum L. The expected angular momentum limits (L_f=90 given by the fission barrier equal to zero), the residual nucleus ^{126}Ba angular momentum (L_{res}=73) and the yrast crossing (L_y=70) are indicated by vertical lines.

to 76 $\hbar$. The influence of an increase of the width of the low GDR component from 4.2 to 6.2 MeV due to the Coriolis effects was also tested and show that only the HD structures close to yrast increase a factor of 2, in contrast the HD continuum which essentially remain unchanged.

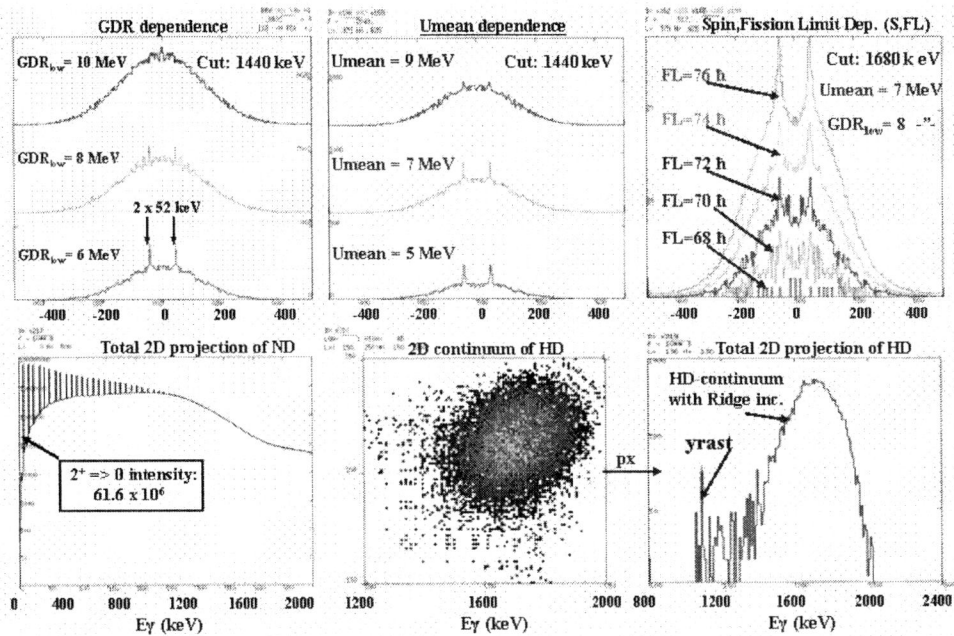

FIGURE 4. Simulations of transitions in the HD potential are displayed in the 3 upper panels as perpendicular cuts on 2D spectra. They show the dependence of the low energy component of the GDR built on HD states in the left panel, the dependence on entry energy, Umean, in the middle panel and the dependence on entry spin and fission limit in the right panel. Typical spectra using "best guess" parameters are shown in the 3 lower panels. The number of counts in the ridge structures shown in the upper right panel is 0,16,210,974,8359 with increasing fission limit, respectively.

EXPERIMENTAL SEARCHES IN BA NUCLEI

The experiment with Gammasphere

The first experiment in the Ba region was made using the GS in Berkeley. The cold fusion reaction, ^{64}Ni + ^{64}Ni $\Rightarrow$ ^{128}Ba* was used with a bombarding energy of 265 MeV. At this energy, the 2n evaporation channel leading to ^{126}Ba was populated at the highest spins the nucleus can accommodate, with a very low cross section of only $\approx$ 2% of the total fusion cross section. However, it was possible to suppress the background from the other reaction channels by a factor of 10 by using a newly developed selective filtering technique [22]. Furthermore, rotational energy correlations were selected by applying the relations: $|E_x + E_y - 2E_z| \leq \delta$, and sorting into a rotational plane, (E_x, E_y) [23, 24], with a thickness of $\delta = 8$ keV giving another factor of about 5-6. The result of this double-selection technique can be seen in fig. 5 where ridges from "warm" rotational bands in the strongly deformed nucleus ^{126}Ba are observed (upper curve in the left panel). The distance between the ridges, 4 x 52 keV, corresponds to a moment of inertia of 77 MeV^{-1} $\hbar^2$ compared to 41 MeV^{-1} $\hbar^2$ for the average of all the normal states, as shown in the lower curve of fig. 5. The ridge structure can also be seen very weakly in the E_y versus E_x plot to be rather smooth over the range 1440 $\pm$ 100 keV in the left part of fig. 5, indicating that more than 5 bands contribute to the observed ridge, $N_{path} \geq 5$.

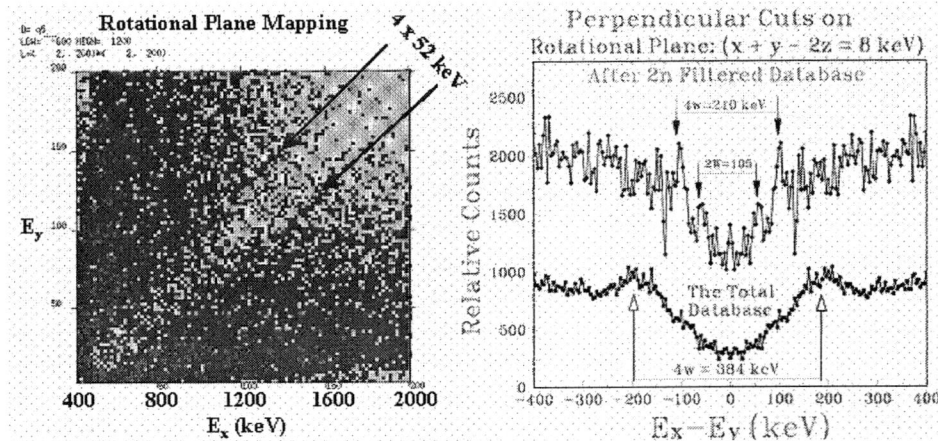

FIGURE 5. Left: The Rotational Plane, for the 2n filtered data. Right: Spectra of ^{126}Ba for perpendicular cuts across the diagonal $E_x = E_y$ of the rotational plane (left), at 1440 $\pm$ 142 keV for the 2n filtered data (upper) and for the full database (lower). The 2n filtered spectrum shows a ridge structure corresponding to a moment of inertia almost twice that extracted from the full dataset. A very weak glimpse of the ridge structure can also be seen in the 2d spectrum to the left.

The first experiment with Euroball IV

The results of the GS experiment stimulated a new attempt with EB, equipped with a BGO inner ball (IB) for detecting the γ-fold distributions, 8 large BaF$_2$ detectors (HECTOR) for detecting high-energy γ-rays from the decay of Giant Dipole Resonances (GDR), and 4 fast BaF$_2$ trigger detectors to ensure that a possible neutron discrimination could be used when needed. We first studied excitation functions for the two reactions, ^{64}Ni + ^{64}Ni $\Rightarrow$ ^{126}Ba + 2n and ^{64}Ni + ^{65}Cu $\Rightarrow$ ^{126}Ba + p2n to make sure that the highest spin region in ^{126}Ba was populated. These results are shown in fig. 6, from the on-line evaluation. From this figure based on spectra of 2n and p2n selected gates, it was concluded on site, that the highest multiplicity was reached for the reaction ^{64}Ni + ^{64}Ni $\Rightarrow$ ^{126}Ba + 2n, using 255 MeV bombarding energy, and it was decided to proceed with this condition for a major part of the available beamtime. This was also in accordance with our initial theoretical estimated shown in fig. 3, where the grazing model [19] had shown that a L_{max} of 76 $\hbar$ may be obtained in the middle of the target, to populate the estimated open spin window of $\approx$ 64-74 $\hbar$ for the 2n channel. However, as we shall see, this turned out not to be valid.

After an extensive and careful analysis of this experiment, we could conclude that the 2n reaction channel was indeed populated most strongly at the highest fold selec-

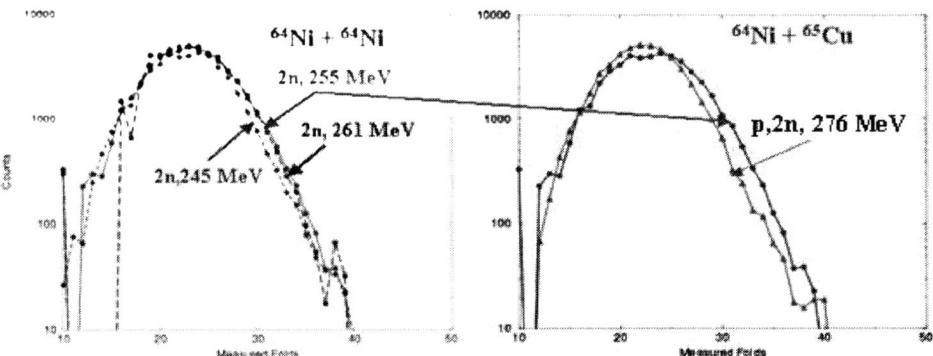

FIGURE 6. Fold distribution functions for the ^{64}Ni + ^{64}Ni $\Rightarrow$ ^{126}Ba + 2n reaction are shown in the left part of the figure. At a bombarding energy of 255 MeV the fold distribution is saturated reaching slightly higher folds than the ^{64}Ni + ^{65}Cu $\Rightarrow$ ^{126}Ba + p2n reaction as shown in the right part of the figure.

tion fold=28, and also that the $\alpha, 2n$ evaporation was rather strong even at fold=26. This pointed to an interesting possibility for "delayed α emission" after decay through strongly deformed rotational bands in the ^{126}Ba continuum, as discussed in greater detail and illustrated in figs. 4 and 5 in ref. [6]. Nevertheless, it was not possible to confirm with reasonable certainty the 4x52 keV ridge structure observed earlier in the GS experiment, displayed in fig. 5.

An Extended experiment with Euroball IV

Gating both by high energy γ-rays, 10-13 MeV, and by high fold with high efficiency due to the inner-ball of EB-IV, showed that even higher sensitivity was needed. This could most probably only be obtained by increasing the running time by a significant factor. Therefore, a new experiment was made, with an extended beamtime of 4 weeks at EB-IV in Strasbourg, Dec. 2002 and Jan. 2003. In this experiment, EB-IV was fully equipped with 4 π of the composite Clusters, Clovers and the Tapered detectors, the 4 π BGO inner ball and in addition the 4 π DIAMANT charged particle detector array. This latter was inspired by the interesting observation of the strongly populated $\alpha, 2n$ reaction channel extending to quite high γ-multiplicity. The beamtime was divided into 2 parts, using 255 MeV and 261 MeV ^{64}Ni beams, resulting in a data sets of 39 and 36 $\times 10^9$ events, respectively, when sorted into fold=10,12,14....48 spike-free 2D-matrices.

The events were sorted into rotational planes as a function of a given "fold plus all higher folds'", independently for the 255 and 261 MeV runs. When studying the highest folds with sufficient statistics, it became clear that the 4 x 52 keV ridge structure observed earlier in the GS experiment at 265 MeV could only be observed in the data from the 261 MeV run. In fact when the 255 MeV data is subtracted after a normalization, all other structures, except the ridge structure vanish as illustrated in fig. 7 (right panel). The intensity of the ridges is very weak, $\approx 5x10^{-5}$ of the 2D projected 256 keV $2^+ \Rightarrow 0^+$ 2n-peak in the 26+higher fold spectrum which again represents a very small fraction ($\approx$ 2-3 %) of the total yield. The statistical significance for this ridge is $\approx$ 4.5 σ. This is good enough to perform a fluctuation analysis and thereby a lower limit for the number of bands passing the observed ridges can be given, $N_{path} \geq 10$. The length of the ridge indicates 4-5 transition in each path. The data were also tested along the ridge structure by an auto-correlation analysis [25], confirming that the bands in the ridges do indeed form sequences with spacing of 52 keV.

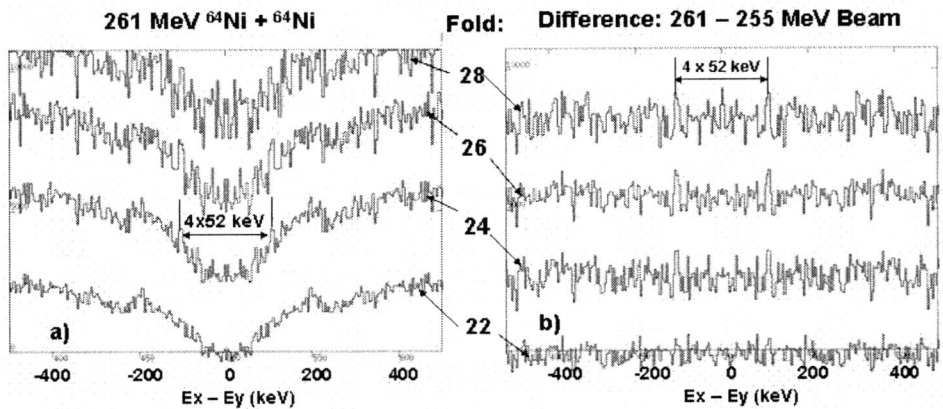

FIGURE 7. 204 keV wide perpendicular cuts at $(E_x + E_y)/2 = 1440$ keV made on rotational planes $(E_x + E_y - 2E_z = \delta) = 12$ keV are shown for folds=22+higher,24+higher,26+higher and 28+higher selections in the left panel for an bombarding energy of 261 MeV. The same spectra are shown in the right panel after subtraction of similar spectra made with 255 MeV, normalized (x 0.76).

The observed strong dependence on the bombarding energy of the ^{64}Ni beam, is indeed very surprising, especially when considering the change of only 3 MeV in excitation energy in the compound nucleus compared to the loss in energy through the target of $\approx$ 12 MeV. This means that the nuclei showing ridge structures are only produced in the first half of the target or perhaps even in a significantly smaller part.

The simulation calculations discussed earlier may guide us towards an explanation of the sensitivity to the bombarding energy, especially the very strong dependence of the ridge structure on angular momentum, as shown in fig. 4 (upper right panel). We therefore tested if a high multiplicity component could be traced in the 2n channel, when comparing the 255 and the 261 MeV data. This is illustrated in fig. 8. In the left hand part of the figure the yields are directly compared as a function of fold for total intensity (upper left: note the scale of 10^{-9}), and for intensities in the peaks from the first exited states in the 2n, 4n and α2n channels (in the other panels: note scale 10^{-6}) all with maximum values for fold $\approx$ 22. The ratio of these yields are shown in the right hand part of the figure, and indeed an increase in yield for the 2n channel can be seen around fold=35. This is significantly higher in fold, although weaker in increased ratio, than the much stronger 4n and $\alpha,2n$ channels, which are known not to saturate within the spin range in the present cases and therefore expected to increase in spin when the bombarding energy is increased. It should be noted that very high multiplicities can be observed from pile-up of two reactions within the time resolution of the electronic system, and can be seen as the second bump on the tails of the distribution functions in the left part of fig. 8, being 3 orders of magnitude smaller than the main intensity. However, a closer look at the functions convinced us that the excess in yield observed as peaks in the ratio spectrum is real, while the increase in the ratio's observed above fold=40 most probably originates from pile-up effects.

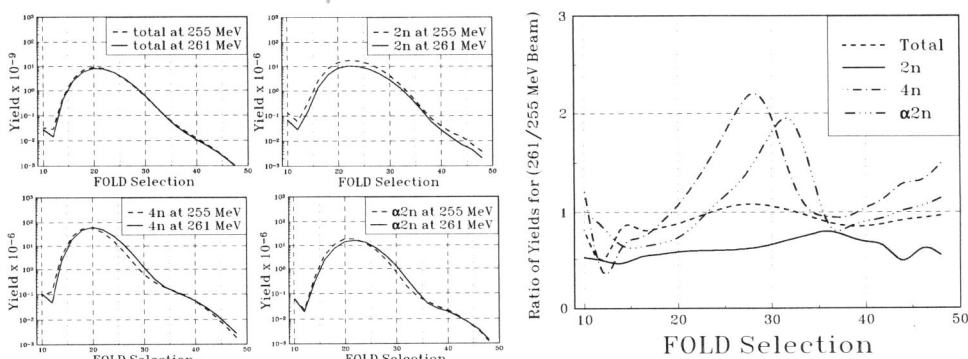

FIGURE 8. The measured yield as function of fold are shown for the total- and the 2n-, 4n- and (α,2n)-intensities, sorted into in 2D spike-free matrices are shown in the 4 panels to the left. The ratio of these intensities are displayed in the right part of the figure as indicated in the legend.

Although it is not easy to translate the increased number of events with high folds $\approx$ 35 with the increase in bombarding energy into spin, it nevertheless points to angular momenta originating from the region of 80 $\hbar$ (before the particle evaporation removes some angular momentum in the decay process). This should be above the fission limit of the residual nucleus discussed earlier and displayed in fig. 3. If significantly higher angular momentum is populated, and fission is delayed to some extent by a complicated equilibration process, then a fraction of the neutron evaporation may be able to compete weakly with fission. Since the simulation shows that the population of the "HD-ridges" change 4 orders of magnitude when increasing the angular momentum by only $8\hbar$, even a very weak population in the $L \approx 76$ region may give the main contribution to the observed ridge structures.

The hard core value for fusion of 261 MeV ^{64}Ni + ^{64}Ni is $L_{max} \approx 71$ $\hbar$, but two recent fusion models have considered i) the collective effects in heavy ion collisions [19] and ii) entrance channel effects caused by a distribution of 3 barriers, which both increase the angular momentum input to the fusion process. In particular, a similar quasi-symmetric systems ^{58}Ni + ^{60}Ni, was investigated in detail by the barrier distribution model and showed spectacular strong entrance channel effects [26].

This latter model was therefore also recently used to calculate the ^{64}Ni + ^{64}Ni reaction for the 3 bombarding energies used in the present experiments [27], after careful fitting

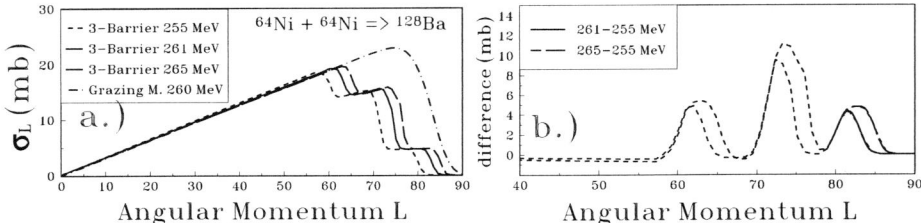

FIGURE 9. The differential fusion cross sections are shown in a) for the reaction ^{64}Ni+^{64}Ni, calculated by the 3-barrier fusion model [27] for the projectile energies used in the present experiments, 255, 261 and 265 MeV, and compared to the grazing model [19] calculations for 260 MeV. Differences expected from the energy changes are shown in b) with full drawn lines for the part of the function which is relevant for the present discussion.

the measured cross section given in [28] for this reaction, from 180 to 220 MeV. The fusion predictions for 255, 261, and 265 MeV are shown in fig. 9a, with a spin population being the weighted sum of the population at the individual barriers. It is also compared to the result of the grazing model for 260 MeV earlier used in fig. 3. The differences obtained by the 3-barrier calculations are displayed in fig, 9b, with the peak at the largest L corresponding to the lowest barriers. In view of the simulations discussed above, the emergence of the ridge might be caused by the addition of cross section around this value of $L \approx 80\hbar$, when increasing the bombarding energy from 255 MeV to 261 MeV, even though fission will remove the major part of this cross section from the fusion channels.

SUMMARY AND CONCLUSIONS

In three experiments we have searched for rotational bands in nuclei of hyperdeformed shapes in ^{126}Ba, using both GS in Berkeley and EB in Strasbourg, trying in different ways to improve the sensitivity for weak structures and high spin, both by hardware and software filters and by extending the beamtime. Although no single discrete band has been found yet, a rotational ridge structure corresponding to a moment of inertia of 77 $\hbar^2 \cdot MeV^{-1}$ extending over 4-5 transitions with an average energy ≈ 1440 keV has clearly been identified in 2 of the 3 experiments. In the last experiment on EB the statistical significance was determined to be 4.5 σ. A lower limit for the number of bands, $N_{path} \geq 10$, was determined from a fluctuation analysis of the rotational ridge structure. The last experiment was made over 2 x 2 weeks of running time with 255 MeV and 261 MeV projectile energy in the fusion reaction ^{64}Ni + ^{64}Ni, respectively, under the same conditions. Surprisingly, the ridge structure was only observed in the data set produced with 261 MeV. Simulation calculations on the population and decay of ^{126}Ba, show an extremely high selectivity to the spin of the entry states for observing the ridge structure. Comparing fold distributions in the 2 reactions show a 15 % increase of the 2n channel in the 261 MeV data at fold ≈ 35, pointing to population from $L \geq 80$. Possible entrance channel effects are discussed in trying to explain the observed very strong influence of the projectile energy on the reaction and decay.

ACKNOWLEDGMENTS

This work was supported by the Danish Science Foundation, by the BMBF, Germany, (contract no. 06 BN 907), by the Polish Committee for Scientific Research (KBN Grant No. 2 P03B 118 22), by the Hungarian Scientific Research Fund, OTKA,(contract number T038404), by the Bolyai János Foundation, by INFN, by the The Swedish Research Council, by DOE, USA, (contract nos. DE-AC03-76SF00098 and FG03-95ER40939), and by the European Commission (contract HPRI-CT-1999-00078).

REFERENCES

1. J. Dudek et al., Phys. Lett. B 211, 252 (1988), see also T.R. Werner and Dudek, Nuclear Data Tables, **50**, (1992).
2. W. D. Myers and W. Swiatecki, Nucl. Phys. **A601**, 141 (1996).
3. K. Pomorski and J. Dudek, nucl-th/0205011; Phys. Rev. **C67**, 044316 (2003).
4. J. Wilson et al. Phys. Rev. **C56**, 2502 (1997).
5. D. Ward, et al., Phys. Rev. **C.66**, 024317 (2002).
6. B. Herskind, et al. Acta Physica Polonica **B34**, 2767 (2003).
7. P.J. Twin, et al. Phys. Rev. Lett. **57**, 8121 (1986).
8. B.M. Nyakó et al. Phys. Rev. Letter **52**, 507 (1984).
9. J.J. Gaardhøje et al., Annu. Rev. Nucl. Part. Sci. **42** 483 (1992)
10. B. Herskind, et al. Phys. Rev. Lett. **59**, 2416 (1987).
11. G. Benzoni, et al. Phys. Lett. **B540**, 199 (2002).
12. S. Cohen, F. Plasil and W. Swiatecki, Ann. Phys (N.Y.), 82:557 (1974).
13. K. Neergård, Phys. Lett. **110B**, (1982).
14. M. Gallardo et al., Nucl. Phys. **A443**, 415 (1985).
15. A. Maj et al, Nucl. Phys. A. in press; nucl-ex/0309018.
16. A. Maj et al. contribution to this proceedings.
17. T. Bengtsson, Nucl. Phys. **A496**, 56 (1989), and Nucl. Phys. A512, 124 (1990).
18. R. Bengtsson, private communication and http://www.matfys.lth.sc/ ragnar/ultimate.html
19. Aa. Winther, Nucl. Phys. **A594**, 203 (1995).
20. K. Shiffer and B. Herskind, Nucl. Phys. **A420**, 521c (1990).
21. S. Leoni et al. Phys. Lett. **B409**, 71 (1997).
22. J. N. Wilson and B. Herskind, NIM **A455**, 612 (2000).
23. B. Mottelson, ANL-PHY-88-2, 1 (1988).
24. B. Herskind, J.J. Gaardhøje and K. Schiffer, ANL-PHY-88-2, 179 (1988).
25. T. Døssing, et al. Phys. Rep., 268 (1996)1
26. A.M Stefanini et al. Phys. Rev. Lett. **74** 864 (1995).
27. N. Rowley, Private communication.
28. M. Beckerman, et al. Phys. Rev. **C23**, 1581 (1981).

Nuclear clusters and structure in light nuclei

Tz. Kokalova*, W. von Oertzen*, S. Thummerer*, H.G. Bohlen*,
M. Milin[†], A. Tumino**, G. de Angelis[‡], E. Farnea[‡], M. Axiotis[‡],
N. Marginean[‡], D.R. Napoli[‡], S.M. Lenzi[§], C. Ur[§], M. Rousseau[¶] and
P. Papka[¶]

Hahn-Meitner-Institut, Glienicker Straße 100, D-14109 Berlin, Germany
[†]*Ruder Bošković Institute, Zagreb, Croatia*
**INFN-Laboratori Nazionali del Sud and Università di Catania, Catania, Italy*
[‡]*INFN-Laboratori Nazionali di Legnaro, Legnaro, Italy*
[§]*Dipartimento di Fisica and INFN, Padova, Italy*
[¶]*Institut de Recherches Subatomiques, IreS, Strasbourg, France*

Abstract. We have studied the γ-decay properties of ^{21}Ne up to the limits of the particle emission thresholds in order to establish the band structure. The GASP γ-ray detector array together with the multi-detector array, ISIS, were used for the selection of the reaction channels. The reaction ^{16}O(^{7}Li,pn)^{21}Ne has been studied at E=29 MeV. The observed decays in ^{21}Ne, support the identification of parity doublets with states of opposite parity connected by strong dipole transitions. The behaviour of the octupole deformed bands in ^{21}Ne is interpreted as consisting of an intrinsic reflection asymmetric (^{4}He+^{16}O)-structure with an additional valence neutron in σ- and π-orbitals. Using the same experimental set-up the emission of the light unbound cluster ^{8}Be has been studied in the reaction ^{18}O+^{13}C$\rightarrow$^{31}Si$\rightarrow$^{23}Ne+^{8}Be. The emission has been studied relative to the sequential emission of 2α-particles.

OCTUPOLE DEFORMATION IN ^{21}NE

The structure of some bands in ^{21}Ne can be interpreted as consisting of an intrinsic asymmetric ^{16}O+α structure combined with a covalent neutron in σ and π orbitals. The parity doublet states observed in ^{21}Ne in the present experiment are discussed and the corresponding band structure for the states up to an excitation energy of 8 MeV in ^{21}Ne is given.

Theoretical consideration and analogy with ^{20}Ne

In the context of covalently bound molecular structures in nuclei, the cluster structure of ^{21}Ne can be based on the underlying structure of ^{20}Ne. The cluster structure of ^{20}Ne has been discussed extensively in Refs. [1, 2, 3, 4]. In Ref. [2] it is shown that a shallow local potential, which is phase equivalent to the deep potential obtained in a double folding model [5], gives an appropriate description of the scattering of α-particles on ^{16}O as well as explaining some of the deformed rotational bands in ^{20}Ne [2, 5, 6]. This "molecular" potential has a strong repulsion at small distances, which can be interpreted (similar to the case of the $\alpha+\alpha$ system) as being due to the Pauli exclusion principle. Calculations based on the antisymmetrised molecular dynamics approach for ^{20}Ne performed by Horiuchi [7, 3] also show the pronounced clustering. In the latter reference it is shown how the ^{16}O+α clustering, related to the octupole degree of freedom, develops with increasing quadrupole deformation; the α-clustering appearing for values of $\beta_2 \approx 0.32$.

The observation of negative parity states due to the intrinsic octupole shapes in the middle of the sd-shell, in conflict with shell-model predictions [8, 9], gives a clue for the existence of these exotic nuclear structures. The observation of parity doublets is also predicted for dinuclear systems (molecules) consisting of two clusters with unequal masses [10]. The phenomenon is well established for ^{20}Ne with bands of $K = 0$. For an odd mass nucleus, like ^{21}Ne with $K \neq 0$ the observation of inversion doublets (both parities for each spin) is predicted. Such shapes, related to the observation of bands of opposite parity connected by strong $E1$ transitions, are well known in heavy odd mass nuclei.

For ^{21}Ne this concept of octupole deformation and a weak coupling of the cluster shape with an extra valence neutron leads to a close analogy with the cluster states in ^{20}Ne, in the form of spin parity doublets with $K = 3/2$ and $K = 1/2$. The existence of such parity doublets, in the context of a reflection asymmetric molecular structure and one covalent molecular orbital for neutrons, are discussed in Ref. [8].

Experimental set up and results

Excited states of ^{21}Ne have been populated in two experiments, a) in the ^{7}Li + ^{16}O reaction from the ^{7}Li + ^{10}BeO experiment and b) in the ^{18}O + ^{13}C reaction, both using the same experimental set up at the Laboratori Nazionali di Legnaro. Targets of $600\,\mu g/cm^2$ ^{10}BeO, layer deposited onto a thick backing (consisting of $4.9\,mg/cm^2$ Pt and $40\,mg/cm^2$ Au) and ^{13}C were used. Gamma-particle coincidence events were collected using the GASP germanium array in coincidence with the ISIS charged-particle detector [11]. This combination enables discrimination between the different reaction channels by selecting the appropriate number of evaporated light charged particles detected in the Si telescopes. Absorber foils of 12 μm thick aluminium were mounted facing the target to prevent scattered beam particles from penetrating the silicon detectors.

In Fig. 1 the observed transitions between states of proposed octupole character populated in the ^{7}Li+^{16}O reaction are shown in the newly ordered band structure. The $K^\pi = 3/2^+$ ground-state band has been observed with intraband transitions up to the $I^\pi = 13/2^+$ state at an excitation energy of 6448 keV. The states in the $K^\pi = 3/2^-$ band have been identified through their *interband* decays to the $K^\pi = 3/2^+$ band. No *intraband* γ-transitions (expected with E2 multipolarity) have been observed in the negative parity band. The lowest state in this band ($3/2^-$, 3664 keV) also feeds the $1/2^-$ state of the $K^\pi = 1/2^-$ band, with an intensity, which is a factor of three lower than the corresponding transition to the $K^\pi = 3/2^+$ band. The fact that the members of the $K^\pi = 3/2^-$ band predominantly decay into states of the $K^\pi = 3/2^+$ band is consistent with strongly enhanced E1 transitions. The existence of an E1 transition connecting the $I^\pi=3/2^-$ bandhead to the ground state could not be clearly established due to the absence

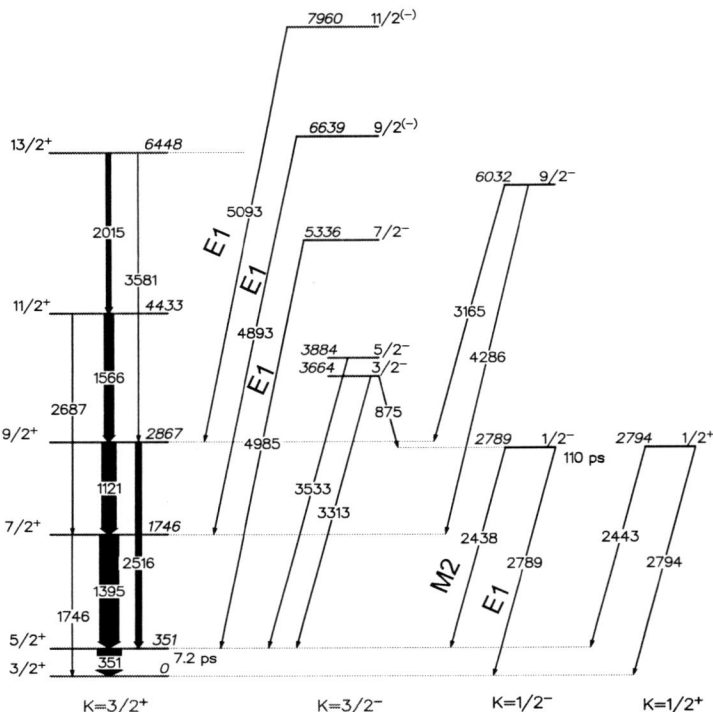

FIGURE 1. Gamma-ray decays observed in the ^{16}O(^{7}Li,np) reaction showing the proposed $K = 3/2$ band of ^{21}Ne. All energies are given in keV. The intensities (widths) of the arrows to the righthand side of the ground-state band are not to scale.

of a feeding transition to gate on and due to the poor signal-to-noise ratio in the ungated spectra.

The transitions de-populating the states differing by intervals of 2 units of spin from the $K = 3/2$ bandhead, namely from $J^\pi = 3/2^-, 7/2^-, 11/2^-$, are clearly visible, whereas the transitions de-populating the interleaved $5/2^-$ and $9/2^-$ states are much weaker, the $13/2^- \to 11/2^+$ transition could not be observed in the present data. This is probably due to the rather low γ-ray branch in this region of excitation energy ($E_x > 9$ MeV), as inferred from very low Γ_γ/Γ-values measured by Billowes *et al.* [12]. (The only exception is the $15/2^-$ level at an excitation energy of 11.988 MeV.) The $5/2^- \to 5/2^+$, 3533 keV transition could not be confirmed here because of its partial overlap with the 3545 keV γ-ray transition de-exciting the state at 6412 keV.

No feeding transitions from higher lying states of the $K = 1/2$ bands were observed. The most striking characteristic of the $K=1/2$ bands, besides the small energy splitting between the members of the parity doublet, is the long lifetime (110 ps) associated with the $1/2^-$ state at 2.79 MeV decaying into the $K = 3/2$ ground-state band. In the cluster description such a transition requires a re-arrangement from a π-type neutron orbital with densities away from the symmetry axis to a σ-type bond [8], where neutrons are concentrated along the symmetry axis. The competing M2 and E1 transitions connecting the $1/2^-$ state to the $5/2^+$ and the $3/2^+$ states in the $K = 3/2$ ground-state band respectively are observed to be retarded by more than three orders of magnitude. The intensity ratio extracted from the data for the two transitions with 2789 keV and 2438 keV is E1/M2 = 3. Warbuton *et al.* [13] have previously measured this ratio and obtained a value of 5.7. Using the systematics of the relative strength of E1 and M2 transitions, the E1 transition would be expected to be stronger than the M2 transition by a factor of 10^6. In the data presented here it is weaker by a factor of 3. The strong suppression of the E1 transition once more indicates that the configurations are not related to single-centre (meanfield) structures, but to two-centre molecular configurations. Two possible explanations were already discussed in Ref. [13]. The explanation of this fact, is that the preferred de-excitation through an M2 transition is related to the change of two quantum numbers going from the $1/2^-$ state ($K^\pi = 1/2^-$ band) to the $5/2^+$ or to the $3/2^+$ state ($K^\pi = 3/2^+$ band), namely the change of the orbital angular momentum and the relative spin orientation, in order to change the K-value. The almost degenerate $I^\pi = 1/2^+$ and $1/2^-$ levels and consequently the small energy splitting in the $K = 1/2$ bands suggests that the internal energy barrier between $\beta_3 < 0$ and $\beta_3 > 0$ for such shapes is very high.

The *non-observation of intraband E2 transitions* in the $K^\pi = 3/2^-$ allows limits to be placed on the observed $B(E1)/B(E2)$ branching ratios. Assuming that for ^{21}Ne, the same value of the measured intrinsic quadrupole moment of ^{20}Ne occurs, $Q_0 = 58(3)$ $e.fm^2$ [14], the intrinsic dipole moment for ^{21}Ne is deduced to be $D_0 > 0.1$ $e.fm$. This large value for D_0 indicates stable octupole deformation, as expected for a reflection asymmetric ^{16}O+n+α structure.

The $K^\pi = 1/2^+$ band was visible in the data only by its weak transitions from the bandhead to the lowest states ($3/2^+$ and $5/2^+$) in the $K^\pi = 3/2^+$ band. No feeding of the $K^\pi = 1/2^+$ bandhead by γ-ray decays from other members of this band could be identified, although its decay is observed.

The de-population of the 6032 keV $9/2^-$ state, which is proposed to be a member of the $K^\pi = 1/2^-$ band, into the $9/2^+$ and $7/2^+$ levels of the $K^\pi = 3/2^+$ band has also been observed. Mixing between the $K^\pi = 3/2^-$, 6639 keV state and the $K^\pi = 1/2^-$ level at 6039 keV, both with spin $9/2^-$, cannot be excluded.

For the $K^\pi = 1/2^-$ bandhead, fed by the $J^\pi = 3/2^-$ ($K^\pi = 3/2^-$) state, only transitions de-populating the $1/2^-$ level were observed. Further feeding of the bandhead was not observed. From this observation and the absence of γ-ray feeding of the $K^\pi = 3/2^-$ band and the $K^\pi = 1/2^+$ bandhead a conclusion could be made that the population patterns observed in this experiment give very strong indications for the direct population of these states by an α-transfer followed by the transfer of a neutron.

The population of the low-spin states of the $K = 1/2$ bands is remarkable and indicates the formation of cluster states, because the (^{7}Li, np) or (^{7}Li, d) reactions favour states with higher spin, due to the angular momentum mismatch between the incoming and outgoing channels.

This result must be contrasted with the *non-observation* of these cluster states in the ^{13}C(^{18}O,2α2n)^{21}Ne reaction. Here the beam energy used was 100 MeV in order to enhance the population of higher spin levels. However, in this purely compound nucleus reaction only yrast and nearly yrast states are strongly populated. The observed decay

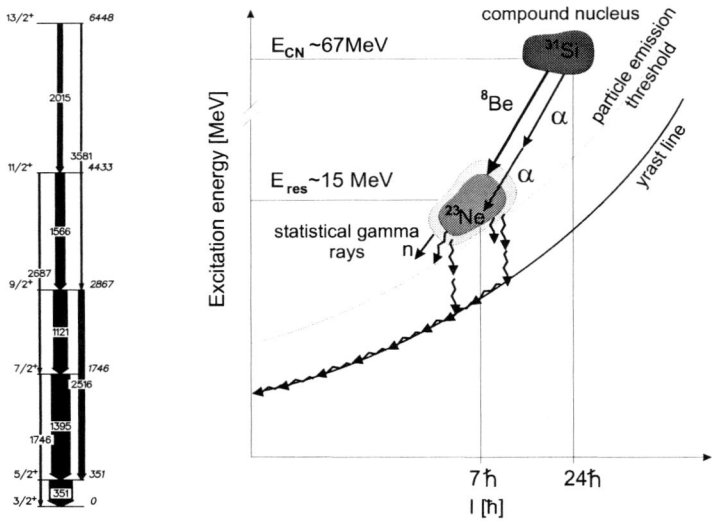

FIGURE 2. Left panel: Partial γ-ray decay scheme for ^{21}Ne as observed in the ^{18}O+^{13}C reaction. All energies are given in keV. Right panel: Diagram of the excitation energy of the compound nucleus decaying by ^{8}Be and 2α emission.

scheme is shown in Fig. 2.

This again shows the importance of the reaction mechanism for populating particular types of states. It is clear, here, that for the investigation of exotic deformed cluster states the use of cluster nuclei for the beam and target, as well as transfer reactions, are required.

EMISSION OF ^{8}BE IN COMPOUND NUCLEUS REACTIONS

In nuclear structure studies the evaporation of light particles following fusion is a decay process which is commonly used to produce residual nuclei which can be subsequently studied via γ-ray spectroscopy. Less commonly used is the emission of heavier charged clusters, a mechanism which is particularly interesting since it may be able to produce nuclei in states of angular momentum and excitation energy that are not normally populated via light-particle emission. Large γ-ray detector arrays like EUROBALL and GASP often make use of light charged-particle triggers to select a particular γ-decaying nucleus. The detection of γ-rays in coincidence with heavier clusters is generally more difficult requiring the integration of large gas counters with close-packed germanium-detector arrays [15]. To overcome this difficulty we have studied the γ-ray decay of compound nuclei in coincidence with reaction products populated in excited states just above the particle threshold, and as a trigger for ^{8}Be we can use, for example, the decay into two α-particles.

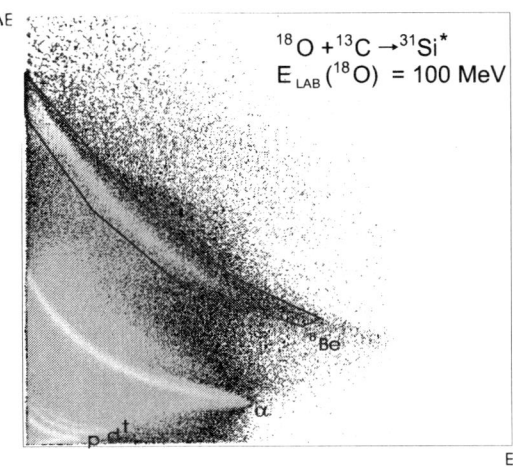

FIGURE 3. A plot of ΔE-E signals from the ISIS telescopes obtained in the ^{18}O+^{13}C experiment. The events representing ^{8}Be are encircled.

Detection of ^{8}Be

The emission of ^{8}Be was studied in the ^{18}O+^{13}C reaction with the same experimental set up as described above. The total solid angle covered by the ISIS detectors is 64% of 4π. Due to the width of the detector frames there is a gap of approximately 6-7° between adjacent detectors. Each ΔE-E telescope covers a solid angle of about 0.20 sr (29° in the reaction plane). Strongly correlated α-particles emitted in the decay of weakly unbound cluster states will be detected in the same Si telescopes and observed through the pile-up of signals. The identification of the decay products, produced by the cluster decay, relies on the former being emitted in a narrow angular cone and a functional dependence following the Bethe-Bloch formula (see Figure 3).

Such a feature is related to the relatively small decay energies of the cluster products as in the case of ^{8}Be emission, the emission cone is approximately 4°. Such kinematic conditions are compatible with the opening angle of 29° of the ISIS ΔE-E telescopes. In this case of ^{8}Be emission it is not possible to distinguish between the ground state, which is unbound by only 91.9 keV and the first excited 2^+-state. The latter, having a width of 1.5 MeV at an excitation energy of 3.04 MeV, will produce an opening cone of more than 24°. Thus a part of the 2α pairs coming from the decay of such an excited state are also registered in one ISIS-telescope.

Plotting the energy signal of the first (thin-ΔE) detector versus the signal of the second (thick-E) detector, the events for each different mZ^2 value, following the Bethe-Bloch formula, will be separated into a distinct banana shaped distribution. The ^{8}Be emission events are registered in the same pattern as multiple hit-signals with 2 times higher values than the original Bethe-Bloch-curves for the single α-particles (see Figure 3). The majority of the 2α-events (97%) due to cluster decay are observed in the first 3 rings of the ISIS detector system. Taking only the events from the first ring of ISIS, which contains 80% of the 2α-event rate, gives us the opportunity to work under the same kinematical conditions and compare the ^{8}Be and 2α-particles events.

Gamma-ray coincidence spectra for different fragments

We are interested in the angular momentum and the energy balance of the binary decay process. The first step is to look into the relevant γ-ray spectra which belong to a given residual nucleus and a given charged-particle trigger. In the off-line analysis the different reaction channels were selected by requiring that only the events corresponding to the detection of the proper number of α-particles in the ΔE-E silicon telescopes were incremented into a symmetrised $E_\gamma - E_\gamma$ matrix. Generally, there are still many reaction channels due to subsequent decays (protons, neutrons, *etc.*) within a charged-particle ISIS-trigger and additional γ-ray gates are needed to identify one residual nucleus uniquely. We consider two possible triggers - 2α particles and ^{8}Be.

The relatively high velocity of the compound nucleus and the larger mass of the emitted fragments, means that the γ-ray spectra must be Doppler corrected. Using the ISIS information a satisfactory Doppler shift correction was obtained, resulting in an energy resolution of about 10 keV FWHM at 1000 keV.

In this reaction, as previously observed for the ^{8}Be emission [16], the binary cluster channel carries less energy and less angular momentum (see the schematical illustration in Fig. 2-right panel) from the residual nucleus than the sequential α-particle emission. Therefore, subsequent neutron and/or proton evaporation becomes more conspicuous (see Fig. 4). The γ-ray spectrum gated by ^{8}Be (see Fig. 4) shows a stronger population for ^{21}Ne, relative to ^{22}Ne, which corresponds to a subsequent neutron evaporation. Comparing the γ-ray spectra gated by 2α-particles (Fold=2), the ratio between the population of ^{22}Ne, N(2α)/N('^{8}Be'), is ≈ 4 and for ^{21}Ne only ≈ 1.5 (from examination of the transition from the first excited state to the ground state in both cases). The appearance of peaks due to transitions which belong to ^{23}Na in the '^{8}Be' gate (see Fig. 4) can not be attributed to the emission of 2α-particles or ^{8}Be. This effect is due to lithium emission [17], overlapping with the ^{8}Be particle gate in the $\Delta E - E$ signals. This effect occurs because the neutron excess gives more favourable Q-values for Li emission. Furthermore, using particle-γ coincidences it was possible to investigate the energy dependence of the relative population of these nuclei by gating on different energies for the '^{8}Be' particles, registered in the same detector.

Although the ^{8}Be events appear along the same event line of the multiple hit events, it was found [18] that the double hit events give only a very small contribution. The latter is explained by the strong spatial correlations of the sequentially emitted α-particles, which only in rare cases are emitted in the same direction in an emission cascade. The cluster fragments, ^{8}Be, carry away less energy from the compound nucleus than the two sequentially emitted α-particles and the emission of another light charged particle from the residual nucleus is possible. The higher probability of sequential α-particle emission than of cluster emission, is in good agreement with the statistical model. Furthermore, in the sequential process the level densities emphasise the emission of a higher total energy for the case of the α-particles. Due to the higher total sum energy carried by the

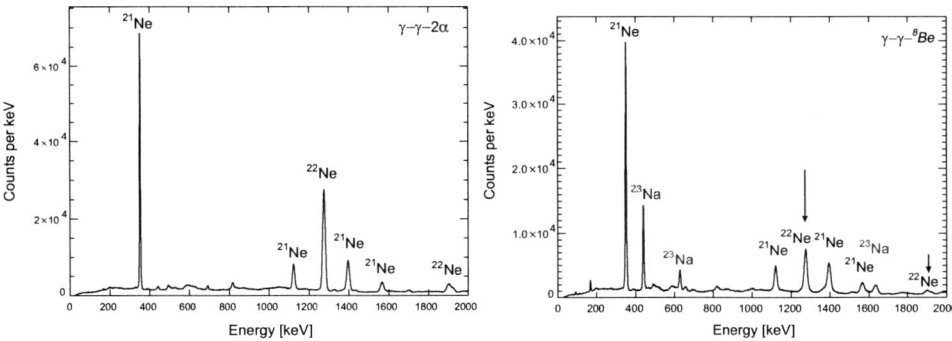

FIGURE 4. Gamma-ray spectra from the ^{18}O+^{13}C reaction gated by different charged particle triggers. Left panel: Doppler corrected spectrum gated by the 2α channel. Right panel: Doppler corrected spectrum gated by the ^{8}Be channel. The peaks corresponding to ^{23}Na are due to the emission of ^{6}Li and ^{7}Li.

α-particles, these events populate states in the compound nucleus closer to the yrast line, than the binary cluster emission process.

ACKNOWLEDGMENTS

This work was supported by the BMBF grant Nr.06-OB-900 and the EU-LSF program (contract Nr.HPRI-CT-1999-00083).

REFERENCES

1. Butler, P., and Nazarewicz, W., *Rev. Mod. Phys*, **68**, 350 (1996).
2. Ohkubo, S., (editor), *Prog. Theor. Phys. Suppl.*, **132** (1998).
3. Kimura, M., Sugawa, Y., and Horiuchi, H., *Prog. Theor. Phys (Kyoto)*, **106**, 1153 (2001).
4. Dufour, M., Descouvemont, P., and Baye, D., *Prog. Theor. Phys.*, **C 50**, 795 (1994).
5. Abele, H., and Staudt, G., *Phys. Rev.*, **C 47**, 742 (1993).
6. Buck, B., Dover, C., and Vary, J., *Phys. Rev.*, **C 11**, 1803 (1975).
7. Horiuchi, H., Ikeda, K., and Suzuki, Y., *Suppl. Prog. Theor. Phys.*, **52**, 89 (1972).
8. von Oertzen, W., *Eur. Phys. J.*, **A 11**, 403 (2001).
9. Descouvemont, P., *Phys. Rev.*, **C 48**, 2746 (1993).
10. Herzberg, G., *Spectra of diatomic molecules*, D. van Nostrand Comp. Inc, Princeton, 1950, vol.i edn.
11. Farnea, E., et al., *Nucl. Instr. and Meth.*, **A 400**, 87 (1997).
12. Billowes, D., et al., *Annual Report, Schuster Laboratory, University of Manchester* (1987).
13. Warburton, E., Olness, J., Engelbertink, G., and Jones, K., *Phys. Rev.*, **C 3**, 2344 (1971).
14. Horikawa, Y., et al., *Phys. Lett.*, **9**, 36B (1971).
15. Thummerer, S., Gebauer, B., Oertzen, W., and Wilpert, T., *Il nuovo cimento*, **111 A**, 1077 (1998).
16. Thummerer, S., *Ph.D Thesis*, Freie Universität and Hahn-Meitner-Institut Berlin, Germany, 1999.
17. Kokalova, T., *Ph.D Thesis*, Freie Universität and Hahn-Meitner-Institut Berlin, Germany, 2003.
18. Kokalova, T., von Oertzen, W., et al., *Eur. Phys. J.*, **A**, to be submitted (2003).

The Neutron Facility at NCSR "Demokritos"-Implementation in the Case of the ^{232}Th(n,2n) Reaction

R. Vlastou[a], C.T. Papadopoulos[a], G. Perdikakis [a,b], M. Kokkoris [a,b], C.A. Kalfas[b], S. Kossionides[b], D. Karamanis[c], P.A. Assimakopoulos[c]

[a] *Department of Physics, National Technical University of Athens, 157 80 Athens, Greece*
[b] *Institute of Nuclear Physics, NCSR "Demokritos", 153 10 Aghia Paraskevi, Greece*
[c] *Department of Physics, The University of Ioannina, 451 10 Ioannina, Greece*

Abstract. The neutron facility at the 5.5 MV Tandem T11/25 Accelerator of NCSR "Demokritos" will be presented. The facility can deliver monoenergetic neutron beams in the energy range 120-650 keV, 4-11.5 MeV and 16-20.5 MeV via the ^{7}Li(p,n), ^{2}H(d,n) and ^{3}H(d,n) reactions, respectively. The corresponding beam energies obtained from the accelerator, are 1.92-2.37 MeV protons, 0.8-9.6 MeV deuterons and 0.8-3.7 MeV deuterons, for the three reactions, respectively. Experimental results for neutron energies up to 11.5 MeV will be presented for the ^{232}Th(n,2n)^{231}Th reaction. In context with the CERN n-TOF collaboration, the cross section of this reaction has been measured relative to the ^{56}Fe(n,p)^{56}Mn and ^{27}Al(n,α)^{24}Na reaction cross sections, by using the activation method. In addition to the experimental work, theoretical Statistical model calculations have also been performed using the computer code STAPRE/F. The results are being compared to the experimental data.

INTRODUCTION

In the last decade, there has been an increasing interest in the study of neutron induced reactions, mainly caused by the extensive programs for various Accelerator Driven projects, which have started recently [1-3] and are aimed at nuclear energy applications as well as medical, astrophysical and fundamental Nuclear Physics purposes.

The main technological applications of these projects are the future production of clean and safe nuclear energy, the incineration of nuclear waste and the radioisotope production for medical applications and dosimetry. All these tasks require knowledge of complete and precise cross sections for neutron induced processes. The existing data and the evaluations, which have been compiled in databases, show so many differences and discrepancies, that they cannot be considered as reliable basis for planning in detail these Accelerator Driven Systems [4,5]. In order to produce reliable and consistent neutron cross section data, there are projects under way, mainly at CERN, LANL and collaborative laboratories with neutron time of flight facilities. Using spallation they produce high intensity neutron beams in the energy interval from 1eV to 250 MeV. These facilities allow the systematic study of neutron-induced reactions for almost any isotope, offering the added feature of an accurate time of flight determination of neutron energy. However, complementary cross section measurements with monoenergetic neutron beams are essential for ensuring reliability. The results of these measurements will provide a complete and reliable database over a wide energy range for nuclear energy research. They will also allow verification of nuclear reaction models and improve the evaluated neutron reaction database.

Apart from the technological motivation, cross section measurements of neutron induced reactions are of special interest to Nuclear Astrophysics, since neutron reactions are responsible for the formation of all elements heavier than iron. Furthermore, spectroscopic studies of reactions between nuclei and fast neutrons can provide valuable information on both nuclear structure and nuclear reactions.

In view of the aforementioned remarks, the neutron facility at the 5.5MV Tandem T11/25 Accelerator of NCSR "Demokritos", has been upgraded and extended to cover broader energy range. Recent developments and techniques used for the production of monoenergetic neutron beams at "Demokritos", will be presented in this report. Within the framework of the collaboration with the n-TOF facility at CERN [1,2], the ^{232}Th(n,2n) reaction has been studied in the energy range 7.5-11.5 MeV [6]. The experimental results and the theoretical investigation of this reaction will also be presented.

THE NEUTRON FACILITY

The production of monoenergetic neutron beams at the 5.5MV Tandem T11/25 Accelerator at the NCSR "Demokritos" can be achieved by using different reactions for different energy regions.

For low energies between 120 and 650 keV, neutrons can be produced via the ^{7}Li(p,n)^{7}Be reaction by using a ^{7}LiF target on Al backing and a proton beam in the energy range 1.9-2.4 MeV. The neutrons thus produced at zero degrees are monoenergetic. At higher proton energies, up to 5.5MeV, the zero degree neutrons can reach an energy of 3.8 MeV, but they are not strictly monoenergetic since they contain a ~10% contribution of neutrons coming from the first excited state of ^{7}Be. The neutron beam flux in all cases is of the order of 10^6 n/cm^{2}sec for a proton beam current ~1 μA.

For the middle energies between 4.0 and 11.5 MeV, neutrons are produced via the ^{2}H(d,n)^{3}He reaction by using deuteron beam in the energy range 0.8-9.6 MeV and a gas cell target 3.7 cm long made of stainless steel, as described in Ref.[7]. The entrance window is a 5μm Mo foil and the beam stops on a 1mm Pt foil. The deuterium gas pressure can be monitored and refilled electronically when the cell pressure falls below a preset level. For deuteron currents up to 5μA and deuterium gas pressures up to 1.5 atm, the neutron flux at 0° is of the order of 10^7 n/cm^{2}sec.

For the higher energies between 16.0 and 20.5 MeV, neutrons are produced via the ^{3}H(d,n)^{4}He reaction by using a deuteron beam in the energy range 0.8-3.7 MeV and a Ti tritiated target of 5Ci activity on a Ag backing for good heat conduction. To avoid heating and outgassing, the target is water-cooled during irradiation with the deuteron beam of ~2μA. The corresponding neutron beam flux at 0° is of the order of 10^6 n/cm^{2}sec.

Neutron energies are calculated from reaction kinematics by taking into account the proton or deuteron energy loss in the target, the cross section of the reaction and the irradiation geometry [8]. Lower energies of the neutron beam can be achieved by placing the target for the reaction under investigation at an angle to the neutron beam, leading to lower flux and lower beam energy resolution. A BF_3 detector whose spectra are stored at regular time intervals, monitors the flux variation of the neutron beam in all three cases. The absolute flux of the beam can be obtained with respect to reference reactions, such as ^{56}Fe(n,p), ^{27}Al(n,α) and ^{93}Nb(n,2n), whose cross sections are well determined and can be found in the literature.

THE ^{232}Th(n,2n) REACTION

In the frame of the n-TOF collaboration, the cross section of the ^{232}Th(n,2n) reaction, which is important for the Th-U cycle, has been measured in the energy range 7.5-11.5 MeV by using the activation method. The β-decay (with a half-life of 25.52 h) of the residual nucleus ^{231}Th leads to the excited states of ^{231}Pa. The de-excitation of ^{231}Pa proceeds via gamma rays, the most prominent of which for the activation measurement is the 25.7 keV transition. Induced gamma activity was measured with an 80 mm^2 Si(Li) detector, continuously for a few days after the exposure to the neutron beam. The experimental details regarding these measurements and the analysis of the data are described in Ref. [6].

The cross section values extracted by the experimental data are presented in Fig. 1 together with the most recent data by Raics et al. [9] and the evaluations from four reference libraries of neutron data. In the overlapping energy region the data of Raics et al. seem to coincide with the results presented here, while the evaluated values from the various databases [10], seem to underestimate the experimental ones at energies above ~10 MeV.

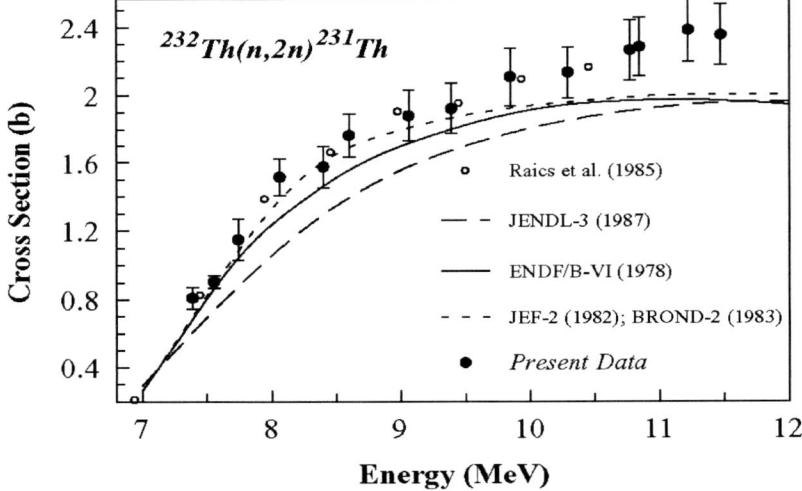

FIGURE 1. Experimental cross sections of the ^{232}Th(n,2n) reaction in comparison with existing data from Ref. [9] and evaluations from different databases [10].

STATISTICAL MODEL CALCULATIONS

The excitation function for the cross sections of the ^{232}Th(n,2n) and fission reactions have been calculated in the framework of the Hauser-Feshbach theory using a modified version of the code STAPRE/F [11]. The code has been designed to estimate energy-averaged cross sections for particle-induced reactions with several emitted particles and gamma-rays under the assumption of sequential evaporation. Discrete energy levels, branching ratios and transmission coefficients required as input parameters in the code, were taken from [12]. Fissioning nuclei level densities at fission saddle deformations were extracted by the generalized superfluid model [12-14]. The values of the parameters used in the calculations are summarized in Table 1.

TABLE 1. Parameters used for the statistical model calculations

Nucleus	α (MeV^{-1})	Δ (MeV)	Fission Barrier Height (MeV)		Fission Barrier Curvature (MeV)		Δ (MeV) For fissioning nuclei
			Inner Saddle	Outer Saddle	Inner Saddle	Outer Saddle	
^{233}Th	24.4	0.786	5.1	6.65	0.7	0.5	0.806
^{232}Th	24.5	0.788	5.8	6.7	0.9	0.6	0.806
^{231}Th	25.5	0.800	6.0	6.7	0.7	0.5	0.830
^{230}Th	27.6	0.791					

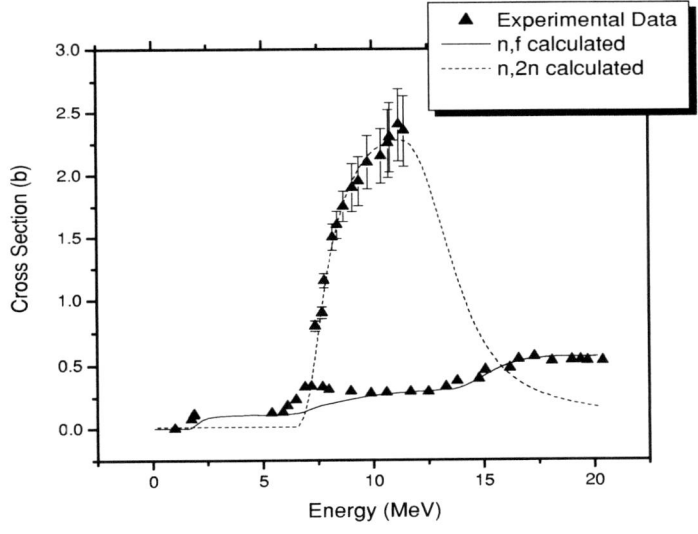

FIGURE 2. Experimental cross sections of the ^{232}Th(n,2n) and ^{232}Th(n,f) reactions in comparison with statistical model calculations. The data for the ^{232}Th(n,f) cross sections were taken from the EXFOR experimental nuclear reaction database.

Statistical model calculations for the ^{232}Th(n,2n) and fission reactions are compared to the data in Fig. 2 and in general are seen to reproduce the overall trend of the excitation function. They seem however, to underestimate the observed fission cross section in the energy region of 7 MeV. Furthermore, it is evident that the lack of experimental data at energies above 12 MeV is of crucial importance for the determination of the parameters in the theoretical calculations. Thus, it is planned to extend the measurements to higher neutron energies, up to 20.5 MeV, by using the ^{3}H(d,n)^{4}He reaction.

SUMMARY

The neutron facility at the 5.5 MV Tandem T11/25 Accelerator of NCSR "Demokritos" can deliver monoenergetic neutron beams in the energy range 120-650 keV, 4-11.5 MeV and 16-20.5 MeV via the ^{7}Li(p,n), ^{2}H(d,n) and ^{3}H(d,n) reactions, respectively, at a flux of the order of 10^6-10^7 n/cm^{2}sec.

In the frame of the CERN n-TOF collaboration, the cross section of the ^{232}Th(n,2n)^{231}Th reaction has been measured between 7.5 and 11.5 MeV, relative to the ^{56}Fe(n,p)^{56}Mn and ^{27}Al(n,α)^{24}Na reaction cross sections, by using the activation method. In addition to the experimental work, theoretical statistical model calculations have also been performed using the computer code STAPRE/F. The results have been compared to the experimental data and appear to be in reasonable agreement. From the behavior of the theoretical predictions, however, the necessity of experimental data at higher energies is evident. Future plans include extending the measurements in the energy region 16.0 and 20.5 MeV, via the ^{3}H(d,n)^{4}He reaction.

REFERENCES

1. Rubbia, C., *International Conference on Accelerator-Driven Transmutation Technologies and Applications*, AIP Conference Proceedings 346, Las Vegas, 1994.
2. Pavlopoulos, P., *et al.*, CERN/SPSC 99-8, 1999.
3. Bowman, C., et al., *Nucl. Instr. Meth.* **A320**, 336 (1992).
4. Koning, A. J., et al., *Nucl. Instr. Meth.* **A414**, 49 (1998).
5. Fessler, A., et al., *Nucl. Sci.Eng.* **134**, 171 (2000).
6. Karamanis, D. et al., *Nucl. Instr. Meth.* **A505**, 381 (2003).
7. Vourvopoulos, G., et al., *Nucl. Instr. Meth.* **A220**, 23 (1984).
8. Doukellis, G., et al., *Nucl. Instr. Meth.* **A327**, 480 (1993).
9. Raics, P., *Phys. Rev.* **C34**, 87(1986).
10. *JENDL-3.2,OECD NEA Data Bank* (1996).
    JEF-2.2, OECD NEA Data Bank (1996).
    ENDF/B-VI, OECD NEA Data Bank (1996).
    BROND-2, OECD NEA Data Bank (1996).
11. Uhl M. and Stohmaier B., *Report IRK-76/01*, 1976.
12. "Handbook for Calculations of Nuclear Reaction Data-Reference Input Parameter Library-Final Report of a Co-Ordinated Research Project *"IAEA-TECDOC*-1034 (August 1998*)*.
13. Ignatyuk A. V., Istekovand K. K., Smirenkin G., N.,*Sov.J.Nucl.Phys.* **29**, 450(1079).
14. Ignatyuk A.V., "Statistical Properties of Excited Atomic Nuclei" (in Russian). *Energoatomizdat*, Moscow 1983, Translated by IAEA Report

Studies Around A~100 Using Binary Reactions

P.H. Regan*, A.D. Yamamoto*, C.Y. Wu[†], A.O. Macchiavelli**, D. Cline[†],
J.F. Smith[‡], S.J. Freeman[‡], J.J. Valiente-Dobón[§], R.S. Chakrawarthy[‡],
M. Cromaz**, P. Fallon**, A. Hayes[†], H. Hua[†], S.D. Langdown*, I-Y. Lee**,
C.J. Pearson[§], Zs. Podolyák[§], R. Teng[†] and C. Wheldon[¶]

**WNSL, Yale University, P.O. Box 208124, 272 Whitney Avenue, New Haven CT 06520-8124, USA
and Dept. of Physics, University of Surrey, GU2 7XH, UK*
[†]*Nuclear Structure Research Laboratory, Department of Physics, University of Rochester,
Rochester, NY 14627, USA*
***Nuclear Science Division, Lawrence Berkeley National Laboratory, Berkeley, CA 94720, USA*
[‡]*Dept. of Physics and Astronomy, The University of Manchester, Manchester, M13 9PL, UK*
[§]*Dept. of Physics, University of Surrey, GU2 7XH, UK*
[¶]*Kernphysik II, GSI, Max-Planck-Straße 1, D-64291 Darmstadt, Germany*

Abstract. The structure of stable and neutron-rich nuclei with Z~42 and N~58 have been studied following binary heavy-ion reactions between a ^{136}Xe beam and a thin, self-supporting ^{100}Mo target. Discrete states at spins of $20\hbar$ and above have been identified and these data allow both gamma-ray fold and binary-fragment particle angular distribiution analyses to be made. On the basis of these results, suggestions are made for future directions of this research extending to higher spin states in this region where more exotic nuclear shapes are predicted.

INTRODUCTION

The study of beta-stable and mildly neutron-rich nuclei at medium to high spins is problematic. This is due to the difficulty in synthesizing such systems with stable beam/target combinations in fusion-evaporation reactions, while spontaneous [1] and heavy-ion-induced fission [2] populates more neutron-rich systems. However, the near-stable nuclei around A~100 present a number of areas of current physics interest, particularly with regard to the evolution of their structure with increasing angular momentum. These include (i) the re-intepretation of backbending in this region as vibration-to-rotational structural changes associated with the population of high-j, low-Ω intruder configurations [3] and (ii) predictions of competing highly collective rotational states states in the vicinity of the 'doubly magic' superdeformed core of $^{100}_{42}$Mo$_{58}$ [4].

Some initial attempts to identify the near-yrast states around the first backbend in this mass region have been made using deep-inelastic reactions [5]. These studies negated the use of any requirement for event-by-event Doppler correction by utilising backed targets and were able to obtain reasonable quality $\gamma-\gamma$ coincidence data using a second generation gamma-ray spectrometer array. While these data provided some limited information on the collective states just above the first backbending in a number of nuclei, the Doppler broadening associated with the decays from the faster, higher-spin states limited the spins identified in this way to $I \leq 14\hbar$.

EXPERIMENTAL DETAILS, DATA ANALYSIS AND RESULTS

This paper describes the results of an investigation using binary collisions between a 700 MeV ^{136}Xe beam and a self-supporting (420μg/cm^2) ^{100}Mo target. Reaction γ rays were detected using the GAMMASPHERE array [7], which consisted of 102 Compton-suppressed hyperpure germanium detectors and was situated at the Lawrence Berkeley National Laboratory. The binary fragments were detected using the position-sensitive gas-filled detector CHICO [8], thus enabling an event-by-event Doppler correction to be applied to the raw γ-ray data. The detection of co-planar events in CHICO allowed the separation of both beam-like (BLF) and target-like fragments (TLF) by the measurement of their position relative to the beam direction. Figure 1 shows the laboratory angle of the emitted fragment relative to the beam direction, versus the time-of-flight difference between the two, co-planar fragments observed by CHICO. The beam-like and target-like fragments are clearly separated by the reaction kinematics and the yields for both are peaked around the expected laboratory grazing angles of $\sim$26° and 48° respectively. The velocity of the target-like fragments was angle dependent and calculated assuming 2-body kinematics to vary between 3% and 11% of the speed of light.

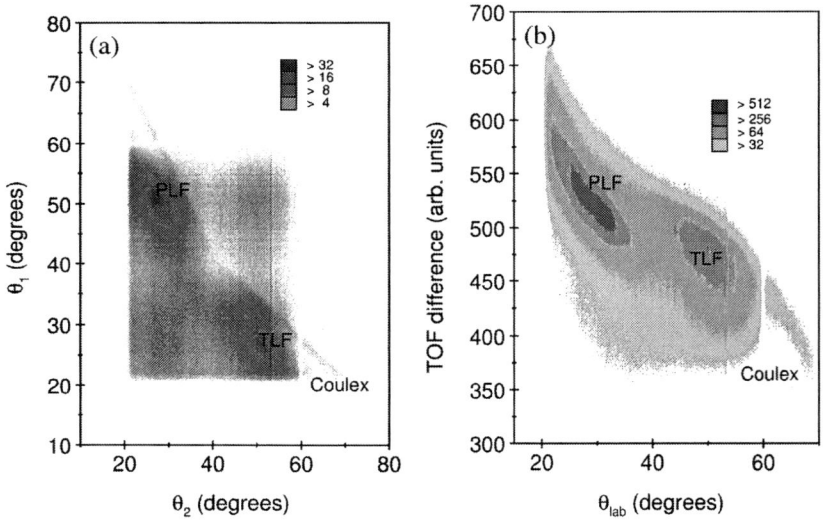

FIGURE 1. (a) CHICO spectrum showing the angular correlations between the two binary fragments. (b) the TOF difference versus laboratory angle for binary fragments in CHICO.

The acquisition master trigger required that at least three prompt, Compton-suppressed γ rays were detected in GAMMASPHERE within approximately 50 ns of each other, together with two, co-planar binary fragments in CHICO. The heavy metal collimators were removed from the GAMMASPHERE BGO suppression shields allowing an improved estimate of the total γ-ray fold to be made. This, in turn, is related to the input angular momenta of the two fragments in the binary reaction. A total of 900×10^6 suppressed germanium triples and higher fold events were detected in coincidence with two, co-planar binary fragments during the course of a four day experiment. Examples of summed double-gated, gamma-ray coincidence spectra for specific target-like fragment nuclei from this analysis are shown in figure 2. The total fold projections gated by double gamma-ray transitions which select specific spins in given nuclei are also shown in figure 2. The yrast decay sequences of selected nuclei around A~100 as observed in the current work are shown in figure 3.

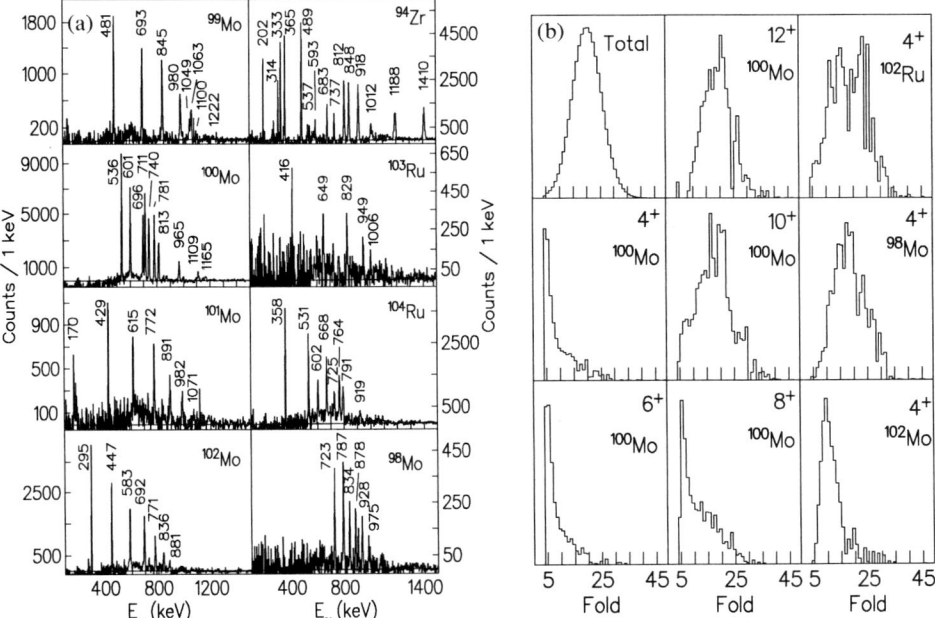

FIGURE 2. (a) Sums of double-gated triples gamma-ray coincidence spectra showing yrast cascades in selected TLFs. (b) Double gamma-gated fold spectra for TLFs from the current work.

Figure 4 shows the particle angular distributions as measured in CHICO for sets of double γ-ray coincidence gates in ^{100}Mo and neighboring nuclei. As can be seen from figures 2b and 4, for spins up to $8\hbar$ in ^{100}Mo the spectra are dominated by a low-fold reaction mechanism (assumed to be Coulomb excitation), while states above the 10^+ level (associated with the aligned $(\nu h_{11/2})^2$ configuration [5]) have distributions shifted to significantly higher folds, presumably associated with deep-inelastic collisions. It is perhaps also worthy of note that the quasi-elastic two-neutron transfer to ^{102}Mo exhibits rather narrow fold and particle angular distributions, intermediate between those observed for Coulomb excitation and deep-inelastic reactions [6].

331

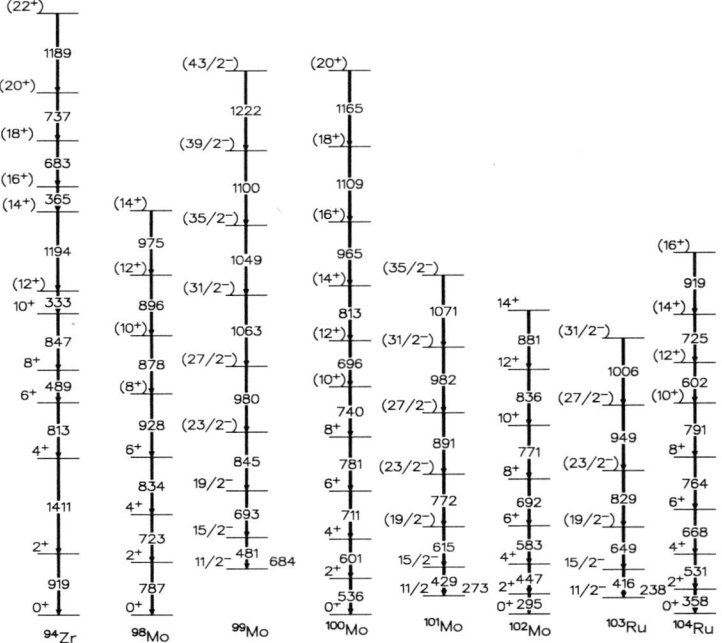

FIGURE 3. Partial decay schemes showing the yrast sequences observed in selected TLFs.

FUTURE DIRECTIONS

There are predictions of superdeformed states approaching the yrast line in ^{100}Mo for spins of approximately $25 \rightarrow 30\hbar$ [4, 6]. The current experiment used an energy of approximately 25% above the nominal Coulomb barrier between the target and projectile, which using the semi-classical 'rolling-mode' approximation [9] corresponds to a maximum expected value of the target-like fragment intrinsic spin of $\sim 25\hbar$. We note that this is close to what is observed discretely in the current experiment. For heavier beams such as ^{208}Pb or ^{238}U at similar relative energies above the Coulomb barrier (i.e., 25-30%), the predicted rolling mode input spin is greater than $30\hbar$, which may be enough to populate the predicted superdeformed minium in both ^{99}Mo and ^{100}Mo.

ACKNOWLEDGMENTS

This work is supported by EPSRC (UK) and the US Department of Energy, under grants DE-FG02-91ER-40609 and DE-AC03-76SF00098 and by the National Science Foundation. PHR acknowledges support from Yale University via both the Flint and the Science Development Funds. ADY, JJVD and SDL acknowledge the receipt of EPSRC postgraduate studentships. ZP acknowledges an EPSRC Advanced Fellowship.

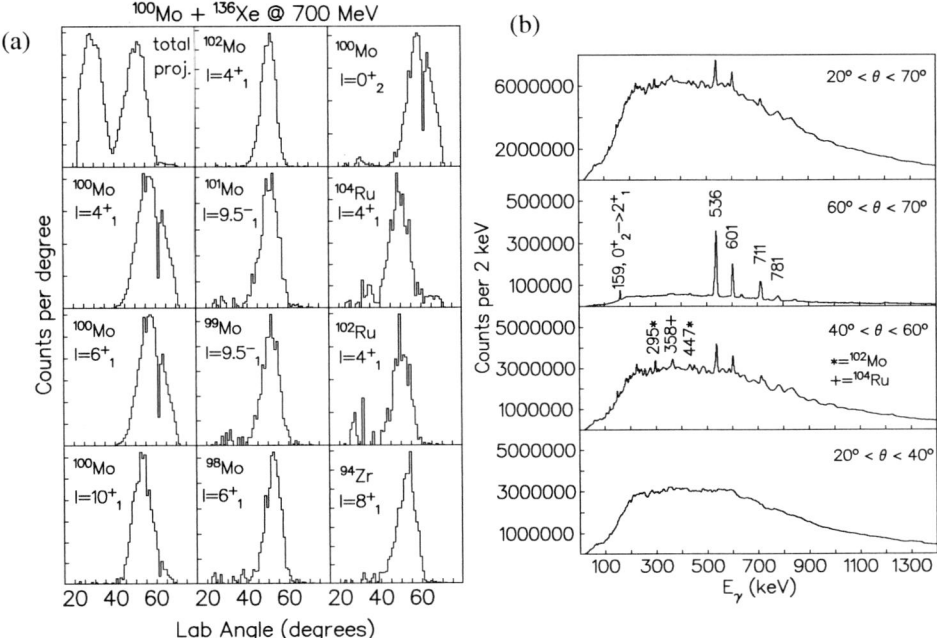

FIGURE 4. (a) laboratory frame angular distributions of gated by discrete decays in various TLFs. (b) Prompt gamma-ray total projections Doppler corrected for TLFs gated by laboratory detection angle of the binary fragment.

REFERENCES

1. J.H. Hamilton et al., *Phys. Rep.* **264**, 215 (1996); I. Hamad and W.R. Phillips, *Rep. Prog. Phys.* **58**, 1415, (1995); H. Hua et al., *Phys. Lett.* **B562**, 201 (2003)
2. R. Krücken et al., *Eur. Phys. J.* **A10**, 151 (2001)
3. P.H. Regan et al., *Phys. Rev. Lett.* **90**, 152502 (2003); P.H. Regan et al., Proceedings of the Frontiers of Nuclear Structure, eds. P. Fallon and R. Clark, AIP Conf. Proc. **656** (2003) p422
4. J. Skalski, S. Mizutori and W. Nazarewicz *Nucl. Phys.* **A617**, 282 (1997)
5. P.H. Regan et al., *Phys. Rev.* **C55**, 2305 (1997)
6. P.H. Regan et al., *Phys. Rev.* **C68**, 044313 (2003)
7. I-Y. Lee *Nucl. Phys.* **A520**, 641c (1990)
8. M.W. Simon et al., *Nucl. Instr. Meth. Phys. Res.* **A452**, 205 (2000)
9. R. Bock *et al.* Nucleonika **22**, 529, (1977)

The radioactive ion beam project SPES at LNL

A. Pisent[1] and A. Bracco[2]

[1] INFN - Laboratori Nazionali di Legnaro, I-35020 Legnaro (PD), Italy
[2] Dipartimento di Fisica and INFN sez. Milano, I-20133 Milano, Italy

Abstract. At LNL it has been proposed an Advanced Exotic Ion Beam facility, that will allow a frontier program in RIBs Physics. In particular beams of neutron rich nuclei in the medium heavy mass region will be produced. The structure properties of these nuclei are interesting also for the understanding of the astrophysical problem of the stellar nucleosyntesis. We illustrate here the concept of the SPES facility, and how this proposal is interconnected with the longer term European project EURISOL. The first part of the SPES facility is presently under construction.

INTRODUCTION

In the last years the availability of new intense radioactive ion beams has been recognized as a fundamental tool for future research in Nuclear Physics. The ISOL (Ion Separation On Line) method, represented in Europe by CERN facility [1] as its most relevant example, is complementary to the fragmentation method, presently used at GSI. For the future development GSI has presented the proposal of a major upgrading of the fragmentation facility [2], while for the ISOL method, after a preliminary feasibility study, a European design study, named EURISOL-DS [3], is under preparation. The size of EURISOL is comparable to the American project RIA [4], but it takes advantage of the complementarily with the GSI present and future capabilities.

The time scale for EURISOL is likely to be behind fifteen years from now. In the meantime an important role for RIB's physics should be played by European national laboratories, with facilities like SPES (Study and Production of Exotic nuclear Species) at LNL (Laboratori Nazionali di Legnaro). This facility, with an intermediate size between existing first generation facilities and EURISOL, will allow, together with an extensive physics program, to boost the technological development (accelerator, production targets, detectors etc) in view of the European project. Indeed at LNL the design work for EURISOL is perfectly integrated in the design and R&D work for SPES, and with the accelerator development for future high intensity linacs (TRASCO [5]). A first design report of SPES was published in '99 [6] while a detailed Technical Design Report has been published in 2002 [7]. In October 2003 the INFN has decided the construction of a first phase of SPES, corresponding to the first part of the driver linac, equipped with a neutron source for interdisciplinary applications.

THE PHYSICS CASE

The study of the nucleus at the limits of isospin, excitation energy and angular momentum has motivated during the last years an extensive experimental program, particularly with the use of large detector arrays for gamma spectroscopy. The future perspectives concerning this activity are presently under discussions and require the use of radioactive beams which will allow investigations of nuclear properties farther away from the stability line. These nuclei which do not exist on earth are instead very abundant in the Cosmos. The known elements have appeared at different stages of the history of the universe and different astrophysical processes have originated them. One of the astrophysics problem particularly complex and intensively investigated is that of the stellar nucleosyntesis (r-process) which concerns the origin of heavy elements (heavier than iron). The development in this field requires progress in experimental nuclear physics for nuclei in the neutron rich side as their properties affect the modeling of the abundance composition of the universe.

The physics program presently under discussion in connection with facility like SPES at LNL and SPIRAL2 at GANIL concerns the study of medium and heavy mass neutron rich nuclei, and it is therefore of particular relevance for the nuclear physics aspects of the astrophysics problem of the stellar nucleosyntesis.

New combinations of proton and neutron numbers are expected to provide new information on the nuclear forces. Among the questions at the forefront there are those related to the shell structure, to the microscopic origin of collective modes, the possibility of new collective modes in weakly bound systems and the mechanisms behind large-amplitude collective motion such as fission or the shape coexistence.

One of the problems currently debated is whether or not the classical shell gaps vanish for neutron rich nuclei and if new magic numbers are present in this situation. Considerable experimental effort has been devoted recently to the study of shell closures at N=40, 50 and 82 in the isotopes around $^{68}Ni_{28}$, $^{100}Sn_{50}$ and $^{132}Sn_{50}$. While the single-particle gap and 2+ excitation energy ($\approx$ 5 MeV and $\approx$ 4 MeV, respectively) could naively recall the situation similar to that of ^{208}Pb ($\approx$ 4 MeV for both quantities), pairing correlation energy (pairing gap) associated with the systems with two neutrons outside closed shell (i.e ^{134}Sn and ^{210}Pb) are approximately 0.7 and 1.4 MeV respectively. Two-particle transfer reactions are the specific tools to probe these correlations and to shed light on the apparent fact that ^{132}Sn seems to be the best available closed shell candidate to date. Absolute cross sections of the order of 0.5-1 mb/sr are expected for these reactions.

New data on the more neutron rich nuclei around the double closed shell ^{132}Sn are expected to be instrumental for the understanding of the effective nuclear interactions. In addition, information as deduced from one and two-particle transfer reactions are necessary to obtain spectroscopic factors and the pairing force. The knowledge of the spectroscopic factors, as obtained from one nucleon transfer reactions at 5-10 MeV/u, provides detailed information on the mixing of single particle states with more complicated configurations. The mixing is expected to be mainly with configurations containing a low-lying surface vibrational mode.

A related topic is the study of the spin-orbit splitting in ^{132}Sn nucleus namely that of the $\nu h_{9/2}$ and the $\nu h_{11/2}$ as well as the $\pi g_{7/2}$ and the $\pi g_{9/2}$ orbits. The spin-orbit splitting plays an important role in the binding and structure of nuclei. In fact, the introduction of spin-orbit splitting of the nucleon orbitals in the nuclear mean-field potential has been originally a landmark in the development of the nuclear shell model. Indeed, depending upon the relative orientation of the spin and the orbital angular momentum, the energy of the associated nuclear state is either pushed up or down. This effect is essential for the understanding of magic number. According to certain model predictions, the energy splitting of these spin-orbit partners should decrease or even vanish far from stability for very neutron-rich isotopes.

Among the other interesting open problems in nuclear structure that can be investigated using neutron rich RIB there is of the existence of hyperdeformed shapes. The use of fusion reactions induced by neutron rich beams is expected to favor the population of these extreme deformations at the very high spins, being the fission barrier higher in the case of neutron rich nuclei. Hyperdeformation is predicted in ^{176}Er, ^{178}Yb, ^{180}Hf that can be produced by ^{130}Cd, ^{132}Sn, and ^{134}Te induced reaction, respectively, on a ^{48}Ca target.

It is important to stress that the developments related to the accelerator systems for the production of intense radioactive beams are accompanied to those of the new instrumentation which is needed to carry out the new experimental program. In this connection it is worth mentioning that gamma-decay measurements will be essential in the investigation of the nuclear structure far from stability and that efforts are being made in a European collaboration to construct an array of new generation (AGATA [8]).

SPES FACILITY AT LNL

SPES is a new generation ISOL facility proposed at LNL, able to represent a competitive intermediate step between the existing facilities, and the longer-range high performance facility EURISOL. Between EURISOL and SPES there will be one or two orders of magnitude, in beam power on target, in the production of specific isotopes and likely in financial investment. The nominal fission rate is 10^{15} Hz for EURISOL and $5 \cdot 10^{13}$ Hz for SPES, while the beam power on target is 5 MW and 100 kW respectively.

The LNL capability to play a role in this research field is related to the presence of the superconducting linac (ALPI), able to accelerate exotic ions above 15 MeV/u, the well consolidated know-how in linac construction, the existing detectors and the related know-how. An other very important requisite for the implementation of the new facility is the real estate, available thanks to the extension of the Laboratory site (more than a factor two in area). New infrastructures (like a 40 MW power station) are being implemented.

An efficient way to produce a large variety of neutron rich isotopes is the use of the nuclear fission induced by high-energy neutrons. The fission target consists of ^{238}U, under uranium carbide form. The neutrons are produced by a primary proton beam impinging on a thick ^{9}Be or ^{13}C target. The main advantage of the two targets method is that the beam power (~ 100 kW) is dissipated (mainly through electromagnetic interactions) in the first target (converter), while the second target (production target) only withstands the fission power (few kW).

The fission rate follows the neutron production, that increases with proton energy, provided that the geometry of the target is optimized for each energy to follow the neutron spatial distribution. We chose a proton energy of 100 MeV, that allows to exceed 10^{13} f/s for a 4 Kg UCx target and a heat dissipation of 100 kW in the converter; this last value corresponds roughly to the limit for solid targets (non liquid metals). One of the most severe problems for this set of parameters is the high power density in the target, since a rather small neutron source is needed for the optimization of the RIB production target. A rotating target is therefore needed, and the option under study at LNL is the use of a ^{13}C target, which resists better to heat avoiding the toxicity of Be. The first results with a ^{13}C graphite are very encouraging.

The driver linac design based on independently phased superconducting cavities (ISCL) and with a capability of 5 mA, is the same as for the corresponding part of EURISOL project. The beam current choice is convenient for the linac, since a power level below 10 kW per amplifier can be achieved with solid state technology. Moreover an RF system of this kind is modular, so that it is proposed to implement the RF system for 3 mA corresponding to one basic RF unit per cavity, leaving the possibility for a future upgrade open.

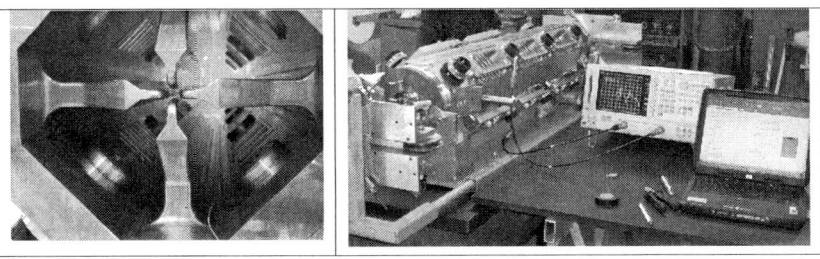

FIGURE 1. TRASCO RFQ: view of the first module before and after brazing.

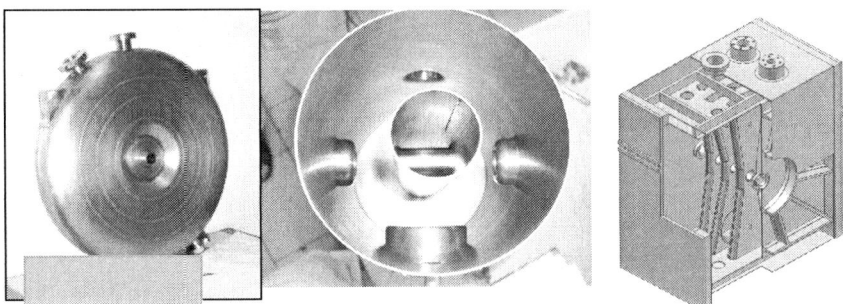

FIGURE 2. Superconducting cavities operating at 352 MHz under development at LNL for the acceleration of protons: reentrant, half wave and ladder prototypes.

The main linac components are the off resonance rf source TRIPS [9], the TRASCO RFQ [10] and the Independently phased Superconducting Cavity Linac (ISCL). The source and the RFQ have been developed within the TRASCO research program, aimed to the development of a high intensity linac for nuclear waste transmutation. The source TRIPS is being commissioned at LNS, while the RFQ is under construction at LNL (Fig. 1). The source and the RFQ, installed at LNL, will represent a unique facility, able to deliver 30 mA 5 MeV beam. The RFQ is a 7.13 m long structure composed by 6 brazed modules, fed by one klystron of LEP kind.

The following ISCL is the natural extension of the superconducting heavy ion linac technology, where LNL has a recognized leading role in Europe. Respect to the accelerator ALPI now operating at LNL [11], there will be a similar scheme, with many independent resonators working in continuous wave mode (CW), but at higher frequency and much higher beam current. As a consequence new RF amplifiers (keeping the reliability of solid state RF sources) and new cavities (with large beam hole and good field quality) are being developed. In fig.2 the reentrant cavity [12], half wave [13] resonator and ladder cavity [14] are shown. All cavities are built with using Nb sheets electron beam welded. The reentrant cavity prototype has already been built and tested up to its design field.

Finally is should be noted that a high current proton beam requires a very good control of beam losses so to avoid an unacceptable activation of the accelerator components. In particular one has to control the mechanisms able to push few particles at a large distance from beam core (the so called beam halo).

The distribution of fission products in A achieved in this scenario is more homogeneous than for a reactor and simulations show for example 2×10^{11} ^{132}Sn per second produced in target. From the UCx target heated to more than 2000 °C the fragments diffuse into the source, where they are ionized by an electron or laser beam.

A first bending magnet selects the ions with a resolution of about 1/300 in q/A and cuts the largest part of unwanted radioactive species. The charge state is therefore boosted in an ion source, of ECR or EBIS type, working in breeder mode. A high-resolution spectrometer, able to resolve isobars in specific cases, can be located at this point. A line of bypass should give the possibility to the user of choosing between higher resolution and higher transmission.

The ions are then bunched and accelerated in an RFQ section, in ALPI and finally impinge into the experimental target. Again for ^{132}Sn typically 10^8 ions/s (0.02 pnA) can be expected. A final energy of about 15 MeV/u can be achieved in ALPI, with the completion of both the low β and the high βsections. A bunching RFQ, plus a couple of superconducting RFQs, identical to PIAVE, have to be installed in the same ALPI vault.

This is not the only scheme that can be used. Indeed protons can produce exotic ions in targets different from U using spallation or direct reactions. Moreover the proton driver linac, thanks to the use of many independent superconducting cavities, is an "open accelerator". Indeed, with an upgraded version of the front-end, the same linac can accelerate ions with A/q up to three with good efficiency (~100 MeV/q). This allows the use of different production reactions, like fusion-evaporation or multinucleon transfer.

The extended capabilities of the linac would also allow the acceleration of deuterons and an increased production of neutrons so that the rate of 10^{14} f/s can be exceeded. On the other hand the acceleration of d is much more difficult for the operation of the linac, due to the neutron activation of the structure including the RFQ. Finally this high intensity ion beam can be directly interesting for experiments.

Fig. 3 shows a schematic layout of the facility, connected with the existing LNL complex. The main new construction is the driver-linac and RIB production building, west of TANDEM-ALPI complex. Integrated in the same building there will be a part dedicated to BNCT, which will follow the related specific requirements.

The three experimental halls presently used at LNL will all be reached by accelerated RIBs. Additional experimental space will be available for traps and other low energy experiments in the region of the mass selection. The upgrading of ALPI, and the installation of the new injector for RIBs, is possible in the existing vault with an upgrade of the cryogenic system and of some infrastructures. In the proposal the acceleration of ^{132}Sn up to 15 MeV/u is considered, corresponding to the installation of cryostats equipped with high performance cavities in all ALPI locations.

In a longer time prospect the proposed facility could be upgraded up to the implementation of EURISOL proposal. An upgrading of the reaccelerator is possible with an intervention in the layout of the complex (ALPI and experimental vaults). Two linac sections, 20 m each, for a final energy of 40 MeV/u (^{132}Sn), could be installed in ALPI vault and in the third experimental hall. As alternative a 100 MeV/u linac, still injected by ALPI, could be installed in a separate tunnel to be built south of ALPI vault. On the other hand the driver proposed for SPES is the initial part of EURISOL driver, and in LNL site there is space for the completion of the 1 GeV linac.

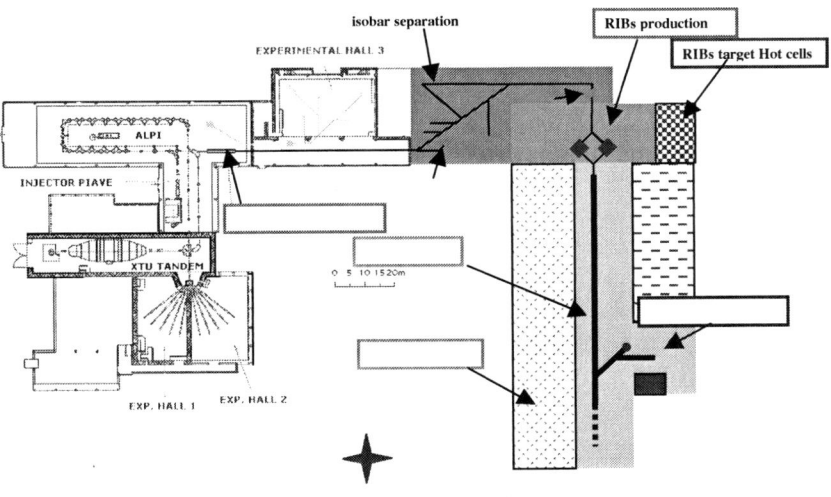

FIGURE 3. Lay out of the SPES facility and integration with the existing PIAVE-ALPI complex.

INTERDISCIPLINARY USERS

The main interdisciplinary user of the facility is the Boron Neutron Capture Therapy (BNCT). The 5 MeV, 30 mA proton beam produced by the first acceleration step, the RFQ, through (p,n) reaction on ^{9}Be and a heavy water moderator, generates the neutron flux of thermal neutrons (about 2.5×10^9/(s cm2)) necessary for patient treatment. A concentration of B in correspondence of neoplastic tissues is achieved pharmacologically, so that the α particles, emitted in correspondence of n capture, determine a very localized damage of these tissues.

In particular, at LNL the BNCT facility is foreseen to explore the cure of extended skin melanoma with this method [15]. An interdisciplinary research group, named GIRA (Gruppo Interdisciplinare Radioterapia Avanzata), formed by medical doctors, biologists, physicists, nuclear engineers, has gathered around SPES. This researchers belongs to different institutions (Istituto Oncologico Veneto, Padova Univ., Politecnico di Milano, Molteni pharmaceutics, ENEA, INFN......) with the common aim to carry out an experimental program, up to the medical application, of advanced radiotherapy topics. The first topic to be attached will be the BNCT treatment of skin melanoma, combined with phototherapy, using SPES beam. The main ingredients of our research program for are the development of the neutron beam, the molecule (boron carrier) and the microdosimetry.

The installation of the high intensity RFQ (TRASCO RFQ) makes possible the reliability and availability tests required by the ADS community in view of the future nuclear waste transmutation plants. The 30 mA, 5 MeV beam will also be available for the tests of new high intensity linac structures.

THE APPROVED INITIAL PHASE OF SPES

Taking into account the boundaries in terms of the limited resources available in Italy for scientific research, and at LNL in particular, and taking advantage from the European framework for the development of RIBs in the other main laboratories, like GSI and GANIL, it was decided to launch SPES FI (fase iniziale) that will be a first significant step in the direction of SPES and EURISOL, a very good test for the high intensity community (ADS), and will be able to serve a community of interdisciplinary physics and medical users.

The specific investment, approved by INFN in September '03, will include:
1. the completion and installation of the 5 MeV 30 mA proton injector,
2. the development and construction of the thermal neutron facility for BNCT,
3. the development and realization of the superconducting p linac up to 20 MeV,
4. the continuation of the R&D program on RIB production targets.

In fig. 4 a schematic lay-out of SPES-FI is shown, with the source TRIP, the TRASCO RFQ, and the transport line that allows to deliver the beam either to the superconducting linac or to the BNCT complex. This lay out is now being tested with extensive beam dynamics simulations.

It should be noted that respect to the SPES TDR proposal the current accelerated by the superconducting linac has been increased from 3 mA to 10 mA. This allows a better use of the intensity available from the RFQ, and makes the linac development relevant for the other high intensity applications, like the accelerator driven systems for waste transmutation.

Moreover for RIB this opens an additional possibility: a 10 mA 30 MeV proton beam hitting a thick ^{13}C or Be target would produce a neutron flux above fission threshold, and therefore a fission rate, comparable with the nominal 100kW 100MeV SPES beam; the main difference is indeed in the difficulty to develop a much more challenging converter. This implies that if the R&D program on solid converters will show that it is sensible to think to very high power density converters, LNL will be able to propose a compact and cost effective RIB source to be coupled to ALPI reaccelerator.

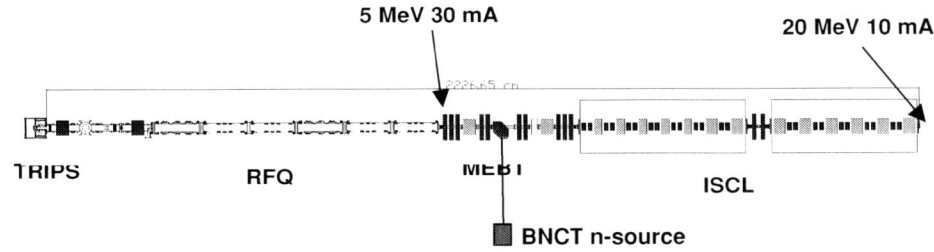

FIGURE 4. Schematic layout of SPES FI; TRIPS (TRASCO Ion Source) is the source, RFQ (Radio Frequency Quadrupole) is the first accelerating element, the MEBT (Medium Energy Beam Transfer) transfer the beam to the ISCL (Independent Superconducting Cavity Linac), composed by two cryostats, or to the BNCT (Boron Neutron Capture Therapy), the thermal neutron source dedicated to the medical experimental program.

SUMMARY

The long range ISOL facility will necessarily be on European scale (EURISOL), based on a high intensity linac as driver and a superconducting linac for the reacceleration. The choice of the accelerators was not generally accepted few years ago, but the flexibility, modularity and cost of superconducting linacs made them preferable respect to cyclotrons or other machines. Moreover the technology of the driver (and of the converter) is common to other applications like spallations sources for material science, nuclear waste transmutation and muon sources for (ν-factory).

It exists an intermediate phase with an essential role for National Laboratories, like SPES project at LNL, described in this paper, or SPIRAL II at GANIL, that is also based on a superconducting driver. There is therefore an integrated plan for the development of ISOL facilities in Europe, and good perspectives for the implementation of the relative Physics programs. The first step in Italy will be the construction of the first phase of SPES in the next five years.

REFERENCES

1. http://isolde.web.cern.ch/ISOLDE/
2. http://www-aix.gsi.de/GSI-Future/project/eng/index.php
3. http://www.ganil.fr/eurisol/index.html
4. http://www.phy.anl.gov/ria/ and http://www.nscl.msu.edu/future/ria/
5. "STATUS OF TRASCO", Internal Report INFN-TC-00-23 21 December 2000 and http://trasco.lnl.infn.it/
6. SPES Study Group "Project Study of an advanced facility for Exotic Beams at LNL" LNL-INFN (REP) 145/99 report
7. SPES Technical design report (A. Bracco and A. Pisent editors) LNL-INFN(REP) 181/2002 and http://www.lnl.infn.it/~spes/
8. contribution of D. Bazzacco to this conference
9. G. Ciavola, L. Celona, S. Gammino: "First Beam from the TRASCO Intense Proton Source (TRIPS) at INFN-LNS" Proceedings of the 2001 Particle Accelerator Conference, Chicago, June 18 - 22, 2001
10. A. Pisent, et al. "TRASCO RFQ", Proceedings of the XX International Linac Conference, Monterey, California, August 21 - 25,p. 902 (2000).
11. A. Porcellato et al., "Operation experience with ALPI resonators", SRF2001 proceedings
12. A . Facco et al. "RF Testing of the TRASCO Superconducting Reentrant Cavity for High Intensity Proton Beams" EPAC2002, Paris (2002).
13. A. Facco, V. Zviagintsev "Study on Beam Steering in Intermediate-β Superconducting Quarter Wave Resonators", Proceedings of the 2001 Particle Accelerator Conference, Chicago, June 18 - 22, 2001
14. V. Andreev et al. "Study of a Novel Superconducting Structure for the very low beta part of high current linacs" EPAC2002 p 2208.
15. S.Agosteo et al. "Advances in the INFN-Legnaro BNCT Project for Skin Melanoma". Proceedings of the international physical and clinical workshop on BNCT. Held in Candiolo (Torino) on the 17th of Febbruary 2001.

The SPIRAL2 project at GANIL

H. Savajols, A.C.C. Villari and W. Mittig for the SPIRAL2 group

GANIL, B.P. 55027 Caen Cedex 5 France

Abstract. - Based on the "LINAG Phase 1" conceptual design, a two years detailed study on a ISOL-type facility for the production of high intensity exotic beams, named SPIRAL2, has been launched. The radioactive isotope beams are produced via the fission process, with the aim of 10^{13} fissions/s at least, induced in a UC_x target either by fast neutrons from a C converter or by direct bombardment of fissile material. Fusion-evaporation residues, using heavy ions beams in different targets can also be produced in this facility. The driver, with an acceleration potential of 40 MV will accelerate deuterons (5 mA) and q/A =3 ions (1 mA) Even heavier ions will be possible at a later stage. The driver consists in high-performance ECR sources, an RFQ cavity and independent phase superconducting resonators. Further upgrade will be possible in the future.

INTRODUCTION

The systematic and very successful use of high-energy fragmentation at GANIL, (the first operational high intensity heavy ion accelerator in the 50-100 MeV/nucleon domain) for exploring the structure of nuclei far from stability, triggered the question how to proceed even further in this domain. The study of nuclei far from stability has become one of the major activities at GANIL, and is one of its domains of excellence. It turns out from the principle of production and separation using a spectrograph – the so-called in-flight method, which corresponds to the use of SISSI or LISE at GANIL – that the optimum efficiency of the process is reached when the radioactive beam has a velocity similar to that of the primary beam. This production process, however, does imply losses in intensity and/or quality of the secondary beam, which become increasingly important as the beam is slowed down. The ISOL (Isotopic Separation On-Line) method, used at SPIRAL [1] since November 2001, provides the production and separation of radioactive ion beams, with subsequent acceleration by a K=265 cyclotron CIME [2] between 1.7A and 25A MeV, thus allowing the study of nuclear reactions around the Coulomb barrier with radioactive ion beams. While SPIRAL is well suited for the production of light masses (A<80), the new project SPIRAL2 [3] will be devoted to the production of fission-like and fusion-evaporation radioactive ion beams, with subsequent acceleration by the CIME cyclotron. An important part of this project is the production of radioactive beams via fission induced by neutrons. The neutrons will be produced by the break-up of deuterons in a carbon wheel converter. The intensity of the deuteron beam (40 MeV and 5 mA) allows to reach 10^{13} fissions per second at least using a standard UC_x target of density 2.3 g/cm^3. A possible encrease of this yield can be obtained by changing the size or the density of the UC_x of the target. If one considers the high density UC_2 with 11 g/cm^3, the yield will be enhanced by the same factor, i.e. a yield of 5 x 10^{13} fissions per second can be reached. In this contribution, other uses of the driver of SPIRAL2, e.g. for production of neutron deficient nuclei via fusion-evaporation reaction or the use of in-flight production for the study of superheavies are not discussed.

The SPIRAL2 detailed design study will lasts up to November 2004 with IN2P3/CNRS, DSM/CEA and the Lower Normandy Region support. In a second phase, the full accelerator LINAG (Linear Accelerator at GANIL) could allow to accelerate light and heavy ions to energies of around 100A MeV at intensities as high as 1 mA, corresponding to an enhancement of the radioactive beam intensities of more than two orders of magnitude in comparison with the present GANIL accelerators.

PRODUCTION OF FISSION FRAGMENT RADIOACTIVE BEAMS

One of the techniques proposed for SPIRAL2 has been already discussed in the EU-RTD report [4], consisting in the use of energetic neutrons to induce fission of ^{238}U. The neutrons are generated by the break-up of deuterons in a thick target, the so-called converter, of sufficient thickness to prevent charged particles from escaping. The energetic forward-going neutrons impinge on a thick production target of fissionable material, i.e. Uranium carbide UC_x. The resulting fission products accumulated in the target diffuse to the surface from which they evaporate, are ionised, mass-selected, eventually charge breeded and finally post-accelerated. This method has several advantages and was suggested by Gerald Nolen [5] in the framework of RIA. The material of the highly activated converter can be chosen to withstand the power of the beam without constraint concerning the diffusion of radioactive atoms. Moreover, the temperature of the converter does not affect the neutron flux. As projectiles, neutrons do not contribute to the heating of the target material directly neither the entrance window, which can be very thick and do not present any special security issue. Neutrons then bombard the target loosing energy only in useful nuclear interactions and having a high penetrating power, which allows very thick targets to be used.

The choice of the Deuteron bombarding energy – 40MeV –has been made taken into consideration basically four main factors:

- the production rate of neutrons at forward angles as a function of energy,
- the angular distribution of the neutrons,
- the excitation energy of uranium, which defines the fission fragment distribution,
- the cost of the project.

The strong forward peaking of the yield of high-energy neutrons from the deuteron break-up at 40 MeV favours a compact geometry, consisting of a converter to produce the neutrons followed by a second target containing the fissionable material. The energy distribution of the neutrons produced in the deuteron break-up (that determines the excitation energy of Uranium) is centred at about 40% of the energy of the incident deuterons and has at 0° a width between one-third and half of the energy of the incident deuterons. This energy is optimal to produce neutron rich fission fragments.

The Rotating Target/Converter

During the studies of SPIRAL2 EU RTD, simulations of neutron spectra for several beam-converter configurations have been performed with the LAHET high-energy transport code combined with Monte Carlo N-Particle MCNP code for low-energy transport were performed. Deuteron Coulomb dissociation has been added to the more standard processes, the forward peaked break-up and direct reactions and the rather isotropic evaporation of low-energy neutrons [6]. A comparison with data, in the energy regions where it was available, has shown an agreement of around 30%. Three different converters were studied: lithium, beryllium and carbon.

The neutron yield is only one of the factors to be taken in consideration for the choice of the converter material. Other aspects in the evaluation are:

- Thermal properties that allow a compact geometry of the converter and the production target
- Toxicity and material properties of the converter;
- Production of long-lived radioactive nuclides;
- Cost of operation.

The conclusions we can infer, regarding these aspects, are summarized below:

A Beryllium converter produces the largest amount of neutrons. However, its low melting point (1278° C) or its location very close to the hot target does not allow high-intensity deuteron beams. Liquid lithium is a more robust converter with respect to deposited beam power than Beryllium or Carbon. However, the flow of hot liquid Lithium containing some amount of radioactive products requires special care of design especially due to safety considerations. A converter designed along the lines originally described by Grand and Goland [7] is probably not to be considered in the context of SPIRAL2, but could be of interest for a next generation facility, e.g. EURISOL since it can stand extremely high beam power.

The above mentioned properties clearly favor Carbon as converter material. It is non-toxic, easy to handle and has a high melting point of 3632 °C. These excellent properties allow high beam intensities with a rotating wheel cooled mainly by thermal radiation.

For SPIRAL2 the main parameters of the rotating Carbon wheel have been obtained by simulations using the code SYSTUS [8]. For the simulations, an infinite rotation velocity was taken, in a first approach. Once the main characteristics chosen, the temperature variation with respect of the beam impact was calculated for a real angular velocity of 1000 RPM. The thermal shock for which graphite achieves the ultimate strength (brake-down condition) is for about 50°C in our conditions, therefore a maximum temperature variation of 20-30°C has been considered.

The evaporation ratio of carbon dependents on the graphite saturated vapor pressure. Experiments performed at GANIL and IPN-Orsay [9] for a specific carbon from POCO and Carbon Lorraine industries show that the evaporation ratio of carbon is in agreement with the values found in literature. The rate for evaporating 1mm thickness of the carbon wheel in a period of 2000 hours is 2.6×10^{-7} Kg/m^{2}s. It corresponds to a temperature of 2085°C. This consideration fixes the sizes of the wheel and the beam spot on the carbon converter; i.e. 350 mm of radius for a beam spot of 10 x 35 mm.

A similar study with equal results [LOG00] has been made in the framework of the SPES project, Legnaro, Italy. The difference between both projects is that, in the latter case, the beam considered were protons of 10 MeV, with a total beam power of 100kW.

The Target And Ion Source Production System

The target and ion source production system is placed just behind the rotating carbon converter. With 1×10^{13} fission per second, the total power produced in the UC_x target is 310W. As mentioned above, the production target perceives no influence on the primary beam. The target temperature is completely controlled by an independent heating system, with a power of about 5 kW.

The Production Target

Two possibilities have been considered in this study for the fissile targets. The first one of UC_x with 2.3 g/cm^3, using the technology developed and used for many years at CERN-ISOLDE [10]. Oxide of uranium is mixed with carbon powder in a small container, pressed and heated at 2000°C during approximately one day. The chips produced have generally around 1 mm thickness and a diameter of 20 mm. Larger diameter do not seem to be a problem, but smaller thickness could be very difficult to produce. We assumed in our calculations of production yields a thickness of 1 mm. Any development in order to reduce this value would be welcome.

The second possibility is the use of high-density UC_2 target (11 g/cm^2) already developed at Gatchina – Russia. The high density UC_2 allows either to reduce considerably the size of the target or to increase by a factor of around 5 the yields for a constant geometry. Preliminary results show that the diffusion properties of high-density UC_2 are similar to the low-density. The following yield estimations were done for both possibilities.

The geometry adopted in the simulations is not the optimum. The best would be to have a UC_x targets of conical shape, with an angle of approximately of 30-40°, as proposed in ref. [11]. In order to simplify the simulations, a simple cylindrical geometry has been adopted in all cases. Moreover, a reasonable size of the UC_x target has been adopted. Therefore, the in-target yields can be considered as **lower limits** for all cases.

In the simulations the beam profile was 2 cm of diameter. The UCx fission target of 6 cm diameter is placed at 2 cm behind the converter. It consists of slices of 1 mm thickness, separated by 0.5 mm, distributed over a length of 8.5 cm, corresponding to 360 g with low-density UC_x or 1.8 kg of uranium for the high-density material. This target could be made of self-supported disks (if mechanically possible) or with a combination of several smaller targets in a suitable geometry. A prototype of both low-density [12] (1.6 cm diameter) and high-density (1 cm diameter) already exists and has been tested at the PARRNe2 and at the Gatchina on-line mass separators.

Ion Sources

The ionisation source will be installed in a module as close as possible to the target. The chemical features of the selected radioactive element will define the type of the ion source regarding its efficiency. The main methods considered are surface ionisation for alkali elements, an electron cyclotron resonance ion source for noble gases or for volatile mono-atomic or molecular elements, and a laser ion source for refractory elements. A new kind of electron beam ion source, developed at Gatchina is another attractive possibility. Correspondingly existing sources for radioactive ion production are respectively described in the references [13, 14, 15, 16].

It will be possible to install any one of the proposed sources in the production module, the ECRIS being the largest one. In all cases, the lifetime to be considered for the production system is 3 months. The replacement of the whole production system will be performed in a hot cell, specially designed for such an operation.

Charge Breeding

The charge breeding (1+/n+ transformation) will increase the charge-state of the singly charged incoming ion to a charge state compatible with the acceleration by the CIME cyclotron.

An ECR charge booster has been developed by the SSI group at ISN Grenoble (France) [17] based on the use of a Phoenix ion source at 14 GHz. The present results with this source are summarised in the table 1:

Table 1: Order of magnitude of charge breeding efficiency with stable elements.

	Charge Breeding Efficiency on the most abundant charge state	Overall Efficiency
Noble gas	10%	50 to 70%
Condensable elements	6%	45%

Other solutions could also be considered, like the use of super-conducting ion sources (SERSE or GYROSERSE) in the future.

Production Rates

Yield calculations of fission fragments with the LAHET+MCNP+CINDER code for ~5 mA deuteron beam of 40 MeV energy in a carbon converter, followed by a UC_x target has been performed.

Two UC_x densities were considered:

- Case 1: An UC_x target with a high density $\rho(^{238}U) = 11 g/cm3$ and 1 atom of U for 2 atoms of C as developed by V. Panteleev, Gatchina.
- Case 2: An UC_x target with a low density $\rho(^{238}U) = 2.3 g/cm^3$ and 1 atom of U for 9 atoms of C as used in the first PARRNe2 experiment.

The following geometry was used :

The deuteron beam of 2 cm diameter hits a carbon converter of 1.8 g/cm^3 density with an effective length of 0.7 cm, in order to stop deuterons and protons from stripping reactions. The UC$_x$ fissile target of 6 cm diameter is placed at 2 cm behind the converter. The target is composed of 56 slices of 1 mm thickness spaced by 0.5 mm, distributed over a length of 8.5 cm. It contains 360g of ^{238}U in the case of the low density $\rho(^{238}U) = 2.3$g/cm^3 and 1800g for higher density $\rho(^{238}U) = 11$g/cm^3. Effectively and for simplicity, the target was assimilated to one cylinder of 8.5cm length with density reduced by a factor 2/3 in the LAHET code. This gives the correct solid angle for neutron impact and the correct ^{238}U quantity. For such a target, 10^{13} fission/s could be obtained with a deuteron beam of 3.8mA for the low-density UC$_x$ target and of 1mA for the higher density, as resumed in table 2. The gain in the number of fission is not exactly proportional to the density ratio. This effect is probably due to high-energy neutrons. These neutrons are produced in the converter and are probably mostly absorbed in the higher U density target. A more important proportion of lower energy neutrons are directly produced inside the UC$_x$ target from fission of U, corresponding to a neutron source inside the target, and are then less sensitive to the effective length of the target. For short living isotopes, a shorter high-density/reduced-volume UC$_x$ target equivalent to 25 mm length will produce 10^{13} fission/s and will ensure faster diffusion out of the target.

Table 2: Required primary beam intensity in order to obtain 10^{13} fission/s and total number of fission for 5 mA deuterons for 2 different densities.

Target density	2.3 g/cm^3	11 g/cm^3
Intensity for 10^{13} fission/s	3780 µA	947 µA
Fission per 5 mA	1.32 x 10^{13}	5.28 x 10^{13}

Release Of The Products

Theoretically, the diffusion, effusion and ionisation processes are well known phenomena, provided the particular properties of the selected element are known. However, despite of a large experimental and theoretical effort, the behaviour of diffusion and effusion of ions implanted into some materials, including the details of their thermal transport or their trapping in the temperature range relevant to ISOL system, remain unknown. In particular, the Arrhenius diffusion coefficients are measured for numerous elements mainly in W, Ta, Re or C-matrix. To our knowledge, these coefficients are not known for different tracers in uranium carbide matrices of different densities. A European RTD project "TARGISOL" N° HPRI-2001-50063 has been proposed and accepted recently in order to progress in this critical field.

The expected radioactive beam intensities (after diffusion, effusion, ionisation and acceleration) are shown in the plots of Fig. 1 for some elements: Zn, Kr, Sr, Sn Sb and Xe. The in-target production yields are those calculated using the 11 g/cm^3 UC$_2$ target in the geometry described in section 2.3. The Arrhenius coefficients used in this calculation were supposed to be the same as for C and Ta, both tabulated in the literature. The assumed 1$^+$ and 1$^+$/N$^+$ ionisation efficiencies are adopted as 90% (1$^+$)

and 12% ($1^+/N^+$) for Kr and Xe, 30% (1^+) and 4% ($1^+/N^+$) for Zn, Sr, Sn, I and Cd. The assumed acceleration efficiency in the CIME cyclotron is 50%.

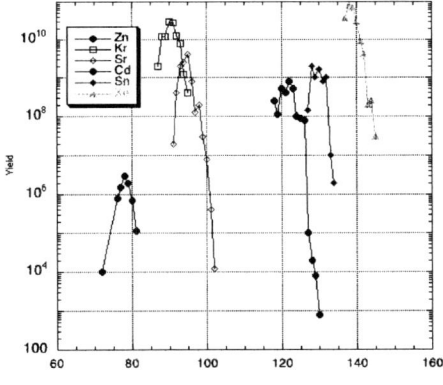

FIGURE 1. Yield after acceleration. See text.

Radio-Protection And Target Handling

The dose rate just after the stop of the primary beam is estimated to be 32 Sv/h at 1 m from the target, after an irradiation time of 3 months. One month later, the radiation rate is still 34mSv/h. This high level of radiation does not permit manual intervention on the target itself. Therefore, automatic handling is necessary. The high level of radiation also imposes strong protection of critical parts of the production system, like insulators, o-rings, etc.

The Plug Solution

The target and the ion source will be placed in a parallelepiped module (called plug) which is surrounded by at least 2 m of concrete and iron shielding. The same principle has been applied on the ISAC facility at TRIUMF (Vancouver, Canada), for 100 µA of 500 MeV protons. The 2 m thickness of concrete reduces the dose rate on the top of the plug down to 7.5 µSv/h when the beam is stopped, allowing people to come and work on the equipment located at this place.

After three month of irradiation, the different connections of the plug (electrical connections, primary pumping, water connections, etc.) can be manually removed. The plug is then remotely removed from the production cave and is evacuated to a storage cell. After a delay and depending on the state of the target ion source system, the plug can be re-used or moved to a hot cell where the elements at the bottom of the module can be replaced with master/slave manipulators.

THE DRIVER ACCELERATOR

The proposed LINAG1 driver for the SPIRAL2 project has the capability to accelerate a 5 mA d$^+$ beam up to 40 MeV; nevertheless, the different parameters are optimised for q/A=1/3 ions up to 14.5A MeV in order to preserve a long term evolution towards an heavy ion driver. It is a continuous wave (CW) mode machine, getting a maximum efficiency in the intensity transmission for heavy ions. It consists in an injector (ECR source + radio-frequency quadrupole), which accelerates the beam up to 0.75 A.MeV, followed by a superconducting linear accelerator based on independently phased resonators.

The RFQ Injector

The RFQ must operate in a CW mode. Its frequency has been chosen equal to 87.5 MHz, sub-harmonic frequency of 350 MHz. This quite low value has been determined for the following reasons:
    - the RF power density is quite low at this frequency, and allows a solution based on a formed metal technology, leading to a cheap mechanical solution.
    - at lower frequency, the inter-vane distance is larger, and allows a higher margin for the mechanical tolerances.

The RFQ output energy, 0.75 A.MeV, has been preliminarily determined by the fact that the first cavities of the SC Linac must allow a possible evolution of the machine for q/A = 1/5 or 1/6 ions, which means that their beta value has to remain quite low ($\approx$ 0.04). This value should already be optimised during the detailed study phase.

Table 3 presents a summary of the main design parameters. In particular, the maximum peak field value is kept to a conservative level, lower than LEDA and Chalk River RFQs, which also work in a CW mode.

Table 3: Main design parameters of RFQ-LINAG

Parameters	Values
Length	5 m
Aperture	8 - 10 mm
Modulation (m)	1 - 2
Frequency	87,5 MHz
Voltage	100 -113 kV
Transmission (q/A = 1/2 and 1/3)	99.9 %

Error simulations have been performed, considering mechanical tolerances of +/- 0.1 mm on the vanes machining and +/- 0.2 mm for misalignments. The results confirm that the deuteron beam transmission remains very close to 100%. This gives a quite comfortable situation from the safety point of view: losses of 3% have been considered for the estimation of the protections.

The Superconducting Linear Accelerator

The Linac must have the capability to accelerate D^+ and $q/A=1/3$ ions with the maximum energy gain, as well as to extend its performances to heavier ions in the future. A Linac based on independent phase superconducting resonators is thus proposed. The Linac design requires accelerating voltages of the order of one MV/cavity and two beta values, around 0.07 and 0.14, at sub-harmonic frequencies of 350 MHz (availability of power sources). The starting frequency is 87.5 MHz, not too high for the lowest beta cavity and not too low for the RFQ. Low beta super-conducting (SC) cavities in the beta range 0.04 to 0.2 are typically quarter wave resonators (QWR), operated at 4.2 °K as the frequency is less than 500 MHz.

The beams dynamics calculations have been performed with the codes TraceWin, PARTRAN and LIONS. Emittances behaviour in the Linac is shown in Figure 2. All these codes are able to use 3D field maps. Meanwhile, it is important to notice that the steering effect, induced by QWRs, is not included in the simulation shown by the Figure 2. This problem is studied at present time using maps computed by the code SOPRANO. The first studies shows that vertical displacements are sufficient to reduce the emittance growth induced by the dependent phase deflection for the first family. For the second family, HWR may be used as a fall back solution if steering of 176 MHz QWR is too important.

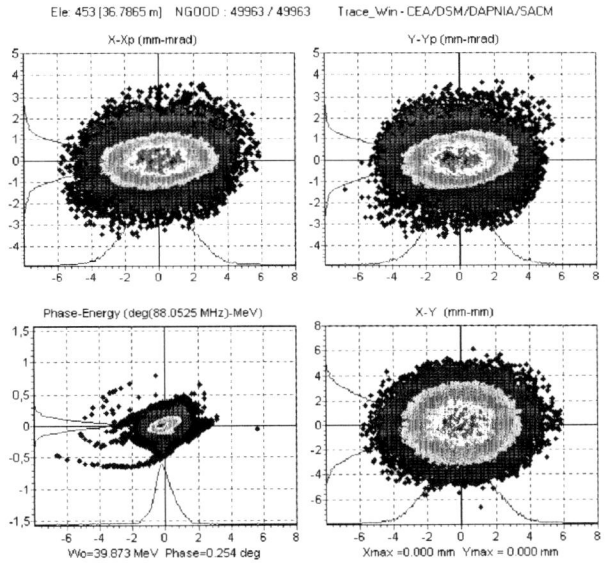

FIGURE 2. Deuterons beam in phase space at Linac exit.

POST ACCELERATION

The recently commissioned cyclotron CIME will perform the post acceleration in the SPIRAL2 project. It allows acceleration of heavy ions in the energy range of 1.7A MeV and 25A MeV, depending on the q/A. For fission fragments and considering presently performances of the charge booster, optimal energies would be of the order of 8A MeV.

SUMMARY

In the present report we have described the technical aspects of the SPIRAL2 project and commented more precisely the aim of adding medium-mass nuclei to those available in the present GANIL. Fission induced by light particles (e, p, d, etc.) was proposed to produce the radioactive ions, **with an aim of 10^{13} fissions/s at least**.

We have shown here that the SPIRAL2 project can reach even higher fission rates using proven technologies. Using a C-converter and a 5 mA deuteron beam, neutron-induced fission will be $1 \cdot 10^{13}$ fissions/s using standard-density UC_x, and 5×10^{13} fission/s for high-density UC_2. For both cases, a very small volume (240-cm^3) ion source was selected, in order to have relatively fast diffusion-effusion times for short-lived nuclei. In principle, larger volumes could result in even higher fission rates, up to 2×10^{14} fissions/s, for which the heat produced by fission in the ion source reaches 6 kW, the present limit for the SPIRAL targets.

The linear accelerator as a driver in this project belongs to the technology of high intensity accelerators, which are of strong current interest for various domains and are a domain of rapid technological development.

As a consequence of the high production rates, the radioprotection constraints become a major factor in the project. This implies a change of technology compared with SPIRAL, with higher costs for the target/ion-source and associated infrastructure. The technology of target "plugs", as used at TRIUMF-ISAC, was chosen. It offers the guarantee of safe handling of the high levels of activity produced.

The SPIRAL2 project is part of a multi-beam policy of GANIL. Another aspect is its possible synergies with EURISOL. We note that the most promising possibility examined for EURISOL for both driver and post-accelerator is a superconducting Linac, precisely the same technology for the Linac as proposed for the SPIRAL2 driver. These two machines could in fact be one and the same, by adding an appropriate RFQ fo higher M/Q ratios. The proposed (final) LINAG driver should be able to accelerate ions up to mass 100 to energy of 100A MeV.

Even though the EURISOL proposal considers MW beam power, the experience with the 200 kW beams proposed for SPIRAL2 would be very relevant. A demonstrated competence in these areas would be a big advantage to any laboratory proposing to host such a facility.

REFERENCES

1. A.C.C. Villari, the SPIRAL group, Nucl. Instr. Meth. B204 (2003) 31.
2. E. Baron, 14th Int. Conf. Cycl. Applic., Cape Town, South Africa (1995), Word Scientific.
3. LINAG Phase I, a technical report, version 1.3, GANIL, June 27, 2002, GANIL R 02 08.
4. M.G. Saint Laurent et al., SPIRAL phase II European RTD report, GANIL R 01-03 2001.
5. J. Nolen, "A target concept for intense radioactive beams in the 132Sn region", Proc. Third Intern. Conf. On Radioactive Nuclear beams, ed. J. Morrissey, East lansing, Mi, May 24-27, 1993.
6. D. Ridikas thesis: Optimisation of beam and target combination for hybrid reactor systems and radioactive ion beam production by fission, GANIL T99-04, 1999.
7. P. Grand and A.N. Goland, Nucl. Instr. Meth. 145 (1977) 49.
8. SYSTUS code, www.esi-group.com/Products/Systus/index_html.
9. J.C. Putaux et al., Nucl.Instr. Meth., B126 (1997) 113.
10. [P.V. Logatchev, L.B. Tecchio et al.; Graphite neutron target for exotic beams, SPES internal report, Legnaro 2000.
11. D. Ridikas, W. Mittig and A.C.C. Villari, Nucl. Phys. A701 (2002) 343c.
12. Roussière, B. et al., Release properties of UCx and molten U targets preprint IPNO-DR-2002-002. - 2002.
13. J-L.Biarotte, IPNO, Private Communication.
14. J.Lesrel, IPNO, Internal Note: Etude préliminaire des systèmes RF de LINAG 1 (2003).
15. L.Dalesio, EPICS: Recent applications and future directions, Proceedings of the 2001 PAC, Chicago, 2001.
16. N. Chauvin, thesis, Université Joseph Fourier Grenoble (2000).

Status Of The EXCYT Facility at INFN-LNS

G. Cuttone, R. Alba, L. Calabretta, L. Celona, F. Chines, L. Cosentino,
P. Finocchiaro, A. Grmek, S. Gammino, M. Menna, G. E. Messina,
G. Raia, S. Passarello, M. Re, D. Rifuggiato, A. Rovelli, S. Russo,
G. Schillaci, V. Scuderi, E. Zappalà

Istituto Nazionale di Fisica Nucleare, Laboratori Nazionali del Sud
Via S.Sofia, 44 . 95123 Catania Italy

Abstract. The EXCYT facility (EXotics with CYclotron and Tandem) at the INFN-LNS is based on a K-800 Superconducting Cyclotron injecting stable heavy-ion beams (up to 80 MeV/amu, 1 eμA) into a target-ion source assembly to produce the required nuclear species, and on a 15 MV Tandem for post-accelerating the radioactive beams. Since December 1999 the Superconducting Cyclotron operates in a stand-alone mode by means of the new axial injection beam line. The primary beam line has been already mounted and tested. The part of mass separator on the two high-voltage platforms together with low intensity diagnostics is already installed while the ancillary items along with the part of mass separator at ground potential will be installed during the next stop of accelerator operations. The target-ion source unit has been successfully tested on-line at GANIL. The goal of such efforts will be represented by the test of the mass separator with stable beams planned at LNS by the end of the year. The commissioning of the EXCYT facility is foreseen in 2004 together with the start of nuclear experiments program.

INTRODUCTION

In order to complete the EXCYT facility [1, 2], a four-months stop from September to December 2003 has been planned. During this period no beam will be delivered to the users to allow the installation of the mass separator and of the ancillary equipment. In addition, several minor tasks already started at the end of 2002 will be completed. During May 2003 the Target-Ion Source assembly (TIS) was successfully tested at GANIL under the same operational conditions that will be initially used at EXCYT. To reach such an important achievement, the TIS had to be modified according to the constraints of the SIRa test bench as described in the following. The commissioning of the Superconducting Cyclotron (CS) as a primary driver started in December 2002 and continued in the first half of 2003. Big efforts were made in the development of the cooled version of the electrostatic deflectors (DEFL) that have been routinely working since last December. The low-energy beam diagnostic devices were positively tested on the Tandem pre-injector platform. Following suggestions by the Referees' and the LNS Research Division, we decided to produce ^{8}Li as the first EXCYT radioactive beam (primary beam ^{13}C, ^{18}O or ^{15}N). This choice also takes in account the availability of Magnex in 2004 as well as the requests and the first results obtained by the Big Bang collaboration. The request to increase the authorised primary-beam intensity from 100 to 500 Watt has been sent. We are confident to get the final authorisation from the national control office (ANPA) within February 2004. We also would like to remark that in the first seven months of 2003 a very intense experimental activity has been successfully carried out both by means of both accelerators, CS and Tandem.

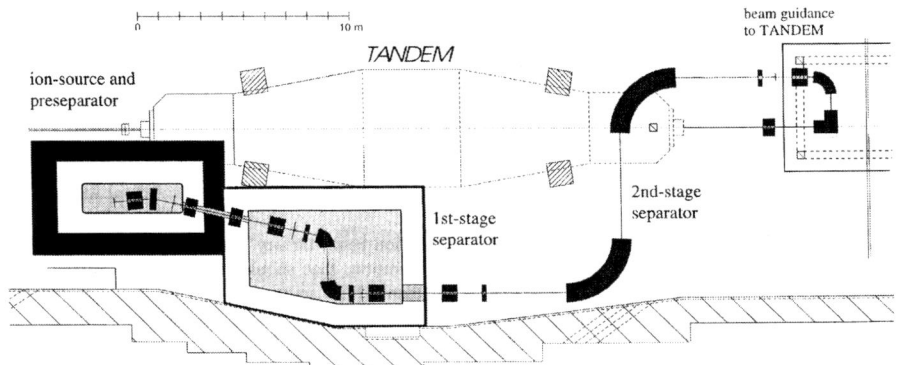

FIGURE 1 Geometry of the EXCYT mass separator.

RESULTS AND STATUS OF EXCYT

Superconducting cyclotron

During last year we improved the performance of the CS [3] by optimising the efficiencies for the injection, the acceleration and the extraction. The ECR sources and the axial injection line were routinely in operation fulfilling the EXCYT requirements without any significant troubles.

At the end of 2002 a water-cooled DEFL E1 was installed in the CS to make the high-intensity test. The first beam tests were performed with $^{20}Ne^{7+}$ accelerated to 45 MeV/u and phase slits to obtain an extraction efficiency of 50%. Tuning was preliminarily done at low intensity (< 5 Watt) by using a chopper system in the injection beam line; its purpose was to decrease the average beam current without changing the number of particles inside a bunch. With this method, it was possible to gradually increase the injected beam without modifying the transverse phase space.

The maximum extracted beam power was 25 W (200 enA). Beyond this value a significant instability of the DEFL caused serious beam fluctuations, while it was ascertained that the RF system and the inflector were not affected by the beam. Therefore it was decided to improve the DEFL features by using a new septum with a V-type notch at the entrance and a decreased thickness around the median plane in order to decrease the beam power dissipated on it.

In this way in April 2003 we easily obtained 40 W (320 enA) of extracted beam current. For higher beam current the expected extracted value was observed only for a short time, and then it quickly dropped down to the previous maximum level of about 320 enA. These effects are not yet completely understood and therefore more investigation is needed. The visual inspection performed after the beam tests showed the septum of the first DEFL to be damaged at the entrance; there was also a hole close to the end of the first section, which clearly indicates that part of the beam has been stopped by the septum and caused its melting.

As a consequence of this result, we decided to install next autumn water-cooled current probes in the CS and after the deflectors; they will allow understanding where and how the beam loss occurs. New tests have to be performed to improve the cooling efficiency of the deflector and we will develop fully stripped ^{13}C and ^{18}O beams in order to avoid probable effects of beam stripping due to high partial pressures in the DEFL gap. Finally, we will check the beam trajectories in order to adjust the deflector shape to the extraction path.

Primary beam line

During the high intensity tests we spent some time to transport the beam along the primary beam line. Despite the diagnostic devices were not fully operational, the beam was delivered just before the 90° magnet to the TIS. The transport efficiency was not measured but the test gave indications on the small adjustments to do on the hardware and on the control system for the final commissioning. Before end of 2003, according to the experience gained so far, the primary beam line will be tested in terms of transport efficiency.

Target-ion source assembly

During last year most of the developments scheduled in the agenda for the TIS were carried out.

The Positive surface Ionisation Source (PIS) was successfully tested in autumn 2001; it is easy to operate and reproducible. Tests confirmed its good quality as already observed at CERN-ISOLDE. The measured off-line intrinsic efficiency for ^{7}Li was greater than 70%, in agreement with values determined from on-line experiments at ISOLDE. The efficiency tests were carried out by implanting $2.76*10^3$ C of ^{7}Li^{3+} (46 MeV) into a graphite target, then inserted in the TIS unit. Further tests with ^{6}Li implanted into W and graphite specimens are in progress. The Negative surface Ionisation Source (NIS) was positively tested in March 2002 for the ionisation of chlorine; however, attempts to determine its ionisation efficiency with chlorine salts failed. Efficiency tests with ^{19}F and ^{37}Cl ions implanted into W and LaB$_6$ specimens are in progress. We obtained from the Hot Plasma Ion Source (HPIS) the ionisation of argon gas with a maximum stable and reproducible current higher than 30 µA. The Kinetic Ejection Negative Ion Source (KENIS), developed at HRIBF, was design and adapted to our target. This modification took longer than expected from our counterpart and the first item will be delivered within 2003. The graphite targets were redesigned and the material was carefully chosen with an high open-closed porosity ratio. Currently they are in stock and the first version of EXCYT target is ready. A slightly modified version has been successfully tested last May at GANIL. The room for the test bench was available only in September 2002 and the services (water and mains) were updated in March 2003. Magnets were delivered in July 2003. The completion is expected before the autumn of 2004.

For the EXCYT commissioning we plan to use the NIS and the KENIS for the negative ionisation of respectively Cl and F. The PIS will be used for positive ionisation of Li, provided that the Charge Exchange Cell (CEC) available in December 2003 will satisfy our requirements.

TIS test at GANIL

The previous TIS assembly was radically modified in its shape, dimensions and geometry. In 2003 a new TIS was tested to verify its behaviour with respect to the mechanical and thermal stresses: we obtained an excellent performance for temperatures up to 2000 °C. Though the first results were encouraging, several other questions had to be answered before on-line operation i.e. if the TIS was able to withstand 400 W of beam power and to produce radioactive ions with the requested efficiency. To answer these questions in a reasonable lapse of time, we asked and obtained 18 Shifts of ^{13}C (60 MeV/amu) from GANIL at the SIRa test bench. The ^{13}C primary beam produced ^{8,9}Li which thermally diffused from our graphite target to the PIS where it was ionised with charge 1^{+}. The primary beam energy was chosen so that more than 60% of the produced Li was stopped close to the surface-end of the target. Since the primary beam at SIRa is horizontal, we had to redesign and test the interface between our assembly and SIRa. This implied rotating by 90° the target, its container and its heater as well as some changes in the mechanical support system, in the thermal and in the electrical insulation.

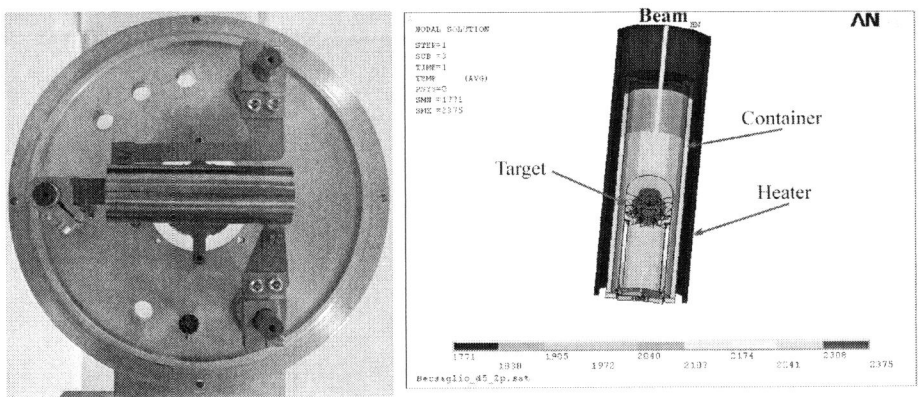

FIGURE 2. a)(left) TIS assembly in the version for GANIL test without the Ta screens. b)(right) cross-section of the TIS model used for the ANSYS simulations (Beam Power: 360 W, Heater Temperature: 1773 K).

In order to have a better understanding of the mechanical and thermal stresses under the simultaneous influence of primary beam and heating power, simulations were carried out with ANSYS and SRIM 2003 codes. The temperatures in the experimental conditions were estimated with a good approximation (± 50 K Standard Deviation) and we can conclude that in working conditions at GANIL we exceeded the required operational target temperature of 2000 K. From the mechanical stresses point of view it turned out that the target was always well below (~10 MPa) its tensile strength (69 MPa), therefore even under higher beam powers the target does not break but sublimates. Under high-power beams (200-400 W) the graphite vapour pressure is higher than 10^{-5} mbar: this leads to a graphite deposition (carbonisation) in the Ta container, thus reducing the target temperature of about 40-60 K. In the simulation we modified the target slope (from 30° to 0°) but this did not significantly affect any mean temperature in our working conditions. We finally checked that the simulation was stable (same results) also when increasing the mesh.

We mounted, outgassed and calibrated our TIS at SIRa and the experiment run in May 2003 was completely successful. In short, our TIS proved to be robust and able to withstand 370 W of beam power even after several accidental fast thermal cycles from 2000 °C to room temperature. The production efficiency of ^{8}Li and ^{9}Li is defined as the ratio between the number of extracted radioactive ions and the number of atoms produced by nuclear reactions in the target core, the latter quantity being estimated by the EPAX code. During the experiment the efficiency increased by passing time; this is mainly due to three factors: the target contained less impurities; the increase of the primary beam power; the increase of the heating temperature. The latter two factors are correlated and currently we are trying to estimate the influence of the second factor upon the third one.

The maximum efficiencies were 0.48% for ^{8}Li and 0.063% for ^{9}Li; these values were limited by the maximum available power heating. The efficiency obtained by GANIL assemblies with ^{36}Ar 95 MeV/amu as primary beam is higher (1.2% for ^{8}Li), but one should bear in mind that the error on the efficiency (coming from the uncertainties in the EPAX code) is about a factor two. Clearly the TIS performance can be still improved but we are confident to reach this goal during routine on-line operation in the near future. This experiment also demonstrates that our TIS fulfils the EXCYT requirements, considering that we were able to produce $6.4*10^6$ pps of ^{8}Li and $1.5*10^5$ pps of ^{9}Li, well sufficient to conduct the first RIB experiments at our facility.

Front-end, HV platform and isobaric mass separator

The front-end has been installed; tests of the positioning and centering system of the electrode source have been carried out while the final part the vertical line is under installation. The driving system and the electronic controls to be mounted on the robot are ready as well as the atmospheric pressure control system (including the outlet gaseous contaminants).

The HV platforms was installed since 1999. Part of the mass separator together with low intensity diagnostics, definition slits, Faraday cups are already installed while the ancillary equipment, the part of the mass separator at ground potential and the isobaric mass separator will be installed during the next stop of accelerator operations. We expect to be ready before end of 2003 for the commissioning with stable beams.

Computer control system

Due to the accelerator activity that exclude operations in the installation area, the main part of the work was devoted to design, construction and laboratory tests. Installation and final commissioning will be performed in the second half of 2003. According to the time schedule, the activity has been oriented to the beam diagnostics for the HV platform, in particular: LEBI for low intensity beams; Faraday cups; slits, collimators and actuators; moving systems DC and STEP; control systems of the power supplies for the HV platform, of the radioactive source, of electrostatic elements, of power supplies for the low energy radioactive beam line. Moreover we developed the local bus inside the HV platform for the beam diagnostics and we upgraded: the computer control LAN to include the new local stations; the local interfaces for the new local stations; the remote interfaces in the main console of all the previous items. These systems are completely tested while the installation and debugging will take at least four months.

Some other systems are partially defined or completely designed: the beam loss monitors for the primary beam line; the security system for the primary beam line; the low intensity beam diagnostics for the high energy radioactive beam line; the control system of the magnetic elements of the spectrometer. This represents the most important part of the activity after the installation of the TIS. The whole system will be ready before spring 2004.

Beam diagnostic

In view of the implementation of a beam-imaging device capable of detecting stable and radioactive low-energy beams (LEBI) [4], we proved that its sensitivity is high enough to display the two-dimensional profile of very low current / low energy beams.

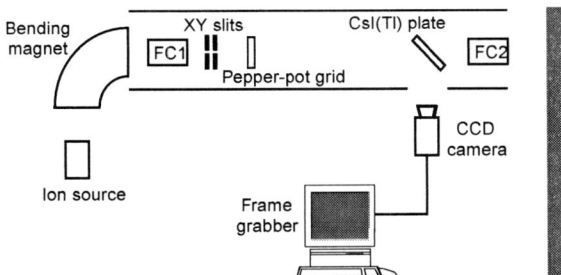
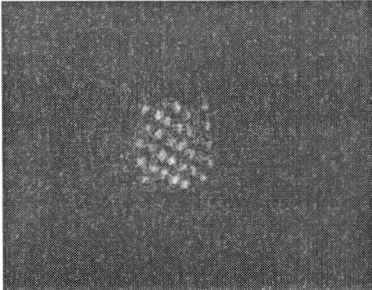

FIGURE 3. a)(left) Sketch of the experimental set-up; b)(right) $^{16}O^-$ beam (0.2 pA, 50 keV) collimated through the pepper-pot grid

In order to test it, we have produced useful beams by employing the ion source of the LNS Tandem accelerator, which can be operated down to 40 kV, spanning the ions from hydrogen to silver [5]. A sketch of the set-up is shown in fig. 3a: downstream the 90° bending magnet there was a standard quality Faraday cup (FC1) for the ion source mass analysis. Then there was a pair of collimators slits, followed by a removable pepper-pot grid for intensity reduction factor of ≈ 0.12. Following this beam attenuator there was our sensor and a high sensitivity Faraday cup (FC2).

The sensor basically consists of a thin CsI(Tl) crystal plate watched by means of a CCD video-camera. In order to prevent the charge-up of the plate, which would scatter away the incoming low-energy ions, we placed a ground-connected grid of thin tungsten wires just on the crystal face (1 mm pitch). The tests were performed in a ion energies range of 40-170 keV; we easily detected corresponding well defined images, with intensities going down to the order of 100 fA ($\approx 10^5$ particles per second) for light ion beams. In table 1 we report a few typical low-intensity beam cases.

TABLE 1. Low intensity beam tests

Mass [u]	Ion	Energy [keV]	Beam Current (FC2) [pA]	Image on CsI(Tl) Scintillator
1	H	170	~0.03	yes
16	O	50	~0.2	Yes (figure 3b)
15	F	170	~0.2	yes
58	Ni	170	~0.2	yes
109	Ag	170	~0.5	yes

ACKNOWLEDGMENTS

We gratefully acknowledge for their help and suggestions: A. Amato, G. Cacopardo, C. D'amato, G. Deluca, G. Gallo, D. Garufi, P. Maniscalco, R. Marletta, S. Marletta, A. Maugeri, G. Panascì, M. Piscopo, S. Tringale, F. Tudisco, A. Seminara, A. Varisano from LNS. Special thanks to A. Villari, M. G. Saint-Laurent and the SIRa group at GANIL for their strong support.

REFERENCES

1. G. Ciavola, D. Rifuggiato, H. Weick, M. Winkler, H. Wollnik, *Nuclear Instruments and Methods in Physics Research* **B 126** (1997) p. 17-21.
2. G. Ciavola, R. Alba, L. Calabretta, L. Celona, G. Casentino, G. Cuttone, P. Finocchiaro, S. Gammino, M. Menna, R. Papaleo, G. Raia, D. Rifuggiato, A. Rovelli, M. Silvestri, D. Vinciguerra, M. Winkler, *Nuclear Physics* **A 701** (2002) pp. 54c-57c.
3. D. Rifuggiato, L. Calabretta, G. Cuttone, *Nukleonika* **48**(s2), (2003) pp.131-134.
4. S. Cappello, L. Cosentino, P. Finocchiaro *Nuclear Instruments and Methods in Physics Research* **A 479** (2002) p. 243-253.
5. L. Cosentino, P. Finocchiaro, *Nuclear Instruments and Methods in Physics Research* **B 211** (2003) p.443-446.

Heavy ion radiative capture: ^{12}C(^{12}C,γ)

D.G. Jenkins[*], B.R. Fulton[*], J. Pearson[*], C.J. Lister[†], M. P. Carpenter[†], S. Freeman[†], N. Hammond[†], R.V.F. Janssens[†], T.L. Khoo[†], T. Lauritsen[†], E.F. Moore[†], A.H. Wuosmaa[†], P. Fallon[**], A. Görgen[**], A. O. Macchiavelli[**], M. McMahan[**], M. Freer[‡] and F. Haas[§]

[*]*Department of Physics, University of York, York, YO10 5DD, UK*
[†]*Physics Division, Argonne National Laboratory, Argonne, IL 60439*
[**]*Nuclear Science Division, Lawrence Berkeley National Laboratory, Berkeley CA 94720*
[‡]*School of Physics and Astronomy, University of Birmingham, Birmingham B15 2TT, UK*
[§]*IReS, Centre National de la Recherche Scientifique, F-67037 Strasbourg Cedex 2, France*

Abstract. The heavy ion radiative capture reaction ^{12}C(^{12}C,γ) has been investigated around a beam energy of 16 MeV. The total cross-section has been measured with the Fragment Mass Analyser at Argonne National Laboratory and found to be somewhat larger than has previously been measured. A subsequent measurement with the Gammasphere array has shown that a considerable proportion of this extra cross-section relates to a highly non-statistical decay through high-lying states in ^{24}Mg

Radiative capture is a common and well-understood process for light nuclei. By contrast, far less is known about the process of radiative capture involving heavy ions. This is unsurprising given the fact that the cross-sections involved are very small and particle emission completely dominates in such systems. The existing knowledge is mostly confined to the ^{12}C(^{12}C,γ) and ^{12}C(^{16}O,γ) systems; a comprehensive review of heavy ion radiative capture (HIRC) up to the 1980s has been presented by Sandorfi [1].

Sandorfi and Nathan [2, 3], and Dechant and Kuhlman [4] demonstrated the direct relationship between the ^{12}C+^{12}C entrance channel and the fused ^{24}Mg compound. Sandorfi employed a large single NaI detector to detect the high energy gamma rays representing direct capture to the first few low-lying states in ^{24}Mg. The peak cross-sections for capture to individual excited states were of the order of 20 nb/sr [1]. The origin of these high energy γ rays was attributed to a coupling to the giant quadrupole resonance strength in ^{24}Mg. Since a single sodium detector was employed in their work, the piling up of low energy gamma rays from particle emission channels in the detector, permitted the observation of only the very highest energy gamma rays to low-lying excited states in ^{24}Mg. This is unfortunate since it did not allow an investigation of a 'doorway' mechanism involving high multiplicity decays passing through high-lying ($E_x > 5$ MeV) states in ^{24}Mg. Such doorway states could, in principle, mediate a large fraction of the total decay width of the capture resonances. Given that the capture resonance has a large deformation ($\varepsilon \sim 1$) associated with a carbon-carbon molecule, the selected doorway states might be states with a large associated deformation, which have a greater overlap with the capture state. A good candidate for such doorway states is the so-called shape isomeric band, whose band is predicted within a number of theoretical prescriptions to be around 10 MeV e.g. refs [5, 6]. Such a structure is yet to be observed

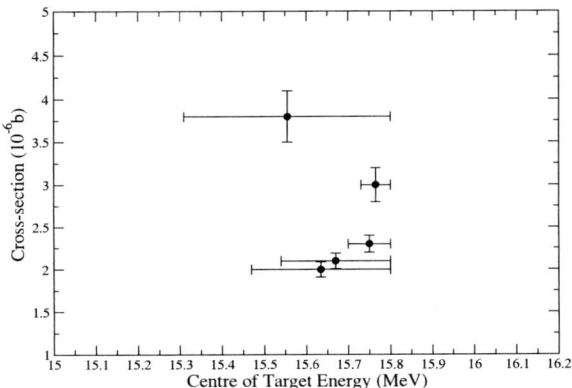

FIGURE 1. Measured cross-sections for $^{12}C(^{12}C,\gamma)$.

experimentally.

In order to determine whether the total capture cross-section was larger than that inferred from earlier measurements of high energy capture gamma rays, an experiment was performed using the Fragment Mass Analyser (FMA) at Argonne National Laboratory to detect ^{24}Mg residues following the $^{12}C(^{12}C,\gamma)$ reaction. We elected to investigate an energy region ($E_{beam} \sim 16$ MeV) where capture resonances had previously been reported by Sandorfi and Nathan [1].

A ^{12}C beam accelerated to 15.8 MeV by the ATLAS accelerator at ANL was incident on thin self-supporting carbon foils with various thicknesses. The FMA was employed to separate fusion residues from the primary beam and to disperse them by mass/charge (M/q) at the focal plane. At the focal plane of the separator the residues passed through a parallel-plate avalanche counter (PPAC), and into an ion chamber, containing isobutane at a pressure of 1.5 torr, and were subsequently implanted into a thick silicon detector at the focal plane. The higher energy loss of the fusion residues in the ion chamber gas allowed them to be cleanly discriminated from scattered beam particles in a plot of energy loss (ΔE) versus energy (E) deposited in the silicon detector. The FMA was set up to focus ^{24}Mg recoils corresponding to radiative capture events, close to the centre of the focal plane. These recoils had a mass/charge ratio, A/q=24/5, and a recoil energy of around half the centre-of-target energy. Residues with different values of A/q were found to be well separated at the focal plane due to the low masses involved. Residues with A=23 were removed by closing physical slits at the focal plane. Calibration of the energy and time of flight verified that the selected residues had A=24; this identification being confirmed by the detection of 1368 keV γ-rays, correlated with the selected recoils, by a single germanium detector at the target position. Cross-sections were derived on the basis of the expected FMA efficiency and are presented in figure 1.

The FMA has a large energy ($\pm 20\%$) and recoil cone acceptance ($\pm 3°$) and so even for radiative capture events where the recoil kick from the emitted photon is largest i.e. a single 22 MeV gamma ray emitted at 90° to the beam axis, the residues fall well within the acceptance of the separator. The FMA efficiency is, therefore, in principle, solely dependent on the fraction of residues in the charge state selected. The ORNL program CHARGE predicts that 17% of residues have q=5, in good correspondence with data tabulated by Shima et al. [7]. The contribution of ^{24}Mg residues produced in reactions with likely target impurities can be rigorously discounted. The reaction Q-values for, for example, ^{13}C(^{12}C,n) and ^{16}O(^{12}C,α) are 8.98 MeV and 6.77 MeV, respectively. This ensures that even for particle emission with the lowest possible kinetic energies, the residues are boosted to recoil energies well outside of the acceptance of the FMA.

Sandorfi measured a total ^{12}C(^{12}C,γ) capture cross-section to the first few excited states of ^{24}Mg of around 1 μb [1]. Since the cross-sections measured with the FMA (figure 1) exceed by a factor of 3 the earlier measurements of Sandorfi, it was clear that the majority of the decay path-ways associated with the capture process must pass through higher-lying states. The challenge, then, was to determine how the capture process was mediated. In order to detect the capture gamma rays with sufficient resolution to allow the decay path-ways to be inferred, the Gammasphere spectrometer, comprising a 4π coverage of 100 high-efficiency HPGe detectors, was employed [8]. Since the Gammasphere array was situated at Lawrence Berkeley National Laboratory where no recoil mass spectrometer was available, it was necessary to devise an alternative methodology for selecting radiative capture events. This new methodology exploited the 4π solid angle of the device as a sum energy spectrometer, summing the energy recorded in each of the germanium crystals and their contiguous BGO shields. Since the Q value for the ^{12}C(^{12}C,γ) reaction is large and positive (+13.93 MeV), and greatly exceeds that associated with competing particle evaporation channels, the end-point of the sum energy spectrum associated with ^{12}C(^{12}C,γ) should be the highest end-point in the spectrum. Selecting the highest sum energy events, therefore, allowed the radiative capture channel to be cleanly selected and subsequently deconvoluted into its constituent gamma rays. This technique has the advantage of guaranteeing that all the gamma rays in the decay cascade must have been observed, if the sum energy is close to the expected end point. The gamma-ray detection efficiency may also be improved at the expense of some resolution by permitting the detection of gamma rays in modules, i.e. summing the energies recorded in the germanium crystal with those recorded in the BGO shield. In practice, the cost in resolution is largely acceptable, since for high energy gamma rays the principal interaction process is pair production, so that the most likely events comprise one or more 511 keV γ rays being recorded in the BGO shield. In the remainder of this discussion, we distinguish between 'clean' modules where the suppression shield did not fire and 'add-back' modules where the summing procedure was applied. Modules where only the BGO shield fired were not considered useful in the analysis of the decay pathways but were included in the construction of the γ-ray sum energy.

In order to find the point of maximum yield, a limited excitation function was performed. A ^{12}C beam accelerated by the 88" cyclotron at LBNL was incident on a 47 μg/cm^2 enriched ^{12}C target. Gamma rays were detected by the Gammasphere array with a trigger condition of one module. A peak in the capture yield, inferred from the number of events with a sum energy above 12 MeV, was found for a centre-of-target energy of 16.1 MeV, which shows qualitative agreement with the cross-section measurement made with the FMA. Having located the point of maximum yield, the array was switched to a coincidence mode where three gamma rays had to be detected in any event and at least two of them in a 'clean' module. The beam current was increased to 100 pnA, and high

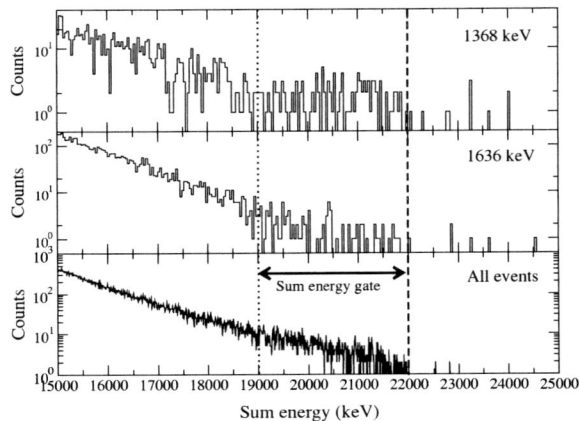

FIGURE 2. Gamma-ray sum energy spectra obtained for the ^{12}C(^{12}C,γ) reaction at a centre-of-target energy of 16.1 MeV: (top) Sum energy spectra for events containing a cleanly detected 1368 keV γ ray. (middle) Sum energy spectrum for events containing a cleanly detected 1634 keV or 1636 keV γ ray - strong transitions in ^{20}Ne and ^{23}Na respectively (bottom) Sum energy spectrum for all events. The expected end-point in sum energy for the radiative capture channel (22.0 MeV) is marked with a dashed line. The high sum energy region from which capture events were selected is marked.

statistics were obtained both for the point of maximum yield (16.1 MeV) and an 'off resonance' position (15.9 MeV). Good separation of radiative capture events and particle emission channels was observed in the sum energy spectrum (see figure 2), with a clear excess of counts at the high sum energies for cascades containing a 1368 keV γ ray - the characteristic $2^+ \rightarrow 0^+$ transition in ^{24}Mg. Moreover, the end point in the sum energy spectrum moved up and down in conformity with changes in beam energy.

In the subsequent analysis, we selected events with sum energies larger than 19 MeV (endpoint is $\sim$22 MeV) and deconvoluted them into their respective cascades. The cascades were sorted into matrices of 'clean' modules vs 'clean' modules and 'clean' modules vs 'add-back' modules each with the relevant sum energy criterion. This coincidence data indicates that the decay following capture does not proceed in a statistical fashion. This is illustrated in figure 3 where cascades containing the 1368 keV γ ray are selected and the remaining gamma rays in both 'clean' and 'add-back' modules are projected. Clear differences are seen between the 'on' and 'off' resonance data, and the population of certain states in the decay of the capture resonance is seen to be enhanced. In particular, in the 16.1 MeV data, the 2^+, 3^+ and 4^+ states in the K=2 rotational band in ^{24}Mg are strongly populated relative to the yrast ground state band. Higher energy transitions around 8-10 MeV are also observed in the gate on the 1368 keV transition, which implies cascades passing through states around 9-11 MeV in ^{24}Mg. This is of particular interest since this is the energy region in which shape isomeric bands are predicted to lie.

It is difficult to extract a total capture cross-section for the ^{12}C(^{12}C,γ) reaction on the basis of the observed decay cascades since the high energy response of Gammasphere is presently unmeasured and the efficiency for each cascade depends, in principle, on that

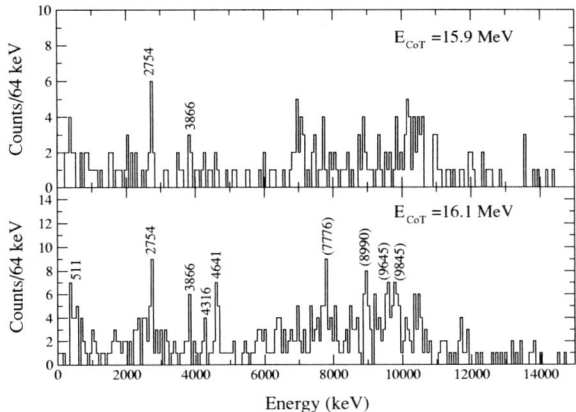

FIGURE 3. Spectra of gamma rays detected in both clean and add-back modules for cascades containing a 1368 keV (detected in a clean or add-back module) with a minimum sum energy requirement of 19 MeV: (top) Spectrum taken from the data taken at a centre-of-target energy of 15.9 MeV (bottom) Spectrum taken from the data taken at a cetre-of-target energy of 16.1 MeV

of its constituent γ rays. A simulation of the high energy response of Gammasphere was performed using the Monte Carlo code MCNP. This simulation reveals that the efficiency for a particular cascade is not especially sensitive to the manner in which the energy is partitioned. Making use of the observed decay patterns, we estimate from the simulated efficiencies a cross-section of between 1 and 10 μb which is in qualitative agreement with the cross-section measured with the FMA.

In summary, we have re-opened research into the area of heavy ion radiative capture as a means of learning more about molecular behaviour in nuclei. Our initial work indicates that the total capture cross-section for the $^{12}C(^{12}C,\gamma)$ reaction is significantly larger than was earlier believed and that the majority of this decay appears to cascade through highly-lying excited states in ^{24}Mg in a non-statistical fashion.

This work was supported in part by the U.S. Department of Energy, Nuclear Science Division, Contract Nos. W-31-109-ENG38 and DE-FG02-95ER40934. DGJ acknowledges receipt of an EPSRC Advanced Fellowship.

REFERENCES

1. A.M.Sandorfi, Treatise on Heavy Ion Science Vol. 2, sec. III (ed. D. Allan Bromley) and references therein.
2. A.M.Nathan, A.M.Sandorfi and T.J.Bowles, Phys. Rev. C **24**, 932 (1981).
3. A.M.Sandorfi and A.M.Nathan, Phys. Rev. Lett. **40**, 1252 (1978).
4. B.Dechant and E.Kuhlman, Z. Phys. A **330**, 93 (1988).
5. H.Flocard, P.-H.Heenen, S.J.Krieger and M.S.Weiss, Prog. Theor. Phys. 72, 1000 (1984).
6. D.Baye and P.-H.Heenen, Phys. Rev. C **29**, 1056 (1984)
7. K. Shima *et al.*, Atomic Data Nucl. Data Tables **51**, 173 (1992).
8. I.Y. Lee, Nucl. Phys. **A520**, 641c (1990).

Chiral Bands and Triaxiality

C.M. Petrache

Department of Physics, University of Camerino, and INFN, Sezione di Perugia, Italy

Abstract. The results obtained with the GASP array in the A=130 mass region are reviewed, emphasizing the discovery excited highly-deformed bands and their decay out, the study of the odd-odd Pr nuclei up to high spins, the discovery of stable triaxial bands in Nd nuclei close to the N=82 shell closure. The very recent studies of nuclei near the proton drip line are described. A discussion of the origin of the various doublet bands observed in odd-odd nuclei of the A=130 mass region is presented.

INTRODUCTION

The spectroscopic study of neutron defficient nuclei in the A~130 mass region undertaken at the Legnaro National Laboratory with the GASP array and the ancillary detectors ISIS - charged particle detector, CAMEL - recoil mass spectrometer and the neutron ring, lead to numerous achievements on various fields of interest in both the high- and low-spin regime. One can mention: the discovery of highly-deformed bands which helped to understand the main excitations at high deformation [1-13], the discovery of the decay out of these bands in a series of six Nd nuclei which allowed the extraction of Δ_v for the highly-deformed nuclei [12,13], the discovery of doublet bands in odd-odd nuclei which were interpreted as the manifestation of chiral or pseudospin symmetry in nuclei [14-17], the evolution of triaxiality which lead to the discovery of stable triaxial bands at the highest spins when approaching the N=82 shell closure [18-20]. For all studied nuclei we established very complete level schemes on the basis of data obtained in heavy ion induced reactions, being able to efficiently identify very weak transitions in the energy range from about 20 keV to several MeV.

THE DECAY OUT OF HIGHLY DEFORMED BANDS AND STABLE TRIAXIALITY IN ND NUCLEI

The most extended decay-out studies have been performed for the Nd nuclei. We observed the linking transitions over a series of six nuclei, from ^{132}Nd to ^{137}Nd, which lead to the determination of the excitation energy, spin and parity of the highly-deformed states. From these studies we could conclude that the highly-deformed bands have different decay-out patterns depending *(i)* on the type of nucleus (even-even or odd-even) and *(ii)* on the parity of the highly-deformed band. The bands in the odd-even Nd nuclei decay out towards low-lying normal-deformed states via E2 transitions ^{133}Nd, E2 plus E1 transitions ^{135}Nd and cascades of E2 plus M1 transitions ^{137}Nd.

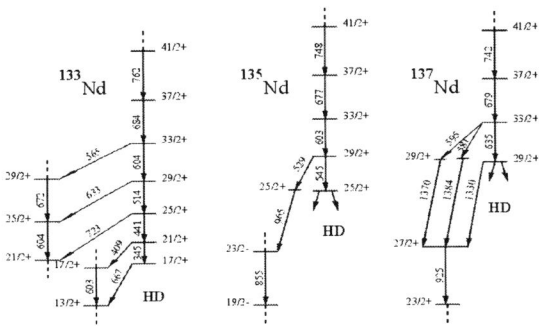

FIGURE 1. Decay out of the highly-deformed bands in odd-even Nd nuclei.

The highly-deformed bands in even-even Nd nuclei, which are based on two quasi-particle neutron excitations, are based on two types of configurations, which lead to either negative parity (the yrast bands observed in ^{132,134,136}Nd which are shown in Fig. 2) or positive parity (the excited bands observed in ^{132,134}Nd).

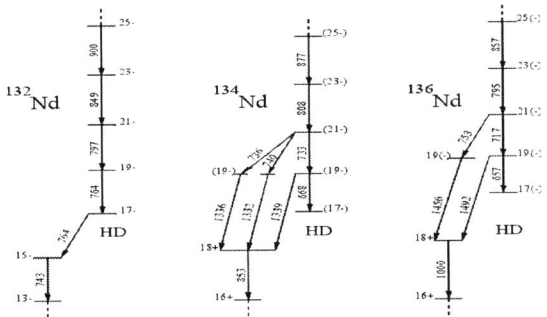

FIGURE 2. Decay out of the negative-parity highly-deformed bands in even-even Nd nuclei.

The negative-parity bands decay in two steps via E2 + enhanced E1 transitions with strengths of $\sim 10^{-3}$ W.u., whereas the positive-parity bands decay in one step via E2 transitions with strengths of ~ 1 W.u. The conclusion was drawn that the mixing with the normal-deformed states plays a more important role than the sliding between different deformations via the pairing interaction. The interaction matrix elements and mixing amplitudes between normal- and highly-deformed bands in Nd nuclei were deduced in the decay-out region and at high spins, showing that the barrier at the point of decay out disappears. The discovery of discrete linking transitions in the series of six Nd nuclei from ^{132}Nd to ^{137}Nd, allowed to estimate the pairing strength in the highly-deformed configuration, from the odd-even mass differences. Using a first-order Taylor expansion [21] and the three-point formula [12], we extracted the experimental neutron pairing gap in the normal-deformed ground state and in the highly-deformed configurations for two odd-even and two even-even Nd nuclei (see Fig. 3).

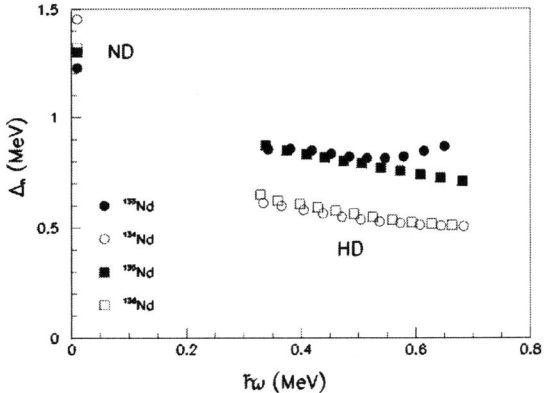

FIGURE 3. Experimental neutron pairing gaps for the normal-deformed ground states and for the highly-deformed configurations in Nd nuclei.

The pairing gap values for the highly-deformed configurations, are reduced by approximately a factor of 2 with respect to the values for the normal-deformed ground states ($\Delta_\nu \sim 0.6 - 0.7$ MeV) and remain rather constant in a wide range of frequencies. In the normal-deformed case, the pairing gaps of the odd-even nuclei are smaller than those of the even-even ones, whereas in the highly-deformed configurations the opposite is true: the pairing gaps of the two-quasiparticle configurations present in even-even nuclei is smaller than the pairing gaps of the one-quasiparticle configurations present in odd-even nuclei. This peculiar feature is due to blocking effects, which are stronger in the two-quasiparticle configurations based on the intruder orbitals $\nu i_{13/2}$ and $\nu f_{7/2}$, than in the one-quasiparticle $\nu i_{13/2}$ configurations.

One of the last achievements we obtained in the study of the Nd nuclei is the discovery of bands with stable triaxiality up to the highest states identified in [138,139]Nd [20]. The observed bands have considerably smaller collectivity than the triaxial bands identified in the other A~130 nuclei, and are based on oblate-triaxial shape with $\gamma \sim$ +30° rather than the prolate-triaxial shape with $\gamma \sim$ -30° present in the lighter nuclei (see Fig. 4). Very recently, exciting results were obtained from an experiment performed with the Euroball array for the study of the [140]Nd nucleus, which is predicted as a good candidate for hyperdeformation. A preliminary analysis performed in various laboratories involved in the collaboration (Bonn, Camerino, Copenhagen) indicates the existence of both superdeformed and triaxial bands in [140]Nd. These results extend the new region of triaxiality consisting of nuclei with N<82 and show that a "bridge" between the A=130 and A=150 regions of superdeformation can be envisaged, leading to a overall understanding of the phenomenon of superdeformation from the A=100 to the A=150 nuclei.

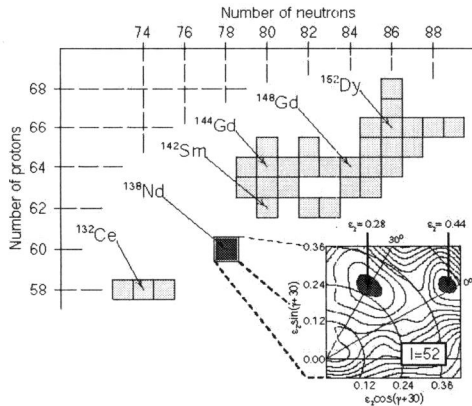

FIGURE 4. Extract of the nuclear (N,Z)-chart indicating those nuclei where superdeformed bands have been observed. Note how the nucleus ^{138}Nd would "fill the gap" between the Ce and the Sm/Eu regions of superdeformation. The calculated potential energy surface for this nucleus (with a contour line separation of 0.5 MeV) shows one minimum at $\varepsilon_2 = 0.28$ ($\gamma \sim 30°$), corresponding to presently observed highly-deformed bands in Nd nuclei, and then the superdeformed minimum at $\varepsilon_2 = 0.44$.

STUDY OF ODD-ODD PR NUCLEI

In the odd-odd nuclei of the A=130 mass region the prevalent structures are semi-decoupled deformed rotational bands built on $\pi h_{11/2} - \nu h_{11/2}$, $\pi g_{7/2}/d_{5/2} - \nu h_{11/2}$ and $\pi h_{11/2} - \nu g_{7/2}$ configurations [14-17,22]. However, there are several cases in which also weakly populated double-decoupled bands were observed, with assigned configurations involving the $\pi g_{7/2}$ and $\nu f_{7/2}/h_{9/2}$, $\nu i_{13/2}$ configurations [10,11,14,15,17]. From the study of the odd-odd Pr nuclei we obtained important information on the signature inversion phenomenon through the interpretation in the framework of the Interacting Boson-Fermion-Fermion model [15,16,22], on the competition between prolate and oblate shapes in the γ-soft nuclei by means of Total Routhian Surface calculations [11,14,15], and more recently, on the possible existence of chiral doublet bands predicted by 3D Tilted Axis Cranking calculations [17]. A discussion of the observed doublet bands in ^{134}Pr and ^{128}Pr will be given in the following sections.

STUDY OF NUCLEI CLOSE TO THE PROTON DRIP LINE

Recently, our interest was focused on the study of nuclei close to the proton drip line in the A~130 mass region, with the aim to determine their lowest lying configurations. The study was motivated by the need to have experimental data in nuclei as close as possible to the unknown Pr and Pm proton emitters in this mass region, in order to make realistic predictions for the lifetimes of the emitting states. An experiment was performed using a self-supporting 0.5 mg/cm^2 thick ^{92}Mo target with a 190 MeV ^{40}Ca beam of 5 pnA intensity. The beam was provided by the XTU Tandem accelerator of the Laboratori Nazionali di Legnaro and the experimental setup consisted of the GASP array for γ-ray detection and the ISIS ball for charged particle detection. New results were obtained on many nuclei close to the proton drip line, in which we were able to identify excited states up to unexpected high spins. We observed for the first time excited states in ^{126}Pr [22], constructed a detailed level scheme and identify a pair of bands with nearly degenerate levels in ^{128}Pr [17], constructed a detailed level scheme of ^{122}Ba [23], identify several rotational bands in ^{128}Nd [24], constructed detailed level schemes for ^{125}Ce [25] and ^{126}Ce [26]. The band structures observed in ^{126}Pr, ^{125}Ce and ^{126}Ce were interpreted with the Interacting Boson Model + Broken Pairs model, which resulted to be successful in describing the high-spin states the light A=130 deformed nuclei (see Fig. 5 taken from ref. [25]).

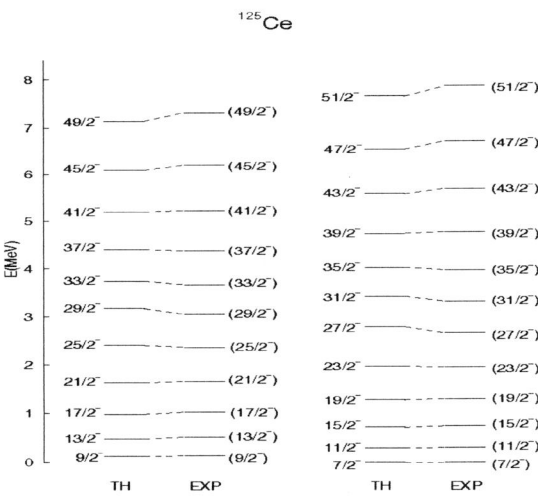

FIGURE 5. The negative parity states in ^{125}Ce compared with the results of the IBFBPM calculations.

THE ORIGIN OF THE DOUBLET BANDS IN ODD-ODD PR NUCLEI

The possibility of chirality in triaxial rotating nuclei predicted theoretically by Frauendorf [27,28], results from a combination of geometry (triaxial nucleus) and dynamics (total angular momentum). The angular momentum vector introduces chirality by selecting one of the octants of the space, giving rise to two degenerate rotational bands. This nice model brought the neutron-deficient odd-odd Pr nuclei and the neighboring nuclei into the focus of experimental studies, since one of the best examples of broken chiral symmetry up to date is ^{134}Pr. Extensive studies were performed in several laboratories around the world, and other candidates for chiral doublet bands were proposed [29]. However, the ^{134}Pr nucleus remained the only case where the behavior of the doublet bands was considered to approach the model predictions.

In the present discussion I would like to remind the main features of the observed doublet bands in ^{134}Pr and stress that we have not yet a good example of broken chiral symmetry in nuclei. In the following we will call band 1 and band 2 the lowest observed bands in ^{134}Pr [14]. The lowest observed state of band 2 at spin 10^+ lies 337 keV above the 10^+ state of band 1. Band 2 crosses band 1 at spin 10^+, which corresponds to a rotational frequency of around 0.45 MeV. As one can see in Fig. 6, the two bands are nearly degenerate only on a very limited spin range, as is the case of many rotational bands built on different shapes or configurations.

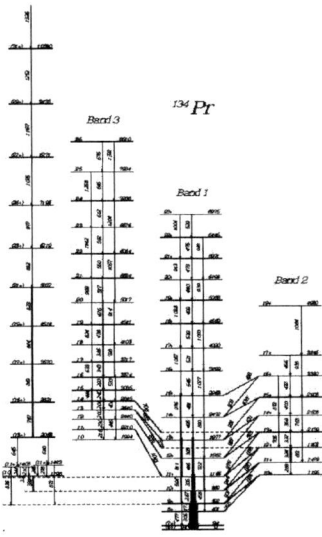

FIGURE 6. Decay scheme of ^{134}Pr.

The single particle alignment of band 2 is larger by 2 units of spin than that of band 1 (see Fig. 7), as one would expect for example for a γ-vibration built on the configuration of band 1. Band 1 exhibits a backbending at a rotational frequency of about 0.5 MeV, which strongly disturbs the regularity of the band at high spins. This has as consequence a very limited spin range in which the properties of the possible chiral candidates remain undisturbed. In particular, this is the reason for which the B(M1)/B(E2) predictions of the 3D TAC model cannot be compared in a straightforward manner to the experimental values.

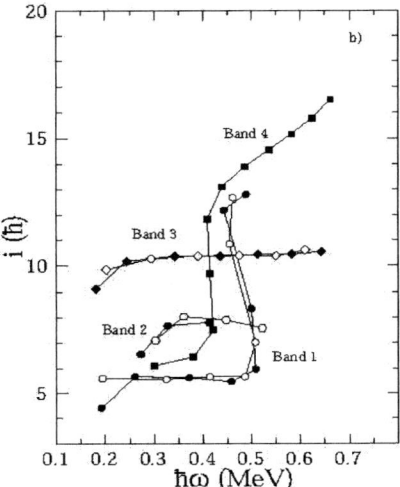

FIGURE 7. Experimental alignments for the observed bands in ^{134}Pr. The bands with signature $\alpha=0$ ($\alpha=1$) are drawn with filled (open) symbols.

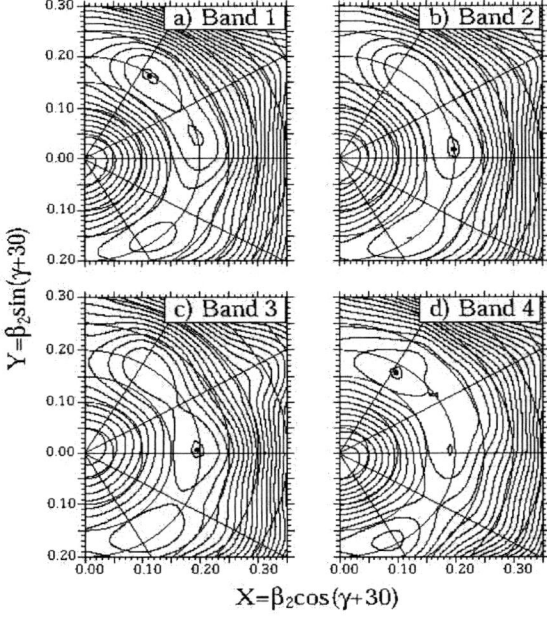

FIGURE 8. Total routhian surface (TRS) calculations for the different configurations assigned to the observed bands in ^{134}Pr for $\omega=0.112$ MeV/h.

In the tentative to understand the structure of the two bands, we performed Total Routhian Surface calculations based on the cranked shell model with a Woods-Saxon potential and two-particle plus triaxial rotor calculations [14]. As can be seen in Fig. 8, the minima are very flat in the γ degree of freedom, the barrier between the minima with positive and negative γ being of the order of 200 keV. Possible descriptions of the two bands involve either shape coexistence in terms of different γ-deformation of the two structures or the coupling to the γ-phonon. Within the mentioned calculations we were not able to find a consistent interpretation of bands 1 and 2 of ^{134}Pr. However, one cannot be content with the description in terms of chiral doublet bands, because the experimental features of the bands can be due to a subtle superposition of various effects, leading to a simple near degeneracy of a few levels in the region of band crossing. It is therefore necessary to continue the search for doublet bands in odd-odd nuclei, to find a good example of broken chiral symmetry in nuclei.

Another origin of doublet bands in odd-odd nuclei, with characteristics slightly different that the claimed chiral doublets, is the pseudospin symmetry [30]. In the pseudospin formalism one assigns new spin and orbital momentum labels to the single-particle levels in accordance with an observed near degeneracy of certain normal-parity eigenstates (pseudospin doublets) of the spherical nuclear mean field. It could be that for particular nucleon numbers the Fermi level lies just between the two members of the pseudospin members, giving rise to bands with levels which are nearly degenerate. One such case is suggested to occur in the odd-odd ^{128}Pr nucleus [17], in which the proton Fermi level is expected to lie close to the [420]1/2$^+$ - [422]3/2$^+$ doublet of mixed $d_{5/2}/g_{7/2}$ composition. However, both in a Tilted Axis Cranking and a Two-quasiparticle Rotor calculations including the residual n-p interaction, it turned out to be impossible to achieve the observed small level spacing and the staggering. If the observed close spacing is not restricted to ^{128}Pr, but appears in a systematic way, it could be due to some accidental combination of effects not taken into account in the standard models and may be a hint of a hidden symmetry to be revealed.

CONCLUSIONS

One conclusion of our recent studies with GASP is that very high spins can be populated in nuclei close to the proton drip line, allowing a detailed spectroscopy much far that previously thought. The study of the doublet bands in the odd-odd nuclei showed that their origin can be traced back to either chiral or pseudospin symmetry. The need to search for good examples of chiral doublet bands in odd-odd nuclei remains an imperative for future studies. However, one should not limit our interest only to odd-odd nuclei, because in principle the chiral regime can be achieved also in even-even or odd-even nuclei.

ACKNOWLEDGMENTS

I am highly indebted to the GASP group which assured the smooth operation of the setup during the experiments we performed in the last decade and to the accelerator group of LNL for delivering high quality beams. I would like to mention the essential contribution to the understanding of the obtained results of Prof. Ramon Wyss and Ingemar Ragnarsson from Sweden, as well as Prof. Slobodan Brant and Dario Vretenar from Croatia.

REFERENCES

1. D. Bazzacco et al., *Phys. Lett.* **B309**, 235 (1993).
2. C.M. Petrache et al., *Phys. Lett.* **B335**, 307 (1994).
3. S. Lunardi et al., *Phys. Rev.* **C52**, R6 (1995).
4. C.M. Petrache et al., *Phys. Rev. Lett.* **77**, 239 (1996).
5. C.M. Petrache et al., *Phys. Let.* **B383**, 145 (1996).
6. C.M. Petrache et al., *Phys. Lett.* **B387**, 31 (1996).
7. C.M. Petrache et al., *Phys. Lett.* **B415**, 223 (1997).
8. C.M. Petrache, *Zeit. Phys.* **A358**, 225 (1997).
9. C.M. Petrache et al., *Phys. Rev.* **C57**, R10 (1998).
10. M.N. Rao et al., *Phys. Rev.* **C58**, R1367 (1998).
11. C.M. Petrache et al., *Phys. Rev.* **C58**, R611 (1998).
12. S. Perriès et al., *Phys. Rev.* **C60**, 064313 (1999).
13. C.M. Petrache et al., *Nuclear Structure 98*, edited by C. Baktash, AIP Conf. Proc. No. 481 (AIP, Woodbury, NY, 1999), p. 31.
14. C.M. Petrache et al., *Nucl. Phys.* **A597**, 106 (1996).
15. C.M. Petrache et al., *Nucl. Phys.* **A603**, 50 (1996).
16. C.M. Petrache et al., *Nucl. Phys.* **A635**, 361 (1998).
17. C.M. Petrache et al., *Phys. Rev.* **C65**, 054324 (2002).
18. C.M. Petrache et al., *Phys. Lett.* **B373**, 275 (1996).
19. C.M. Petrache et al., *Nucl. Phys.* **A617**, 228 (1997).
20. C.M. Petrache et al., *Phys. Rev.* **C61**, 011305 (2000).
21. D. Madland and J. Nix, *Nucl. Phys.* **A476**, 1 (1988).
22. C.M. Petrache et al., *Phys. Rev.* **C64**, 044303 (2001).
23. C.M. Petrache et al., *Eur. Phys. J.* **A12**, 135 (2001).
24. C.M. Petrache et al., *Eur. Phys. J.* **A12**, 139 (2001).
25. C.M. Petrache et al., *Eur. Phys. J.* **A14**, 439 (2002).
26. C.M. Petrache et al., *Eur. Phys. J.*, submitted.
27. S. Frauendorf and J. Meng, *Nucl. Phys.* **A617**, 131 (1997).
28. V.I. Dimitrov, S. Frauendorf, F. D\"onau, *Phys. Rev. Lett.* **84**, 5732 (2000).
29. K. Starosta et al., *Nucl. Phys.* **A682**, 375c (2001).
30. A. Bohr, I. Hamamoto and B. Mottelson, *Phys. Scripta* **26**, 267 (1982).

Supersymmetry in nuclei

A. Algora

IFIC, CSIC-Univ. Valencia, Valencia, Spain
Institute of Nuclear Research, Debrecen, Hungary

Abstract. Supersymmetry, first introduced in high energy particle physics studies; found its first experimental confirmation in nuclear physics. Following the pioneering work of Iachello [1] several applications of this concept to the description of the structure of nuclei have been studied in different degrees of complexity [2]. In this work an overview of recent developments in the application of the supersymmetry concept to nuclear structure will be presented. Special emphasis will be given to our study of the ^{73}As, ^{74}Se supermultiplet in the framework of the $U(6/12)$ symmetry scheme [3].

INTRODUCTION

Greece, and in particular Crete, the place were the most ancient european civilization was born, is an inspiring place to talk about symmetries. The word symmetry originates from the greek "sum metria" which means "same measure". A greek sculptor from the fifth century B. C., Polykleitos, is apparently one of the first to have used the term. It is clear why a sculptor was one of the first persons to use the word, since proportion and symmetry in an object seem to be essential for our perception of beauty. The recognition of something having a pattern seems to bring an intuitive or instinctive aesthetic response. This general feature of our perception of beauty is also valid in physics.

Symmetry has been widely used in physics as a powerful tool of classification and simplification. Supersymmetry in particular is a complex type of symmetry, which distinguishes itself from other types of symmetries by including operations that transform bosons into fermions and viceversa. Although supersymmetry was originally introduced in Particle Physics, it has shown its potential in nuclear physics applications. Dynamical supersymmetries were introduced in nuclear physics by Iachello in the framework of the Interacting Boson Model(IBM) and its extensions. The IBM describes the collective excitations of even-even nuclei in terms of bosons with angular momentum $l = 0, 2$. The s bosons ($l = 0$) are analogous to Cooper pairs and the d bosons ($l = 2$) can be viewed as a generalization of this. The number of bosons in the model is related to the number of valence proton and neutron pairs counted from the nearest closed shells. Since by definition the IBM is only applicable to even-even nuclei it is necessary to include other degrees of freedom in the model to increase its versatility. One natural extension is the inclusion of single-particle degrees of freedom [4]. The Interacting Boson Fermion Model (IBFM) has as its building blocks N bosons with $l = 0, 2$ and $M = 1$ fermions with j. Within the framework of the IBFM, a supersymmetry occurs whenever: a) the boson and fermion single particle states are related by a supersymmetry transformation; and b) the coefficients in the Hamiltonian H describing the bosonic and fermionic degrees of freedom are also related by a supersymmetry transformation. The mathematical framework needed to describe supersymmetry is called superalgebra. The IBM and IBFM can be unified into a superalgebra $U(n/m)$ where $n = \sum_l (2l+1)$ and $m = \sum_j (2j+1)$ are the dimensions of the boson and fermion spaces respectively. The even-even and odd-mass nuclei described in this context are called a supermultiplet. The supermultiplet is characterized by the $\mathcal{N} = N + M$ number, the total number of bosons and fermions which is the same for the members of the supermultiplet. This scheme can be further generalized by including the proton-neutron degree of freedom [5]. In this case instead of describing "magic rectangles" we describe "magic squares" or supersymmetric quartets consisting of an even-even, an odd-even, an even-odd and an odd-odd nucleus.

After postulating the concept of supersymmetry in nuclei, as the unifying symmetry that establishes precise links among the spectroscopic properties of certain neighboring nuclei one question arises. Does this symmetry exist in nature even if it is partially broken?. As mentioned by Casten [6] there are two levels to this question. The first is: given a particular supersymmetry scheme, can one find a set of data that can be predicted by the theory?. The second level: how can one demonstrate the existence of such a symmetry?. The second question is more difficult to address and requires great effort both theoretically and experimentally, therefore most test of supersymmetry have attempted to answer the first level of the question.

In the first part of this contribution, we will present a recent application of the U(6/12) supersymmetry scheme to the ^{74}Se and ^{73}As nuclei in great detail. In the second part and only briefly, other recent applications of the supersymmetry concept to nuclei will be summarized. Most of the applications presented here concern the first level of the question addressed before.

THE U(6/12) SUPERSYMMETRY SCHEME: APPLICATION TO THE ^{74}SE, ^{73}AS SUPERMULTIPLET

The detailed description of the U(6/12) supersymmetry in its U(5) limit can be found in refs. [2, 7, 8, 9]. Here we restrict ourselves to a brief discussion of the main ideas. The group structures associated with the IBFM hamiltonian of a nucleus whose valence shells contain the single particle orbits $j = 1/2, 3/2, 5/2$ are $U^F(\sum_j(2j+1)) = U_F(12)$ and $U^B(6)$, where $U^F(12)$ is the fermion group related to the single particle degrees of freedom, and $U^B(6)$ is the boson group necessary to describe the collective excitations. Since there is no group isomorphism between the complete fermion space and boson space to construct spinor groups, the pseudo-spin symmetry is used to deduce the boson-fermion symmetry. The single particle orbits with $j = 1/2, 3/2, 5/2$ can be considered as arising from a combination of a pseudo-orbital part $l = 0, 2$ and a pseudo-spin part $s = 1/2$. This corresponds to the $U^F(12) \supset U^F(6) \times SU^F(2)$ reduction in group theoretical language.

The group chain for the above mentioned problem reads [9]:

$$U^B(6) \times U^F(12) \supset U^B(6) \times U^F(6) \times SU^F(2) \supset U^{BF}(6) \times SU^F(2) \quad (1)$$
$$\supset U^{BF}(5) \times SU^F(2) \supset O^{BF}(5) \times SU^F(2)$$
$$\supset O^{BF}(3) \times SU^F(2) \supset Spin(3)$$

The related classification schemes can be applied to even-even or odd nuclei depending on which representations of $U^B(6) \times U^F(12)$ are used. A connection between these nuclei is obtained by embedding the group chain into the $U(6/12)$ supergroup. The U(6/12) supergroup allows the simultaneous description of the even-even and odd nuclei (usually called supermultiplet) in the framework of the SUSY concept.

The associated Hamiltonian of the dynamical symmetry can be found in refs. [2, 8, 9]. Diagonalization of the Hamiltonian using the basis states obtained from the classification schemes leads to the following analytical expression for the eigenvalues:

$$E = A[N_1(N_1+5) + N_2(N_2+3)] + B_1(n_1+n_2) + B_2[n_1(n_1+4) + n_2(n_2+2)] \quad (2)$$
$$+ C[v_1(v_1+3) + v_2(v_2+1)] + DL(L+1) + EJ(J+1),$$

where $N_1, N_2, n_1, n_2, v_1, v_2, L, J$ are the quantum numbers characterizing the states of the members of the supermultiplet and A, B_1, B_2, C, D and E are free parameters (which are the same for the supermultiplet as imposed by SUSY).

The U(6/12) dynamical supersymmetry (in its vibrational limit) has been used previously in the arsenic region to describe the states of ^{76}Se, ^{75}As [8]. Necessary but certainly not sufficient conditions for the applicability of this approximation are [7]: i) the even-even nucleus can be described by the U(5) dynamical symmetry of the IBM, ii) the available single particle orbits which can be occupied by the odd nucleon in the odd-even nucleus have angular momenta $j = 1/2, 3/2, 5/2$.

Conditions i)-ii) are partially fulfilled by ^{74}Se, ^{73}As. Previous studies of ^{74}Se by means of the IBM have shown that this nucleus can be considered to lie between the U(5) and SU(3) symmetry limits [10] with a strong U(5) character. Being interested in the description of the negative parity states of ^{73}As, condition ii) is trivially fulfilled since the orbitals available for the odd nucleon are $2p_{1/2}, 2p_{3/2}$ and $1f_{5/2}$ in the 28 - 50 valence shell.

Since we have an analytical solution of the eigenvalue problem, the simplest way to test the $U(6/12)$ SUSY model is to perform a fit procedure to the spectra of ^{74}Se, ^{73}As using (2) in order to determine the A, B_1, B_2, C, D and E parameters, and then with the resulting parameters, the SUSY spectra can be generated. The first and at the same time the most crucial step in such a procedure is the proper identification of the states. As in Ref. [11] the quantum numbers were assigned to the levels on the basis of the pattern of the experimental energy spectra, the decay properties of the levels, $U(6/12)$ wave functions [9], and single particle transfer reaction spectroscopic factors. In the case of ^{74}Se the one, two and three phonon states can be easily identified on the basis of energies and $B(E2)$ values between the low-lying states, so the assignment of the quantum numbers to the states is straightforward. The identification of the ^{73}As states is not so simple. Here we have followed a procedure similar to that of Ref. [11], based mainly on the decay pattern of the levels, the result of former IBFM calculations [12, 13], and single particle transfer reaction results [14, 15]. There are a few differences with respect to Ref. [11] due to the new experimental results.

The quality of the fitting procedure was quantified using the values ϕ and σ [8]:

$$\phi = (\sum_i |E_i^{exp} - E_i^{the}| / \sum_i E_i^{exp})(\%) \quad (3)$$

$$\sigma = [\sum_i (E_i^{exp} - E_i^{the})^2 / (n-k)]^{1/2} (keV) \quad (4)$$

where n is the number of levels included in the fit and k is the number of parameters and $(n-k)$ is the number of degrees of freedom.

In Fig. 1 we compare the experimental spectra of ^{74}Se and ^{73}As with the calculated ones using (2). Also in this figure the proposed classification of the states of ^{74}Se and ^{73}As in terms of $U(6/12)$ can be found. The results presented in Fig. 1 were obtained using the following parametrization: $A = 43$, $B_1 = 569$, $B_2 = 3$, $C = -1$, $D = -26$, $E = 39$ (all in keV). This parametrization follows the trend of the results of Vervier *et al.* for ^{75}As [8] showing only differences in the B_1 and B_2 parameters. The quality of the fit is quantified by the values of $\phi = 3\%$ and $\sigma = 105$ keV, which represent a smaller breaking of the SUSY than in the ^{76}Se, ^{75}As case ($\phi = 6\%$ and $\sigma = 117$ keV in Ref. [8]).

A reasonable reproduction of the energy spectrum, though essential, is not sufficient for testing the applicability of the supersymmetry scheme. In addition it is necessary to compare observables other than the energies, such as electromagnetic transition probabilities, branching ratios and one nucleon transfer intensities in order to assess the goodness of the description with more sensitive probes. For that reason we have calculated B(M1), B(E2) and one-nucleon transfer probabilities using the $U(6/12)$ wave functions and the following operators:

$$T(E2) = e_b(s^+\tilde{d}+d^+s)^{(2)} + e'_b(d^+\tilde{d})^2 \quad (5)$$
$$+ e_f[-\sqrt{4/5}(a^+_{1/2}\tilde{a}_{3/2})^{(2)} - \sqrt{6/5}(a^+_{1/2}\tilde{a}_{5/2})^{(2)} + h.c.]$$
$$+ e'_f[-\sqrt{14/25}(a^+_{3/2}\tilde{a}_{3/2})^{(2)} - \sqrt{24/25}(a^+_{5/2}\tilde{a}_{5/2})^{(2)}$$
$$+ \sqrt{6/25}(a^+_{3/2}\tilde{a}_{5/2} - a^+_{5/2}\tilde{a}_{3/2})^{(2)}]$$

$$T(M1) = \sqrt{3/4\pi}[g_b\sqrt{10}(d^+\tilde{d})^{(1)} \quad (6)$$
$$- \sum (3)^{-1/2} \langle l1/2j|g_l l + g_s s + g_T(Y_2 \times s)|l'1/2j'\rangle$$
$$(a^+_j \tilde{a}_{j'})^{(1)}(u_j u_{j'} + (v_j v_{j'})]$$

$$P^+_j = \xi_j a^+_j \quad (7)$$

In the application of the T(E2) operator one normally takes $e_b = e_f$ and $e'_b = e'_f$, because then the transition operator becomes a generator of the $U^{BF}(6)$ group. This not only reduces the number of free parameters but also leads to selection rules. For instance the operator cannot connect states with different $[N_1, N_2]$ quantum numbers [8]. Following this apparently drastic approach the only free parameters of the T(E2) operator (e_b and e'_b) can be obtained from a fit to the $B(E2; 0^+_1 \to 2^+_1)$ and to the quadrupole moment of the ground state of ^{74}Se [16]. Then the obtained $e_b = 0.097$ and $e'_b = 0.212$ parameters can be used to calculate the $B(E2)$s of transitions in ^{74}Se and ^{73}As and we can test how good the agreement is with the known experimental values according to the SUSY prescription. The results obtained for ^{74}Se show good agreement between experiment and theory [3]. The $B(E2; 5/2^-_1 \to 3/2^-_1)_{exp} = 0.0072^{+108}_{-54} e^2b^2$ [17] and $Q_{exp}(5/2^-_1) = \pm 0.356(12)e^2b^2$ [17] of ^{73}As are also well reproduced in the calculations with this parametrization: $B(E2; 5/2^-_1 \to 3/2^-_1)_{SUSY} = 0.009 e^2b^2$, $Q_{SUSY}(5/2^-_1) = 0.359 e^2 b^2$. As for the IBFM results, the recently measured $B(E2)$s of Bucurescu *et al*

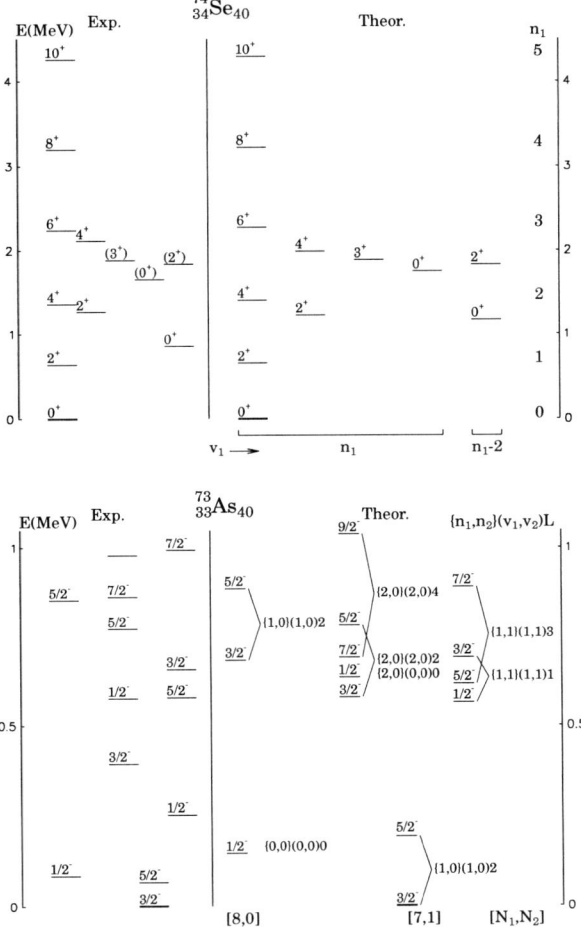

FIGURE 1. Experimental energy levels of ^{74}Se [16] and ^{73}As [12] in comparison with the results of U(6/12) SUSY calculations.

[18] for the decay of the high lying $7/2_2^-$ state are not well reproduced in the SUSY framework ($B(E2; 7/2_2^- \rightarrow 3/2_2^-)_{exp} = 0.1740(765)e^2b^2$, $B(E2; 7/2_2^- \rightarrow 3/2_1^-)_{exp} = 0.0390(163)e^2b^2$, $B(E2; 7/2_2^- \rightarrow 3/2_2^-)_{SUSY} = 0.00e^2b^2$, $B(E2; 7/2_2^- \rightarrow 3/2_1^-)_{SUSY} = 0.0075e^2b^2$). As stated in [18] the experimental $B(E2)$ values for the decays from the

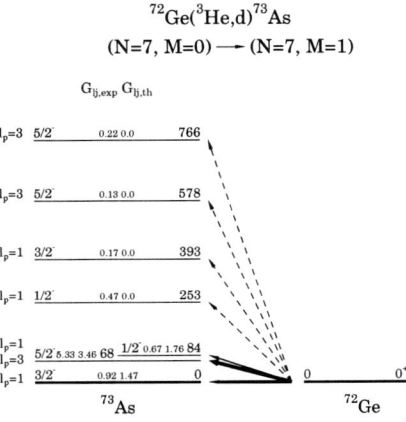

FIGURE 2. Comparison between the experimental [15] and calculated one-proton transfer intensities between ^{72}Ge and ^{73}As. The theoretical allowed transitions are marked with continuous lines and the dotted lines represent tilted transitions in first order.

$7/2_2$ state have very large uncertainties due to the low recoil velocities in the $(p, n\gamma)$ measurement, but this can not account for the bad description of the $B(E2)$ values in both models (IBFM and SUSY).

The choice of the T(M1) operator was similar to that used in Ref. [19]. The parameters for the T(M1) operator were taken in accordance with reference [13]. The values of the effective boson charge and gyromagnetic ratios were: $g_b = 0.452, g_l = 1, g_s = 0.7 g_s^{free} = 3.91, g_T = \frac{1}{25} \langle r^2 \rangle g_s^{free} = 3.372$. The v_j values used were $v(\pi p_{1/2}) = 0.344$, $v(\pi p_{3/2}) = 0.761$, and $v(\pi f_{5/2}) = 0.584$ as in Ref. [13].

Comparison of the calculated and experimental branching ratios supports most of the primary assignments of Fig.1. This result also confirms the approximate validity of the $U(6/12)$ classification scheme for the majority of the states of ^{73}As. The overall agreement of calculated and experimental results is remarkable: for 6 states the leading branches are well reproduced (states $5/2_1, 1/2_1, 1/2_2, 3/2_2, 5/2_2, 3/2_3$), and for states $5/2_3, 7/2_1, 7/2_2$ the quality of the results is comparable with the more involved IBFM calculations. Only the branches for states $1/2_3, 5/2_4$ are poorly reproduced indicating that the weight of the different components of the wave functions is not appropriate or that they should include additional components to obtain a better agreement. Also in Ref. [3] it was shown that the choice of the T(E2) operator as the sum of two $U^{BF}(6)$ generators is not unrealistic for this case. Most of the transitions that connect states with different $[N_1, N_2]$ numbers have weak branchings (see transitions $1/2_2 \rightarrow 1/2_1; 3/2_2 \rightarrow 1/2_1; 5/2_2 \rightarrow 1/2_1; 5/2_3 \rightarrow 1/2_1; 5/2_4 \rightarrow 1/2_2$). At this point it is worth mentioning that using a slightly different parametrization than the one used in [12] Bucurescu *et al* [18] obtained an IBFM description of ^{73}As which for some levels gives a slightly better agreement with the experimental branching ratios than the results of [12]. The quality of the results obtained for the branching ratios of ^{73}As in the SUSY calculations is in any case comparable with the better IBFM results of [18].

In addition to E2 and M1 transitions one-nucleon transfer reactions provide an important tool to study the structure of states of odd-even nuclei. For transfer reactions between even-even nuclei with N bosons and odd-even nuclei with N bosons and M=1 fermion the simplest operator that can be used has the form of (7) [7]. In this case the transfer intensities for the $(N, M = 0) \to (N, M = 1)$ reaction can be calculated easily using the explicit expression for the wave functions of Ref. [9]. In Fig. 2 we present a comparison of the results obtained with the experimental data of Ref. [15]. As can be seen from this figure, the stronger one-nucleon transfer intensities as well as the general trend of the experimental data are well reproduced using this very simple operator, but it is also clear that to obtain a quantitatively better description of the states $1/2_2, 3/2_2, 5/2_2, 5/2_3$, which are in first order tilted transitions, a more complex one-nucleon transfer operator is required. Notice that the semi-microscopic operator used in [20] does not induce fragmentation when a pure $U(5)$ limit is assumed, because it relies on the annihilation of a d-boson in the ground state of an even-even nucleus.

OTHER RECENT APPLICATIONS

Without the aim of completeness I summarize here some recent results obtained in the application of the supersymmetry concept to nuclei.

Probably one of the most important recent results is the confirmation of the validity of the classification schemes for the ^{196}Au case, which is considered the best example available [20, 21, 22]. In this case the supersymmetry scheme applied is the $U_\pi(6/4) \times U_\nu(6/12)$, which was used successfully to describe the magic square formed by the ^{194}Pt, ^{195}Pt, ^{195}Au and ^{196}Au nuclei.

Another recent result is related to the extension of the application of supersymmetry concept without the use of dynamical symmetry, "SUSY without chains" as it was called by A. Frank and coworkers. In an earlier work [23] they described some Ru and Rh isotopes in a transition between two dynamical symmetry limits ($U(5)$ and $O(6)$). There are new developments in this direction using what is called "generic SUSY" [23]. In [24] some preliminary results are presented for W and Hf using this concept. The importance of these generalizations is that they open the possibility of testing SUSY in wider nuclear regions, since dynamical symmetries are scarce.

There is also another interesting result from the same group [25, 26]. Most tests of supersymmetry do not probe directly the correlations present in the "quartets" wave functions. Usually the wave functions are tested from "outside" the quartet. In these recent works they propose to probe the fermionic sector of the SUSY supermultiplets using transfer reactions that connect the members of the supermultiplet. These tests may provide the most stringent checks of nuclear supersymmetry. It is still an open question whether the correlations predicted by SUSY are indeed verified by experiment.

Most supersymmetry applications to nuclear structure are based on the extensions of the IBM (IBFM, IBFFM). But there are also other possible generalizations. In a recent work Levai *et al.* [27] have studied a supersymmetry application to nuclear cluster systems. In this application the bosonic sector is identified with the relative motion of the clusters (U(4) group of the vibron model), while the fermions are defined as holes in the p shell. Within this framework Levai et al. were able to describe the α-cluster states in ^{20}Ne and ^{19}F.

There are a few topics of research on supersymmetry not covered in this talk because of the lack of time. They are related to research on the microscopic framework for the existence of SUSY in nuclei (see some recent work of Cejnar and coworkers, Jolos and Brentano, etc.), as well as some applications to the description of superdeformed bands (see some recent work of Yu-xin Liu et al.). The interested reader is encouraged to look for the original publications.

Summarizing, we have shown another example of the application of the SUSY concept in detail (^{74}Se, ^{73}As) and some other examples of recent work in the field. Considering these examples and the work performed in the past we can say that supersymmetry is well established in nuclear structure, there are still new results coming out and there is still a future for new applications. This work was supported partially by the OTKA research grant T037502, the Swiss National Science Foundation, and the MCYT(Spain) under contract No. FPA2002-04181-C04-03. Special thanks are devoted to Prof. J. Jolie, Dr. D. Sohler, Dr. Zs. Podolyák, Prof. T. Fényes, and Dr. Zs. Dombrádi who made the ^{73}As study possible. We are also indebted to Prof. J. Cseh, Dr. G. Levai, Prof. A. Frank, Dr. R. Bijker and Dr. J. Barea for their contributions to this work.

REFERENCES

1. Iachello, F., *Phys. Rev. Lett.*, **44**, 772 (1980).
2. Iachello, F., and Van Isacker, P., *The interacting boson-fermion model*, Cambridge University, Cambridge, 1991.
3. Algora, A. et al., *Phys. Rev. C*, **67**, 044303 (2003).
4. Iachello, F., and Scholten, O., *Phys. Rev. Lett.*, **43**, 679 (1979).
5. Van Isacker, P., Jolie, J., Heyde, K., and Frank, A., *Phys. Rev. Lett.*, **54**, 653 (1985).
6. Casten, R. F., and Feng, D. H., *Physics Today*, **November**, 26 (1984).
7. Bijker, O., Ph.D. thesis, University of Groningen (1984).
8. Vervier, J., Van Isacker, P., Jolie, J., Kota, V., and Bijker, R., *Phys. Rev. C*, **32**, 1406 (1985).
9. Van Isacker, P., Frank, A., and Sun, H.-Z., *Ann. Phys. (N.Y.)*, **157**, 183 (1984).
10. Tokunaga, Y. et al., *Nucl. Phys. A*, **430**, 269 (1984).
11. Algora, A. et al., *Z. Phys. A*, **352**, 25 (1995).
12. Sohler, D. et al., *Nucl. Phys. A*, **618**, 35 (1997).
13. Algora, A. et al., *Nucl. Phys. A*, **588**, 399 (1995).
14. Rotbard, G. et al., *Phys. Rev. C*, **14**, 58 (1976).
15. Schrader, M. et al., *Phys. Rev. C*, **19**, 1236 (1979).
16. Singh, B., and Viggars, D., *Nucl. Data Sheets*, **51**, 225 (1987).
17. King, M., and W.-T., C., *Nucl. Data Sheets*, **69**, 857 (1993).
18. Bucurescu, D. et al., *Int. J. Mod. Phys. E*, **8**, 17 (1999).
19. Jolie, J., *Habilitation, Rijksuniversiteit Gent* (1992).
20. Metz, A. et al., *Phys. Rev. Lett*, **83**, 1542 (1999).
21. Gröger, et al., *Phys. Rev. C*, **62**, 064304 (2000).
22. Metz, A. et al., *Phys. Rev. C*, **61**, 064313 (2000).
23. Frank, A., Van Isacker, P., and Warner, D. D., *Phys. Lett. B*, **197**, 474 (1987).
24. Frank, A. et al., *Proc. of the Conf. "Symmetries in Nuclear Structure", Erice, March 23-30* (2003).
25. Barea, J., Bijker, R., Frank, A., and Loyola, G., *Phys. Rev. C*, **64**, 064313 (2001).
26. Bijker, R. et al., *Proc. of the Conf. "Symmetries in Nuclear Structure", Erice, March 23-30* (2003).
27. Levai, G., Cseh, J., and Van Isacker, P., *Eur. Phys. J. A*, **12**, 305 (2001).

Chirality in the A~100 region

P. Joshi*, D.G. Jenkins*, P.M. Raddon*, A.J. Simons*, R. Wadsworth*,
T. Wilkinson*, D. B. Fossan†, K. Starosta†, C. Vaman†, J. Timár**,
Zs. Dombrádi**, A. Krasznahorkay**, J. Molnár**, D. Sohler**, L. Zolnai**,
A. Algora‡**, E. S. Paul§, G. Rainovski§, J. Gizon¶, A. Gizon¶,
P. Bednarczyk‖, D. Curien‖, G. Duchene‖, N. Fotiades.†† and
J. N. Scheurer‡‡

*Department of Physics, University of York, York, YO10 5DD, UK
†Department of Physics and Astronomy, SUNY, Stony Brook, New York, 11794-3800, USA
**Institute of Nuclear Research, Pf. 51, 4001 Debrecen, Hungary
‡Instituto de Fisica Corpuscular, Valencia, Spain
§Oliver Lodge Laboratory, Department of Physics, University of Liverpool, UK
¶ISN, IN2P3-CNRS/UJF, F-38026 Grenoble-Cedex, France
‖IReS, 23 rue du Loess, Strasbourg, 67037, France
††Los Alamos National Laboratory, Los Alamos, New Mexico 87545, USA
‡‡Université de Bordeaux, F-33175, Gradignan Cèdex, France

Abstract. Evidence for chirality in nuclei has been found in N ~ 75 isotopes in the A ~ 130 mass region. This phenomenon is a signature of triaxiality in nuclei and there is a clear need to study other regions of the Segre chart to see if further examples can be found. Potential Energy Surface (PES) calculations suggest that the N ~ 57-63, Z ~ 43-45 region is another promising island of triaxiality. The present experimental study with the Euroball γ-ray array, using the reaction ^{96}Zr(^{13}C, p2n) at 51 MeV is aimed to search for chirality in ^{106}Rh. Channel selection of ^{106}Rh was done using the charged-particle array DIAMANT. The yrast level scheme, studied previously in fusion-fission work, has been confirmed and extended in the present study. In addition, a new strongly coupled band, lying at an excitation energy of ~ 300 keV above the yrast band, has been found. The two structures show the characteristic properties of the chiral phenomenon.

INTRODUCTION

Considerable effort has gone into searching for convincing evidence of triaxial shapes in nuclei. The presence of such nuclear shapes is difficult to prove, however two key indications of triaxiality in nuclei have been revealed in recent years. These are the observation of wobbling bands [1] and chirality [2]. To date, evidence for the wobbling phenomenon has been observed at high spins in nuclei which are highly deformed in the A ~ 160 region (e.g. see [1, 3, 4]). In contrast chirality was first predicted to exist in normal deformed nuclei at low to medium spins in the A ~ 130,190 mass regions [5, 6]. Experimental evidence has also been obtained [7-10] for the presence of this phenomenon in the light rare earth nuclei. Very recent work has also provided the first evidence for this phenomenon in ^{104}Rh [11, 12].

The properties of structures generated by the rotation of deformed triaxial nuclei about a tilted axis have been the focus of much attention in the recent past. In triaxial odd-

odd nuclei which have configurations involving high-j proton (neutron) holes and high-j neutron (proton) particles, it is found that the total angular momentum vector **J** does not lie in one of the principal planes of the mean field and the invariance with respect to the compound operation of time reversal and rotation, $TR_x(\pi)$, is broken. The rotation which then occurs may result in two nearly degenerate $\Delta I = 1$ bands, having the same parity.

Two primary requirements for observing the chiral phenomenon are to have a triaxial shape of the nucleus and the occupation of a high-j proton (neutron) hole and a high-j neutron (proton) particle orbit coupled to this triaxial core. This situation is realized in the A~130 region with the occupation of the $\pi h_{11/2} \times \nu h_{11/2}^{-1}$ configuration for the odd proton and neutron, in a triaxial mean-field. Another mass region where a unique parity hole-particle type configuration is observed is the A ~ 100 region, where the proton hole from the $g_{9/2}$ sub shell aligns its angular momentum vector along the long axis whilst the $h_{11/2}$ neutron particle aligns its angular momentum vector along the short axis of the triaxial mean field. The occupation of this configuration has been established from near yrast spectroscopy experiments in odd-odd Rh isotopes with A=100-106 [11-16]. The triaxial nature of the mean-field for nuclei in this region is also suggested from macroscopic-microscopic calculations. Recent studies of ^{104}Rh [11, 12] have shown the first evidence for chirality in this region of the Segre chart. These results, coupled with the results of the PES calculations, prompted us to select ^{106}Rh for further investigation.

EXPERIMENTAL DETAILS AND DATA ANALYSIS

The fusion evaporation reaction ^{96}Zr(^{13}C, p2n) was used for populating the high spin states of ^{106}Rh. The reaction channel had a small yield (2-4 mb) for this channel. However, using the high resolution Euroball γ-ray array in conjunction with the highly efficient charged-particle tagging device DIAMANT [17], which consisted of 88 CsI detectors, it was possible to study this weak reaction channel. The experiment involved a ^{13}C beam impinging upon a stack of two targets of 85% enriched ^{96}Zr, each of 560 μgm/cm^2 thickness. The experiment was performed at two different beam energies 51 MeV and 58 MeV. The data from the 51 MeV part of the experiment were found to be the best for investigating the nucleus ^{106}Rh. The rare events involving the γ-rays associated with the decay of the ^{106}Rh nucleus were identified by the detection of a proton in the DIAMANT array.

The list mode data were pre-sorted with the condition of a proton being detected in the DIAMANT array in coincidence with a γ-ray event. The proton and alpha channels were clearly separated using gating on the two dimensional spectrum with the energy plotted along one axis and the PID (Particle Identification) plotted along the other axis. The PID is an electronically generated signal from the DIAMANT electronics [18] which utilises a combination of zero cross over timing and the ballistic deficit techniques. The proton gated events with three and higher Ge folds were then sorted into a Radware cube for the further analysis.

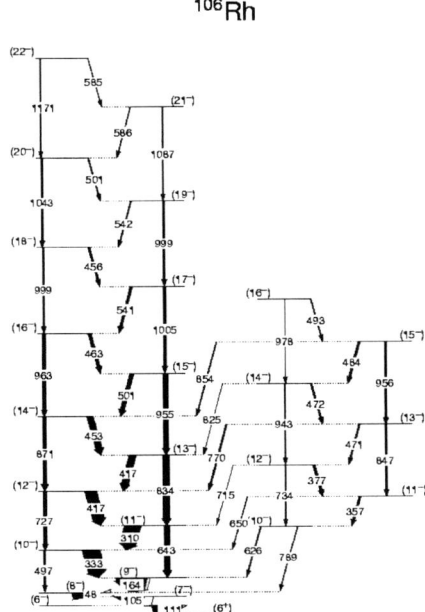

FIGURE 1. Partial level scheme of ^{106}Rh studied from the present work

RESULTS AND DISCUSSION

The analysis performed has enabled the yrast band to be extended up to spin 22^-. This is close to the maximum spin which could be populated using this reaction. The decay of the levels in this yrast band (Fig. 1) proceed with the emission of strong M1 transitions with E2 cross-over transitions of weak or medium intensity compared to the M1 transitions. The preliminary analysis of the angular correlation and polarisation data for the transitions in this band has confirmed the M1 nature of the interconnecting transitions and the E2 nature of the cross-over transitions. The low spin transitions of this band have been studied by N. Fotiades et al. and M. Porquet et al. [14, 15, 16] using a fusion-fission reaction. They were able to observe this sequence up to spin 16^- and have interpreted it in terms of a $\pi g_{9/2}^{-1} \times \nu h_{11/2}$ configuration. This configuration has indeed been understood to be responsible for the negative parity yrast sequences in all the odd-odd Rh nuclei in A>100 region [11-16].

A new strongly coupled band has been identified in the present study (Fig. 1). This lies at an excitation of ∼300 keV above the yrast sequence. This band, which is linked to to the yrast band via several transitions, also consists of an inter-connecting sequence of M1 transitions along with weak E2 cross-over transitions. The ongoing angular correlation analysis suggests a mixed M1/E2 nature for the linking transitions, which sets the parity of this band to be the same as that of the yrast band (i.e. negative). The very recent observation of chiral partner bands in the nearby isotope ^{104}Rh based on

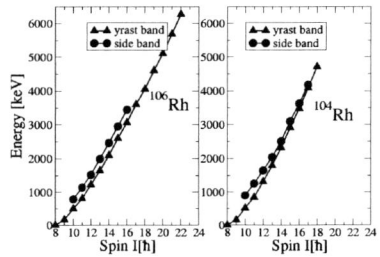

FIGURE 2. Energy vs spin plot for ^{106,104}Rh

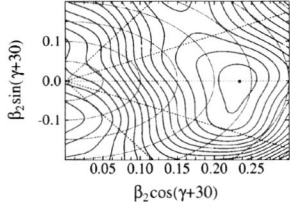

FIGURE 3. TRS plot for the ^{106}Rh nucleus with yrast configuration at a rotation of $\hbar\omega$=0.4 MeV

the $\pi g_{9/2}^{-1} \times \nu h_{11/2}$ configuration has provided strong evidence for the existence of a new region of chirality. The present results for ^{106}Rh suggest that the structures can be interpreted in a similar manner. One principle difference between the chiral partner band in ^{104}Rh and that seen in the present studies in ^{106}Rh is that in ^{104}Rh this band becomes degenerate with the yrast band at spin 17^- (see Fig. 2) (i.e. in a manner similar to the case of chiral bands in ^{134}Pr [19]) whilst in ^{106}Rh it lies at an approximately constant energy of $\sim$300 keV above the yrast band. This is similar to what is observed in many of the other isotopes and isotones of ^{134}Pr.

In order to test whether a triaxial mean-field might exist in ^{106}Rh, we have performed Total Routhian Surface calculations using the macroscopic-microscopic formalism. The calculations used the Woods-Saxon potential with the Universal parameterisation of the mean field and monopole pairing between the pairs of like-nucleons was included as the residual interaction. The results of these calculations (Fig. 3) at a rotational frequency of 0.4 MeV/$\hbar$ show that the nucleus ^{106}Rh prefers a triaxial shape, for the $\pi g_{9/2}^{-1} \times \nu h_{11/2}$ configuration, with the value of $\gamma \sim -30^0$. It is also clear from this figure that the triaxial minimum is γ-soft. This could explain why the partner bands do not become degenerate, but give rise to what has been called as chiral vibrations [2].

Part of the previous work on ^{104}Rh involved a detailed theoretical investigation [11] of the conditions that had to be satisfied for chirality to be observed in nuclei. One of the important tests was that the parameter S(I) = [E(I) - E(I-1)]/2I should be independent of the spin. Fig. 4 shows a plot of this quantity for ^{106}Rh. The results clearly satisfy the expectations. A further criterion which has been established for the existence of chirality is a staggering of the B(M1)/B(E2) values within each ΔI=1 sequence and also a similar staggering, with the same phase for the B(M1)$_{in}$/B(M1)$_{out}$ ratios. These features are

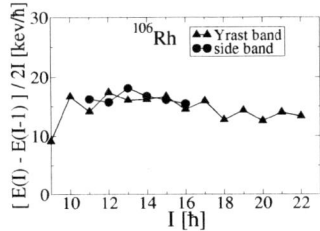

FIGURE 4. The S(I) plot for the doublet bands in ^{106}Rh

observed in both ^{106}Rh and ^{104}Rh.

The present results, along with those for ^{104}Rh strongly suggest that a new and important region for chiral studies has been found. They also provide evidence for the existence of triaxial deformation at low - medium spins in these nuclei.

ACKNOWLEDGMENTS

The authors would like to acknowledge the EPSRC in the UK, the European Community - Access to Research Infrastructures action of the Improving Human Potential Programme (contract EUROVIV: HPRI-CT-1999-00078) and the Hungarian Scientific Research Fund, OTKA (contract numbers T038404 and D 34587) as well as the Bolyai János Foundation, for the financial support.

REFERENCES

1. S.W. Odegard *et el.*, Phys.Rev.Lett. **86**, 5866 (2001).
2. K. Starosta *et el.*, Phys.Rev.Lett. **86**, 971 (2001).
3. H. Amro *et el.*, Phys.Lett. **506B**, 39(2001).
4. D.R. Jensen *et al.*, Phys.Rev.Lett. **89**, 142503 (2002).
5. V. Dimitrov *et al.*, Phys.Rev.Lett. **84**, 5732 (2000)
6. S. Frauendorf and J. Meng, Nucl. Phys. **A617**, (1997), 131.
7. T. Koike *et al.*, Phys. Rev. **C63** (2001), 061302(R)/1-4.
8. A.A. Hecht *et al.*, Phys. Rev. **C63** (2001), 051302(R)/1-4.
9. D. Hartley *et al.*, Phys. Rev. **C64** (2001), 031304(R)/1-4.
10. K. Starosta *et el.*, Nucl. Phys. **A682**, (2001) 375c.
11. T. Koike *et al.*, Frontiers of nuclear structure, Berkeley, (29 July - 2 August, 2002), AIP conference proceedings vol. 656, p. 160, (AIP New York, 2003).
12. C. Vaman *et al.*, submitted to PRL.
13. A. Gizon *et al.*, Eur. Phys. J. **A 2**, 325 (1998).
14. N. Fotiades *et at.*, conference proceeding, International conference on nuclear structure, Grand Teton National Park, Wyoming, USA, (May 22-25, 2002).
15. N. Fotiades *et at.*, Phys.Rev. **C67**, 064304 (2003).
16. M.-G. Porquet *et al.*, Eur. Phys. J. **A 15**, 463 (2002).
17. J.N. Scheurer *et al.*, Nucl. Instr. and Meth. **A 385**, (1997) 501.
18. J. Gál *et al.*, Nucl. Instr. and Meth. A, in print.
19. C.M. Petrache *et al.*, Nuc. Phys. **A597**, 106 (1996).

Author Index

A

Åberg, S., 164, 213, 289
Abu Saleem, K. S., 230
Achouri, N. L., 31
Adimi, N., 222
Adrich, P., 174
Agodi, C., 68
Ahmad, I., 230
Airoldi, A., 104
Akkus, B., 82
Alba, R., 68, 354
Algora, A., 252, 303, 375, 383
Al-Khatib, A., 303
Allia, M.C., 54
Amon, L., 82
Amorini, F., 77
Andreoiu, C., 208, 213, 289
Angélique, J.C., 31
Aoyama, S., 49
Appelbe, D. E., 208, 222
Aritomo, Y., 200
Asahi, K., 135
Assimakopoulos, P. A., 324
Astabatyan, R., 169
Astier, A., 208
Aumann, T., 87, 112, 174
Axiotis, M., 316
Azaiez, F., 31, 303
Azhari, A., 244

B

Baby, L. T., 169
Balabanski, D. L., 169
Bardayan, D. W., 73
Barranco, F., 19
Bauer, H., 39
Baumann, T., 112
Bazin, D., 26
Bazzacco, D., 39
Beck, F. A., 222
Becker, F., 31
Becker, J. A., 60
Bednarczyk, P., 104, 222, 273, 303, 383
Bélier, G., 169

Belleguic, M., 31
Benlliure, J., 112
Benzoni, G., 104, 157, 303
Bernstein, L. A., 60
Binder, B., 39
Blackmon, J. C., 73
Blasi, N., 104
Bogatchev, A., 200
Bohlen, H.G., 316
Borcea, C., 31
Boretzky, K., 174
Borge, M. J. G., 112, 252
Borremans, D., 169
Bortignon, P. F., 19
Boston, A. J., 208
Bouchat, V., 200
Bourgeois, C., 31
Bracco, A., 104, 157, 222, 303
Brambilla, S., 104
Brekiesz, M., 104, 303
Bringel, P., 303
Broglia, R. A., 19, 117
Brondi, A., 208
Brown, B. A., 31
Bürger, A., 303
Byrski, T., 222, 273, 303

C

Calabretta, L., 68, 354
Camera, F., 95, 104, 303
Campbell, C. M., 26
Cano-Ott, D., 252
Cappuzzello, F., 54
Cardella, G., 77
Carpenter, M. P., 230, 361
Casandjian, J. M., 208
Cederwall, B., 222, 303
Celona, L., 354
Cherubini, S., 68
Chines, F., 354
Chulkov, L.V., 112
Church, J. A., 26
Cizewski, J. A., 73
Clark, R., 230
Cline, D., 329

Colò, G. L., 19
Coraggio, L., 149
Cortina-Gil, D., 112, 174
Cosentino, L., 68, 354
Courtin, S., 222, 252
Covello, A., 149
Cromaz, M., 230, 329
Cullen, D., 208
Cullen, D. M., 222
Cunsolo, A., 54
Curien, D., 104, 222, 273, 303, 383
Cuttone, G., 354

D

Dakov, O., 303
Datta Pramanik, U., 112, 174
Daugas, J. M., 31, 169
D'Auria, J. M., 1
de Angelis, G., 39, 104, 273, 303, 316
de France, G., 127, 208, 222
Deloncle, I., 208
Del Zoppo, A., 68
de Oliveira Santos, F., 31, 169
Descovich, M., 208
Dessagne, P., 252
Devlin, M., 60
Dietrich, A., 39
Dinca, D. C., 26
Di Pietro, A., 68, 77
Dlouhý, Z., 31
Dobaczewski, J., 273
Dombrádi, Z., 31, 208, 303, 383
Donzaud, C., 31
Dorvaux, O., 200, 222, 273
Døssing, T., 157, 164, 213, 230, 289, 303
Dubray, N., 104
Duchêne, G., 222, 273, 303, 383
Dudek, J., 303
Duprat, J., 31

E

Elze, T. W., 174
Emling, H., 174
Erduran, M. N., 82
Ertürk, S., 222

Escrig, D., 39, 252
Evans, A. O., 208, 303

F

Fahlander, C., 213, 289
Fallica, G., 77
Fallon, P., 230, 295, 303, 329, 361
Farnea, E., 39, 104, 257, 273, 316
Fernandez-Vazquez, J., 112
Figuera, P., 68, 77
Finocchiaro, P., 354
Forssén, C., 112
Fossan, D. B., 383
Foti, A., 54
Fotiades, N., 60, 383
Fraile, L. M., 112, 252
Freeman, S., 361
Freeman, S. J., 329
Freer, M., 361
Fulton, B. R., 361
Fülöp, Z., 31

G

Gade, A., 26
Gadea, A., 39, 273
Gagliardi, C.A., 244
Gál, J., 208, 303
Galindo-Uribarri, A., 10
Gall, B., 222, 273, 303
Gammino, S., 354
Gargano, A., 149
Garrett, P. E., 60
Gast, W., 303
Geissel, H., 112, 174
Gelletly, W., 252
Georgiev, G., 169
Gerl, J., 112, 280
Girod, M., 208
Gizon, A., 383
Gizon, J., 383
Gladnishki, K. A., 280
Glasmacher, T., 26
Goergen, A., 303
Goldring, G., 169
Goodman, A.L., 285
Görgen, A., 361

Gori, G., 19
Górska, M., 143, 280
Goutte, H., 169
Grębosz, J., 104, 273
Greene, J. P., 230
Greife, U., 73
Grévy, S., 31
Grmek, A., 354
Gros, S., 208
Gross, C. J., 73
Guillemaud-Mueller, D., 31
Guinet, D., 208
Gund, C., 39

H

Haas, F., 361
Hagemann, G. B., 157, 303
Hammache, F., 112
Hammond, N., 361
Hanappe, F., 200
Hannachi, F., 230, 303
Hardy, J.C., 244
Härtlein, T., 39
Hass, M., 169
Hauschild, K., 303
Hayes, A., 329
Heinz, A. M., 230
Hellström, M., 174, 280
Herskind, B., 104, 157, 230, 303
Himpe, P., 169
Hua, H., 329
Hübel, H., 303

I

Iacob, V. E., 244
Ibrahim, F., 31
Ikeda, K., 44
Imai, N., 135
Ishihara, M., 135
Isocrate, R., 273
Itaco, N., 149
Itagaki, N., 44, 49
Itahashi, K., 112
Iwasa, N., 174

J

Jäger, H., 303
Janik, R., 112
Janssens, R. V. F., 230, 361
Jenkins, D. G., 230, 361, 383
Jones, K. L., 73, 174
Jonson, B., 112
Joshi, P., 208, 222, 383
Joss, D., 208
Juhász, K., 208, 303
Julin, R., 184
Jungclaus, A., 39, 252

K

Kalfas, C. A., 324
Kalinka, G., 208, 303
Kameda, D., 135
Karamanis, D., 324
Kerek, A., 31
Khalfallah, F., 303
Khiem, Le Hong, 174
Khoo, T. L., 157, 164, 230, 361
Kicińska-Habior, M., 104
King, S. L., 222
Kinnard, V., 200
Kmiecik, M., 104, 303
Kobayashi, Y., 135
Kokalova, T., 316
Kokkoris, M., 324
Kondev, F. G., 230
Kopatch, Y., 280
Korichi, A., 222, 230, 303
Kossionides, S., 324
Kozub, R. L., 73
Krasznahorkay, A., 31, 383
Kratz, J. V., 174
Kulessa, R., 174

L

Lagergren, K., 222, 303
Lamia, L., 68
Lane, G., 230
Langdown, S. D., 329
La Rana, G., 208, 303
Lauritsen, T., 164, 230, 361

Lautesse, P., 208
Lazzaro, A., 54
Lee, I-Y., 303, 329
Lee, Y., 208
Leenhardt, S., 31
Leistenschneider, A., 174
Lemmon, R., 208
Lenzi, S. M., 316
Leoni, S., 104, 157, 222, 303
Le Scornet, G., 252
Lewitowicz, M., 31, 169
Liang, J. F., 73
Lieder, R. M., 238, 303
Lisle, J. C., 303
Lister, C. J., 230, 361
Livesay, R. J., 73
Lo Bianco, G., 222
Lopez-Jimenez, M. J., 31
Lopez-Martens, A., 157, 164, 222, 230, 303
Lukyanov, S. M., 31, 169
Lunardi, S., 39, 222

M

Ma, Z., 73
Macchiavelli, A. O., 230, 329, 361
Maj, A., 104, 273, 303
Mandal, S., 31, 112, 280
Maréchal, F., 252
Marginean, N., 316
Markenroth, K., 112
Mason, P., 303
Matea, I., 169
Materna, T., 200
Matsuo, M., 157
Mayes, V. E., 244
Mayet, P., 31
McMahan, M., 361
Męczyński, W., 104, 273
Meister, M., 112
Menna, M., 354
Méot, V., 169
Merdinger, J. C., 273
Messina, G. E., 354
Meyer, M., 208
Miehé, C., 252
Milin, M., 316
Million, B., 104, 222, 303

Mittig, W., 31, 343
Miyoshi, H., 135
Mocko, M., 112
Molnár, J., 208, 303, 383
Moore, E. F., 361
Moro, R., 208, 303
Mrázek, J., 31
Mueller, W. F., 26
Mullins, S. M., 82
Münzenberg, G., 174
Musumarra, A., 68, 77

N

Nácher, E., 252
Nakatsukasa, T., 179
Napoli, D. R., 39, 104, 316
Negoita, F., 31
Neilson, R. G., 244
Nelson, R. O., 60
Nesaraja, C. D., 73
Neusser, A., 303
Neyens, G., 169
Nociforo, C., 54, 174
Nolan, P. J., 208
Norman, J., 208
Nourreddine, A., 222
Nyakó, B. M., 208, 303
Nyberg, J., 104

O

Odahara, A., 222
Ødegård, S., 303
Ogawa, H., 135
Ohta, M., 200
Ohtsubo, T., 112
Olliver, H., 26
Olofsson, H., 213
Orrigo, S. E. A., 54
Otsuka, T., 44
Ozawa, A., 112

P

Pachoud, E., 222
Page, R. D., 208

Paleni, A., 303
Palit, R., 174
Pansegrau, D., 39
Papa, M., 77
Papadopoulos, C. T., 324
Papka, P., 316
Pappalardo, G., 77
Pappalardo, L., 68
Passarello, S., 354
Pasternak, A. A., 238
Patel, S. B., 303
Paul, E. S., 208, 222, 383
Pearson, C. J., 329
Pearson, J., 361
Pellegriti, M. G., 68
Penionzhkevich, Y. E., 31, 169
Perdikakis, G., 324
Petrache, C. M., 104, 222, 303, 366
Petrache, D., 104, 303
Pignanelli, M., 104
Piqueras, I., 222, 303
Pisent, A., 334
Pizzone, R. G., 68
Podolyák, Z., 31, 280, 329
Podsvirova, E. O., 238, 303
Poirier, E., 252
Pomorski, K., 104
Porquet, M. G., 31, 208
Pougheon, F., 31
Prévost, A., 208, 273
Prezado, Y., 112
Pribora, V., 112
Prokhorova, E., 200
Provasi, D., 117

R

Raddon, P. M., 208, 383
Ragnarsson, I., 213, 289
Raia, G., 354
Rainovski, G., 208, 303, 383
Re, M., 354
Redon, N., 208, 222, 273, 303
Regan, P. H., 280, 329
Reiter, P., 174, 230
Rifuggiato, D., 354
Riisager, K., 112
Rinollo, A., 68
Rizzo, F., 77

Robin, J., 222, 273, 303
Roccaz, J., 303
Roig, O., 169
Rolfs, C., 68
Romano, S., 68
Rønn Hansen, C., 303
Rossé, B., 208
Rossi-Alvarez, C., 39
Rousseau, M., 316
Rovelli, A., 354
Rowley, N., 303
Rubio, B., 252
Rudolph, D., 213, 289
Russo, S., 354

S

Saint Laurent, M. G., 31
Saltarelli, A., 222
Sampson, J., 208
Sanchez-Vega, M., 244
Sato, W., 135
Savajols, H., 31, 343
Sawicka, M., 169
Scheidenberger, C., 174
Scheit, H., 112, 174
Scheurer, J. N., 208, 303, 383
Schillaci, G., 354
Schmidt, K.-H., 280
Schmidt-Ott, W.-D., 135
Schmitt, C., 200
Schneider, R., 112
Schrieder, G., 112
Schwalm, D., 39
Scuderi, V., 354
Seweryniak, D., 230
Shapira, D., 73
Sherrill, B. M., 26
Shimada, K., 135
Siem, S., 230, 303
Simon, H., 112, 174
Simons, A. J., 383
Simpson, J., 208, 222
Singh, A. K., 303
Sitar, B., 112
Siwek-Wilczynska, K., 200
Sletten, G., 31, 303
Smith, J. F., 208, 329
Smith, M. S., 73

Sobolev, Y., 31
Sohler, D., 31, 303, 383
Sorlin, O., 31
Spitaleri, C., 68
Stanoiu, M., 31
Starosta, K., 383
Stézowski, O., 208, 222, 273
Stodel, C., 31
Stolz, A., 112
Stracener, D. W., 73
Strieder, F., 68
Strmen, P., 112
Stuttgé, L., 200
Styczeń, J., 104, 273, 303
Sümmerer, K., 112, 174
Szarka, I., 112

T

Taín, J. L., 252
Tavukcu, E., 60
Teng, R., 329
Tengblad, O., 252
Theisen, C., 208
Thiamova, G., 44
Thomas, J. S., 73
Thummerer, S., 316
Tian, W., 77
Tiana, G., 117
Timár, J., 31, 208, 303, 383
Trache, L., 244
Tribble, R.E., 244
Tudisco, S., 68
Tumino, A., 68, 316
Twin, P. J., 222
Typel, S., 174

U

Ueda, M., 179
Ueno, H., 135
Ur, C., 39, 316

V

Valiente-Dobón, J. J., 329
Valvo, G., 77

Vaman, C., 383
Van der Marel, H., 31
Vardacci, E., 208
Venturelli, R., 222
Vigezzi, E., 19, 157
Villari, A. C. C., 343
Vivien, J. P., 222, 273
Vlastou, R., 324
Vondrasek, R. C., 230
von Oertzen, W., 316

W

Wadsworth, R., 208, 383
Wajda, E., 174
Walus, W., 174
Ward, D., 230
Watanabe, H., 135
Weick, H., 112, 174
Wheldon, C., 329
Wiedenhöver, I., 230
Wieland, O., 104, 303
Wilkinson, A., 208
Wilkinson, T., 383
Wilson, J. N., 303
Winfield, J. S., 54
Wollersheim, H. J., 280
Woods, P. J., 192
Wu, C. Y., 329
Wuosmaa, A. H., 361

Y

Yabana, K., 179
Yamamoto, A D., 31, 329
Yoneda, K., 135
Yordanov, O., 280
Yoshimi, A., 135
Younes, W., 60
Yurkewicz, K. L., 26

Z

Zappalà, E., 354
Ziębliński, M., 104
Zolnai, L., 208, 303, 383
Zuber, K., 104, 222, 273, 303